PHYSICS LIBRARY

Asymptotic Methods in Electromagnetics

Springer

Berlin
Heidelberg
New York
Barcelona
Budapest
Hong Kong
London
Milan
Paris
Santa Clara
Singapore
Tokyo

Daniel Bouche
Frédéric Molinet
Raj Mittra

Asymptotic Methods in Electromagnetics

 Springer

Authors

Dr. Daniel Bouche
CEA, CELV
F-94195 Villeneuve St. Georges Cedex, France

Dr. Frédéric Molinet
Mothesim, La Boursidière
F-92357 Le Plessis, Robinson, France

Professor Raj Mittra
Electrical and Computer Engineering Department
University of Illinois
1406 W. Green St., Urbana, IL 61801-2991, USA

This book is a revised translation of the French edition: Daniel Bouche, Frédéric Molinet, Méthodes asymptotiques en électromagnétisme – Mathématiques & Applications Vol. 16, © Springer-Verlag Berlin Heidelberg 1994.

Library of Congress Cataloging-in-Publication Data

Bouche, Daniel, 1958 –
[Méthodes asymptotiques en électromagnetisme. English]
Asymptotic methods in electromagnetics.
p. cm.
Includes bibliographical references (p.   ) and index.
ISBN 3-540-61574-1 (alk. paper)
1. Electromagnetic waves – Diffraction magnetics.   2. Asymptotic expansions.   I. Title.
QC665.D5B68   1997   539.2–dc21   96-39966   CIP

ISBN 3-540-61574-1 Springer-Verlag Berlin Heidelberg New York

© Springer-Verlag Berlin Heidelberg 1997
Printed in Germany

The use of general descriptive names, registered names, trademarks, etc. in this publication does not imply, even in the absence of a specific statement, that such names are exempt from the relevant protective laws and regulations and therefore free for general use.

Typesetting: Data conversion by Springer-Verlag
Cover design: *design & production* GmbH, Heidelberg
SPIN 10508432   55/3144 – 5 4 3 2 1 0 – Printed on acid-free paper

Preface

The subject of this book is the diffraction of high-frequency electromagnetic waves by conducting bodies, or scatterers coated with dielectric or magnetic materials. For four decades, this field has been the subject of intensive research, with its primary focus directed toward the prediction of the scattered field by a target illuminated by a radar wave. As an outgrowth of this research, many methods have been developed for the computation of the radar cross-section (RCS) of complex targets, e.g., aircraft and missiles, and these methods are being widely used by engineers today. Notable among these are the Geometrical Theory of Diffraction (GTD) introduced by J. B. Keller; the Physical Theory of Diffraction (PTD) developed by P. Y. Ufimtsev; the Uniform Asymptotic Theory (UAT) and the Uniform Theory of Diffraction (UTD) formulated by Lewis et al. and Kouyoumjian and Pathak; and, the Spectral Theory of Diffraction (STD) introduced by Mittra and others. It is interesting to note that mathematicians specializing in the theory of partial differential equations have concurrently developed powerful tools for the study of diffraction problems. For instance, V. M. Babitch and his collaborators have not only adapted but have also provided a firm theoretical foundation for the so-called "*matched asymptotic expansion methods*," which were initially developed in fluid mechanics. V. Maslov and L. Hörmander have established the theory of the Fourier-Maslov-Hörmander operators, which provide integral representations of the diffracted field, although they have not provided any explicit results that are readily usable for electromagnetic diffraction problems.

The objectives of this treatise are to present a comprehensive review of various methods for computing the asymptotic diffracted field, to lay the mathematical foundations of these techniques, and to provide a cohesive exposition of the underlying principles.

As mentioned above, our focus in this book will be the study of plane-wave diffraction by perfectly conducting scatterers, which may or may not be coated with dielectric materials. Mathematically speaking, our objective will be to extract solutions of Maxwell's equations, subject to appropriate boundary conditions that depend upon the nature of the object under consideration.

The problem at hand, as enunciated above, is well-defined in the theory of partial differential equations. It can be shown that a unique solution exists in a prescribed function space whose characteristics depend upon the regularity of the boundary surfaces. The choice of the function space is typically dictated by the condition that the energy associated with the field solution be locally bounded. This choice is particularly important, because it governs the existence and uniqueness of the solution, especially when the boundary surfaces contain edges or tips.

It should be realized, however, that the demonstration of the existence and uniqueness of the solution alone does not generate the explicit solution for the problem at hand. One could, of course, resort to techniques such as the finite element method or the method of moments and derive a strictly numerical solution to the problem. However, these numerical methods require, as an initial step, a discretization of the surface of the scatterer or the entire computational volume, with a typical cell size on the order of $\lambda/10$ or less, where λ is the wavelength of the incident field in the medium. It is evident that the number of degrees of freedom rapidly increases with frequency and that we are typically limited, at high frequencies, by the availability of computer time and memory in our ability to solve the problems associated with many practical structures of interest.

In contrast, the accuracy of the asymptotic methods improves with the increase in frequency as these methods provide a representation of the solution in the form of a series of inverse powers (integral or fractional) of the wave number k, of which only the first term is typically retained. We will typically deal with lossy coating materials with a relatively high index, and for the boundary condition on the surface of a scatterer with such a coating, we will employ an impedance condition that relates the tangential electric and magnetic fields on the surface of the scatterer.

Historically, the asymptotic method employed to describe the interaction of electromagnetic waves with objects was geometrical optics (GO), which is based upon the intuitive notion of rays that obey the laws of reflection and refraction in accordance with Fermat's principle, which mandates that the optical trajectories be stationary. Interestingly, the foundations of GO were established long before the advent of Maxwell's equations or even the introduction of the mathematical formalism of partial differential equations.

The principal drawback of GO is that, in contradiction to experimental observations, it predicts vanishing fields in the geometrical shadow regions. For instance, the theory of geometrical optics is unable to reproduce the interference fringes in Young's experiment that are observed experimentally.

In the 1960s, a second method called the geometrical theory of diffraction (GTD) was introduced to overcome this shortcoming of GO. It augmented the fields by adding the contribution of diffracted rays, especially those diffracted by the edges, that penetrate in the shadow zone. These rays are derived via a generalization of the Fermat's principle and, as in GO, the diffracted GTD field is generally a superposition of a multiplicity of ray contributions. In common with GO, GTD provides an intuitive interpretation of the scattering phenomena and, in most cases, enables us to handle the problem of computing the fields diffracted by complex objects.

This theory of GTD is based upon intuitive principles rather than rigorous mathematical proofs. In common with GO, the phase varies linearly along a ray in GTD and the power is conserved in a tube of rays. The diffracted field carried by a ray depends only upon the local properties of the incident field and that of the object where it intercepts the ray. This postulate enables us to calculate each contribution to the diffraction by replacing the original scatterer with a canonical object for which the solution to the diffraction problem can be derived with relative ease. The key, and often difficult step in the GTD, is to interpret the solution to the canonical problem in terms of rays. This interpretation is essential because it enables us to derive for later use the quantity termed the diffraction coefficient, which is defined as the ratio of the diffracted and incident ray fields. At this stage, it may be useful to point out that, in the practical application of the GTD, it is only necessary to identify the rays that are the predominant contributors to the diffracted field and to evaluate the field along each ray using the GTD diffraction coefficients in accordance with the principles enunciated above. The point we wish to emphasize here is that by using the GTD, we can conveniently elude the mathematical complexities associated with the scattering problem, since the mechanics of applying the GTD procedure are relatively straightforward once the diffraction coefficient has been derived. Chapter 1 provides an introduction to the Geometrical Theory of Diffraction. Section 1.2 enunciates the principal postulates of GTD: High frequency diffraction is a localized phenomenon; the field is expressed as a sum of rays that are given by the generalized Fermat's principle; these rays follow, when propagating in the free space, the same laws as do the ordinary rays of Geometrical Optics. Section 1.2 explains the different types of rays determined by using the generalized Fermat's principle. Using the aforementioned laws, we derive in Sect. 1.3 the expressions for the field carried by an arbitrary ray. These formulas contain certain geometrical factors that take into account the divergence of the ray pencil and the diffraction coefficients. The geometrical factors are derived in Sect. 1.4, and the most useful diffraction coefficients are given in Sect. 1.5. The derivation of some of the diffraction coefficients is provided in Appendix 1, while the examples of Sect. 1.7 are designed to assist the reader in obtaining some familiarity with the GTD procedure. It is hoped that the exposition of GTD presented in Chap. 1 will enable the reader to not only solve a wide variety of diffraction problems, but also to acquire physical and intuitive understanding of the various mechanisms that govern the interaction of an incident field with a given complex object.

In spite of the fact that GTD is useful for the various field computations described above, the reader may not be thoroughly satisfied with it for two main reasons. First, the GTD relies on some postulates, and, to this date, no demonstration of these postulates has been provided. Second, the field computed by GTD is infinite on the envelope of the rays, namely caustics, and is discontinuous on the light-shadow boundaries. These unphysical results, reviewed in Sect. 1.6, can be attributed to the fact that the GTD is invalid in these zones. To fill these gaps, the remaining chapters of the book provide the mathematical foundations of the GTD, specify the zones in which it is valid, and, finally, present the methods as well as useful formulas for computing the field in regions where the GTD does not yield valid results.

A complete understanding of the mathematical foundations of GTD can only be developed by expanding Maxwell's equations in terms of the wavelength λ and by investigating the limits as this parameter tends to zero. This limiting process can be carried out systematically within the framework of asymptotic expansion methods and is discussed in Chap. 2. These methods are usually based upon an ansatz that postulates a representation of the solution in the form of a product of a rapidly varying exponential of a phase function $S(x)$ and a slowly variable amplitude function $A(x)$. We will see in Chap. 2 that such a representation enables us to recover all of the laws of Geometrical Optics. In particular, the rays can be identified as the characteristic curves of the eikonal equation, which is obtained as a first-order (in k) approximation of the solution of Maxwell's equations. By generalizing the ansatz upon which GO is based, it is possible to clarify the mathematical foundations of the GTD and to define its range of validity. More precisely, Chap. 2 provides the ansatz upon which both the GO rays as well as different kinds of diffracted rays are based. The GO and the GTD solutions are identified as a first-order (in k) approximation of the solution, which is given as a series in terms of entire or fractional powers of $1/k$. This method allows us to demonstrate that the GO and GTD rays follow similar laws, viz., that the phase varies linearly along the ray and that the power is conserved in a tube of rays. These laws were enunciated as postulates in Chap. 1. GO and GTD, identified as first-order approximations in k of the solution to the diffraction problem, are valid only if this approximation is valid, i.e., if the other neglected turns, with higher power of $1/k$, are small enough to be discarded, implying that the amplitude $A(2)$ does not vary too rapidly.

Hence, the range of applicability of the GTD is limited to the regions in which the field is locally similar to a plane wave, and where it propagates along a direction which coincides with the gradient of the phase, with a slow variation of the amplitude of the field in a direction orthogonal to the phase gradient. However, during the course of the application of the GTD, we frequently encounter the situation in which the point of observation is located within a region where the field amplitude varies rapidly in the direction orthogonal to the phase gradient. These regions are referred to as the boundary layers, a terminology commonly used in fluid mechanics. These boundary layers are generally located close to the surface of the scatterer, in the vicinity of the light-shadow boundaries, or near the caustics, and the thickness of these layers tends to zero as the wavelength is decreased. Techniques for dealing with the boundary layers will be presented in Chap. 3, where we will show that the discussion not only contributes to a better understanding of the mathematical foundations of the GTD but also yields the appropriate diffraction coefficients without the need to revert back to the canonical problems. More precisely, Chap. 3 explains how to use the boundary layer method, which is an efficient tool for computing the field in the boundary layers, where the GTD fails. The idea underlying this method is quite simple: in an ordinary point in space, the amplitude $A(a)$ of the grid varies with the ordinary "slow" space scale, whereas the phase $k\,S(x)$ varies with a "fast" scale, namely the space scale multiplied by k. However, in a boundary layer, some intermediate scales appear, such as $k^{1/3}$, $k^{1/2}$, or $k^{2/3}$ times the space scales. Therefore, when computing the field in the boundary layers, one has to explicitly introduce these scales in the solution, i.e., modify the ansatz to include these scales. Although the concept of the boundary layer is quite simple, the application of this method is somewhat involved. We believe that the best way to learn how to use this powerful technique is to work through some examples and to follow the details of the computation. Chapter 3 provides a number of examples of this type of computation, beginning with simple boundary layers, such as the neighborhood of the caustic, and continuing with more complicated boundary layers, such as the neighborhood of the points where the incident ray field is tangent to the object. Finally, the method of matching the solution in the boundary layer to the ray field outside the boundary layer is explained and illustrated via some examples. The boundary layer method is a powerful technique, but it must be complemented by other methods so that the field can be computed most conveniently at every point in space. First, we must determine how the boundary layer field is diffracted; this problem is addressed in Chap. 4. Second, the boundary layer method leads to different forms of ansatz and, therefore, of solution, inside and outside the boundary layer. This feature can be somewhat inconvenient. The generation of a uniform solution, one that is valid in both zones, will be addressed in Chap. 5. Finally, the boundary layer method is not the only technique that can be used to generate the solution in the regions where the field is not a ray field, and where, consequently, both the GO and GTD solutions fail. Another powerful technique consists of deriving an integral representation for the field, which is valid everywhere in space. This technique can, in some cases, provide the result with less effort than the boundary layer method or uniform theories of diffraction, and is detailed in Chaps. 6 and 7.

We will now review the contents of Chaps. 4 through 7. As mentioned above, conventional GTD methods must be suitably modified when the incident field is no longer a ray field or when the boundary layer fields are diffracted by a scatterer. The spectral theory of diffraction (STD), described in Chap. 4, is a powerful and general technique for handling the interaction of these complex fields with diffracting structures. The underlying concept of this theory is that a general incident field can be represented as a superposition of plane waves, both the homogeneous and inhomogeneous types. The diffracted field is expressed as a superposition of the diffracted fields originating from these incident plane waves. Chapter 4 provides the plane wave spectrum representations of some of the commonly encountered boundary layer fields and some examples of their diffraction by discontinuities. For example, the diffraction of a creeping ray by a wedge is presented, and the heuristic formula obtained in Chap. 1 is justified.

The boundary layer method, mentioned above, typically leads to different forms of solutions inside and outside the boundary layer. In contrast, the so-called *uniform methods* provide a unique representation of the solution throughout the entire domain including the transition region. These theories play an extremely important role in practical applications and have been the focus of attention of a vast number of researchers. No unique approach exists for extracting the uniform solutions, and a variety of different techniques, including matched asymptotic expansions, have been developed for electromagnetics and fluid dynamic applications. In Chap. 5, we will present a unified approach to the uniform theories, including the uniform asymptotic theory (UAT), the uniform theory of diffraction (UTD), and their relationships with the spectral theory of diffraction (STD). In addition, this chapter will include the most useful uniform solutions, viz., diffraction by an edge, a curved surface, and a wedge with curved faces. More precisely, Sects. 5.1 and 5.2 provide the definition and general material on uniform asymptotic expansions. Section 5.3 deals with uniform solutions across the light-shadow boundaries of a wedge. Section 5.4 gives a uniform solution for the discontinuity in the curvature, Sect. 5.5 deals with the smooth surface. For the case of diffraction by a wedge with curved faces near grazing incidence on one face, there is an overlap of light-shadow boundaries of wedge type, treated in Sect. 5.3, and of smooth surface type presented in Sect. 5.5. This, in turn, leads to a new type of uniform solution presented in Sect. 5.6. Finally, a uniform solution at a caustic is presented in Sect. 5.7. This chapter is designed to provide the reader with a good understanding of the different methods used to derive uniform solutions, of the relationships between the various theories, and convenient formulas for the most frequently encountered diffraction problems. With this chapter, we conclude the coverage of the asymptotic expansion techniques that provide a convenient way of computing the diffracted field for the majority of practical cases. However, the field integral representation approach presented in Chaps. 6 and 7 provides a very interesting and complementary insight into the subject and even enables us to compute the field in a more convenient and accurate way than do the field solutions.

Although, in principle, the matched asymptotic expansions enable us to calculate the field everywhere in space, its application becomes very involved, if not cumbersome, in regions where the boundary layers overlap. In this event, it becomes more convenient, instead, to employ an integral representation of the solution in these regions. Although the choice of the integral representations is not unique, they must lead to results that agree with those derived by the ray method when the stationary phase approximation is applied to the integral representation. Mathematicians specializing in the theory of partial differential equations have made significant advances in this area, especially in the field of the integral operators of Fourier-Maslov-Hörmander. In Chap. 6, we present the Maslov version of the integral representation method as well as other techniques for deriving these integral representations. More specifically, Chap. 6 is principally devoted to deriving integral representations of the field in the vicinity of the caustics. Section 6.1 presents the foundations and some applications of Maslov's method, which provides a representation of this field in terms of its plane wave spectrum. For example, the field representation in the vicinity of a caustic, which is derived in Chap. 3 by using the boundary layer method, is rederived in this section via the Maslov method. As another example of the capabilities of the method, the field in the vicinity of the cusp of a caustic, which is difficult to compute by using the techniques of the preceding chapters, is derived in Sect. 6.1. Section 6.2 shows another method for computing the integral representation of the field near the caustic, using the field at the wavefront far away from the caustic and Green's formula.

Yet another approach to deriving an integral representation of the diffracted field is to go through an intermediate step of introducing a surface field and to evaluate the space field from the radiation of the surface currents. The latter approach leads to the physical theory of diffraction (PTD) and its extensions, and these are described in Chap. 7. The PTD turns out to be a very efficient tool to compute the field diffracted by complex objects, because it has the desirable property that it yields bounded results at any point in space, including the points where the boundary layers overlap, which is why this method is widely used on engineering applications.

After reading this introduction, the reader may conclude that there are too many methods for computing the diffracted field, and may find it somewhat difficult to understand how these various methods are interrelated.

To shed a little light on the subject, we offer the following thoughts. As explained above, at an ordinary point in space where the field is a ray field, i.e., it does not differ much from a plane wave, or from a superposition of a few plane waves, GO and GTD analyses yield the correct results for the diffracted field. More precisely, these methods are valid when the phase of the field varies relatively rapidly, because it is proportional to the wavenumber k, whereas its amplitude varies relatively slowly, since it is a function only of the observation coordinator. However, difficulties appear in these methods when these conditions are not fulfilled and the asymptotic expansion methods of Chaps. 3 and 5 circumvent the problem by introducing intermediate scales, and, therefore, different ansatz than employed in GO and GTD. The solution is expressed by using special functions, for example, the Fresnel or the Airy, and the GO-GTD results are recovered by replacing these functions with their asymptotic expansions for large arguments. The boundary layers are the regions where the arguments of these functions are of $O(1)$. The methods described in Chaps. 6 and 7 provide integral representations of the field, which is uniformly valid at all observation points. All of these integrals have phase functions with several stationary and, more generally, critical points. The GO-GTD results are recovered by computing the integrals via the method of stationary phase. When two critical points are close to each other, the stationary phase method is not valid, and we must evaluate the integral with great care. In fact, we may again use the boundary layer approach. Then the evaluation of the integral reproduces the results derived via the asymptotic expansion method. Hence, the asymptotic expansion and integral representation methods are closely related, and yield the same results when properly used. These two methods are, in fact, only two different routes to the solution to the problem at hand, and whether one is more convenient than the other depends upon the problem itself.

Chapter 8 presents the foundations of the impedance boundary conditions that are used to handle the case of material layer coatings on conducting scatterers and whose use enables us to derive a closed-form solution for many diffraction coefficients. Finally, for the convenience of the reader, we have grouped together a summary of the frequently used techniques in the context of GTD in the Appendices. These include the canonical problems for diffraction coefficient computation (Appendix 1); differential geometry (Appendix 2); complex rays (Appendix 3); and asymptotic expansion of integrals (Appendix 4). For convenience, we have also included, in Appendix 5, a summary of Fock functions, which are often used in the context of diffraction by a smooth object, and have included a brief discussion of the reciprocity principles. We have provided a summary of the technique and an extensive list of references.

Although the space does not permit us to include a description of other approaches to the field solution than those presented herein, we nonetheless hope that the book will be helpful to the reader interested in asymptotic methods in electromagnetic theory.

Acknowledgement
The authors wish to thank gratefully Patricia and Daniel Gogny for translating the original text.

February 1997   *D. Bouche, F. Molinet, R. Mittra*

Contents

1. Ray Optics

1.1 Introduction

The ray optical technique, comprising of geometrical optics and its extensions, provides simple and physical approaches to the description of the diffraction of an electromagnetic wave by an object. The technique relies upon a number of principles, which, though they may not be rigorous from a mathematical point of view, nonetheless provide approximate solutions to problems that would otherwise be intractable by using analytical or numerically rigorous techniques. In this chapter, we begin by enunciating the principles of ray optics and show how they enable us to extract the essential characteristics of various scattering phenomena.

1.1.1 The principle of localization

a) Physical nature of the rays

At high frequencies, the field diffracted by a scatterer and observed at a given point does not depend upon the field at every point on the surface of the scatterer, but only upon the field in the vicinity of certain points of the object called *diffraction points*. Thus, at high frequencies, the diffraction appears essentially as a *local phenomenon,* as though the field emanates from the vicinity of the diffraction points.

A very familiar example of this phenomenon is the specular reflection of a light source from a polished metallic surface, that produces a *bright spot* on the surface (Fig. 1.1). Another and perhaps less typical example is that of an opaque disc obstructing light source, that causes a halo to appear at the edge of the disc (Fig. 1.2). These two examples show that the diffraction is often local, originating from a specular point in the first case, and from an edge in the second.

We will define the *ray* as the trajectory between a point of diffraction and the point of observation. Along the direction of propagation of a ray, the field *looks like* a plane wave, that is to say its phase variation is essentially the same as that of a plane wave propagating in the direction of the ray. Furthermore, the amplitude and phase variation of the field is relatively slow in the transverse direction, i.e., in the direction perpendicular to the ray (Fig. 1.3). Whenever the field under consideration satisfies these criteria, we will refer to it as a *ray field*.

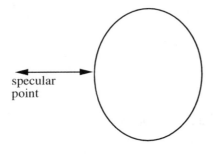

Fig. 1.1. Reflection at a specular point

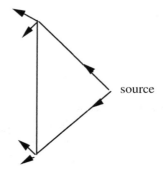

Fig. 1.2. Diffracted rays

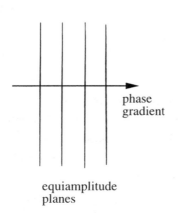

Fig. 1.3. Equiamplitude contours and the phase gradient

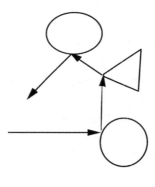

Fig. 1.4. Multiply diffracted rays

It is evident that a field may encounter multiple diffractions as it propagates around a complex object, or diffracts around a collage of scatterers, as in Fig. 1.4. Ultimately, the total field diffracted by an object can be viewed as the superposition of individual ray contributions that may be either simply- or multiply-diffracted by the object. Note that we are using the terminology *diffracted* in a broad sense, and are including all possible interactions, including specular reflections.

b) Local treatment of diffraction

The localization principle can be used in a relatively simple manner to relate the amplitude of the diffracted field to that of the incident one. The field diffracted by an object uniquely depends upon the electromagnetic properties and local geometry of the surface, and also upon the local configuration of the field in the vicinity of the diffraction point. As a consequence, it becomes possible to treat each diffraction type of interaction between the illuminating field and the object in a localized form.

In summary, the localization principle enables us to resolve the original global diffraction problem into three simpler ones that entails:

(i)   determination of the rays

(ii)   calculation of the fields propagating along the rays

(iii)   solution of local diffraction problems.

We will now elaborate on these steps involved in the computation of ray fields in the paragraphs below.

1.1.2 Determination of the rays and generalization of Fermat's principle

The geometrical theory of diffraction is an extension of geometrical optics (GO), and is designed to account for the penetration of the GO field in the shadow zones, where the field is associated with diffracted rays. The propagation of geometrical optical rays is governed by the *Fermat's principle*, which dictates that the ray paths follow the trajectory of minimal length between two points. In free space, Fermat's principle implies that the rays must be straight lines and, as will be seen in Sect. 1.2, that they must satisfy Descartes' law when the ray undergoes reflection.

The diffracted rays, on the other hand, are determined by a generalization of the Fermat's principle, viz., *generalized Fermat's principle*, which has been enunciated by Keller [1, 2] as follows: *The diffracted rays are the trajectories of minimal length between two points, but they are among a restricted class of trajectories subjected to some constraints.* This generalization of the variational principle leads to new solutions and, consequently, to a new class of rays. We will see in Sect. 1.2 that, the *rays diffracted by the edges and discontinuities*, as well as the *creeping rays* on the surface of an object, augment the geometrical optics rays to produce the total field and that the former two types of rays account for the penetration of the field in the shadow zone.

Once all of the different species of rays have been identified and their respective paths have been determined, one can go on to calculate the phase, amplitude and polarization of the field along the rays, by using the postulate that all of the rays satisfy the laws of the geometrical optics as described in the paragraph below.

1.1.3 The laws of geometrical optics

a) Propagation of the phase

The phase variation along a ray propagating between two points is expressed as the product of the wave number k and the distance between these two points. Furthermore, the phase must be continuous at the point of diffraction, except at the point of contact of the ray with its envelope, called a caustic, where it exhibits a jump discontinuity.

b) Power conservation

The power carried by a tube of rays is proportional to the square of the field amplitude and is conserved along a tube of rays. The amplitude of the ray field is thus inversely proportional to the square root of the cross-sectional area of the tube of rays.

c) Conservation of the polarization

The polarization of the field is also conserved along a ray. The laws of geometrical optics enable us to calculate the field *at each point* of a ray once it is known *at the point of diffraction*. At this point, the diffracted field can be expressed as a vector which, in turn, can be written as a 2×2 matrix operating on the incident field vector. The determination of this matrix, which is called the diffraction coefficient, will be discussed next.

1.1.4 Determination of the diffraction coefficient

The diffraction coefficient is calculated by invoking the localization principle as follows. First, the original object is replaced by a *canonical object*, one whose shape closely resembles that of the original object in the vicinity of the diffraction point, and for which the diffraction problem can be solved exactly. The detailed procedure for the determining the diffraction coefficients will be presented in Sect. 1.5.

1.1.5 Summary: The GTD as a ray method

Thanks to the *localization principle*, it is possible to resolve the global problem of the diffraction of a given incident field illuminating an object as a superposition of distinct contributions associated with diffracted rays that satisfy the *generalized Fermat's principle*. The ray paths are determined by invoking this principle, and the amplitude and phase along these rays are calculated according to the laws of *geometrical optics*. The diffraction coefficients, that relate the amplitude of the incident and diffracted fields are obtained by utilizing the *localization principle*, that enables us to replace the original object with a *canonical object* in the neighborhood of the diffraction point.

When placed in the context of a ray approach, the GTD provides an intuitive picture of the diffraction phenomenon and helps one visualize the various mechanisms of diffraction. Furthermore, it provides explicit and fairly accurate formulas for the fields diffracted by objects whose shapes are not overly complex. Nonetheless, GTD is not without its limitations; for instance, it is based upon principles that are not universal in nature, and it is unable to deal with fields that do not fall within the category of ray fields whose definition was given earlier. However, even in these regions, the GTD provides at least some approximate results that can often be refined by using more sophisticated techniques.

In this chapter, we will restrict to the implementation of GTD in the context of a ray technique and defer the discussion of refinement issues to a later chapter.

1.2 Ray tracing using generalized Fermat's principle

1.2.1 Condition for a path L to be a ray

A generalization of the Fermat's principle can be carried out in the following way. Let us consider a path T connecting the points M_0 and M_{N+1}. Let T be composed of N regular segments T_i, and let the connection points be M_i, $i = 1, N$, where either the direction or the curvature of T changes. The segments T_i may reside either in the space outside the object, or be located on the surface. The points M_i, in turn, are located either on the surface, on edges, or on the tips (see Fig. 1.5). N represents the total number of interactions of the path T with the object.

Fig. 1.5. General path between the two extremities of a ray

Let us designate $\hat{t}_i$ as the tangent at M_i to the segment T_i and let $\hat{t}_i'$ be the tangent at M_i to the segment T_{i-1}, where $\hat{t}_i$ is not equal to $\hat{t}_i'$ in general.

The optical path is defined as the curvilinear integral

$$L(T) = \int n\, ds,$$

where ds is the differential arc length along the curve T and n is the refraction index of the medium through which the ray traverses and is simply equal to 1 when the object is located in free space. Then, the optical path simply is the length of the path T.

$$L(T) = \int_T ds \qquad (1.1)$$

This is a functional expression defined on the entire set of compatible paths on the surface, that link the points M_0 and M_{N+1} in a way such that the intermediate points M_i located on the surface, the edge, or the tip, remain there and the segments T_i on the object remain on the surface of the object.

Fermat's generalized principle is stated as follows: *T is a ray if and only if the length of T is stationary for all paths satisfying the connections on the surface.*

Next, one applies the technique of the calculus of variations to express the infinitesimal variation $\delta(L(T))$ when each point undergoes an incremental displacement $\vec{\delta M}$ compatible with the connections on the surface.

The variation δ of L may be related to the variation $\vec{\delta M}$ (s) of M as follows:

$$\delta(L(T)) = \int_T \hat{t} \cdot d(\vec{\delta M}), \qquad (1.2)$$

where $\hat{t}$ is the tangent to T at point M.

Integrating by parts, (2) becomes

$$\delta(L(T)) = \sum_{i=1}^{N} (\hat{t}_i' - \hat{t}_i) \cdot \vec{\delta M}_i - \sum_{i=1}^{N} \int_{T_i} \vec{\delta M} \cdot d\hat{t}. \qquad (1.3)$$

We can now define a ray as follows: *T is a ray if $\delta(L(T)) = 0$, for all compatible $\vec{\delta M}$ with the constraints imposed on the ray segments.* This definition leads to the following two types of conditions:

(i) $N+1$ conditions characterizing the $N+1$ elementary segments T_i

$$\int_{T_i} \vec{\delta M} \cdot d\hat{t} = 0. \qquad (1.4a)$$

(ii) N conditions of transition associated with the diffraction points

$$(\hat{t}_i' - \hat{t}_i) \cdot \vec{\delta M}_i = 0. \qquad (1.4b)$$

As we will see below, the conditions (1.4) embodies all of the laws of ray propagation, and enables us to construct the optical as well as the diffracted rays.

1.2.2 Laws for diffracted rays (Fig. 1.6)

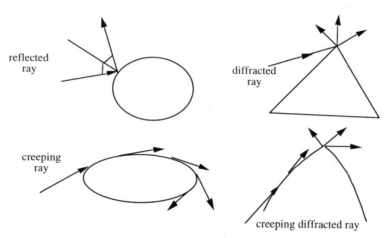

reflected ray

diffracted ray

creeping ray

creeping diffracted ray

Fig. 1.6. Different types of rays

1.2.2.1 Segments in free space

In free space, $\vec{\delta M}$ is arbitrary and has three degrees of freedom. According to (1.4a), $\delta \hat{t} = 0$, hence $\hat{t} = \hat{t}_i = \hat{t}'_{i+1}$ = constant, and, as a result (1.4b) is also satisfied.

This leads us to the first law of geometrical optics, which is simply stated as: *In free space, the rays are straight lines.*

1.2.2.2 Reflection from a smooth surface

For a smooth surface, both the incoming and outgoing rays corresponding to a reflection point M are straight lines. Let us denote the unit vector along the incident ray as $\hat{t}'_i = \hat{i}$, and the corresponding unit vector along the reflected ray as $\hat{t}_i = \hat{r}$. Since the reflection point M is on the surface S, $\vec{\delta M}$ must be a vector in the tangent plane P at the reflection point M. From (1.4b), it appears that $\hat{i} - \hat{r}$ must be normal to S at M, i.e.,

$$\hat{i} - \hat{r} = \lambda \hat{n} , \qquad (1.5)$$

where λ is a scalar, $\hat{n}$ is the normal to S at M_1, pointing outward.

Let us denote θ_i as the incident angle. Then,

$$\hat{i} \cdot \hat{n} = -cos\theta_i . \qquad (1.6)$$

In view of (1.5), and utilizing the fact that $\hat{r}$ is a unit vector, we obtain

$$\hat{r} = \hat{i} + 2 \cos \theta_i \, \hat{n}. \tag{1.7}$$

The equation above represents the law of reflection. It is often stated as follows: *The reflected ray is in the plane of incidence, which is defined by the surface normal, and the propagation vector of the incident ray, and the angle of reflection is equal to the angle of incidence.*

The two conditions, viz., *that the rays are straight lines*, and the law of reflection, comprise the laws of geometrical optics. Our next concern is the development of the rules that govern the propagation of the diffracted rays.

1.2.2.3 Diffraction by an edge

In common with the reflected rays, the diffracted rays from an object, emanating from a diffraction point M, also are straight lines. The paths between the diffraction and observation points are such that $\vec{\delta M}$ is along the tangent $\hat{s}$ to the diffracting edge. Denoting $\hat{d}$ as the unit vector on the diffracted ray, we obtain from (1.4b)

$$(\hat{i} - \hat{d}) \cdot \hat{s} = 0. \tag{1.8}$$

Equation (1.8) defines a cone of diffracted rays, the axis of the cone being tangent to the edge, and its semi-angle β at the vertex of the cone being the angle between $\hat{i}$ and $\hat{s}$. *This cone of diffraction is called the Keller's cone.* The law of diffraction from an edge can be simply stated as: *All of the diffracted rays originating from a diffraction principle must reside on the Keller's cone.* The law, stated above, is valid for any line discontinuity of the curvature or higher-order derivatives of the surface.

1.2.2.4 Diffraction from a tip or a singular point

Finally, returning to (1.4b), if we have the condition $\vec{\delta M} = 0$, which is true for a ray diffracted by a tip, the above equation is satisfied for all directions of the diffracted ray. This, in turn, implies the rule: *A tip or a singular point diffracts in all directions.* This rule is also valid for the intersection of two line discontinuities.

1.2.2.5 Surface ray

If T_i is a surface ray, $\vec{\delta M}$ lies in the plane tangent to the surface. Then, according to (1.4a), $d\hat{t} \cdot \vec{\delta M}$ must vanish, and this leads to the condition:

$$\frac{d\hat{t}}{ds} = \lambda \hat{n} \,. \tag{1.9}$$

The normal to the surface ray coincides with the normal to the surface of the object. This property is characteristic of the geodesics on a surface. This leads us to the following rule governing the propagation of rays on a surface: *The surface rays follow the geodesics of the surface.* The surface rays are also referred to as *creeping rays*.

1.2.2.6 Creeping ray launched by a space ray on a smooth surface

At the intersection of the two rays the point M can move in the plane tangent to the surface. Let $\hat{t}_i' = \hat{i}$ be the unit vector along the space ray and $\hat{t}_i = \hat{r}$ along the creeping ray. Then, from (1.4b), we have

$$(\hat{i} - \hat{r}) = \lambda \hat{n} \,. \tag{1.10}$$

Since $\hat{r} \cdot \hat{n} = 0$, it follows from (1.6) and (1.7) that the angle of incidence $\theta_i = \pi/2$. This implies that $\hat{i}$ is along the direction of grazing incidence and, consequently, the point M is located on the light-shadow boundary. In addition, we have $\hat{i} = \hat{r}$, i.e., the creeping ray exists along the direction of the space ray. In accordance with the laws of reflection, it is also possible to obtain a situation where the space ray is reflected in the incident direction. In this event, the energy of the incident ray is shared between a space ray, which is an extension of the incident ray, and a surface ray.

This leads us to the following rule: *The space rays launch creeping rays at the light-shadow boundaries and the tangent to the creeping ray is along the incident space ray.*

1.2.2.7 Emission of a space ray induced by a creeping ray

This situation is the reciprocal to the previous one. The creeping ray emits space rays along its tangent. The energy of the creeping ray splits into two parts, with one part proceeding along the geodesic and the balance emitting in the form of a space ray.

We can now state the following rule: *The creeping rays launch space rays along their tangent.* In particular, this rule implies that the creeping rays attenuate due to loss of energy via radiation.

1.2.2.8 Creeping rays emanating from an edge
As for the diffraction from an edge, it follows from (1.4b) that the projection of the emitted ray on the tangent to the edge must be equal to that of the incident ray. This leads us to the following rule: *At the edges of curved surfaces, the incident rays generate creeping rays in the direction of the Keller cone.*

1.2.2.9 Creeping ray diffracted into a space ray
The situation of creeping rays diffracting on a surface is reciprocal to the previous one. The space rays, which are diffracted, are on the Keller's cone defined by the tangent to the diffracted edge and that of the incident creeping ray.

1.2.2.10 Creeping rays emanating from a tip
Creeping rays are emitted along each generator of the cone tangent to the tip.

1.2.2.11 Diffraction of a creeping ray by a tip
For the case of the tip, the creeping ray diffracts in all directions of space outside the tip.

1.2.2.12 Ray along an edge or a wire
A class of rays that are called *traveling waves* for wires are associated with rays that attach and detach themselves from the wire either tangentially, or at one end of the wire. Similar rays for the edges are called the *edge waves*.

1.2.3 Conclusions
The generalized Fermat's principle enables us to derive, in a simple and uniform manner, all of the laws that govern the propagation of both the geometrical optics and diffracted rays. The phase, amplitude, and the polarization of the field along these rays can all be calculated once the ray paths have been determined.

We show below how the laws of geometrical optics can be employed to determine all of the desired characteristics at each and every point of the ray, once they are known at one single point along the ray.

1.3 Calculation of the field along a ray

Since the geometrical theory of diffraction is based on the premise that all rays in space obey the laws of geometrical optics, knowing the reflected or diffracted field at any one point along a space ray enables us to determine the field at any other point along the ray. The treatment of creeping rays is somewhat different, however. As mentioned earlier, these rays shed space rays along the direction of their tangent and, consequently, they attenuate by virtue of radiation. As a result, the wave numbers associated with these rays must be complex to account for this damping (1.3.1.2). In addition, the creeping rays are surface rays that require a special formulation to ensure the satisfaction of the principle of conservation of power. Finally, and perhaps most importantly, it must be understood that the field computed by using the GTD along a creeping ray on the object is not physical but fictitious, and we will refer to it hereafter as the fictitious field of a surface ray. We will return to this point in Sect. 1.3.4.

In Sects. 1.3.1, 1.3.2, and 1.3.3, respectively, we will deal with the propagation of the phase, the conservation of the power in a tube of rays, and the evolution of the polarization along a ray. For the reasons explained above, we will treat separately, in each of these three sections, the case of the creeping rays. Finally, explicit formulas will be provided for the different types of ray in Sect. 1.3.4.

1.3.1 Phase propagation along a ray

1.3.1.1 Space ray

According to the laws of geometrical optics, the gradient of the phase is directed along the ray; hence, the surfaces that are orthogonal to the rays are the surfaces of constant phase. Again, invoking the laws of geometrical optics, the variation of the phase along a ray is given by $k\,ds$, where k is the wave number, and s is the curvilinear abscissa along the ray. The phase difference ΔS between the points O along and M separated by the distance s, is then given simply by

$$\Delta S = S(M) - S(O) = ks . \qquad (1.11)$$

1.3.1.2 Creeping ray

To determine the propagation characteristics of the creeping rays, we first solve a canonical problem, e.g., diffraction by a cylinder. The results reveal that the creeping rays suffer an attenuation as they propagate and that their velocities are slightly less than those of the space rays which propagate at the velocity of light. Thus, the phase of a creeping ray can be interpreted as a sum of two contributions.

The first of these is *ks, s* being the curvilinear abscissa of the creeping ray. This term is purely geometrical and is similar in nature to that associated with the space rays.

The second contribution to the phase, which is complex, is determined by invoking the localization principle that entails replacing the surface of the scatterer, locally, by that of a canonical object. The increment of phase shift suffered by a ray as it traverses the distance *ds* is given by $Re\{\alpha(s)ds\}$, which is positive, where α is the complex wave number. Locally, the phase velocity of the creeping ray is given by $k+Re\alpha(s)/k$. The total variation of the phase between the points *O* and *M*, separated by the distance *s*, is given by $ks + \int_0^s \alpha(s)$. The imaginary part of α, which is also positive, is responsible for the attenuation of the creeping ray.

The attenuation suffered by the ray field is expressed as:

$$exp\left(-\int_0^s Im\ \alpha(s)\ ds\right).$$

The quantity $\alpha(s)$ depends only upon the local parameters of the surface *S* and upon the nature of the creeping ray. Specifically it depends upon the following:

(i)   geometrical parameters characterizing the surface, primarily the curvature of the geodesic path traversed by the creeping ray

(ii)   electromagnetic properties of the surface

(iii)   polarization of the creeping ray

(iv)   the order *p* of the ray as defined below.

In principle, there exists an infinite number of distinct creeping rays, labeled according to their order *p*. In practice, the attenuation increases rapidly with the order and, consequently, it is sufficient to retain only the first few terms of the series.

1.3.1.3 Phase-shift through caustics

In this section, we will merely state the rule that governs the phase-shift suffered by a ray as it goes through a caustic, viz., that the phase of the field jumps by $-\pi/2$, when the time convention is *exp (–i ωt)*.

Notice that the caustic appears to shorten the path traveled by the ray, contrary to what we might have guessed intuitively, because the relative phase delay at a point located at a distance *s* from a caustic is $(iks - i\pi/2)$.

The above rule is valid for all ordinary points on the caustic, namely for points where only one of the radii of curvatures changes sign (see Sect. 1.3.2.1). For exceptional points, where both the radii of curvature change sign, the phase-shift is $-\pi$.

We will return once more to this rule in Sect. 1.3.2.1, where we will present a heuristic justification of it, and in Chaps. 3 and 6 where we will demonstrate this result.

The rule for evaluating the phase-shift at caustics applies to all types of rays, be they space or surface rays.

1.3.2 Conservation of the power flux in a tube of rays

1.3.2.1 Space ray

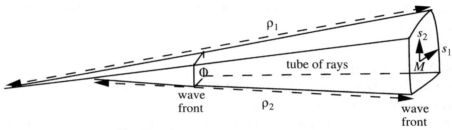

Fig. 1.7. Infinitesimal tube of rays and wavefronts

According to the laws of geometrical optics, the power, which is proportional to the square of the amplitude of the field, is conserved in a tube of rays. Let us consider an infinitesimal tube of rays, as shown on Fig. 1.7, and choose the origin of the abscissa at the point O on the ray. The wavefront at point O is a surface $F(O)$ whose normal is in the direction of the ray. In the vicinity of the point O, which is assumed to be an ordinary point on $F(O)$, i.e., neither an inflection point or an umbilic, the curvature lines of $F(O)$ define an orthogonal set of coordinates (s_1, s_2). Let ρ_1 and ρ_2 be the radii of curvature of $F(O)$ and let us denote $\delta s_1 \cdot \delta s_2 = \delta A(0)$ as the area of a curvilinear rectangle with sides δs_1 and δs_2. Now consider a point M located on the ray with abscissa s and the corresponding wavefront $F(M)$. The infinitesimal tube of rays intersects the surface $F(M)$ according to an infinitesimal curvilinear rectangle with sides $(\rho_1 + s/\rho_1)\, \delta s_1$ and $(\rho_2 + s/\rho_2)\, \delta s_2$ whose area is $((\rho_1 + s)\,(\rho_2 + s)/\rho_1\rho_2)\, \delta s_1\, \delta s_2 = \delta A(M)$. The conservation equation for the power in a tube of rays reads:

$$|E(M)|^2\, \delta A(M) = |E(0)|^2\, \delta A(0),$$

which leads to the well-known formula of geometrical optics, viz.,

$$|E(M)| = |E(0)|\, \sqrt{\frac{\rho_1\rho_2}{(\rho_1 + s)(\rho_2 + s)}}. \qquad (1.12)$$

Equation (1.3) is not only useful for determining the ray amplitude but also for providing a heuristic justification of the phase-shift rule discussed in Sect. 1.3.1.3. It is evident from (1.3) that the caustics are located at $s = -\rho_1$ and $s = -\rho_2$. Thus, when passing through the caustic located at $-\rho_1$, the quantity $s + \rho_1$ changes sign, and its square root is then multiplied by $\pm i$. The ray undergoes a $(\pm \pi/2)$ phase shift at the caustic; the choice of the sign is not obvious and requires a more refined study.

At an ordinary point on the caustic, $\rho_1 \neq \rho_2$, the two caustics are crossed successively by the ray. In contrast, if $\rho_1 = \rho_2$, the two caustic surfaces have a common point and the ray crosses the two caustics simultaneously, generating a phase shift of $-\pi$ in the process.

1.3.2.2 Creeping ray

There are two ways of viewing a creeping ray, viz., as a surface ray without any thickness or as a volumetric ray with a small thickness. The analysis of the canonical problem of a cylinder reveals that there exists a boundary layer with thickness proportional to $\rho^{1/3}$ in the vicinity of the surface, where ρ is the radius of curvature of the geodesic path traversed by the creeping ray.

The two different interpretations, given above, lead to two different results derived from the application of the conservation of the energy flux. Consider a pencil of infinitesimal creeping rays of width $d\eta$ at point 0, and $d\eta'$ at point M. The first assumption yields

$$ |\vec{E}(M)| = |\vec{E}(0)| \left(\frac{d\eta}{d\eta'}\right)^{1/2} , \tag{1.13a} $$

while the corresponding formula obtained from the second interpretation reads

$$ |\vec{E}(M)| = |\vec{E}(0)| \left(\frac{d\eta}{d\eta'}\right)^{1/2} \left(\frac{\rho}{\rho'}\right)^{1/6} . \tag{1.13b} $$

The above two formulas are not contradictory, however, because the former applies to the fictitious field carried by the surface ray, while the latter to the real surface field on the object. In practical calculations, the fictitious field, as given by (1.13), is used because it is easier to manipulate.

1.3.3 Polarization conservation

In free space, the polarization of space rays is conserved, which means that the reference frame, defined with $(\vec{E}, \vec{H})$ and the direction of the ray, are conserved. Also, when the normalized surface impedance differs sufficiently from unity, the conservation of the polarization can be extended to the creeping waves, in a manner explained below.

First, we define two types of rays (see Fig. 1.8) as follows. The first type of ray, which we will refer to as the *electric creeping ray* has its electric field oriented along the binormal to the geodesic and the magnetic field along the normal to the surface. The second type, termed the *magnetic creeping ray*, has its magnetic field pointed along the binormal and its electric field along the normal.

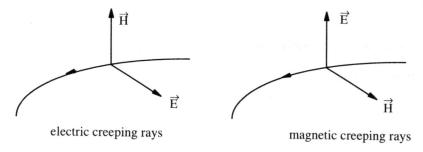

electric creeping rays   magnetic creeping rays

Fig. 1.8. Electric and magnetic fields associated with creeping rays

Note that, for a perfect conductor, the field carried by the creeping ray is obviously fictitious, since the tangential electric field vanishes on the surface of a perfect conductor, which we have assumed the scatterer to be. When the normalized surface impedance differs sufficiently from unity, the electric and magnetic creeping rays propagate independently from one another. The electric (magnetic) field of the electric (magnetic) creeping ray then remains in the direction of the binormal to the geodesic. The polarization is then conserved in the sense that the tangential fields are transported without a change in their orientation along the geodesic. The laws of amplitude and phase propagation, and those that enforce the conservation of power along a tube of rays, enable us to calculate the field at any point M along a ray from the field and the principal radii of curvature of the wavefront at the diffraction point O. The relationship between the field at points O and M for space rays takes the form:

$$\vec{E}^d(M) = \vec{E}^d(0) \sqrt{\frac{\rho_1\rho_2}{(\rho_1 + s)(\rho_2 + s)}} \, e^{iks}, \qquad (1.14)$$

where ρ_1 and ρ_2 are the principal radii of curvature at the point O. The above relationship becomes more complex for the creeping rays but $\vec{E}^d(M)$ is still the product of $\vec{E}^d(0)$ and a transmission coefficient from O to M, where the latter is calculated by using the laws given in this section.

Finally, to evaluate $\vec{E}^d(M)$ it only remains to derive the expression of a diffraction coefficient $\underline{D}$ that relates the diffracted field $\vec{E}^d(0)$ at O to the incident field $\vec{E}^i(0)$ via the equation $\vec{E}^d(M) = \underline{D}\,\vec{E}^i(0)$. According to the principle of localization, $\underline{D}$ will depend only on the geometry of the surface in the vicinity of the point O.

Next we will apply the above procedure to the computation of amplitude of reflected and diffracted rays, as well as to the creeping and diffracted creeping rays, and we will provide explicit formulas for each.

1.3.4 Final formulas for the ray computation of fields

1.3.4.1 Reflected field

Consider a ray impinging upon a smooth surface S (Fig. 1.9) at the point O.

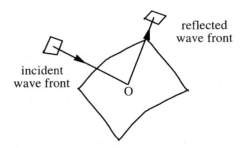

incident wave front

reflected wave front

Fig. 1.9. Reflection of array from a smooth surface

According to the law of reflection given in (1.7) of Sect. 1.2, it generates a reflected ray $\vec{E}^r(M)$ which can be written, using (1.14), in the form

$$\vec{E}^r(M) = \vec{E}^r(0)\sqrt{\frac{\rho_1^r\rho_2^r}{(\rho_1^r + d)(\rho_2^r + d)}}\; e^{ikd}, \tag{1.15}$$

where:

$\vec{E}^r(M)$ is the reflected field at point O

d is the distance OM

ρ_1^r and ρ_2^r are the principal radii of curvature of the wavefront reflected at the point O that are given in Sect. 1.4.

Using the localization principle, we can express $\vec{E}^{r}(0)$ in terms of the incident field $E_i(0)$ as

$$\vec{E}_r(0) = \vec{E}_i(0) \cdot \underline{R} \ . \tag{1.16}$$

Hence the final expression for the reflected field can be expressed as

$$\vec{E}_r(M) = \vec{E}_r(0) \cdot \underline{R} \sqrt{\frac{\rho_1^r \rho_2^r}{(\rho_1^r + d)(\rho_2^r + d)}} \, e^{ikd} , \tag{1.17}$$

where $\underline{R}$ is a dyadic reflection coefficient given by:

$$\underline{R} = R_{TE} \, \hat{e}_{\shortparallel}^i \, \hat{e}_{\shortparallel}^r + R_{TM} \, \hat{e}_{\perp} \hat{e}_{\perp}, \tag{1.18}$$

where TE and TM refer to transverse electric and magnetic with respect to the z-axis.

The right-hand reference frames, given by $\left(\hat{i}, \hat{e}_{\perp}, \hat{e}_{\shortparallel}^i \right)$ and $\left(\hat{r}, \hat{e}_{\perp}, \hat{e}_{\shortparallel}^r \right)$, are attached to the incident and reflected rays, respectively. The vector $\hat{i}(\hat{r})$ is along the ray, $\hat{e}_{\perp}$ is the vector normal to the incidence plane, as defined by $\hat{i}(\hat{r})$ and the normal $\hat{n}$ to the surface [Fr].

R_{TE} and R_{TM} depend on the impedance of S at 0. They are extracted from the canonical problem associated with the reflection of a plane wave on an infinite plane (see Sect. 1.5) in accordance with the localization principle. For reasons given at the end of the next section, these coefficients are also sometimes denoted as R_h and R_s.

1.3.4.2 Field diffracted by an edge (Fig. 10)

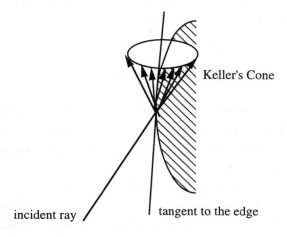

Keller's Cone

incident ray

tangent to the edge

Fig. 1.10. Diffraction of a ray at an edge

Consider a ray impinging upon an edge at the point 0. According to the laws of diffraction from an edge (Sect. 1.2.2.3), such a ray generates a cone of diffracted rays called the Keller cone. All of the diffracted rays emanate from the edge, hence it is a caustic of the diffracted rays. One of the radii of curvature of the diffracted wavefront is the distance of the point of observation to the edge, while the other is denoted by ρ_d. Although it is not possible to use the formula in (1.14) at the point 0, it can still be employed to calculate the diffracted field at a point P, which is on the diffracted ray located at a distance σ from 0. The diffracted field at M and P are related by

$$\vec{E}^d(M) = \vec{E}^d(P) \sqrt{\frac{\sigma}{d+\sigma}} \sqrt{\frac{\rho_d + \sigma}{\rho_d + \sigma + d}} \, e^{ik(d-\sigma)}. \tag{1.19}$$

It can be shown that the $\lim\limits_{\sigma \to 0} E^d(P) \sqrt{\sigma}$ exists, and the diffraction coefficient $\underline{D}$ can be defined in terms of limit as:

$$\lim_{\substack{\sigma \to 0 \\ P \to 0}} E^d(P) \sqrt{\sigma} = E^i(0) \, \underline{\dot{D}}. \tag{1.20}$$

Using (1.19) and (1.20), one finally gets

$$E^d(M) = E^i(0) \, \underline{D} \sqrt{\frac{\rho_d}{d(\rho_d + d)}} \, e^{ikd}. \tag{1.21}$$

The quantity $E^i(0) \, \underline{D}$ will be referred to as the *diffracted field at point 0*.

The diffraction coefficient, $\underline{D}$ is a dyadic, and, for *a perfect conductor*, it is given by

$$\underline{D} = -\hat{\beta}' \, \hat{\beta} \, D_s - \hat{\Phi}' \, \hat{\Phi} \, D_h \tag{1.22a}$$

The right-hand reference frames $(\hat{\imath}, \hat{\Phi}', \hat{\beta}')$ and $(\hat{d}, \hat{\Phi}, \hat{\beta})$ are associated with the incident and diffracted rays, respectively. $\hat{\imath}(\hat{d})$ is in the direction of the incident and diffracted rays; $\hat{\beta}'(\hat{\beta})$ is in the plane $(\hat{\imath}, \hat{t})(\hat{d}, \hat{t})$; $\hat{\Phi}'(\hat{\Phi})$ is perpendicular to it; and, $\hat{t}$ is the tangent to the edge at O.

The previous formulas apply to the line discontinuities as well. Let D_s denote the diffraction coefficient of the wedge that is tangent to the edge for TM-polarization, i.e., with the magnetic field orthogonal to the edge. For a TE-polarized field for which the electric field is orthogonal to the edge, the diffraction coefficient is denoted by D_h. The subscripts s and h, associated with the TM and TE diffraction coefficients, represent soft and hard boundary conditions. The use of this notation is prompted by the fact that for the case of a conducting wedge, the TM and TE problems reduce themselves to the solution of an unknown scalar function p satisfying the Dirichlet and Neuman conditions as follows:

TM Problem: $(\nabla^2 + k^2)\, p = 0$ in $\mathbb{R}^2 - \Omega$ and $p = 0$ on $\partial\Omega = \Gamma,$

TE Problem: $(\nabla^2 + k^2)\, p = 0$ in $\mathbb{R}^2 - \Omega$ and $\partial p/\partial n = 0$ on $\partial\Omega = \Gamma.$

One can now recognize the resemblance between the electromagnetic and its equivalent acoustic problem by identifying p with the acoustic pressure. The condition $p = 0$ corresponds to a *soft* surface, while $\partial p/\partial n = 0$ implies the vanishing of the normal component of the velocity on an acoustically *hard* surface. This terminology, borrowed from acoustics, is frequently used in electromagnetics. It should be noted that all of the results of this paragraph apply to an arbitrary line discontinuity such as, for instance, a curvature discontinuity.

When an impedance boundary condition is employed on the surface of a scatterer, the diffraction coefficient, given in (1.22a), no longer remains valid. In fact, the TE and TM problems become coupled in this case, and the two problems are no longer as distinct as they are for a perfect conductor. As a consequence, $\underline{D}$ loses its diagonal nature and (1.22a) must be replaced by

$$\underline{D} = \hat{\beta}'\, \hat{\beta}\, D_s - \hat{\Phi}'\, \hat{\Phi}\, D_h - \hat{\Phi}'\, \hat{\beta}\, D_{sh} - \hat{\beta}'\, \hat{\Phi}\, D_{hs}. \tag{1.22b}$$

The quantities D_{sh} and D_{hs} are the so-called cross-polarization terms, as they account for the conversion of a TE (TM) ray into a TM (TE) ray. This expression is somewhat formal because, as will be seen in Sect. 1.5, no analytical expressions are available, except for some special cases, for the coefficients D_s, D_h, D_{sh} and D_{hs} when the incidence is oblique on a wedge with a surface impedance.

1.3.4.3 Diffraction field by a tip (Fig. 1.11)

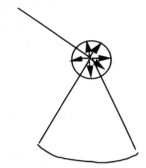

Fig. 1.11. Diffraction of a ray by a tip

When a ray diffracts from a tip, all of the diffracted rays emanate from it. Once again, it is possible to express the diffracted field for this case in the form

$$\vec{E}^d(M) = \underline{D}^p \, \vec{E}^i(0) \frac{e^{ikd}}{d} , \tag{1.23}$$

where $\underline{D}^p$ is the dyadic diffraction coefficient for the tip. The above expression applies equally well to any point discontinuity of the derivative, for instance of the curvature, or to the intersection of line discontinuities of higher-order derivatives.

1.3.4.4 Diffracted field by a smooth surface - Creeping ray

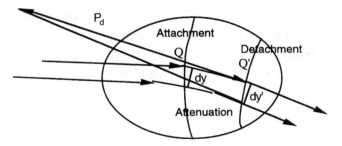

Fig. 1.12. Attachment, propagation, and detachment of creeping rays

Up until now, we have considered only the diffraction processes in which the incident field interacts but once with the scatterer. However, when dealing with the problem of diffraction by a smooth surface (see Fig. 1.12), the process of applying the GTD to the problem can be divided into three distinct steps, viz., (i) attachment at point Q; (ii) propagation of a surface ray; and, (iii) detachment at point Q'.

We will now illustrate the application of the GTD principles to the calculation of the field diffracted by a smooth surface. For simplicity, we will consider an incident field directed along the normal $\hat{n}$ at point Q. The first step, namely the attachment at the point Q will now be detailed.

Attachment at point Q

According to the localization principle, the diffracted field at Q satisfies the following equation for a creeping ray of order p.

$$\vec{E}(Q) = \vec{E}^i(Q) \, A_h^p(Q)\hat{n} \qquad (1.24)$$

The quantity $A_h^p(Q)$ is a scalar function which is called an attachment coefficient. Note that $\vec{E}(Q)$ is a fictitious field carried by the surface ray.

Propagation from Q to Q'

The knowledge of the field $\vec{E}(Q)$ of the surface ray, the laws governing the phase propagation and the conservation of the power and polarization, enable us to calculate the field $\vec{E}(Q')$ at point Q' of the surface ray.

In fact, as was pointed out in Sect. 1.3.1.2, the phase delay between Q and Q' is given by the sum of the following two terms:

   (i)   the first one of these is kl, where l is the length of QQ'.
   (ii) the second one, which corresponds to the phase $\alpha(s)$, is given simply
     by $\int_Q^{Q'} \alpha_h^p(s)\,ds$.

Let us now turn to the problem of amplitude computation. In accordance with the conservation of the power (see Sect. 1.3.2.2) the field $\vec{E}(Q)$ should be multiplied by the factor $\sqrt{d\eta(Q)/d\eta(Q')}$ to derive the field amplitude at Q', where $d\eta(Q)$ and $d\eta(Q')$ denote the cross-sections of the tube of the surface rays at Q and Q', respectively. Finally, according to Sect. 1.3.3, the field $\vec{E}$ remains normal to the surface and, consequently, is directed along the normal $\hat{n}\,'$ at Q'.

Putting all of this together we arrive at the result that the GTD field associated with the fundamental mode at Q' takes the form

$$\vec{E}(Q') = \vec{E}(Q) \, exp \, i\left(k\ell + \int_Q^{Q'} \alpha_h^p(s)\,ds\right)\sqrt{\frac{d\eta(Q)}{d\eta(Q')}} \, \hat{n}'. \qquad (1.25)$$

Detachment at Q'

The last step in the field computation at a given observation point in space is to take into account of the detachment of the ray. For the purpose of this computation, the field $\vec{E}(Q')$ is taken to be the incident field at Q'. The field denoted as $\vec{E}^d(Q')$ is the product of $\vec{E}(Q')$ and a diffraction coefficient $\underline{D}_h^p$ called the detachment coefficient, i.e.,

$$\vec{E}^d(Q') = \vec{E}(Q')\ D_h^p(Q').$$

The diffracted field at point M is then calculated by using the above expression for $\vec{E}^d(Q')$ in formula (1.21). The surface of the scatterer is a caustic for the diffracted rays and one of the radii of curvature of the wavefront vanishes at Q'. Consequently, we obtain

$$\vec{E}^d(M) = \vec{E}(Q')\,D_h^p(Q')\sqrt{\frac{\rho_d}{d(\rho_d + d)}}\,e^{ikd}, \qquad (1.26)$$

where, as in the case of diffraction by an edge, ρ_d is the non-vanishing radius of curvature of the diffracted wavefront. It can also be defined as the distance between the point Q' and the second caustic.

Using (1.24), (1.25) and (1.26) we express the diffracted field at point M as function of the incident field at point Q as follows

$$\vec{E}^d(M) = \vec{E}^i(Q)A_h^p(Q)\,exp\left\{i\left(k\ell + \int_Q^{Q'}\alpha_h^p(s)ds\right)\right\}\sqrt{\frac{d\eta(Q)}{d\eta(Q')}}\,D_h^p(Q')\sqrt{\frac{\rho_d}{d(\rho_d + d)}}\,e^{ikd}\,\hat{n}'.$$
$$(1.27)$$

When the incident electric field is along the binormal vector $\hat{b}(\hat{b} = \hat{i}\hat{n})$, we can proceed to compute the field in a similar manner. An expression similar to the one in (1.27) can be derived for the diffracted field; however, we must now use new diffraction coefficients denoted by A_s^p, D_s^p and α_s^p.

For the general case, the incident field, orthogonal to the direction $\hat{i}$ of propagation of the ray, is in the plane defined by $\hat{n}$, $\hat{b}$ and it can be resolved as $\vec{E}^i(Q) = (\vec{E}^i(Q)\cdot\hat{n})\hat{n} + (\vec{E}^i(Q)\cdot\hat{b})\hat{b}$. By treating each component separately, we obtain the final result which reads

$$\vec{E}^d(M) = \vec{E}^i(Q)(\hat{b}\hat{b}'T_s^p + \hat{n}\hat{n}'T_h^p)e^{ik(\ell + d)}\sqrt{\frac{d\eta(Q)}{d\eta(Q')}}\sqrt{\frac{\rho_d}{d(\rho_d + d)}}, \qquad (1.28)$$

where

$$T_s^P = A_s^P(Q)\,exp\left(i\int_Q^{Q'} \alpha_s^P(s)ds\right) D_s^P(Q'),\qquad(1.29)$$

and

$$T_h^P = A_h^P(Q)\,exp\left(i\int_Q^{Q'} \alpha_h^P(s)ds\right) D_h^P(Q').\qquad(1.30)$$

Let us now make a few observations regarding the above formulas for GTD field computation. First of all, we note that the GTD procedure can be viewed as a block-build-up approach to constructing the diffracted field. It consists of an application of the laws of geometrical optics along the rays and the multiplication of the local field with a diffraction coefficient at each stage. However, it must be kept in mind that one does not work, in the context of GTD, with the *physical field* but instead with a *fictitious* field carried by the ray. For a space ray, the physical and fictitious fields are identical at points far away from singularities, e.g., caustics and shadow boundaries. However, the same can not be said for a surface ray, and this is particularly evident for an electric type of creeping ray propagating on a perfect conductor. Although the physical electric field along the binormal $\hat{b}$ must vanish, the fictitious electric field employed in the GTD computation is non-zero as can be seen on (1.28), (1.29), and (1.30). We will see, in Sect. 1.3.4.5, how the physical field can be deduced from the fictitious GTD surface field.

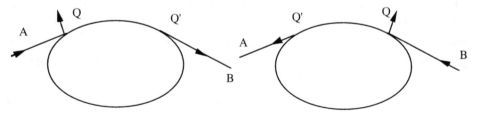

Fig. 1.13. Application of the reciprocity principle to creeping rays

The second observation we make is with regard to the satisfaction of the reciprocity relationship by the fields derived by using the GTD. In most of the publications on GTD, the reciprocity theorem is invoked by asserting that D_s^P and D_h^P are equal to A_s^P and A_h^P, respectively. This is not strictly correct, however, and we will demonstrate this fact by applying the reciprocity theorem to the problem shown in Fig.

1.13. The formulas (1.28), (1.29), and (1.30), applied to the situations shown in Fig. 1.13, yield the following two expressions for the field E normal to the ray and induced by the creeping ray.

Problem (1) :

$$E(B) = \frac{e^{ikr_1}}{\sqrt{8\pi kr_1}} e^{i\pi/4} A_h^p(Q) exp\left(i\int_Q^{Q'} \alpha_h^p(s)ds\right) D_h^p(Q') \frac{e^{ik(\ell+r_2)}}{\sqrt{r_2}} \qquad (1.31)$$

Problem (2) :

$$E(A) = \frac{e^{ikr_2}}{\sqrt{8\pi kr_2}} e^{i\pi/4} A_h^p(Q') exp\left(i\int_Q^{Q'} \alpha_h^p(s)ds\right) D_h^p(Q) \frac{e^{ik(\ell+r_1)}}{\sqrt{r_1}} \qquad (1.32)$$

From the reciprocity theorem it follows that $E(A) = E(B)$, and since (1.31) and (1.32) can be derived from each other merely by interchanging Q and Q', we obtain

$$A_h^p(Q) \, D_h^p(Q') = A_h^p(Q') \, D_h^p(Q), \qquad (1.33)$$

or, equivalently,

$$\frac{D_h^p(Q')}{A_h^p(Q')} = \frac{D_h^p(Q)}{A_h^p(Q)} = C, \qquad (1.34)$$

where C is a constant. We conclude, therefore, that D_h^p and A_h^p are proportional to each other and are equal only to within an arbitrary constant C. Keller [K1, K2], and the majority of authors have chosen the constant C to equal 1 in their works. For 2D problems, Albertsen and Christiansen [AC] decompose the process of detachment into two, the first one being equivalent to an attachment, and the other associated with the radiation by a line source. One then has $D_h^p = C A_h^p$, where C is the coefficient relating the fictitious field at the emission point to the intensity I of the source.

The field at a large distance r from the emission point is given by

$$E(r) = E(0) \frac{e^{ikr}}{\sqrt{r}} = C I \frac{e^{ikr}}{\sqrt{r}}. \qquad (1.35)$$

The constant C is equal to $e^{i\pi/4} \sqrt{8\pi k}$ for a line source radiating in free space. This value of C was chosen by Albertsen and Christiansen.

As we will see in Sect. 1.5, D_h^p, D_s^p, α_s, and α_h are obtained by solving an associated canonical problem, the simplest one being that of scattering by a circular cylinder with a normally incident wave. Since the cylinder is a developable surface, the geodesic paths followed by creeping rays along the surface are readily obtained. However, at normal incidence, the cylinder is not a representative geometry for supporting creeping rays with the following characteristics: a torsion of the geodesic; the transverse curvature of the surface in the direction orthogonal to the geodesic; or, the derivative of the curvature of the geodesic with respect to the curvilinear abscissa. Unfortunately, there is no one single canonical problem that takes into account all of these features simultaneously, although there are a number of geometries which do support creeping rays with curvature and/or one of the other features listed above. A partial list of these canonical problems include:

(i)   cylinder with oblique incidence for the torsion of the geodesic
(ii) the sphere for the transverse curvature
(iii) the parabolic or elliptic cylinder for the derivative of the curvature.

The above situation leads us to introduce the notion of *partial* canonical problems that support creeping rays with only some of the characteristics. To construct the solution for a general surface, one can then attempt to piece together the solutions of these partial canonical problems and synthesize a solution in this manner. Historically, the problem of a cylinder with a normal incidence was chosen to be the basic canonical problem for smooth surfaces. Later, the study of more elaborate canonical problems, in particular by Pathak and his co-workers, have made it possible for us to refine the calculation of the diffraction coefficients, and to recognize new physical effects. The canonical problems in question were essentially the study of diffraction by a sphere or a cylinder with an obliquely incident field.

Finally, we observe that (1.28), (1.29), and (1.30) provide the expressions for the diffracted field for a creeping ray of order p. In principle, the total result can be obtained by summing up the contributions of the different orders; however, in practice, it is usually sufficient to include only the first mode because the contributions of those with higher orders attenuate rather rapidly.

1.3.4.5 Diffraction of a creeping ray by an edge

When a creeping ray strikes an edge with curved faces, it generates, as we have seen in Sect. 1.2, a cone of diffracted rays. The GTD provides an expression for the field at the diffraction point, but this field is fictitious and one must not continue to use it for computing subsequent diffraction effects as though it were a physical field. Physically, the incoming field at the edge has approximately the same phase as that of a plane wave propagating along the tangent to the creeping ray, but has a variation along the normal described by a special function as explained in Chap. 3. However, the localization principle still applies, and the diffracted field at point O is the product of a fictitious surface field carried by the creeping ray at O and a *hybrid* diffraction coefficient denoted by $D_{sr,r}$ (see [AC]). The calculation of $D_{sr,r}$ is based upon the following two concepts: first, because the ray has an *exp(iks)* behavior, it looks like a plane wave at grazing incidence; second, since the field must satisfy the boundary condition on the surface of the scatterer, it is comprised of a superposition of an incident field and a field reflected by the surface. To calculate the diffraction of the creeping ray by an edge, one needs only consider the incident field.

For the sake of simplicity, let us restrict ourselves, initially, to two-dimensional problems. We will regard the creeping ray as one that impinges upon the edge at a grazing incidence. The amplitude of the incident field may be estimated as follows. From the fictitious field of the surface ray, denoted by the subscript sr, as in reference [AC], one computes the surface field denoted by the subscript s. These two fields are related by the equation

$$E_s = D_{sr,s} \, E_{sr} , \qquad (1.36)$$

where $D_{sr,s}$ is a diffraction coefficient, which depends upon the polarization of the electric and magnetic creeping rays that are designated, respectively, with the subscripts s and h. The above equation enables one to relate the real field E_s to the fictitious field of surface ray E_{sr} . The surface field E_s is the sum of the incident and reflected fields, the latter being R times the incident field where R is the reflection coefficient. Accordingly, one must divide the surface field by $(1+R)$ in order to obtain the incident field on the edge.

Finally, temporarily omitting the subscripts s and h designating the polarization, the incident field may be written as

$$E^i = \frac{D_{sr,s}}{(1+R)} E_{sr} . \qquad (1.37)$$

The fictitious field, E_{sr}, carried by the surface ray, is calculated via the GTD using the method described in Sect. 1.3.4.4. For a magnetic creeping ray on the surface of a perfect conductor, $R = 1$, and the incident field is one half of the surface field. For a surface with impedance Z, the reflection coefficient R is given by $(\sin \theta - Z)/(\sin \theta + Z)$ for the TE polarization, and $(Z \sin \theta - 1)/(Z \sin \theta + 1)$ for the TM case, where θ is the angle of incidence. We note that $(1 + R) \to 0$ as θ approaches the grazing angle. However, the results for the diffracted field remain finite in this limit, as we will demonstrate shortly.

Next, we turn to the problem of diffraction of the field E^i, given by (1.37). To this end, we seek a representative canonical problem for the diffraction by an edge. The only possible choice is a wedge with planar faces, since the problem of a wedge with curved faces is not soluble in a convenient form that enables one to extract a diffraction coefficient.

Thus, we consider the problem of diffraction of the field E^i by a wedge with planar faces. The diffracted field at point O is given by $E^i D$ where D is the diffraction coefficient of a wedge with planar faces, with the appropriate coefficient D_s or D_h, chosen in accordance with the polarization of the field.

Using the value of E^i, given in (1.37), we obtain the following expression for the diffracted field at the point O

$$E^d(0) = D_{sr,s} \lim_{\theta \to 0}\left(\frac{D}{1 + R}\right) E_{sr}(0), \qquad (1.38)$$

where the coefficient $D_{sr,s}$ is chosen to be either $D_{sr,s}^s$ or $D_{sr,s}^h$ depending upon the polarization. In order to derive the limit appearing in (1.38), one takes into account the fact that both $(1 + R)$ and $D \to 0$ as $\theta \to 0$, and the limit of $D/(1 + R)$ is finite at $\theta = 0$.

Finally, the diffracted field at the point O is related to the fictitious field incident upon the surface through a *hybrid* diffraction coefficient $D_{sr,s}$. The same notation is used throughout for the subscripts. For instance, the subscript sr implies that the incident field is a fictitious field of a surface ray whereas r is associated with the diffracted field of a space ray.

According to (1.38), $D_{sr,r}$ is given by

$$D_{sr,r} = D_{sr,s} \lim_{\theta \to 0}\left(\frac{D}{1 + R}\right). \qquad (1.39)$$

The above result is readily extendable to the three-dimensional case by using a dyadic diffraction coefficient which is obtained from the previous scalar coefficients $D_{sr,r}^s$ and $D_{sr,r}^h$. The dyadic diffraction coefficient that relates the fictitious electric field on the surface to the physical surface electric field is given by

$$\underline{D}_{sr,s} = \hat{b}\hat{b}\, D_{sr,s}^s + \hat{n}\,\hat{n}\, D_{sr,s}^h. \qquad (1.40)$$

The two vectors $\hat{\beta}'$ and $\hat{\Phi}'$, defined in Sect. 1.3.4.2 in connection with the dyadic coefficient of diffraction D of the wedge with plane faces, are quite simple for the particular case of grazing incidence. The vector $\hat{\beta}'$ is orthogonal to $\hat{\imath}$ and lies in the plane $(\hat{\imath},\ \hat{\imath}\ )$, and is the binormal vector to the incident creeping ray at the diffraction point. As for the vector $\hat{\Phi}'$, it is normal to the incident creeping ray at the diffraction point. Note that it is also normal to the curved face since the creeping ray follows a geodesic path.

The frame attached to the edge, as defined in Sect. 1.3.4.2, coincides with Frenet's frame attached to the creeping ray at the diffraction point.

The dyadic diffraction coefficient relating $E_{sr}(0)$ to $E^d(0)$, and denoted by $\underline{D}_{sr,r}$ is given by (1.39), with the conventional products now replaced by dyadic products. The diffraction coefficient is given by

$$\underline{D}_{sr,r} = -\hat{\beta}'\hat{\beta}\ D^s_{sr,s}\ \lim_{\theta \to 0}\ \frac{D_s}{1+R_s} - \hat{\Phi}'\hat{\Phi}\ D^h_{sr,s}\ \lim_{\theta \to 0}\ \frac{D_h}{1+R_h},$$
$$-\hat{\Phi}'\hat{\beta}\ D^s_{sr,s}\ \lim_{\theta \to 0}\ \frac{D_{sh}}{1+R_s} - \hat{\beta}'\hat{\Phi}\ D^h_{sr,s}\ \lim_{\theta \to 0}\ \frac{D_{hs}}{1+R_h}. \qquad (1.41)$$

The expressions given previously must be modified when considering electric creeping rays on the surface of a perfect conductor, since, in this case, the surface field vanishes along the binormal $\hat{b}$, although its normal derivative, being proportional to the current, does not. Rather than working with the usual normal derivative $\partial/\partial n$, it is convenient to introduce the normal derivative with respect to the dimensionless variable kn. This derivative, i.e., $\partial/\partial kn$, whose dimension is the same as that of the field, will still be referred to as the *normal derivative* in what follows. For the sake of simplicity, let us return for a moment to the two-dimensional problem, and consider a creeping ray propagating along one of the faces of the wedge with curved faces.

Again, with the notation of reference [AC], let $D_{sr,s}$ · designate the diffraction coefficient relating the fictitious field carried by the surface ray to the normal derivative of the surface field. Then,

$$\frac{\partial E}{\partial (kn)} = D_{sr,s'}\ E_{sr}. \qquad (1.42)$$

The derivative of the surface field includes, as in the previous case discussed above, an incident and a reflected part. The reflection coefficient of the field is -1, and the corresponding one for the normal derivative is $+1$. It is then necessary to divide the normal derivative given by (1.42) by two in order to retain only the incident part. Thus, we have

$$\left(\frac{\partial E}{\partial (kn)}\right)^i = \frac{1}{2} D_{sr,s'} E_{sr}$$

(1.43)

We now consider the diffraction of this field when incident upon a wedge with curved faces. For the canonical problem, we use the case of diffraction of an incident field by a wedge with planar faces. The incident field vanishes on the edge but its normal derivative to the edge is nonzero. The diffraction coefficient relating the normal derivative to the field on the diffracted ray at point O, sometimes called the *slope diffraction coefficient*, is given by

$$i\frac{\partial D_s}{\partial \theta}.$$

(1.44)

Using (1.43) and (1.44), the diffracted field at point O may be written as

$$E^d(0) = \frac{D_{sr,s'}}{2} i\frac{\partial D_s}{\partial \theta} E_{sr}.$$

(1.45)

The generalization of this case to three dimensions can be formally carried out in a manner similar to the one followed for the case of an object with a surface impedance. The dyadic diffraction coefficient $D_{sr,r}$ reads

$$\underline{D}_{sr,r} = -\hat{\beta}'\hat{\beta}\frac{i}{2} D_{sr,s'}\frac{\partial D_s}{\partial \theta} - \hat{\Phi}'\hat{\Phi}\frac{1}{2} D_{sr,s}^h.$$

(1.46)

The use of the diffraction coefficient $D_{sr,s'}$, given in (1.45), enable us to convert the fictitious field of surface ray into the normal derivative of the electric surface field carried by the electric creeping ray. We will see in Chap. 3 that, in the vicinity of the surface, this normal derivative is essentially directed along the binormal to the creeping ray and, consequently, along the same direction as the fictitious surface field. In (1.46), $D_{sr,s'}$ is simply the diffraction coefficient extracted from the solution of a two-dimensional problem. However, a problem arises in the use of this coefficient because, as we will see in Chap. 3, the surface magnetic field of a magnetic creeping ray has a component along the binormal to the ray, proportional to $\tau\rho$, where τ is the torsion and ρ the radius of curvature of the geodesic followed by the ray. The normal derivative of the surface field does not completely define the field of the incident creeping ray. The expression in (1.46), which takes into account only the normal derivative, does not adequately reflect the effect of the torsion of the incident creeping ray on the diffracted field.

However, this may not present too serious a problem in practice because the electric creeping rays attenuate quite rapidly as they propagate. As a consequence, the second term in (1.45) which corresponds to the diffraction of the magnetic component of the incident creeping ray, becomes the predominant contributor to the field.

In the next section, we will discuss the reciprocal problem of the excitation of creeping rays as induced by an incident field on a curved edge.

1.3.4.6 Excitation of creeping waves on an edge with curved faces

According to the localization principle, the field of a surface ray is given by

$$E^{sr}(0) = E^i(0) D_{r,sr}, \qquad (1.47)$$

where $D_{r,sr}$ is a diffraction coefficient relating, at point O, the incident field $E^i(0)$ of the space ray to the fictitious surface field carried by one or two creeping diffracted rays.

It can be shown [AC] by using the reciprocity theorem that $D_{r,sr}^{s(resp.h)}$ is the ratio of the diffraction coefficient $D_{sr,r}^{s(resp.h)}$, introduced in the previous section, and the coefficient of excitation C.

The extrapolation to the three-dimensional case can be carried out in a straightforward manner. $D_{r,sr}$ is given by (1.41) for the case of a surface impedance (after dividing by the factor C) and for the case of a perfect conductor by (1.46), again after C is factored out. Note that $\hat{\beta}$ is nothing but the binormal and $\hat{\Phi}$ the normal vector to the creeping diffracted ray.

Once E^{sr} is known, the fictitious surface field on the diffracted creeping ray may be calculated by following exactly the same procedure as employed for the computation of the usual creeping ray induced on a smooth surface. The only difference is that the attachment coefficient $A = \hat{b}\hat{b} A_s^p + \hat{n}\hat{n} A_h^p$ is now replaced by the hybrid diffraction coefficient $\underline{D}_{r,sr}$.

Finally, it should be mentioned before closing that the method described in this paragraph may not apply to the excitation of the electric creeping rays when the torsion at the diffraction point is large.

1.3.4.7 Creeping ray on an edge with curved faces

The problem of a creeping ray traveling through an edge with curved faces can be handled by combining the procedure for the two cases we just discussed above. There are two diffracted creeping rays, one on each face. The ratio $D_{sr,sr}$ between the fictitious surface field of the incident ray and the fictitious surface field of one of the two diffracted rays is given by [AC]

$$D_{sr,sr} = \lim_{\substack{\theta_i \to 0 \\ \theta_d \to 0}} \frac{D_{sr,s}}{1 + R(\theta_i)} \frac{D_{sr,s}}{1 + R(\theta_d)} \frac{D(\theta_i, \theta_d)}{C}, \qquad (1.48)$$

where $E^i_{sr}(0)$ is the fictitious field on the incident surface ray.

If the incident ray is TM polarized and it propagates on the face of a perfect conductor, the previous formula must be replaced by the following

$$D_{sr,sr} = \lim_{\theta_d \to 0} \frac{D_{sr,s'}}{2} i \frac{\partial D_s}{\partial \theta_i} \frac{D_{sr,s}}{1 + R(\theta_d)} \frac{1}{C}. \qquad (1.49)$$

A similar formula exists for the TM-polarized diffracted creeping ray as it propagates on a perfect conductor. Finally, and again for TM polarization, the formula given below applies when the two faces are perfect conductors, viz.,

$$D_{sr,sr} = -\frac{D_{sr,s'}}{2} \frac{D_{sr,s'}}{2} \frac{\partial^2 D_s}{\partial \theta_i \partial \theta_d} \frac{1}{C} E^i_{sr}(0). \qquad (1.50)$$

As in the previous sections, the transition from 2D to the 3D case is readily achieved in the present case. We denote, with the indices s and h, the scalar diffraction coefficients for the TM and TE polarizations, respectively. The dyadic diffraction coefficient for a perfect conductor can be written as

$$\underline{D}_{sr,sr} = \hat{b}\hat{b}' D^s_{sr,sr} + \hat{n}\hat{n}' D^h_{sr,sr}.$$

It follows that the propagation of a creeping ray along an edge can be described in terms of a diffraction coefficient multiplying the fictitious surface field.

We should also mention that for the case of an impedance condition it becomes necessary to include cross terms in the diffraction coefficient D, which no longer remains diagonal.

1.3.4.8 Launching and diffraction of creeping waves by a tip with curved faces

The method of calculating the hybrid diffraction coefficient can be formally extended to the case when the object is a tip.

An incident creeping wave striking a tip generates diffracted waves in all directions in the space outside of the tip.

The diffracted field at a distance r from the tip is given by

$$E^d(M) = E_{sr}(0) \, \underline{\underline{D}}_{sr,s} \, \lim_{\theta \to 0} \left(\frac{D}{1+R} \right) \frac{e^{ikr}}{r}, \tag{1.51a}$$

where $\underline{\underline{D}}$ is the diffraction coefficient of the tip at grazing incidence in the direction of the creeping ray and the observation in the direction of M.

The above formula for the tip is derived in the same manner as that used earlier to treat the wedge. An important case occurs in practice when the tip is that of a circular cone with a small angle, and is perfectly conducting. This particular example falls among a few handful of cases for which it is possible to obtain the diffraction coefficient. It so happens that the diffraction coefficient becomes large along certain directions, and the contribution of the diffracted field may be sizable in those directions. Under these conditions, the formula in (1.51) can be reduced to

$$E^d(M) = E_{sr}(0) \, \underline{\underline{D}}_{sr,s} \, \frac{D}{2} \frac{e^{ikr}}{r}. \tag{1.51b}$$

From now on we will restrict ourselves to this case in the following.

In accordance with the principle of reciprocity, an incident wave striking a tip with curved faces generates creeping rays on the tip. We will only calculate the fictitious surface field due to the magnetic creeping rays on the tip. It is evaluated from (1.51b) by invoking reciprocity, and is given by

$$E_{sr}(0) = E^{inc}(0) \, \underline{\underline{D}}_{sr,s} \, \frac{D}{2C} \tag{1.52}$$

where D is the tip diffraction coefficient for observation along the tangent to the creeping ray. Only the magnetic field parallel to the surface is taken into account in D since the electric creeping waves are negligible.

Finally, an incident creeping ray generates diffracted creeping rays on a tip with curved faces. Again, restricting ourselves to magnetic creeping rays on a perfectly conducting tip, we can obtain the following expression for the fictitious surface field

$$\vec{E}_{sr}(O) = \vec{E}^{inc}(O) \cdot \frac{\underline{\underline{D}}_{sr,s}^2}{4} \frac{\underline{\underline{D}}}{C} \tag{1.53}$$

where D is the diffraction coefficient of the tip, for grazing incidence and observation, for an incident wave with a magnetic field parallel to the surface.

Note that, in (1.51), the diffraction coefficient D for a circular conical tip depends upon two parameters characterizing the direction of the diffracted (incident) ray with respect to the direction of the incident (diffracted) creeping ray. In (1.53), D depends upon the angular separation between the incident and the diffracted creeping rays.

In all of the previous discussion, we have assumed that the scatterer was illuminated by an incident field due to a distant source. However, in some problems, the incident field is generated by a source situated on the surface of the object itself, that not only emits space rays but launches creeping rays as well. The space rays interact with the object just as though there were incident fields from distant sources and, consequently, they can be handled by using the methods described earlier. The creeping rays, on the other hand, must be handled differently, and the phenomenon of launching these rays is discussed in the next section.

1.3.4.9 Excitation of creeping waves by a source located on the surface of the scatterer

We have assumed, thus far, that the incident field was a ray field. The case of more general fields cannot be handled directly by using the GTD, and one must appeal to the spectral theory of diffraction discussed in Chap. 4. However, when the source is located precisely on the surface, the GTD can still be used in accordance with the localization principle. This, in turn, leads to the assumption that the launching coefficient of the creeping ray depends only upon the geometry of the surface in the vicinity of the source. The fictitious surface field on the creeping ray at the point S, where the source is located when the source is a magnetic dipole $\vec{S}$ lying in the plane tangent to the surface. It is given by

$$\vec{E}(S) = \frac{ik}{4\pi} \vec{S} \cdot [(\hat{b}\hat{n})L_h + (\hat{b}\hat{b})L + (\hat{t}\hat{b})L_s + (\hat{t}\hat{n})L'], \qquad (1.54)$$

where the coefficient $ik/4\pi$ is referred to as the source factor. The first term represents the intensity of the magnetic creeping ray launched by the source whose direction is along the binormal to the creeping ray, while the third represents the intensity of the electric creeping ray launched by a source whose direction is that of the tangent to the creeping ray. Until the 1970s, the second term, which provides the intensity of the electric creeping ray launched by a source directed along the binormal, was missing from the literature on GTD. In the case of a perfect conductor, the first term is dominant, while the other two are of the $O(k^{-1/3})$ relative to this term. There also exists a magnetic creeping ray launched by a source directed along the normal to the creeping ray. It is associated with the fourth term $(\hat{t} \cdot \hat{n})L'$ occurring in the dyadic coefficient of diffraction. However, it is of $O(k^{-2/3})$ with respect to the first term mentioned above, and is typically neglected.

For an electric dipole source in the direction of the normal to the surface $\vec{S} = S\hat{n}$, the fictitious surface field takes the form

$$\vec{E}(S) = \frac{ik}{4\pi} Z_0 S \left[\hat{n} M + \hat{b} N \right]. \tag{1.55}$$

The fictitious field at any point along the creeping ray is deduced from the field at point S by following a procedure which is the same as that for a classical creeping ray (see Sect. 1.3.4.4). The only difference is the geometrical factor resulting from the application of the power conservation in a tube of rays. For the case of the source, the power is conserved in an *infinitesimal* pencil of rays, characterized by its initial angular width $d\psi$. After propagating on the surface, the width of the pencil becomes $d\eta$ and the geometrical factor to be used is then $\sqrt{d\psi / d\eta}$.

We will now illustrate, via an example, the application of the procedure we just described. Let us calculate the field radiated by a magnetic dipole source of unit intensity and tangent to the surface. Furthermore, let us carry out the calculation for the case where the creeping ray is launched parallel to the source, $\vec{S} = S\hat{t}$.

According to (1.54), the initial field S is given by

$$\vec{E}(S) = \frac{ik}{4\pi} L_s \, \hat{b}. \tag{1.56}$$

This field propagates, with accompanied attenuation, until it reaches the detachment point Q' where it is given by

$$\vec{E}(Q') = \frac{ik}{4\pi} L_s \, exp \, i\left(k\ell + \int_Q^{Q'} \alpha_s ds \right) \sqrt{\frac{d\psi(S)}{d\eta(Q')}} \, \hat{b}'. \tag{1.57}$$

It then propagates from Q' to the observation point M where it can be expressed in terms of $E(Q')$ as

$$\vec{E}(M) = \vec{E}(Q') \, D_s(Q') \sqrt{\frac{\rho_d}{s(\rho_d + s)}} \, e^{iks}. \tag{1.58}$$

Notice that this procedure is exactly the same as the one employed earlier for the case of a creeping ray launched by an incident field from a distant source.

1.3.4.10 Launching and diffraction of traveling waves and edge waves

We have seen in Sect. 1.2.2.12, that Fermat's generalized principle predicts the existence of waves propagating along the discontinuity lines in general, and along the wires and edges in particular.

We will use the terminology *traveling waves* to describe waves that propagate along a wire, or more generally along a slender body whose longitudinal dimension is much larger than its transverse cross-section. Traveling waves are very well known in antenna theory, where one makes a clear distinction between the class of antennas operating with traveling waves *vs.* standing waves. In principle, it is possible to integrate the traveling waves into the GTD framework. Unfortunately, however, very little attention has been devoted to this subject in the literature, and the necessary diffraction and attenuation coefficients are not all known. The most well-known traveling waves are those propagating on a straight wire, which attenuate slowly with distance, and have been studied by Sommerfeld [St].

The traveling waves propagate on slender bodies by following along the geodesics. The ogival cylinder is an example of such a body. These waves may be viewed as creeping waves in the limit where the curvature transverse to the direction of propagation is very large compared to that in the direction of propagation. To date, the transition between the creeping and traveling waves has not been fully explored.

We will now briefly discuss two cases which have been partially investigated in the literature, viz., waves propagating along wires and the edge waves.

a) Traveling waves along wires

These wire types of traveling waves follow Fermat's generalized principle and these waves can either be generated by: (i) a voltage source located on the wire; (ii) an incident wave whose wave vector is tangent to the wire; and, (iii) a wave impinging on one of the extremities of the wire. These waves detach themselves tangentially from the wire, or diffract from one of the extremities of the wire. All of the canonical problems associated with these waves have not been resolved; in particular, the attenuation coefficients of these waves on a curved wire are not known. Although the straight-wire problems have been investigated more extensively than their curved counterparts, a consensus on the use of the diffraction coefficients has not been reached. We refer the reader to the article by Peters et al. and Shamansky et al. [SD], where more details may be found on an interpretation of the diffraction by a straight wire at high frequencies [SD].

b) Edge waves

As the name implies, the edge waves propagate along the edges of the wedges. They can be generated by a dipole located in the vicinity of the edge [Bu], by an incident wave whose wave vector is along the tangent to the edge, or by a discontinuity, for instance, a corner on the edge. They detach tangentially from the edge, or diffract around the corners. These waves can play an important role, especially on straight edges. For this case, they decrease as $r^{-\pi/\gamma}$, as a function of the distance r from the edge, where γ is the external angle of the wedge [Bu]. Thus, if the wedge is a half-plane, $\gamma = 2\pi$ and the edge wave decreases as $r^{-1/2}$, while a space ray emitted by a dipole behaves as $1/r$. The attenuation coefficients of the waves are not known, because an analytical solution to the canonical problem for a wedge with a curved edge is not available. Additionally, the exact diffraction coefficients of these waves by a corner are not known. Nevertheless, approximate techniques have been proposed for handling the problems of the quarter-plane [Ha1, Ha2] and the polyhedric corner [IM]. An examination of these techniques shows that it is not always possible to incorporate these edge waves into the GTD, even though they play an important role in scattering, in particular, in the investigation of the problem of diffraction on polyhedral structures. Unfortunately, until now, these waves have been neglected in the publications pertaining to the GTD, and it is felt that they should be examined more thoroughly in the future because of their contributions to the diffracted field, which, in many instances, are far from negligible.

1.3.5 Summary

In this section we have shown how to extract, from the postulates of the GTD, the rules that govern the propagation phase (Sect. 1.3.1), amplitude (Sect. 1.3.2), as well as the polarization (Sect. 1.3.3) along a ray. Then, in Sect. 1.3.4, these rules were applied in conjunction with the localization principle to identify and characterize different species of diffracted rays, and have also introduced the notion of *diffracted field* originating at the *point of diffraction*. This notion of the phenomenon of diffraction is not only convenient, especially when multiple interactions are encountered, but it also provides a powerful tool for systematically calculating the field (fictitious) along a ray. The physical diffracted field is ultimately obtained from the fictitious field at the last point of diffraction and from the geometrical parameters of the diffracted field, viz., the radii of curvature of the wavefront, be it reflected or diffracted. The calculation of these various geometrical parameters are discussed in the next section, and the diffraction coefficients that are employed most frequently in GTD are given in Sect. 1.5.

1.4 Calculation of the geometrical factors

The formulas pertaining to the computation of the diffracted field given in Sect. 1.3 depend upon a number of geometrical factors such as the radius of curvature of the reflected wave ρ_1^r and the quantity ρ_2^r defined in Sect. 1.3.4.1; namely, the radius of curvature ρ_d of the diffracted wavefront (1.3.4.2), the ratio $\delta\eta(Q)/\delta\eta(Q')$ of the widths of a infinitesimal pencil of geodesics between the attachment and detachment points of a creeping ray (1.3.4.3), and the ratio $\delta\psi(S)/\delta\eta(Q')$ between the angle at the observation point and the width at the detachment point of an infinitesimal pencil of creeping rays emitted by a source on the surface.

In Sect. 1.4.1, we discuss the evaluation of the geometrical parameters along a ray. We will make a distinction between the case of the space rays, that are characterized by the curvature matrix of the wavefront, and the surface rays (see 1.4.1.2). We will discuss the evaluation of the width and the angular opening of an infinitesimal pencil of surface rays as it propagates. The changes in the geometrical parameters of the ray pencil as it interacts will be discussed in detail for each type of interaction in Sect. 1.4.2.

The ray calculations described in this section rely upon the use of notions in differential geometry, which, though fairly simple in concept, can nonetheless seem to be quite involved when one attempts to implement them in practice. The section will describe the method and will also provide the relevant references where the details of the calculation can be found.

1.4.1 Evolution of the geometrical parameters along a ray

1.4.1.1 Introduction

In this section we will discuss the evolution of the geometrical parameters of the rays as they propagate.

A wavefront is typically described in terms of its principal radii of curvature ρ_1 and ρ_2, and the orientation of the frame of the principal directions. However, we will find it convenient, instead, to employ the frames attached to the diffracting object. For instance, we will use the frame $\left(\hat{\imath}, \hat{e}_\perp, \hat{e}_\shortparallel^i\right)$ for the field incident on a reflecting surface (see 1.3.4.1), and $\left(\hat{\imath}, \hat{\Phi}', \hat{\beta}'\right)$ for the field incident on an edge.

In these frames, the wavefront is represented by a symmetric matrix of curvature Q, which is non-diagonal. More precisely, if s is the coordinate along the ray, and b is the two-component vector of coordinates perpendicular to the ray, the second-order approximation to the wavefront is given by

$$s = -\frac{1}{2}{}^t b\, Q\, b. \tag{1.59}$$

If the phase reference is chosen at the point $(s, b) = (0, 0)$, then its value at a point (s, b) is given by

$$s + \frac{1}{2}{}^t b\, Q\, b, \tag{1.60}$$

to within a second-order approximation. If the curvature matrix Q is diagonal, i.e., if the wavefront is expressed in its principal axes, (1.59) becomes

$$s = -\frac{1}{2}\left(\frac{b_1^2}{\rho_1} + \frac{b_2^2}{\rho_2}\right). \tag{1.61}$$

If one traverses a distance s along the wavefront, the principal axes of the wavefront are conserved along the ray, and the radii of curvature ρ_1 and ρ_2 become $\rho_1 + s$ and $\rho_2 + s$, respectively. Let P be the matrix that transforms the frame chosen to express Q to the principal frame

$$P^{-1}\, Q^{-1}\, (s)\, P = P^{-1}\, Q^{-1}(0)\, P + s\, I, \tag{1.62}$$

which can be rearranged to read

$$Q^{-1}\, (s) = Q^{-1}\, (0) + s\, I. \tag{1.63}$$

Equation (1.63) describes the evolution of Q along the space ray, for a reflected, diffracted, or a creeping ray but not for surface rays. While the wavefront for these rays is described in terms of its radius of curvature, the evolution of ρ is more complicated than that of the space rays. We will discuss this point in some detail in the next section.

1.4.1.2 Infinitesimal width and curvature of the wavefront of a pencil of surface rays
Consider an infinitesimal pencil of surface rays bounded by two geodesics that are infinitesimally close to each other (see Fig. 1.14). This pencil is characterized by its width $d\eta$, i.e., the separation between the two geodesics, and by the angle $d\psi$ between the tangents to the two geodesics.

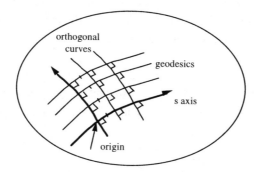

Fig. 1.14. Geodesic coordinate system for creeping rays

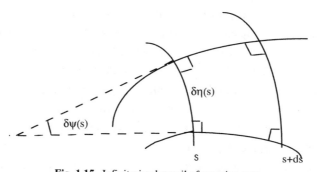

Fig. 1.15. Infinitesimal pencil of creeping rays

Let us now use the geodesic coordinate system defined by the pencil of rays. The essential property of this system (see Fig. 1.15) is that the curvilinear coordinates are equal on all geodesics and, as a consequence, the curvilinear abscissa s can be chosen on any geodesic. The quantities $d\eta$ and $d\psi$ depend on s and it can be shown that they obey the following differential equations [Stru]

$$\frac{d(d\eta)}{ds} = d\psi, \tag{1.64}$$

$$\frac{d(d\psi)}{ds} = -K\,d\eta, \tag{1.65}$$

where K is the Gaussian curvature of the surface.

The radius of curvature of the wavefront ρ, which is also the geodesic radius of curvature of the orthogonal trajectory of the geodesic, is simply the ratio $d\eta/d\psi$.

$$\rho = \frac{d\eta}{d\psi} \tag{1.66}$$

For the coordinate system we have chosen, the first fundamental quadratic form of the surface is diagonal. If α is the coordinate on the curve orthogonal to the geodesics, chosen as the coordinate axes, the quadratic form reads

$$I = d^2 s + g_{\alpha\alpha} \, d^2 \, \alpha, \tag{1.67}$$

where $g_{\alpha\alpha}$ characterizes the widening (or narrowing) of the pencil between the axis of the coordinates α ($s = 0$) and the point with abscissa s and $d\eta$ is proportional to $\sqrt{g_{\alpha\alpha}}$. Equations (1.64) and (1.65) imply that

$$\frac{d^2 \sqrt{g_{\alpha\alpha}}}{ds^2} + K(s)\sqrt{g_{\alpha\alpha}} = 0, \tag{1.68}$$

which is the Jacobi equation [Stru]. We will have the occasion to use the parameter $\sqrt{g_{\alpha\alpha}}$ again in Chap. 3. Equations (1.64) and (1.65) enable us to calculate $d\eta(s)$ and $d\psi(s)$ from the initial conditions $d\eta(0)$ and $d\psi(0)$. Evidently, all these quantities are defined within a scale factor. The geometrical factors entering the calculations are, in effect, the ratios of the infinitesimal quantities $d\eta(Q)/d\eta(Q')$, $\rho_d = d\eta(Q')/d\psi(Q')$, defined in Sect. 1.3.4.2, and $d\psi(S)/d\eta(Q')$, defined in Sect. 1.3.4.9. Thus, a single initial condition is sufficient to calculate all of the geometrical factors. For the creeping rays, or diffracted creeping rays, this initial condition is given by the radius of curvature ρ at the attachment or the diffraction point. As for the rays emitted by a source, the initial condition is given by $d\eta(S)/d\psi(S) = 0$.

For an arbitrary general surface, the system of Eqs. (1.64) and (1.65), that are equivalent to (1.68), can only be solved numerically, unless K is a constant, in which event an analytical solution becomes possible. As an illustrative example, we consider the simple problem of the diffraction of a plane wave by a sphere. The Gaussian curvature K is equal to $1/R^2$, where R is the radius of the sphere. Since the incident field is a plane wave the radius of curvature at the attachment point is infinite and the initial condition becomes $d\psi/d\eta = 0$, or alternatively

$$d\psi(0) = 0, \tag{1.69a}$$

$$d\eta(0) = d\eta(Q). \tag{1.69b}$$

Equations (1.64) and (1.65), when combined with (1.69), lead to the result that at the point M on the geodesic QQ'

$$d\eta(M) = \cos\left(\frac{s}{R}\right) d\eta(Q), \qquad (1.70)$$

$$d\psi(M) = \sin\left(\frac{s}{R}\right) \frac{d\eta(Q)}{R}, \qquad (1.71)$$

which leads to

$$\rho = \frac{d\eta(M)}{d\psi(M)} = R \cot g\left(\frac{s}{R}\right), \qquad (1.72)$$

and

$$\frac{d\eta(Q)}{d\eta(Q')} = \frac{1}{\cos\left(\frac{s}{R}\right)}. \qquad (1.73)$$

Notice that this factor becomes infinite when $s = \pi/2\,R$, that is to say at the pole in the shadow of the sphere, as this pole is the focus of the creeping rays.

The two previous paragraphs described a procedure for calculating the evolution of the geometrical parameters along a ray. Next, we will investigate how these parameters are transformed as the ray interacts with the surface.

1.4.2   Transformation of the geometrical factors in the course of interaction of the ray with the surface

1.4.2.1 Calculation of the curvature matrix of the front wave reflected at point 0

The curvature matrix Q^r of the wavefront, reflected at a point 0, depends upon the following:

(i)   the curvature matrix Q^i of the incident wavefront
(ii)   the curvature matrix of the reflecting surface
(iii)   the reflection angle θ.

Several methods are available for performing the calculation of the curvature matrix. An efficient way of doing this is to equate the phase of the reflected and incident wavefronts, or more exactly their quadratic approximations, on the reflecting surface [Ja], the reference frames defined in Sect. 1.3.4.1.

If we take the first coordinate along $\hat{e}_\perp$ and the second along $\hat{e}_\shortparallel^i$, the curvature matrix of the reflected wavefront becomes

$$\underline{Q}^i = \begin{pmatrix} Q_{11}^i & Q_{12}^i \\ Q_{12}^i & Q_{22}^i \end{pmatrix}. \tag{1.74}$$

Using the coordinate system $\hat{e}_\perp$ and $\hat{n}$ ä $\hat{e}_\ddot{o}$, where $\hat{n}$ is the normal to the surface at the reflection point O, the curvature matrix of the surface can be written in the form

$$\underline{C} = \begin{pmatrix} C_{11} & C_{12} \\ C_{12} & C_{22} \end{pmatrix}. \tag{1.75}$$

Thus, referring to n (resp. x_1, x_2) as the coordinate along $\hat{n}$ (resp. $\hat{e}_\perp, \hat{n} \times \hat{e}_\perp$) the osculating paraboloid to the reflecting object can be expressed as

$$\mathrm{n} = -\frac{1}{2}(C_{11}x_1^2 + 2C_{12}x_1x_2 + C_{22}x_2^2). \tag{1.76}$$

By using (1.60) in the triad $\left(i, \hat{e}_\perp, \hat{e}_\shortparallel^i\right)$, the phase of the incident wavefront can be written in the coordinate system (i, x_1^i, x_2^i) as

$$\varphi^i = i + x^i Q^i x^i. \tag{1.77}$$

Similarly, by using the coordinate system for the reflected field we get

$$\varphi^r = r + x^r Q^r x^r. \tag{1.78}$$

We can now perform a transformation such that we can express (1.77) and (1.78) in the coordinate system $(\hat{n}, \hat{e}_\perp, \hat{n} \times \hat{e}_\perp)$ attached to the surface. The equalities of φ^i and φ^r, given by (1.77) and (1.78), provide a way to extract the transformation matrix Q_r. The details of the calculations may be found in reference [Ja] and we will merely present the results here using the same notations. Expressed in the coordinate system, x_1^r, x_2^r, the matrix Q takes the simple form

$$Q^r = \begin{pmatrix} 2C_{11}\sin\theta + Q_{11}^i & 2C_{12} - Q_{12}^i \\ 2C_{12} - Q_{12}^i & 2C_{22}(\sin\theta)^{-1} + Q_{22}^i \end{pmatrix}. \tag{1.79}$$

The principal radii of curvature ρ_1^r and ρ_2^r of the wavefront and its principal directions are obtained by diagonalizing the matrix Q^r.

1.4.2.2 Calculation of the radius of curvature ρ_d of the wavefront of the ray diffracted by an edge

The definition of ρ_d, the non-vanishing radius of curvature of the wavefront of the ray diffracted by an edge has been given in Sect. 1.3.4.2. The radius of curvature ρ_d is oriented along the direction of the vector $\hat{\beta}$, while the one in the $\hat{\Phi}$ direction is zero. Using (1.63), the curvature matrix can be expressed in the coordinate system $(\hat{\beta}, \hat{\Phi})$, at a distance s from the point 0, as

$$\begin{pmatrix} \dfrac{1}{\rho_d + s} & 0 \\ 0 & \dfrac{1}{s} \end{pmatrix}.$$

Following a similar procedure as outlined previously, the following expression for the radius of curvature ρ_d is obtained [Ja] by expressing on the edge the equality of the phases of the incident and diffracted waves.

$$\frac{1}{\rho_d} = Q_{11}^i + \frac{1}{\rho_a \sin^2 \beta}(\hat{i} \cdot \hat{n}_a - \hat{d} \cdot \hat{n}_a), \qquad (1.80)$$

where $\hat{i}$ and $\hat{d}$ are, as in Sect. 1.3.4.2, the unit vectors of the incident ray and the diffracted ray, respectively; β is the angle between $\hat{i}$ (or $\hat{d}$) and the tangent to the edge; $\hat{n}_a$ is the vector normal to the edge; and, ρ_a the radius of curvature of the edge, all of which are expressed at the diffraction point 0. The quantity Q_{11}^i is the first element in the curvature matrix of the incident wavefront, in the coordinate system $\hat{\beta}'$ and $\hat{\Phi}'$. It is simply the curvature of the normal section of the wavefront, with the plane defined by the incident ray vector $\hat{i}$ and the tangent to the edge $\hat{t}$.

1.4.2.3 Diffraction by a tip

As we saw in Sect. 1.3.4.3, the wavefront diffracted by a tip is spherical, with its center located at the diffracting tip. Consequently, for this case the need to evaluate any geometrical parameters is obviated.

1.4.2.4 Creeping ray initiated on a smooth surface

The initial geodesic radius of curvature of the wavefront of the creeping ray is simply the curvature of the normal section of the incident wavefront and the tangent plane to the surface at the attachment point. This radius gives the initial condition to be used in the system of Eqs. (1.64) and (1.65). As we saw in Sect. 1.4.1.2, it is possible to calculate $d\eta$ and $d\psi$ at every point along the ray and consequently, the broadening $d\eta(Q')/d\eta(Q)$ of the ray pencil between the attachment point Q and the detachment point Q' can be readily obtained. The non-vanishing radius of curvature, ρ_d, of the diffracted wavefront is then given by

$$\rho_d = \frac{d\eta(Q')}{d\psi(Q')} .$$

(1.81)

In the above equation, ρ_d is the radius of curvature of the normal section of the diffracted wavefront with the tangent plane at the surface, i.e., the radius of curvature along the binormal to the creeping ray; the other radius of curvature is zero. At a distance s from the diffraction point, the curvature matrix of the diffracted wavefront, expressed in the coordinate system $(\hat{b}, \hat{n})$, is given by

$$\begin{pmatrix} \dfrac{1}{\rho_d + s} & 0 \\ 0 & \dfrac{1}{s} \end{pmatrix} .$$

Note the similarity between this case and that of the ray diffracted by an edge.

1.4.2.5. Creeping ray diffracted by an edge

The radius of curvature of the diffracted wavefront for an edge is extracted from the results for the edge diffraction, given above, by simply inserting the value of the incident angle at grazing in (1.80). This yields

$$\frac{1}{\rho_d} = \frac{1}{\rho} + \frac{1}{\rho_a \sin^2 \theta}(\hat{i} \cdot \hat{n}_a - \hat{d} \cdot \hat{n}_a) .$$

(1.82)

In (1.82), ρ is the radius of curvature of the surface wavefront at the diffraction point. The expression for $\hat{n}_a/\rho_a$ is

$$\frac{\hat{n}_a}{\rho_a} = \frac{\hat{b}_a}{\rho_g} + \frac{\hat{N}}{\rho_n} .$$

(1.83)

The coordinate system $(\hat{t}_a, \hat{b}_a, \hat{N})$ defines Darboux's frame of reference [Stru] for the edge, where $\hat{t}_a$ is tangent to the edge, and $\hat{N}$ is the normal to the surface.

$$\frac{\hat{i} \cdot \hat{n}_a}{\rho_a} = \frac{\hat{i} \cdot \hat{b}_a}{\rho_g} = -\frac{\sin \beta}{\rho_g} \qquad (1.84)$$

1.4.2.6 Excitation of creeping rays on a wedge with curved faces

The geodesic radius of curvature of the surface wavefront is deduced from the results for the diffraction by an edge, by considering the case of the grazing angle of observation in (1.80).

1.4.2.7 Creeping ray traveling on an edge with curved faces

For the case in which both the incidence and the observation angles are grazing, we may use (1.80) and (1.83), to derive the following expressions

$$\frac{1}{\rho_r} = \frac{1}{\rho_i} - \frac{2}{\rho_{g_1} \sin \beta}, \qquad (1.88)$$

$$\frac{1}{\rho_t} = \frac{1}{\rho_i} - \frac{1}{\sin \theta} \left(\frac{1}{\rho_{g_1}} - \frac{1}{\rho_{g_2}} \right), \qquad (1.89)$$

where we denote the geodesic radius of curvature of the incident wavefront by ρ_i, that of the reflected wavefront by ρ_r, that of the transmitted wavefront by ρ_t, and the geodesic radius of curvature of the edge on the face 1 by ρ_{g_1}.

Notice that the expression in (1.88) could also be derived by considering the two dimensional problem of the reflection by a cylinder with radius ρ_g.

1.4.2.8 Launching and diffraction of creeping rays by a tip with curved faces

A creeping ray incident on a tip generates a spherically-diffracted wavefront. The initial condition for this case is the same as in the case of a source, namely $d\eta(0) = 0$.

1.4.2.9 Excitation of creeping rays by a source on the surface

As we saw in Sect. 1.4.1.2, the geometrical factor $\sqrt{d\psi / d\eta}$ for the field amplitude is deduced from the system of Eqs. (1.64) and (1.65) with the initial condition $d\eta(S) = 0$.

1.4.2.10 Wire waves and edge waves

The case of wire and edge wave propagation is another example where no geometrical factors need be calculated. In fact, the rays associated with progressive and edge waves do not diverge, and the wavefront of the diffracted space rays is entirely spherical.

1.4.3 Conclusions

The results given in Sect. 1.4.1 enable us to track the evolution of the geometrical parameters of the wavefronts along the rays. Likewise, those in Sect. 1.4.2 help us compute the transformation of these parameters during the course of the interaction of the ray with the surface of the object. It is thus possible to calculate, recursively, the geometrical parameters of a ray undergoing any number of transformations induced by these interactions. Once these geometrical parameters have been determined, the field can be computed by employing the diffraction coefficient as discussed in Sect. 1.3.4. The method for calculating the diffraction coefficients will be the subject of discussion of the next section.

1.5 Calculation of the diffraction coefficients

In Sect. 1.1, it was shown how the diffraction coefficient establishes the link between the incident and the diffracted field at the diffraction point. According to the localization principle, the diffraction coefficient depends only upon the local geometry of the surface in the neighborhood of the diffraction point. Consequently, we replace the original geometry of the scatterer at the diffraction point by that of a canonical object whose geometrical characteristics bear a close resemblance to those of the object under consideration. It is evident that the canonical object must be chosen such that it permits the construction of an exact solution to the diffraction problem for this object in a form that is convenient for the derivation of the diffraction coefficient. As we have seen already (see Sect. 1.3.4.4), there is no unique definition of the canonical object. For instance, for the case of a smooth surface, one can choose the canonical object to be a cylinder with a normally or obliquely incident wave; however, the choice of a sphere or an elliptic cylinder is also permissible for this example. Frequently, it is necessary to use a combination of several different canonical objects in order to adequately reproduce the various characteristics of the original object. Examples of such a procedure abound in the literature, and, in this text, we will merely list the diffraction coefficients without detailing their derivation. We will present some physical interpretations of the results extracted from the solution of the canonical problems, but refer the reader interested in the details of the calculations to the relevant publications or to the appendix.

Finally, we recall that the localization principle applies equally well to the incident field, and, in view of this, we will replace a ray field by a plane wave for computational purposes in the discussion below. We will defer until in Chap. 4 to deal with the situation where the incident field no longer satisfies the criterion for ray fields.

We will proceed next with the presentation and discussion of the most commonly-used diffraction coefficients in GTD.

1.5.1 Reflection coefficients

The reflection coefficients R_{TE} and R_{TM} for the TE and TM waves, respectively, are evaluated by solving the canonical problem associated with the reflection of an incident wave by a planar surface. The reflection coefficient is calculated by simply imposing the condition that the total field satisfy an impedance condition on the surface for a source impedance Z, and an incidence angle θ with respect to the plane ($\theta = 0$ at grazing incidence). These reflection coefficients are given by

$$R_{TE} = \frac{\sin\theta - Z}{\sin\theta + Z},$$
(1.90)

$$R_{TM} = \frac{Z\sin\theta - 1}{Z\sin\theta + 1}.$$
(1.91)

For a perfect conductor, $Z = 0$, and the expressions given above tend to the limits of $R_{TE} = 1$ and $R_{TM} = -1$, implying that the reflected and the incident waves are equal in amplitude, a result that is not totally unexpected.

1.5.2 Diffraction by an edge or a line discontinuity

The problem of diffraction by a line discontinuity is handled in the same manner as that of the diffraction by an edge, except for the modification of the diffraction coefficients D_s and D_h. In what follows, we will first address the problem of diffraction by an edge, and subsequently discuss the topic of diffraction by other types of line discontinuities.

1.5.2.1 Diffraction by an edge

The canonical problem associated with edge diffraction is that of a plane wave diffracted by an infinite wedge. For a perfect conductor, the solution to this problem can be readily constructed by using the well-known method of separation of variables [SV]. Initially, it is convenient to construct the solutions for normal incidence and the boundary conditions $u = 0$, i.e., the so-called soft boundary case, as well as for the hard boundary condition for which $\partial u/\partial n = 0$. The solution for the case of oblique incidence can then be readily constructed from the knowledge of the above two types of solutions.

The situation becomes somewhat more involved however, when an impedance condition is used. In a classic paper, Maliuzhinets [Ma], has provided the solution for normal incidence, for the problem of diffraction by a wedge whose surfaces satisfy impedance boundary conditions that may, in general, be different for the two faces of the edge. To date, the problem of diffraction by an impedance wedge at oblique incidence remains unresolved, except for some specific angles of the wedge and for particular values of impedances. The solution may be expressed in a series form, an integral form, or be represented in terms of special functions. A ray interpretation of the asymptotic limit of the above solution must be found in order for it to be useful in practice.

The asymptotic representation, in turn, depends upon the location of the point of observation. The geometrical optics solution for the wedge suggests a partitioning of the observation space into the following three zones

(i)   Zone 1 : $u^r \neq 0$ and $u^i \neq 0$

(ii)   Zone 2 : $u^r = 0$, $u^i \neq 0$

(iii) Zone 3 : $u^r = u^i = 0$

that are defined by the combination of the incident field u^i and the reflected one u^r (see Fig. 1.16).

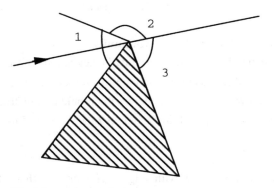

Fig. 1.16. Light and shadow zones in wedge diffraction

The asymptotic expansions of the field in these three zones are given by

$$\text{Zone 1}: u \approx u^r + u^i + u^d$$
$$\text{Zone 2}: u \approx u^i + u^d$$
$$\text{Zone 3}: u \approx u^d$$

The diffracted field u^d is $O(k^{-1/2})$ with respect to the incident and the reflected field. The residual terms are $O(k^{-3/2})$. The diffracted field decreases with r as $1/\sqrt{kr}$. For TM or *soft* polarization, it reads

$$u = D_s / \sqrt{r} + O(k^{-3/2})$$

Similarly, for the TE polarization, the diffracted field is expressed as

$$u = D_h / \sqrt{r} + O(k^{-3/2})$$

D_s and D_h are the desired diffraction coefficients in the two expressions above.

When the angle of incidence is oblique, and the wedge is perfectly conducting, the corresponding diffraction coefficient can be extracted from the expression for the normal incidence. The result is

$$D_h^s = \frac{e^{i\pi+4} \sin \frac{\pi}{n}}{n\sqrt{2\pi k}\,\sin \beta} \left(\frac{1}{\cos \frac{\pi}{n} - \cos\left(\frac{\Phi-\Phi'}{n}\right)} \mp \frac{1}{\cos \frac{\pi}{n} - \cos\left(\frac{\Phi+\Phi'}{n}\right)} \right), \qquad (1.92)$$

$$n = \frac{\gamma}{\pi}.$$

where γ is the exterior angle of the wedge, β is the angle between the tangent to the edge and the incident wave vector, and Φ is the angle with respect to the illuminated face of the wedge of the diffracted wave (see Fig. 1.17).

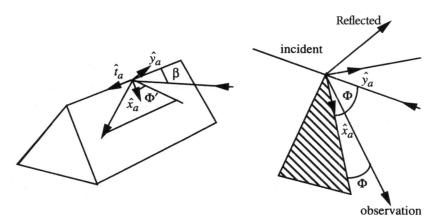

Fig. 1.17. Diffraction by a wedge

More precisely, let $\hat{t}_a$ be the tangent to the edge, $\hat{x}_a$ be the vector residing on the plane tangent to the illuminated face, pointing towards this face and orthogonal to $\hat{t}_a$. Let us consider the right-handed coordinate system $T = (0, \hat{x}_a, \hat{y}_a, \hat{t}_a)$, and let β and Φ be the Euler angles of the diffracted vector with respect to T.

An examination of the diffraction coefficients reveals that D_h^s in (1.92) becomes infinite when

$(\Phi - \Phi') = \pi$, i.e., at the shadow boundary of the incident field

$(\Phi + \Phi') = \pi$, i.e., at the shadow boundary of the reflected fields.

This behavior of the diffraction coefficient is not totally unexpected however, since, strictly speaking, in the neighborhood of these shadow boundaries the field is no longer expressible as a sum of geometrical optics and diffracted fields because the former is discontinuous at the shadow boundaries, while the total field must be continuous everywhere. Thus, in a *boundary layer* in the neighborhood of the boundaries, it becomes necessary to employ alternate representation of the diffracted field so that it behaves correctly at the shadow boundaries. The correction of the singularity of the diffraction coefficient at the shadow boundaries will be addressed in Chap. 5, where the so-called uniform theories will be discussed. We should mention that the diffraction coefficient D_h^s becomes infinite when $\beta = 0$, i.e., for paraxial incidence. However, for this case, an edge wave is launched and, consequently, a field description in terms of diffracted rays is no longer valid.

When the illuminated face $\Phi = 0$ is characterized by a surface impedance Z_0, and the face in the shadow zone $\Phi = n\pi$ with an impedance Z_n, the diffraction coefficient at normal incidence is given by

$$D_h = \frac{e^{i\pi/4}}{\sqrt{2\pi k}}\, \frac{1}{\Psi\left(\Phi' - \frac{n\pi}{2}\right)} \left(\frac{\Psi\left(\Phi - \frac{n\pi}{2} - \pi\right)}{\cos\left(\frac{\pi+\Phi}{n}\right) - \cos\frac{\Phi'}{n}} - \frac{\Psi\left(\Phi - \frac{n\pi}{2} + \pi\right)}{\cos\left(\frac{\pi-\Phi}{n}\right) - \cos\frac{\Phi'}{n}} \right), \qquad (1.93)$$

where ψ is a special function which is fairly involved, because it is not only a function of its argument, but also depends upon the impedance of the faces as well as the angle of the wedge. The special function takes the form

$$\Psi(\alpha) = \Psi_n\left(\alpha + \frac{n\pi}{2} + \frac{\pi}{2} - \theta_0\right) \Psi_n\left(\alpha + \frac{n\pi}{2} - \frac{\pi}{2} + \theta_n\right)$$

$$\Psi_n\left(\alpha - \frac{n\pi}{2} + \frac{\pi}{2} - \theta_n\right) \Psi_n\left(\alpha - \frac{n\pi}{2} - \frac{\pi}{2} - \theta_0\right), \qquad (1.94)$$

where Ψ_n is a special meromorphic function defined by Maliuzhinets [Ma]. Molinet [Mo], and Volakis and Senior [VS], have provided a simplified expression for Ψ_n which reads

$$\Psi_n(u) = \sqrt{\cos\left(\frac{u}{2n}\right)}\, exp\left(\frac{n}{\pi} \int_0^1 \frac{Log\left(1 + w^2 tg^2\left(\frac{u}{2n}\right)\right) dw}{(1 - w^2)\, ch(2nArgthw)} \right). \qquad (1.95)$$

In (1.94), θ_0 is the Brewster angle of the face O and θ_n. More precisely, for the s (or TM) polarization, these angles are given by

$$\sin\theta_0 = 1/Z_0 \quad , \quad \sin\theta_n = 1/Z_n,$$

while for the h (or TE) polarization, the corresponding expressions are

$$\sin\theta_0 = Z_0 \quad , \quad \sin\theta_n = Z_n.$$

Thus θ_0 and θ_n are angles leading to a reflection coefficient equal to zero on each face.

In general, the solution for the impedance wedge also includes surface waves, but they play a significant role only when one of the Zs is almost purely imaginary and negative.

Finally, the case of oblique incidence for an impedance wedge has remained elusive to-date and only some special cases have been treated. They include: (i) plane with an impedance discontinuity [Va]; (ii) half-plane [BF]; (iii) right angle wedge, one of whose faces is a perfect electric or magnetic conductor [Va, Se3]; (iv) wedge with an arbitrary angle and impedance [Be2].

Starting from these particular solutions, Syed and Volakis [Sy, V] have extracted an approximate solution for the general case. However, no correct general solution has been published for the general case of a wedge illuminated at an oblique incidence.

1.5.2.2 Diffraction by a discontinuity in the curvature

The problem of discontinuity in curvature has been solved by Weston [W] and Senior [Se2] for a perfect conductor, and by Kaminetzky and Keller [KK] for an impedance wedge. Weston and Senior have used an approximation of the integral equation on two parabolic semi-cylinders that are joined together such that imposing their tangents are continuous, whereas Kaminetzky and Keller have used a boundary layer method. For a perfect conductor, the diffraction coefficients D_s and D_h, take the form

$$D_s = A \frac{\sin \Phi' \sin \Phi}{(\cos \Phi' + \cos \Phi)^3 \sin^2 \beta}, \qquad (1.96)$$

$$D_h = -A \frac{1 + \cos \Phi' \cos \Phi}{(\cos \Phi' + \cos \Phi)^3 \sin^2 \beta}, \qquad (1.97)$$

where A is defined by the expression

$$A = e^{-i\pi/4} \sqrt{\frac{2}{\pi k}} \left(\frac{1}{ka_1} - \frac{1}{ka_2} \right). \qquad (1.98)$$

In the above, a_1 denotes the radius of curvature of the face $\Phi = 0$, illuminated by the incident ray (the notation is the same as that for the wedge) and a_2 is the corresponding curvature of the face $n = \pi$.

We notice that D_s and D_h diverge for the wedge on the shadow boundary of the reflected field as defined by $\Phi + \Phi' = \pi$. For this case, the incident shadow boundary is located in the interior of the obstacle. The divergence, which has a $(\pi - (\Phi + \Phi'))^3$ behavior, is more rapid than for the wedge at $\beta = 0$.

For the case of a surface impedance, the expression for the diffraction coefficient is, for the acoustic case

$$D_h = -A \frac{\sin \Phi' \sin \Phi (\sin^2 \beta (1 + \cos \Phi' \cos \Phi) - Z^2)}{(\sin \beta \sin \Phi' + Z)(\sin \beta \sin \Phi + Z)(\cos \Phi' + \cos \Phi)^3 \sin^2 \beta}, \qquad (1.99)$$

$$D_s(Z) = D_h(1/Z), \qquad (1.100)$$

where, as previously, Z is the impedance of the diffracting surface relative to free space.

The diffraction contribution due to a curvature discontinuity is weaker than that of the diffraction by an edge, since the diffraction coefficients of the former behave as $k^{-3/2}$, rather than as $k^{-1/2}$ corresponding to the wedge. Kaminetzky and Keller have studied other types of weak discontinuities associated with higher-order derivatives of the surface and derivatives of the impedance. The results for these discontinuities will be presented in the next section.

1.5.2.3 Diffraction by higher-order discontinuities

For higher-order discontinuities, Kaminetzky and Keller employ the same boundary layer method as is used for curvature discontinuities. For a discontinuity $[f^{(j)}]$ of the jth surface derivative and a discontinuity $[Z^{(j-1)}]$ of the $(j-1)^{th}$ derivative of the surface impedance, with $j \geq 2$, the diffraction coefficients read, for acoustics

$$D_h(Z) = \frac{(-1)^{j+1}}{i^{j-2}} \frac{e^{-i\pi/4}}{k^{j-1/2}} ,$$

$$\sqrt{\frac{2}{\pi}} \frac{\sin \Phi' \sin \Phi \{(\sin^2 \beta(1+\cos \Phi' \cos \Phi) - Z^2)[f^{(j)}] + \sin \beta(\cos \Phi' + \cos \Phi)[Z^{(j-1)}]\}}{(\sin \beta \sin \Phi' + Z)(\sin \beta \sin \Phi + Z)(\cos \Phi' + \cos \Phi)^{\gamma+1}(\sin \beta)^{\gamma}} ,$$

and
(1.101)

$$D_s(Z) = D_h\left(\frac{1}{Z}\right).$$
(1.102)

The expressions given in (1.101) and (1.102) are general formulas for the diffraction coefficients, and they include the particular cases discussed in the previous section. We notice, once again, that D_s and D_h diverge on the shadow boundary of the reflected field and at $\beta = 0$, and the singularity increases with increasing j. The diffraction coefficients behave as $k^{-j+1/2}$, and, consequently, become very weak for large j. For these two reasons, the coefficients as given in (1.101) and (1.102) are rarely used in the range $j > 2$.

An interesting case is that of a the linear beveled edge of a high index material backed by a conductor, that corresponds to a discontinuity in the derivative of the impedance. For $\beta = \pi/2$, the results for the diffraction coefficient are

$$D_h = -\frac{e^{-i\pi/4}}{k^{3/2}} \sqrt{\frac{2}{\pi}} \frac{[Z']}{(\cos \Phi + \cos \Phi')^2} ,$$
(1.103)

$$D_s = \frac{e^{-i\pi/4}}{k^{3/2}} \sqrt{\frac{2}{\pi}} \frac{\sin \Phi \sin \Phi' [Z']}{(\cos \Phi + \cos \Phi')^2} ,$$
(1.104)

where $[Z']$ designates the jump discontinuity in the derivative of the surface impedance.

We now have in our repertoire all of the diffraction coefficients needed to deal with the following geometries: (i) line discontinuities on a surface, e.g., discontinuity of the tangent plane and surface impedance discussed in Sect. 1.5.2.1; (ii) discontinuity of the curvature given in Sect. 1.5.2.2; and, (iii) discontinuity of the impedance derivatives and of the higher-order derivatives of the surface presented in Sect. 1.5.2.3. We turn next to the only remaining geometry, viz., the wire configuration, which can also be treated as a line discontinuity.

1.5.2.4 Diffraction by a metallic wire

The problem of diffraction by a metallic wire is discussed in this section. We assume in what follows that the product of the wire radius b and the wave number k is small, and thus we are justified in neglecting the terms $O((kb)^2) \ell n\ (kb)$. Using this approximation [26] we obtain

$$D_s = e^{i\pi/4}\left(\frac{\pi}{2k\sin^2\beta}\right)^{1/2}\frac{1}{\left(\gamma + log\left(\frac{kb\sin\beta}{2i}\right)\right)}, \tag{1.105}$$

$$D_h \approx 0, \tag{1.106}$$

where γ is the Euler constant $\gamma \approx 0.577$.

More accurate results for the diffraction coefficients have been provided by Keller [KA]. Note that as in the case of edge diffraction, the diffraction coefficient for a wire behaves as $k^{-1/2}$ and, as a consequence, can be sizable even at high frequencies.

We have catalogued above some of the most commonly-encountered diffraction coefficients for line discontinuities. A search through the literature is likely to reveal diffraction coefficients for other geometries, and this topic is by no means exhausted. However, we will now move on to the discussion of a different type of geometry, namely the tip, which often plays an important role in diffraction theory.

1.5.3 Diffraction by a tip

There are even fewer results for the tip diffraction problem than there are for the line discontinuities. The tip is first approximated by its tangent cone and the canonical problem to solve, then, is that of diffraction of an electromagnetic wave by a cone.

If the cross-section of the cone is arbitrary, the canonical problem is not soluble. If the cone is elliptical and conducting, a solution can be derived using the method of the separation of variables; however, even for this case it has not been possible to extract a diffraction coefficient from the above solution. Only for the case of a circular and perfectly conducting cone it is possible to determine the diffraction coefficient using an asymptotic expansion of the exact solution. For a cone with an arbitrary angle, this coefficient can only be expressed in an integral form [BS], and further simplification does not appear possible, except in the limit of either a small or a large cone angle. We will restrict ourselves here to the small-angle case and discuss it in Sect. 1.5.3.1.

Another example of the tip diffraction problem, which is exactly soluble, is that of a semi infinite straight wire, and its diffraction coefficient is given in Sect. 1.5.3.2. To our knowledge, there does not exist other exact diffraction coefficients that are available in closed forms. The need to predict the RCS of structures with planar faces has led researchers to derive approximate diffraction coefficients, whose extraction relies upon various approximations for the induced currents (see for instance Sect. 1.5.3.3).

Finally, some results have been obtained by Kaminetzky and Keller for point and weak singularities, i.e., those due to discontinuities in the derivatives of the impedance, or to higher-order derivatives of the surface. They have also given results for singularities in line discontinuities [KK], but only for the acoustic case (see Sect. 1.5.3.4.)

1.5.3.1 Diffraction by a perfectly conducting circular cone with a small angle

The problem of diffraction by a perfectly conducting circular cone has been solved by Felsen [Fe1]. The dyadic diffraction coefficient for the cone problem (see Fig. 1.18) has appeared in [Fe2]. The expression reads

$$D_p = k^{-1} (D_{\theta\theta} \, \hat{\theta}\hat{\theta}_0 + D_{\theta\varphi} \, \hat{\varphi}\hat{\theta}_0 + D_{\varphi\theta} \, \hat{\theta}\hat{\varphi}_0 + D_{\varphi\varphi} \, \hat{\varphi}\hat{\varphi}_0), \qquad (1.107)$$

with

$$D_{\theta\theta} = \frac{i}{\log \sin^2 \frac{\delta}{2}} \frac{tg\frac{\theta}{2} tg\frac{\theta_0}{2}}{\cos\theta + \cos\theta_0}, \qquad (1.108)$$

$$-D_{\theta\varphi} = D_{\varphi\theta} = \frac{4i \sin^2 \frac{\delta}{2} \sin(\Phi - \Phi_0)}{(\cos\theta + \cos\theta_0)^2}, \qquad (1.109)$$

$$D_{\varphi\varphi} = \frac{-2i \sin^2 \frac{\delta}{2}}{(\cos\theta + \cos\theta_0)^3} (\sin\theta \sin\theta_0 + 2\cos(\Phi - \Phi_0)(1 + \cos\theta \cos\theta_0)), \quad (1.110)$$

where (θ_0, φ_0) and (θ, φ) are the Euler angles of the incident and observation directions, respectively, and δ is the semi-angle of the cone.

The formulas given above were established for the angular range $\theta + \theta_0 < \prod - 2\delta$ [Fe2], i.e., outside the region in which the rays reflected by the cone are present (see Fig. 1.19).

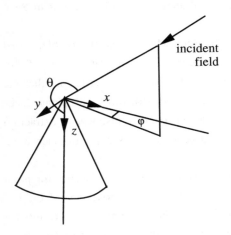

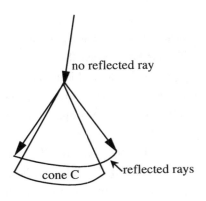

Fig. 1.18. Diffraction by a cone

Fig. 1.19. Cone separating zones with and without diffracted rays

We note that $D_{\theta\varphi}$, $D_{\varphi\theta}$ and $D_{\varphi\varphi}$ are small for a cone angle δ since they behave as k^{-1} and, furthermore, are proportional to $\sin^2 \delta/2$, i.e., to $\delta^2/4$. One finds that the diffracted field is proportional to δ^2, if one assumes that the currents on the surface of the cone are the physical optics currents. However, the dependence of $D_{\theta\theta}$ on θ is logarithmic in nature and, consequently, its contribution can be rather large. Furthermore, its expression contains the term $\left\{ tg\frac{\theta}{2} tg\frac{\theta_0}{2} \right\} / (cos\theta + cos\theta_0)$, which can be recognized as being identical to the one occurring in the expression for the diffraction by a wire (see the following section). This leads us to conclude that $D_{\theta\theta}$ accounts for certain physical effects, for instance the launching of progressive waves on the cone. Owing to the presence of the factor $tg\ \theta/2\ tg\ \theta_0/2$, the coefficient $D_{\theta\theta}$ diverges when the direction of incidence or observation angle corresponds to $\theta \approx \pi$, i.e., for grazing angles of incidence or observation.

Let us now turn to the zone which contains the reflected rays. Excluding the case of grazing rays, it can be shown [Sh] that at large distances (i.e., when $kr \to +\infty$) the diffracted field by any cone C is the sum of the geometrical optics and diffracted fields, and the latter behaves as $1/r$, except in a set of singular directions. This set is simply defined by the cone S, which is the reflection shadow boundary that separates the region containing the reflected rays from the complementary region beyond the access of these rays. The cone S is derived by diffracting the incident wave from the tip of the cone C and obtaining a reflected vector for each generator of the diffracted cone characterized by its normal vector. The cone S is formed by the entire set of these reflected vectors as shown in Fig. 1.19. A simple approach to building the trace of the separator cone on the unit sphere has been provided in reference [Sh]. Let M_0 be the trace on the unit sphere of the incidence direction. One plots the great circle path starting from M_0 up to the trace ∂C of the cone C on the sphere. Consider now the great reflected circle, i.e., the one whose angle with the normal to C is the same as that of the previous great circle. The trace $\partial \delta$ corresponds to the points that satisfy the criterion that the sum of the two circles is equal to M. When one neglects the semiangle δ of the cone C and replaces it by a wire, the cone δ simply is the cone with the top at 0 and a half-angle of θ_0. One then expects the diffraction coefficient to diverge for the angle, $\theta = \prod - \theta_0$. This divergence does occur, in fact, due to the presence of the factor $1/(\cos\theta + \cos\theta_0)$ that enters into the definition of the diffraction coefficient D.

Except for the singular directions of the cone S, and the directions of the grazing rays, the field is given by the sum of the geometrical optics and diffracted fields, and the latter behaves as $1/r$. Felsen [Fe1, Fe2] has postulated that the diffracted field has the same form in the entire space. Using this postulate and the formulas in (1.107) through (1.109), one can calculate the field even in the region where reflected rays exist, provided the following two conditions are met: (i) the point of observation is sufficiently far from S; and, (ii) it is not in the vicinity of the grazing rays. One will note that the formulas in (1.107) through (1.109) are singular not on the exact cone S, but on the approximate one obtained by substituting a wire for the diffracting cone. This difference can lead to erroneous results in some cases, especially near the singular zone.

We now proceed to the study of the diffraction by one extremity of the wire.

1.5.3.2 Diffraction by the extremity of a wire
Ufimtsev [U] has presented an approximate expression for the diffraction of a wire by extracting it from an approximate current induced on the wire. The diffraction coefficient is given by

$$D = k^{-1}\hat{\theta}\hat{\theta}_0 \frac{i}{2\ell n \frac{2i}{\gamma kb \sin\theta_0} \ell n \frac{2i}{\gamma kb \sin\theta}}$$

$$\ell n\left(\frac{i}{\gamma kb \sin\frac{\theta_0}{2}\sin\frac{\theta}{2}}\right) \frac{tg\frac{\theta_0}{2} tg\frac{\theta}{2}}{(\cos\theta + \cos\theta_0)} \,. \tag{1.111}$$

As in Sect. 1.5.2.4, b is the radius of the wire, γ is the Euler constant, θ_0 and θ are the incidence and observation angles, respectively. One may note the presence of the factor $\left\{tg\frac{\theta_0}{2} tg\frac{\theta}{2}\right\}/(\cos\theta + \cos\theta_0)$ in the above expression for D, which occurs in the coefficient $D_{\theta\theta}$ for the thin circular cone as well. The diffraction processes for these two cases are somewhat similar, as are the two geometries, and the fact that progressive waves are launched at the tip in both cases. Equation (1.111) is valid outside the cone $\theta = \pi - \theta_0$, which is the cone S for the wire, and for non-grazing incidence as well (observation θ_0 or $\theta = \pi$). The progressive waves dominate in these directions because they do not decay as $1/r$, and D becomes infinite.

We will now present some approximate results for an arbitrary tip geometry.

1.5.3.3   Approximate calculation of the diffraction coefficient for an arbitrary tip geometry

a) Regular cone

For a regular cone, i.e., one whose cross-section is a smooth curve, it is natural to approximate the induced currents by using the physical optics $J = 2\,\hat{n}\times\vec{H}i$, where $\hat{n}$ is the normal to the cone and $\vec{H}_i$ is the incident field. The diffracted field produced by the physical optics currents located on a cone's generator can be represented in the form of an integral over the angle of the generator [TP].

For the case of the circular cone illuminated along its axis, Gorianov has compared the physical optics result with the exact result [G'O]. In the backscattering direction, the diffraction coefficient derived from the physical optics approximation becomes identical to the exact result, and it never deviates by more than 10 % in the other directions. Although the physical optics diffraction coefficient provides a good approximation to the exact solution for paraxial incidence, we recall from Sect. 1.5.3.1 that the coefficient $D_{\theta\theta}$ did account for certain physical phenomena not included in the physical optics approximation. In view of this, some differences between the approximate and the exact coefficient would naturally be expected. For a cone with a semi-angle of 12.5°, Fig. 1.20 shows the comparison between the physical optics diffraction coefficient with that suggested by Felsen and given in (1.107)–(1.109), when the incidence is close to backside. A significant difference between the two results is evident for this case.

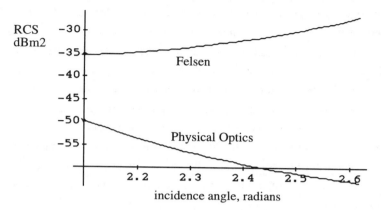

Fig. 1.20. Comparison of Felsen (exact) and PO diffraction coefficient

We conclude, therefore, that the physical optics approximation provides satisfactory results for the paraxial incidence, but that the PO results deteriorate with increasing angle of incidence, with the error rising significantly for incidence from the rear.

b) Polyhedron

When dealing with a polyhedron rather than with a rectangular cone, the current induced on such a structure is well approximated via a superposition (see Fig. 1.21) of three different current components, viz., the physical optics current, the so-called fringe currents launched by the edges, and the edge waves currents launched by the tip.

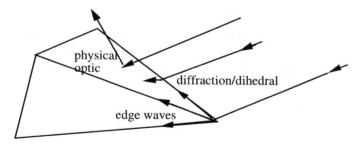

Fig. 1.21. Diffraction by the tip of a polyhedron

A numerical study has been performed by Hansen [Ha1, Ha2] for the quarter plane using the above approximation. He has shown that the currents induced by the edge waves are concentrated in the vicinity of the edges and that they decrease as $r^{-1/2}$ away from the edges. Hansen has also provided explicit formulas for the different components of the diffraction coefficient. In this approach, one calculates, as a first step, the exact contribution to the diffraction coefficient by the physical optics current. Next, the equivalent fringe currents due to the edges of the quarter plane are computed by using Michaeli's fringe currents [Mi] for the half-plane. Such an approximation forms the basis of the physical theory of diffraction (PTD), detailed in Chap. 7. Finally, Hansen provides an approximate expression of the coefficient of diffraction due to edge waves, whose derivation is based on numerical results. He shows that it is essential to account for this component of the diffraction coefficient in order to obtain accurate results. Indeed its omission can lead to errors in the diffraction coefficient, reaching up to 7 dB.

Hill [Hi] has studied the angular sector with an arbitrary angle by taking into account both the physical optics current and the fringe currents launched by the edges. Hill's work provides uniform results for diffraction by an angular sector. Finally, Ivrissimtzis has investigated the corner of a polyhedron, by using the same type of approximations for the currents as those employed by Hill [IM]. However, the shortcoming of both of these two approaches is the omission of the currents induced by the edge waves.

c) Conclusions

The evaluation of the diffraction coefficients of a tip requires, as a first step, an approximation of the currents induced on the tip. The most natural approximations, employing the physical optics currents for the regular cone that are complemented by the fringe currents on the polyhedral corners, frequently produce the correct result, though the accuracy of this procedure cannot generally be guaranteed. Consequently, it often becomes essential to take into account the currents launched by the tip; however, this can only be accomplished using numerical methods.

1.5.3.4 Weak singularities

Let us now turn to the discussion of weak singularities, which have been studied by Kaminetzky and Keller [KK] for the acoustic case by using the boundary layer method which could also be extended to the electromagnetic case. Kaminetzky and Keller have paid particular attention to the point-type discontinuities and to the intersections of the derivatives of higher discontinuity of the line surface and of impedance derivatives. Kaminetzky has also studied the diffraction by a curvature discontinuity on the edge of a wedge. However, these weak singularities lead to very small diffraction coefficients, and can be neglected in general.

1.5.4 Attachment coefficients, detachment coefficients, and attenuation coefficients of creeping rays

Let us consider the canonical problem of the diffraction of a TM plane wave ($E = \exp(ikx)$) by a conducting circular cylinder (see Fig. 1.22). The asymptotic expansion of the exact solution in the shadow zone can be written in the form [Ja]

$$E(\rho, \varphi) = \sum_{p=1}^{p}(D_s^p)^2\left\{exp(ika + i\alpha_s^p)\theta_1 + exp(ika + i\alpha_s^p)\theta_2\right\}\frac{exp(iks)}{\sqrt{s}} \quad (1.112)$$

The above expression is useful for extracting the *soft* diffraction coefficients, defined in Sect. 1.3.4.4. To obtain the *hard* coefficients, it is necessary to solve the canonical problem associated with the diffraction of a TE plane wave instead.

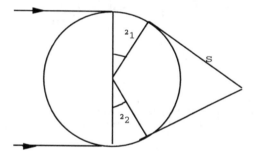

Fig. 1.22. Creeping rays on a circular cylinder

As mentioned in Sect. 1.3.4.4, the circular cylinder is not the only choice for the geometry to evaluate the different diffraction coefficients. Voltmer [Vo] has also investigated the problem of diffraction by a sphere in order to estimate the effect of the radius of curvature ρ_t transverse to the creeping ray.

The effect of a variable curvature has been studied by Keller and Levy [KL] and by Franz and Klante [FK]. Keller and Levy have used canonical geometries with variable curvature, for instance the elliptic cylinder, and Franz and Klante have derived an approximate solution to the magnetic field integral equation for the case of a cylinder illuminated by a TE (hard) polarized wave. Finally, Hong [Ho] has also used the magnetic field integral equation to solve a convex object of revolution, a geometry that enabled him to simultaneously account for the effect of the transverse radius of curvature and the variation of the curvature.

Although Hong's results are restricted to the case of perfect conductors, they do provide accurate expressions for the diffraction coefficients for perfectly conducting scatterers. In practice, the effect of the variation of the curvature is usually not significant. Furthermore, it is only justifiable to include it when this variation is relatively slow, since the accuracy of the results tends to deteriorate [Mo] if this condition is not met. In contrast, the inclusion of the term associated with the transverse curvature, as implemented in the formulas below, may significantly improve the accuracy of the calculations.

We now provide a listing of some of the useful diffraction coefficients, in the following where we will use the convention $A = D$ (or $C = 1$), followed by Keller.

Attenuation coefficient α

Conductor (electric creeping ray):

$$\alpha_s^p \approx q_s^p \frac{m}{\rho} e^{i\pi/3} \left(1 + \frac{1}{m^2} \frac{q_s^p}{60} \right).$$

(1.113)

Conductor (magnetic creeping ray):

$$\alpha_h^p \approx q_h^p \frac{m}{\rho} e^{i\pi/3} \left(1 + \frac{1}{m^2} \left(\frac{q_s^p}{60} + \frac{1}{(q_h^p)^2} \left(\frac{1}{10} - \frac{\rho}{4\rho_t} \right) \right) \right).$$

(1.114)

Impedance (magnetic creeping ray):

$$\alpha_h^p(Z) \approx q_h^p \frac{m}{\rho} e^{i\pi/3}.$$

(1.115)

Impedance (electric creeping ray):

$$\alpha_s^p \approx q_h^p \left(\frac{1}{Z} \right).$$

(1.116)

The coefficient m, known as the Fock parameter is equal to $(k\rho/2)^{1/3}$. The coefficients q_s^P (resp. q_h^P) designate the opposite of the zero of the Airy function (resp. derivative of the Airy function). The $q_h^P(Z)$ is the opposite of the root of equation:

$$A_i'(-q_h^P(Z)) - mZ\,e^{-i\pi/6}A_i(-q_h^P(Z)) = 0. \tag{1.117}$$

Setting $Z = 0$ or $Z = \infty$ in (1.117), we recover the magnetic or electric creeping ray, respectively, on a conductor.

Detachment coefficient

The detachment coefficient can be expressed as

$$(D_s^P)^2 \approx (2\pi k)^{-1/2}m\,\frac{e^{i\pi/12}}{[A_i'(-q_s^P)]^2}\left(1 + \frac{1}{m^2}q_s^P\left(\frac{1}{30} + \frac{\rho}{4\rho_t}\right)\right), \tag{1.118}$$

$$(D_h^P)^2 \approx (2\pi k)^{-1/2}m\,\frac{e^{i\pi/12}}{q_h^P[A_i(-q_h^P)]^2}\left(1 + \frac{1}{m^2}q_h^P\left(\frac{1}{30} + \frac{\rho}{4\rho_t}\right) - \frac{1}{[q_h^P]^2}\left(\frac{1}{10} - \frac{\rho}{4\rho_t}\right)\right), \tag{1.119}$$

$$(D_h^P(Z))^2 \approx (2\pi k)^{-1/2}m\,\frac{e^{i\pi/12}}{q_h^P A_i^2(-q_h^P) + A_i'^2(-q_h^P)}, \tag{1.120}$$

where $q_h^P = q_h^P(Z)$ and

$$D_h^P(Z) = D_h^P(1/Z). \tag{1.121}$$

The following comments are in order.

(i) In practice, only the zero creeping wave mode, i.e., the one corresponding to the first zero of Eq. (1.117) plays a significant role as long as the surface impedance of the scatterer is not too close to a purely imaginary and negative value.

(ii) the electric creeping ray on the conductor suffers considerably more attenuation than its magnetic counterpart. For instance, $q_s^1 \approx 2.338$ while $q_h^1 \approx 1.019$.

(iii) In view of the above, it is possible to obtain fairly good results for the scattered fields of large objects by considering only the first magnetic creeping wave mode.

(iv) The attenuation and detachment coefficients are modified by the surface impedance Z, which therefore controls the amplitude of the creeping rays. The derivation of (1.117) and some of the main results derived from it are given in Appendix 1.

(v) The principal effect of the transverse curvature is to diminish the attenuation of the magnetic creeping ray because of the term $-(1/m^2 (q_h^p)^2) (\rho/4\rho_t)$ occurring in Eq. (1.114). This term is valid provided that ρ/ρ_t is not too large.

The formulas (1.114) to (1.121) provide all the diffraction coefficients needed to calculate the creeping wave contributions to the diffracted field. Next, we will present the diffraction coefficients $D_{sr,s}$, which enable us to make the transition from the fictitious field on the surface ray to the surface field. As seen in Sect. 1.3.4.5, these coefficients associated with the diffraction by discontinuities provide the way to calculate processes of hybrid diffraction.

1.5.5 Calculation of coefficients $D_{sr,s}$

To calculate the diffraction coefficients $D_{sr,s}$, we consider, once again, the canonical problem of the diffraction of a plane wave by a cylinder. In the shadow zone, the surface field is written as a superposition of creeping rays and the coefficient $D_{sr,s}$ is obtained from the ratio of the physical surface field and the fictitious surface field. For a perfect conductor and a magnetic creeping ray, the result is

$$(D_{sr,s}^h)^2 = \frac{1}{mq^h} e^{-i\pi/12} (2\pi k)^{1/2}. \tag{1.122}$$

For a surface with an impedance Z, the above equation becomes

$$(D_{sr,s}^h)^2 = \frac{1}{m} \frac{A_i^2(-q^h)}{A_i'^2(-q^h) + q^h A_i^2(-q^h)} e^{-i\pi/12} (2\pi k)^{1/2} \tag{1.123a}$$

$$= \frac{1}{m} \frac{1}{m^2 Z^2 \exp(-i\pi/3) + q^h} e^{-i\pi/12} (2\pi k)^{1/2}. \tag{1.123b}$$

We remind the reader that in order to obtain the electric field normal to the surface, we must multiply the normal electric field carried by the surface ray with the diffraction coefficient $D_{sr,s}^h$. We also point out that the difference between (1.122) and Albertsen's result [AC] can be explained by noting that he sets the constant C to equal $e^{i\pi/4}/\sqrt{8\pi k}$ instead of 1; hence, one can retrieve Albertsen's coefficient for a perfect conductor from (1.122) by multiplying $D_{sr,s}^h$ with $C^{1/2}$. In addition, Albertsen assumes $Z = O(1)$ for the case of an impedance surface, while we have set $mZ = O(1)$ in (1.123).

Again, we recover Albertsen's result by neglecting q^h in front of $m^2 Z^2$ in the denominator of (1.123b). The reason for the choice $mZ = O(1)$ is that it provides a uniform result when $Z \to 0$, and, in effect, (1.123) reduces to (1.122) when $Z = 0$.

For a surface with an impedance, $D^s_{sr,s}$ which relates the tangential electric field carried by the surface ray to the tangential electric field on the surface, is given by

$$D^s_{sr,s}(Z) = D^h_{sr,s}(1/Z). \qquad (1.124)$$

Finally, the coefficient $D_{sr,s'}$ for a perfect conductor, which relates the tangential electric field carried by the surface ray to the normal derivative $\_/\_kn$ of the tangential electric field on the surface, takes the form

$$D_{sr,s'} = e^{-3i\pi/8} \sqrt{\frac{2}{ka}} (2\pi k)^{1/4} \qquad (1.125)$$

The expressions in (1.122) to (1.125) provide the diffraction coefficients which relate the surface field to the fictitious field on the surface ray. Thus, it is possible to calculate the surface field in the context of the GTD using these coefficients. Furthermore, it was seen in Sects. 1.3.4.5 and 1.3.4.8 that when the above coefficients are combined with the corresponding diffraction coefficients associated with the discontinuities, the result is a hybrid diffraction coefficient which accounts for the diffraction or the launching of creeping rays on edges or tips with curved faces.

We proceed next to present the excitation coefficients of creeping rays by a source located on the surface of the scatterer.

1.5.6 Calculation of the launching coefficients of creeping rays due to a source located on the surface

When the source is located on the surface of the scatterer, we encounter a situation which is reciprocal to the previous one. The diffraction coefficients of the type $D_{sr,s}$ provide the surface field due to a source located far away from the scatterer, while the coefficients like L_h, L_s provide the far fields radiated by a source located on the scatterer surface. Fortunately, the latter can be extracted from the former by involving the reciprocity theorem. For an object with a surface impedance, the result is

$$L_h = D^h_{sr,s} = \frac{1}{m^{1/2}} \frac{1}{(m^2 Z^2 \ exp \ (-i\pi/3) + q_h)^{1/2}} \ e^{-i\pi/24} (2\pi k)^{1/4}, \quad (1.126)$$

$$L_s = \frac{1}{Z} D^s_{sr,s} = \frac{1}{m^{1/2}} \frac{1}{(m^2 \ exp \ (-i\pi/3) + Z^2 q_s)^{1/2}} \ e^{-i\pi/24} (2\pi k)^{1/4}. \quad (1.127)$$

When the impedance $Z \to 0$, (1.126) and (1.127) reduce to the well-known formulas [Pa] for the perfect conductor, which are

$$L_h = e^{-i\pi/24} \left(\frac{2}{ka}\right)^{1/6} (2\pi k)^{1/4} (q_h)^{-1/2}, \quad (1.128)$$

and

$$L_s = e^{i\pi/8} \sqrt{\frac{2}{ka}} \ (2\pi k)^{1/4}. \quad (1.129)$$

We note that L_s is of the order of $1/m$ (we recall that $m = (ka/2)^{1/3}$) with respect to L_h for the case of the perfect conductor. Furthermore, for the case of a perfect conductor, the electric creeping rays attenuate much more rapidly than their magnetic counterparts. Thus, the dominant diffraction process is associated with the launching of the magnetic creeping rays by the magnetic dipole components perpendicular to the geodesic.

For the case of a surface impedance, the magnetic (electric) rays launched by the perpendicular (parallel) component of the magnetic dipole dominates for low (high) impedances. When the impedances are very large, i.e., $|Z| \to +\infty$, L_h and L_s tend toward 0, which is a logical consequence of the fact that the surface becomes a magnetic conductor in this limit.

The case of a electric dipole parallel to the surface is readily deduced from the previous one by using the following transpositions: $E \to H$, $H \to -E$, $J \to M$, $M \to -J$, and $Z \to 1/Z$. Evidently, we revert to the launching coefficients that are zero for the perfect conductors.

Some comments and observations are relevant at this point with regard to the term L in (1.54). In effect, if we choose the diffraction by a cylinder at normal incidence for our canonical problem, we obtain, for the TE polarization, a surface magnetic field parallel to the generators (and thus perpendicular the geodesic when it is a circle) along which the creeping ray propagates. The terminology TE polarization is used here to identify the case where the magnetic field is oriented such that it is parallel to the generators of the cylinder. Correspondingly, for the TM polarization, the surface magnetic field is perpendicular to the generators and thus parallel to the geodesic along which the creeping ray propagates.

It is then tempting to conclude that the surface magnetic field carried by a magnetic (electric) creeping ray is orthogonal (parallel) to the geodesic and that, in accordance with the reciprocity principle L and L' should be identically equal to 0. Indeed, such an assumption was used in all of the publications until the year 1980.

If we consider the canonical problem of a perfectly conducting cylinder for the oblique incidence case, we find that the geodesics now become helices. For the TE polarization, the diffracted magnetic field is oriented principally in the direction of the geodesic [Iv]; however for the TM case, the magnetic field always remains, in view of the definition of the TM field, perpendicular to the generators. Consequently, it has a perpendicular component $H_\perp$ to the geodesic. For the cylinder case, the ratio of $H_\perp$ and $H_{,,}$ is just the cotangent of the angle of the helix with the generator of the cylinder. This cotangent is equal to the product of the torsion τ and the radius of curvature ρ of the helix. Again, invoking the localization principle, we conclude that, for any surface, the ratio $H_\perp / H_{,,}$ is equal to the product $\tau\rho$, a result that will be demonstrated in Chap. 3. Using the reciprocity theorem, we conclude that a magnetic dipole orthogonal to a geodesic launches an electric creeping ray and for a perfect conductor we have the relation

$$L = \tau\rho \, L_s, \tag{1.130}$$

where L_s was given in (1.129).

It is also possible to obtain the above result from the canonical problem of the magnetic dipole located on the cylinder. For a surface with an impedance Z, which is sufficiently different from unity such that the electric and magnetic creeping rays propagate essentially independently, we can obtain

$$L = (\tau\rho) \, \frac{1}{1-Z^2} L_s, \tag{1.131}$$

by using the boundary layer method (see Chap. 3).

As we will also see in Chap. 3, when $Z \cong 1$, it is no longer possible to distinguish the two types of creeping rays and the previous formalism can no longer be used. When $Z = 0$, we revert back to the result for a perfect conductor. When $Z \to \infty$, $L \to 0$, since the surface now becomes a magnetic conductor. The coefficients M and N are evaluated in the same manner as L_h, L_s and L, by first starting from the canonical problem of the diffraction of a plane wave by a cylinder and then using the reciprocity theorem

$$M = L_h, \tag{1.132}$$

$$N = L. \tag{1.133}$$

The case of the magnetic dipole perpendicular to the surface is deduced through the substitutions $E \rightarrow H$, $H \rightarrow -E$, $J \rightarrow M$, $M \rightarrow -J$, $Z \rightarrow 1/Z$.

1.5.7 Coefficients of diffraction for progressive waves and edge waves

To date, the diffraction coefficients for progressive and edge waves have not been determined, because the canonical problem is not solvable in closed form.

1.5.8 Conclusions

Let us now summarize the three key steps involved in the solution of a diffraction problem.

(i) Search for the rays according to Fermat's generalized principle (Sect. 1.2). For a simple object, the search for the rays can be carried out analytically; however, to deal with more complex geometries it may be necessary to use numerical methods for tracing rays.

(ii) Determine the most important rays, although this is somewhat of a nebulous statement. For scatterers with simple shapes, the choice of the rays to be accounted for depends upon the level of accuracy we seek, and this accuracy is typically enhanced by the inclusion of the higher order diffraction phenomenon.

(iii) Calculate the fields along each ray by using the general formulas of Sect. 1.3, in conjunction with the geometrical factors given in Sect. 1.4 and the diffraction coefficients catalogued in Sect. 1.5.

(iv) Add the various contributions to extract the total field diffracted by the scatterer.

Before proceeding to Sect. 1.7 where we provide some simple illustrative examples, it is important to recognize some of the limitations of the ray methods. A discussion of this type appears in the next section.

1.6 Some limitations of the ray method

The geometrical theory of diffraction, and more generally all of the ray methods, have certain limitations. They predict non-physical jumps in the solution at the light-shadow boundaries (see Sect. 1.6.1); they lead to infinite results on the caustics, i.e., at the envelopes of the rays (see Sect. 1.6.2); and, a vanishing field in the shadow zone of the caustic (see Sect. 1.6.3).

The method of the asymptotic expansions, described in Chap. 2, provides a better foundation for the GTD, and, yet it enables us to calculate the field in zones where the GTD fails. However, before discussing this method, it would be worthwhile for us to present some example cases and some simple techniques for circumventing, at least in some cases, the difficulties encountered with the GTD.

1.6.1 The jump discontinuity at the light-shadow boundary

Let us consider a half-plane illuminated by a plane wave at normal incidence (see Fig. 1.23). The field of an incident plane wave in terms of rays is a set (more precisely a congruence) of straight lines. The quarter plane, defined by $x > 0$ and $y > 0$, is the optical shadow zone and the light-shadow boundary is the semi-infinite straight line $y = 0$ and $x > 0$. For the simple case of a half-plane at normal incidence, the Keller cone degenerates into a plane. Also, the edge-diffracted rays reach all points of the optical shadow zone. The field predicted by GTD in the shadow zone is given by

$$u\,(r,\,\theta) = D(\theta)\frac{e^{ikr}}{\sqrt{r}}\,,\tag{1.134}$$

where $(r,\,\theta)$ are the polar coordinates of the observation point.

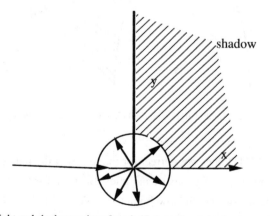

Fig. 1.23. Light and shadow regions for a half plane illuminated at normal incidence

Recall that $D(\theta)$ was given in Sect. 1.5 (the half-plane is a wedge whose exterior angle is equal to 2π), and for TM polarization, for which u designates H, it is given by

$$D(\theta) = \frac{e^{i\pi/4}}{2\sqrt{2\pi k}}\left(\frac{1}{sin(\theta/2)} - \frac{1}{cos(\theta/2)}\right).\tag{1.135}$$

For a point slightly above the shadow boundary, i.e., $\theta = 0$, the total field is given by (1.134) and (1.135). However, for a point slightly below, we must add the incident field of the plane wave to this field.

Two problems arise in following the procedure for field computation described above:

(i)   above the light-shadow boundary, the amplitude decreases as $1/\sqrt{r}$, while it is constant below.

(ii)   the diffraction coefficient becomes infinite at $\theta = 0$, i.e., on the light-shadow boundary.

Consider now a parabolic cylinder illuminated by a normally-incident plane wave (see Fig. 1.24). Slightly below the light-shadow boundary, the field contribution is due to a creeping ray and decreases as $1/\sqrt{r}$, however, the field is constant slightly above the boundary.

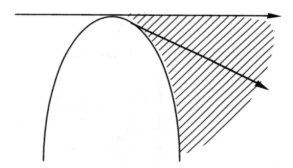

Fig. 1.24. Fictitious jump of the field at light shadow boundary for a parabolic cylinder illuminated at normal incidence

Thus, the usual GTD does not correct for the jumps at the boundary between the illuminated and shadow zones of geometrical optics. Besides, for the case of an edge, it predicts an infinite result on the shadow boundary. We will see, in Chap. 2, that the failure of the GTD stems from the fact that in the neighborhood of the shadow boundary the field is not a ray field, which means that it cannot be approximated by a plane wave. We will introduce, in Chap. 5, the so-called uniform theories which enable us to correct for the deficiencies of the GTD. Finally, we will see in Chap. 7, that the Physical Theory of Diffraction (PTD) also provides a way to circumvent the difficulties on the shadow boundary. The choice of the method for handling the shadow boundary problem depends upon the problem to be solved and we hope that this will be evident after we have described these methods in detail in Chaps. 7 and 5. For now, we turn to the subject of the difficulties encountered in the neighborhood of the caustics.

1.6.2 Infinite results on the caustics

The caustics, as was mentioned earlier, are the envelopes of the rays and they have been so designated because the concentrated light *burns* on the caustics. The example shown on the figure just below gives a simple explanation why the GTD yields infinite results on the caustics. According to the laws of geometrical optics (see Sect. 1.2) the energy is conserved in a tube of rays. On a circular caustic, for instance, (see Fig. 1.25) the cross-section of a tube of rays whose initial section was $d\theta$ is constricted to one which is $O(d\theta)^2$. Hence, the field calculated with the ray approach becomes infinite at the caustic.

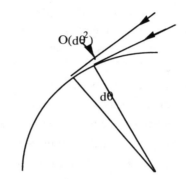

$$O(d\theta^2)$$

$$d\theta$$

Fig. 1.25. Infinitesimal tube of rays at a caustic

An arbitrary ray passes through two caustics located, as we saw in Sect. 1.3.2.1, at the points whose abscissa are $-\rho_1$ and $-\rho_2$, where ρ_1 and ρ_2 are the two radii of curvature of the wave front. From (1.12) of Sect. 1.3.2.1 we have

$$|E(M)| = |E(0)| \sqrt{\frac{\rho_1 \rho_2}{(\rho_1 + s)(\rho_2 + s)}} \; .$$

It is evident that the above expression would yield infinite results at points $s = -\rho_1$ and $s = -\rho_2$ located on the caustic.

A particularly difficult situation arises in RCS computation when either ρ_1 or ρ_2 becomes infinite. This situation is referred to as the *caustic at infinity*. It occurs when a wave reflects from a surface one of whose radii of curvature vanishes, or, alternatively, the wave diffracts from an edge along the direction of the binormal to the edge (see Fig. 1.26).

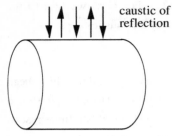

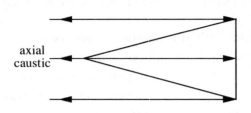

Fig. 1.26. Caustic of reflection on a circular cylinder

Fig. 1.27. Axial caustic for an axis illumination of a circular cone

The *caustic at infinity* frequently occurs when an object of revolution is illuminated along its axis, which is a caustic of rays, and the GTD predicts an infinite result along the axial direction, i.e., at the axial caustic (see Fig. 1.27).

The field in the vicinity of a caustic is not a ray field and, as mentioned earlier, the GTD predicts an infinite field there. In Chap. 2, we will see how to evaluate, in general, the field near the caustics using the asymptotic expansions. Maslov's method (Chap. 6) and the PTD (Chap. 7) also enables us to extract the field near the caustics. Thus, there are several approaches to calculating the field at the caustics, and some are more convenient than others. All of these methods provide a way to calculate the field in the shadow zone of the caustic, a region which cannot be reached by any physical ray. However, for this case, we can still apply the GTD to compute the field provided that we employ the notion of complex rays. We will now present a simple example of the circular caustic, and that of the application of the method of complex rays to the evaluation of the field on the shadow side of the caustic.

1.6.3 Calculation of the field in the shadow zone of the caustic

Let us consider a system of rays which has a circular caustic [K1] (see Fig. 1.28). If $r > a$, two physical rays whose tangential points have the polar angles given by the expression

$$\theta_\pm = \theta \pm cos^{-1}\frac{a}{r},\tag{1.136}$$

pass through the point M whose coordinates are (r, l). The distance of the point M to one of the tangential points is $\sqrt{r^2 - a^2}$.

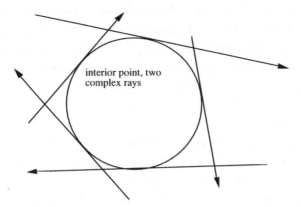

interior point, two
complex rays

Fig. 1.28. Circular caustic

If the point M is inside the caustic, no physical rays exist, but there are complex rays whose tangential points have the polar angles satisfying the equation

$$\theta_{\pm} = \theta \pm i \; cosh^{-1}\left(\frac{a}{r}\right), \qquad (1.137)$$

and the distance of the point M to one of the tangential points becomes $i \sqrt{r^2 - a^2}$. It is then possible to calculate the field at a point inside the caustic by considering these two complex rays. The problem of computing the field directly at the caustic itself is not solved by this procedure, since the field diverges as $(r^2 - a^2)^{-1/4}$ on each side of the caustic. Nonetheless, sufficiently far away from the caustic, we can use this formalism of complex rays to calculate the field. In addition, if the solution on the lit side of the caustic can be written as

$$u = (r^2 - a^2)^{-1/4}\left(A(\theta_+)\, e^{ik\sqrt{r^2-a^2}} - iA(\theta_-)\, e^{ik\sqrt{r^2-a^2}}\right), \qquad (1.138)$$

where A is an analytical function and $\theta_{\pm}$ is given by (1.136), then the solution on the shadow side can be obtained via an analytical continuation of this expression. This simply entails the replacement of $\theta_{\pm}$ by expressions in (1.137), and $\sqrt{a^2 - r^2}$ by $i\sqrt{a^2 - r^2}$ in (1.138). It is then possible, for simple cases, to obtain the field in the shadow zone by extending the notion of a ray in the complex field. The field on the shadow side is a ray field in the sense that it appears as a sum of two inhomogeneous plane waves. In different parts of this work, we will again have occasions to use this method of complex rays.

This example just discussed shows both the advantages as well as the limitations of the techniques discussed. It appears that complex rays provide an elegant way of handling the problem when the geometry is simple; however, the analytical continuation is not achieved easily, or may even be impossible for more complicated cases, thus rendering this method essentially useless for such geometries.

1.6.4 Conclusions

GTD is a powerful tool for field computation in regions where we deal with the ray field, i.e., where it is can be accurately described, locally, with a finite number of plane waves. When this is no longer true, e.g., in the neighborhood of the boundaries between the illuminated and shadow zones, or near the caustics, the GTD does not provide correct results, and we must seek alternate approaches (described in Chaps. 2, 3, and 4) to handling the problem of field computation.

1.7 Examples

In this section, we will present a number of examples to illustrate the usefulness of the techniques discussed earlier in this chapter on some concrete examples. First, in Sect. 1.7.1, the orders of magnitude of the various phenomena contributing to the RCS of an object will be given. Let us recall that the RCS is defined by

$$\lim_{d \to \infty} 4\pi d^2 \ \vec{E}^d \cdot \vec{E}^{*d} / \vec{E}^i \cdot \vec{E}^{*i},$$

where $\vec{E}^d$ is the diffracted field, and $\vec{E}^i$ the incident field. For two-dimensional scatterers, the definition of the RCS becomes

$$\lim_{d \to \infty} 2\pi d \ \vec{E}^d \cdot \vec{E}^{*d} / \vec{E}^i \cdot \vec{E}^{*i}, \text{ for TM polarization,}$$

and

$$\lim_{d \to \infty} 2\pi d \ \vec{H}^d \cdot \vec{H}^{*d} / \vec{H}^i \cdot \vec{H}^{*i}, \text{ for TE polarization.}$$

The principal objective of this section is to assist the reader in choosing the rays that contribute the most to the RCS of a given object with a general shape.

In Sect. 1.7.2, the RCS of some of the objects will be listed. We will restrict ourselves to some simple examples for which the essential part of the calculations can be carried out analytically. We refer the reader to the literature, especially to reference [Ja] given in this book, for additional examples.

1.7.1 Orders of magnitude of the contributions to the RCS

1.7.1.1 Specular point

The specular points, i.e., the points from which the reflection of the rays occurs are the dominant contributions to the RCS of smooth scatterers. The field reflected by the specular point S is given by (1.17) of Sect. 1.3.4.1. The radii of curvature ρ_1^r and ρ_2^r of the reflected wavefront are given by (1.79) Sect. of 1.4.2.1. Since the incident wave is a planar, the curvature matrix of the wavefront vanishes and, consequently, the curvature matrix of the reflected wavefront given by (1.79) becomes

$$\underline{Q}^r = 2\,\underline{C}\,, \tag{1.139}$$

where $\underline{C}$ is the matrix of the surface curvature at the reflection point. Then, the radii of curvature of the reflected wavefront are given simply by

$$\rho_1^r = \rho_1\,/2 \text{ and } \rho_2^r = \rho_2\,/2, \tag{1.140}$$

where ρ_1 and ρ_2 are the curvature radii of the scatterer at point S. For the case of specular reflection, the incident wave vector is directed along the normal to the object so that it is completely defined within a rotation. Furthermore, according to (1.90) and (1.91) of Sect. 1.5.1, the reflection coefficients are given by

$$-R_{\text{TE}} = R_{\text{TM}} = \frac{Z-1}{Z+1}, \tag{1.141}$$

and, since $\hat{e}_\perp^i = -\,\hat{e}_\perp^r$, the expression in (1.18) of Sect. 1.3.4.1 becomes

$$\underline{R} = \frac{Z-1}{Z+1}\underline{I}\,, \tag{1.142}$$

where $\underline{I}$ is the identity matrix.

At a point M, which is located at a large distance d from the specular point, we obtain the following result for the diffracted field

$$\vec{E}^d\,(M) = \frac{(\rho_1\rho_2)^{1/2}}{2d}\,\frac{Z-1}{Z+1}\,\vec{E}^i\,(S)\,. \tag{1.143}$$

This leads us to the following result for the contribution to the RCS by the specular point.

$$RCS = \pi \rho_1 \rho_2 \left| \frac{Z-1}{Z+1} \right|^2 . \tag{1.144}$$

The above RCS does not depend on the frequency, although it increases with the square of the characteristic dimension of the object. For a perfectly conducting object ($Z = 0$), whose surface radii of curvature are both equal to 1 cm, we obtain an RCS of – 27 dB/m^2, and it increases to –7 dB/m^2 when the curvature increases to 10 cm. If the impedance of the object at the reflecting point is 0.9, which corresponds to a reflection coefficient of approximately 1/100 in terms of energy, the RCS values, given above, are reduced by 20 dB. It is evident from the above that it is crucial to eliminate the specular point reflection from an object to reduce its RCS. Also we note that the calculated RCS becomes infinite when one of the two radii of curvature vanishes, which corresponds to the case of a caustic at infinity. The most efficient method for calculating the field at a point located on this type of caustic is the Physical Theory of Diffraction, which will be discussed in Chap. 7.

In summary, we have seen in this section that the contribution of the specular point to the RCS is important. Next, we will discuss the calculation of the RCS due to a diffracting point located on a line discontinuity. At high frequencies, the contribution of such line discontinuities is generally weaker than that of a specular point.

1.7.1.2 Diffraction by a point located on a discontinuity line

In the backscattering direction, the points contributing to the diffraction are those where the tangent to the line discontinuity is perpendicular to the direction of incidence, and the expression in (1.21) of Sect. 1.3.4.2 provides the diffracted field in this case. We note that this field is in the direction of $\vec{E}^i \underline{D}$, which, in general, does not coincide with the direction of the incident field. Also, unlike the specular reflection case, the diffraction by an edge modifies the polarization of the incident field. Let us assume that the direction of the incident electric field is along the edge. The RCS due to the diffracting point is given by

$$RCS = 4\pi D^2 \rho^d , \tag{1.145}$$

where $D = D_s$ is the diffraction coefficient of the line discontinuity for the TM polarization. If the incident electric field is perpendicular to the edge, the RCS is still given by (1.145), provided that we replace D by D_h which is the diffraction coefficient for the TE polarization. Also, ρ^d in (1.145) is the non-vanishing radius of curvature of the diffracted wavefront given by (1.80) of Sect. 1.4.2.2. Thus, we have

$$\frac{1}{\rho^d} = \frac{1}{\rho^a}(\hat{\imath}\cdot\hat{n}_a - \hat{d}\cdot\hat{n}_a), \tag{1.146}$$

or

$$\rho^d = \rho^a/2\,\hat{\imath}\cdot\hat{n}_a. \tag{1.147}$$

Thus, the RCS is given by

$$RCS = 4\pi D^2\,\rho^a/2\,\hat{\imath}\cdot\hat{n}_a. \tag{1.148}$$

The dependence of the diffraction coefficients on k, given in Sect. 1.5, determine the variation of the RCS with respect to k. For a sharp edge, the coefficient D decreases as $k^{-1/2}$, and, consequently, the RCS behaves as $1/k$. Also, the RCS increases with the characteristic dimension of the object and some numerical values are reported in Sect. 1.7.2. For a discontinuity in the curvature (see Sect. 1.5.2.2), the RCS behaves as k^{-3}, and for a discontinuity of the derivative of order n, the behavior is k^{-2n+1}. The contribution to the RCS from the higher-order discontinuity derivatives decreases quickly with an increase in the frequency. In practice, except for objects whose RCS is very low, the contributions of the discontinuities of orders greater than 4 are negligible.

We note that the RCS, as given by (1.148), diverges if the radius of curvature ρ^a of the edge at the diffraction point is infinite, or if the direction of incidence is perpendicular to the osculator plane of the edge. Specifically, this situation occurs when we calculate the diffracted field by a cone with sharp edges for the case of axial incidence, and we encounter the caustics of the diffracted rays at infinity. The most efficient method for calculating the RCS on these caustics is provided by the method of equivalent currents described in Chap. 7.

In the next section, we will study a diffraction phenomena which is weaker than the diffraction by an edge, viz., the diffraction by a tip.

1.7.1.3 Diffraction by a tip

For tip diffraction, the diffracted field is given by the expression in (1.23), given in Sect. 1.3.4.3. For this case the RCS is given by

$$RCS = 4\pi\,|\vec{E}^i\cdot\underline{D}|^2/|\vec{E}^i|^2. \tag{1.149}$$

As we saw in Sect. 1.5, the diffraction coefficient of a circular conical tip has a k^{-1} dependence; consequently, its RCS behaves as k^{-2}. This result still holds true for an arbitrary type of tip. In effect, this result can be derived with simple arguments of dimensional analysis as follows. The RCS is measured in m^2, a tip has no dimension and the wave number is the inverse of tip length (m^{-1}). For the RCS to be $O(m^2)$, RCS must be proportional to k^{-2}. Thus, the contribution of a tip to the RCS decreases rapidly as a function of the frequency. It is generally weak, except in the direction where the diffraction coefficient is large. For the case of a small-angle, perfectly conducting circular cone, we have seen in Sect. 1.5 that this coefficient becomes large near the grazing incidence. We note that the RCS does depend on the characteristic dimensions of the object.

In the above we have reviewed the diffraction phenomenon associated with the singularities. Next, we move on to the problem of evaluating the RCS due to creeping rays on an object.

1.7.1.4 Creeping ray

For the perfect conducting case, the magnetic creeping ray attenuates as $(\sin(\pi/3) \, m/\rho)$ q_h^1, in accordance with the expression in (1.114) of 1.5. Here the quantity q_h^1, which is the zero of the derivative of the Airy function with the opposite sign, equals 1.02, while the electric creeping ray attenuates as $(\sin(\pi/3)m/\rho) \, q_s^1$, where q_s^1, the first zero of the Airy function equals 2.34. Thus, in practice, only the magnetic creeping ray contributes to the RCS. Let us assume that the electric field is directed along the normal to the object at the detachment point of the creeping ray. The diffracted field is then given by the expression in (1.27) of Sect. 1.3. According to (1.114) and (1.119) of Sect. 1.5, the RCS due to the creeping ray is then given by

$$\text{RCS} = 2mm'k^{-1} \rho_d \, (q_h^1 Ai^2 \, (-q_h^1))^{-2} \exp\left(-\int_Q^{Q'} 2(\sin \pi/3)(m/\rho)q_h^1 ds\right). \quad (1.150)$$

To compute the contribution of the creeping rays, it becomes necessary to carry out a numerical integration on the surface of the object. A rough estimation of the RCS due to the creeping ray can be obtained by approximating the path QQ' with a semi-circle of radius R, where R represents one of the characteristic dimensions of the object, and also by replacing the various geometrical parameters occurring in (1.150) by R. This leads to the expression

$$\text{RCS} \approx 2^{1/3} \, k^{-1/3} \, R^{5/3} \, (q_h^1 \, Ai^2 \, (-q_h^1))^{-2} \exp(-2(\sin \pi/3)q_h^1 \, \pi(kR/2)^{1/3}), \quad (1.151)$$

where $q_h^1 \approx 1.019$, and $Ai(-q_h^1) \approx 0.536$. The RCS of the creeping rays exhibits a decay which is greater than any algebraic power of kR; hence, they contribute little to the overall RCS of large scatterers. However, where the scatterer has no specular points or sharp edges, the creeping ray contribution is not negligible even when the object is moderately large.

This concludes the presentation of the topic of evaluation of the various contributions of the rays that undergo only one type of interaction with the scatterer. Next, we will conclude this section with some observations regarding the multiply-diffracted rays.

1.7.1.5 Multiply-diffracted rays

Besides the effect of the caustic, the diffraction by edges introduce a $1/k$ factor in the k-dependence of the RCS; hence, the multiple diffractions can frequently be neglected. This is even more so if these diffractions are induced by tips or discontinuities in the curvature, which introduce $1/k^2$ and $1/k^3$, respectively. Furthermore, as we have seen in the previous section, there is a rapid decrease of the creeping ray field, as a function of the frequency because of its propagation on the surface of the object. In general, it is legitimate to neglect the rays beyond the first few diffractions or circumnavigation of the object. We should point out, however, that the situation is different when we are dealing with multiple reflections, since in effect, the reflections do not modify the k-dependence. Thus we must allow for the fact that the rays undergoing an arbitrary number of reflections may have the same importance as those undergoing either a single or no reflections at all. If these reflections occur on convex surfaces, the radii of curvature of the wavefront increase with each reflection; consequently, the rays undergoing a sufficiently large number of reflections can be neglected for this case. However, this is no longer true for concave surfaces, for instance inside a cavity, and we must take into account reflections of arbitrary order for such surfaces.

1.7.1.6 Conclusions

Several different types of rays must in general be accounted for in the study of the diffraction by a scatterer, viz., reflected, diffracted, and creeping rays, and all of these types of rays can, in principle, be involved in all of the various elementary types of interactions. We have provided above some estimates of the orders of magnitude, as well as some guidelines that would hopefully help the reader weigh the importance of the contributions of the various rays. We will now present some examples that would serve to illustrate this. We should point out, however, that all of the chosen examples are simple, and the reader must understand that the calculation of more complex objects can only be achieved by numerical mean.

1.7.2 Case examples of diffraction calculation

1.7.2.1 Diffraction by a strip

Let us consider a strip illuminated by a TM plane wave (see Fig. 1.29), whose amplitude is unity and the angle of incidence is $\theta \neq \pi/2$. Then the incident field can be written as $\vec{E}^{inc} = \exp(ik(x\cos\theta - y\sin\theta))$.

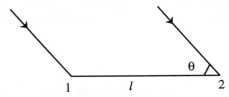

Fig. 1.29. Diffraction by a strip

It is evident that there are no specular reflected rays for this geometry and the essential contributions are the diffracted rays. For this polarization, the double-diffracted rays contribute very little because the diffracted field vanishes on the scatterer. The double-diffraction manifests itself only through the slope diffraction coefficient (to be discussed in Chap. 5), and it can therefore be neglected. We can thus obtain an accurate result by including just the simple diffraction from the two edges. The diffraction coefficient is given by (1.92) of Sect. 1.5, and we note that the angle external to the half-plane is equal to 2π for this case, which implies, in turn, that $n = 2$. Thus the diffraction coefficient becomes

$$D_s = -\frac{e^{i\pi/4}}{2\sqrt{2\pi k}}\left(1 - \frac{1}{\cos\Phi}\right),\tag{1.152}$$

with $\Phi = \pi - \theta$ for edge-1 and $\Phi = \theta$ for edge-2.

The field diffracted by edge-1 is then given by

$$E_1^d = -e^{ikr}\frac{e^{i\pi/4}}{2\sqrt{2\pi kr}}\left(1 + \frac{1}{\cos\theta}\right).\tag{1.153}$$

Likewise, for edge-2 we have

$$E_2^d = -e^{ikr+2ikl\cos\theta}\frac{e^{i\pi/4}}{2\sqrt{2\pi kr}}\left(1 - \frac{1}{\cos\theta}\right).\tag{1.154}$$

Hence the RCS (two-dimensional) is given by

$$\text{RCS} = \frac{1}{k}\left(\cos^2(kl\cos\theta) + \cos^{-2}(\theta)\sin^2(kl\cos\theta)\right).\tag{1.155}$$

We observe that this RCS does not diverge, although the fields given by (1.153) and (1.154) do diverge when $\theta = \pi/2$, i.e., when the shadow boundary coincides with the direction of observation, and this is because the two singularities cancel each other out. Thus, we see that in certain situations we can continue to use the GTD to calculate the RCS, even on the caustics. However, this type of cancellation occurs if and only if the two diffracted rays, whose contributions become infinite, are identical in nature, as for instance diffractions from two edges or discontinuities in the curvature. It should be realized, however, that the cancellations of this type, involving two large quantities, are not numerically stable. For these reasons, the uniform theories presented in Chap. 5, and the physical theory of diffraction detailed in Chap. 7, are two techniques that are most widely employed in practice for the calculation of the RCSes of complex objects.

In the next section we consider the example of a slightly more complicated scatterer, namely the triangular-shaped cylinder, or a prism wedge.

1.7.2.2 Diffraction by a triangular-shaped cylinder or a prism

Consider the problem of diffraction by a triangular-shaped cylinder or a prism when the polarization is TE. We choose this polarization because for a TM wave, the diffraction coefficient vanishes along the surface, and, as a result, the doubly diffracted rays are very weak. The singly and doubly diffracted rays that are taken into account are shown on Fig. 1.30.

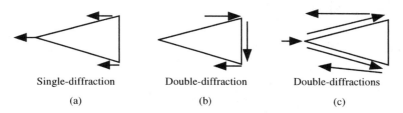

Single-diffraction	Double-diffraction	Double-diffractions
(a)	(b)	(c)

Fig. 1.30. Rays taken into account in the calculation of the diffracted field.

The calculations only require the expression for the diffracted field by a two-dimensional wedge with the phase origin on the edge. The diffracted field is given by

$$H^d = H^i \frac{D^h}{\sqrt{d}}, \qquad (1.156)$$

where the expression for D^h may be found in (1.92) of Sect. 1.5. Their derivation is left as an exercise to the reader.

To evaluate its accuracy, we have compared the GTD results to those obtained via the integral equation code SHF2D, developed by P. Bonnemason and B. Stupfel of CEA-Limeil. For the case of a prism whose angle at the vertex is 30°, we observe significant differences at low frequencies although the agreement becomes much better with the increase in frequency (see Fig. 1.31). For the case of a prism whose cross-section is an equilateral triangle, the agreement is good for the entire range of frequencies considered (see Fig. 1.32). For the first example, the discrepancies at low frequencies stem from the fact that sufficient care was not exercised in computing the GTD results for the doubly diffracted rays. It should be realized that the rays first diffract off the tip and then from the near edge (see Fig. 1.30 (c)), and for the range of frequencies we are considering, we are still in the transition zone of the reflected field. A more accurate solution to this problem would require the use of the spectral theory of diffraction, which will be presented in Chap. 4.

In the next section we discuss the case of the ogive-shaped cylinder which supports diffracted creeping rays that we have not considered heretofore.

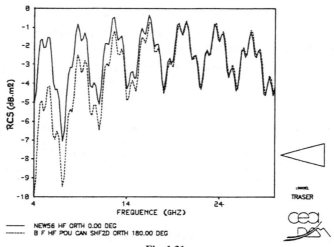

Fig. 1.31.

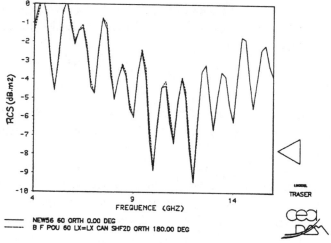

NEW56 60 ORTH 0.00 DEG
B F POU 60 LX=LX CAN SHF2D ORTH 180.00 DEG

Fig. 1.32.

1.7.2.3 Diffraction by an ogive-shaped cylinder

For the ogive-shaped cylinder we take into account the following contributions to the RCS. As shown in Fig. 1.33, the contributions are: the reflected ray at S (designated as #1); the diffracted ray at A (#2); and the diffracted ray at B (#7); the creeping rays that attach at Q_1 or Q_2, diffract at B, and detach at Q_1 (#3) or Q_2 (#4); and finally, the creeping ray that attaches at Q_1 diffracts at B, and detaches at Q_2 (#5). The latter contribution must be doubled because one must take into account the reverse path as well. The rays contributing to the RCS are shown in Fig. 1.33, and the contributions of each ray to the diffracted field are shown in Fig. 1.34. For the TE polarization case these contributions are calculated as follows:

(i)   for the ray #1, the two-dimensional equivalent of (1.143) is used. If O is the phase origin and ρ is the radius of curvature at point S, then we have

$$H^d = \exp\left(2i\,\vec{k}\cdot\vec{OS}\right) H^i \sqrt{\frac{\rho}{2d}}\,, \qquad\qquad (1.157)$$

(ii)   for the rays #2 and #7, (1.156) is used. For instance, for the ray #2 we obtain

$$H^d = \exp\left(2i\,\vec{k}\cdot\vec{OS}\right) H^i \frac{D^h}{\sqrt{d}}\,, \qquad\qquad (1.158)$$

(iii)   for the rays #3, #4 and #5, we can make use of the results given in Chap. 4 for
the diffraction by a wedge with curved faces. For instance, the contribution of
the ray #7 takes the form

$$H^d = \exp{(2i\ \vec{k}\cdot\vec{OQ_1} + 2ikl)}\ H^i \frac{g^2(x)}{4} \frac{D^h}{\sqrt{d}}, \tag{1.159a}$$

where l is the length of the arc Q_1B, g is the magnetic Fock function (see the
appendices). The argument x of g equals $(k\rho/2)^{1/3}\theta$, where θ is the angle of the Q_1B
and equals l/ρ.

The hybrid diffraction coefficients are useful only when the path length of the
creeping ray between the attachment point and the diffracting edge is sufficiently large,
i.e., when one can replace g by its asymptotic expansion for large positive arguments.
In this event, the expression in (1.159) reduces to the result obtained by using the
hybrid coefficients given in Sects. 1.3.4.5 and 1.3.4.6, which leads us to the expression

$$H^d = \exp{(2i\ \vec{k}\cdot\vec{OQ_1} + 2ikl)}\ H^i \left\{ \exp(-2e^{i\pi/6}q_h^1\theta(k\rho/2)^{1/3})\left(D^h_{sr,s}\right)^2\left(D^1_h\right)^2\right\} \frac{D^h}{4\sqrt{d}}$$

$$(1.159b)$$

where $\left(D^h_{sr,s}\right)^2$ and $\left(D^1_h\right)^2$ are given by (1.119) and (1.123) of Sect. 1.5. It can be
verified that the expression in (1.159b) is indeed the square root of the asymptotic
expansion of the Fock function $g(x)$; thus we conclude that (1.159b) is equivalent to
(1.159a) for large x. For TM-polarization, the method we follow is similar, but the
diffracted creeping rays as described by an electric Fock function and the slope diffraction coefficients are much weaker (see Sect. 1.3.4.5 and Chap. 4).

Recently, in the 1993 JINA (Journées internationales d'antennes) workshop
held in Nice, the GTD solution was compared to those derived from the integral
equation methods, and good agreement to within a fraction of a decibel was found.

Finally, we consider as a last example the three-dimensional problem of diffraction by a finite circular cylinder. For this example, we will start with the impulse
response obtained by using the integral equation method. We are now going to illustrate how the GTD can be used to provide a physical interpretation of this response.

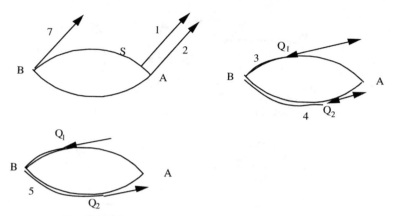

Fig. 1.33. Schematic showing the radius of the ogive cylinder

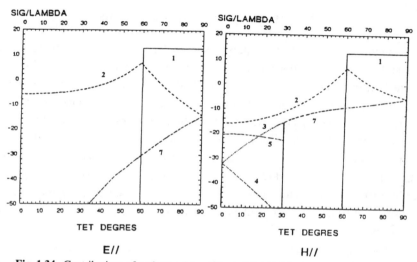

Fig. 1.34. Contributions of each ray of the diffracted field for TM and TE polarization

1.7.2.4 Diffraction by a finite circular cylinder

We have calculated the RCS of a finite circular cylinder illuminated by a plane wave, with an incident angle of 45° and a frequency range of 4 GHz to 16 GHz, by means of an integral equation code developed by P. Bonnemason and B. Stupfel of the CEA/ CELV Laboratoire in Limeil, France. We have subsequently computed the impulse response of the scattered field to extract the different contributions to the RCS. The locations of the different peaks of this response yield the lengths of the rays, and their heights provide their amplitudes averaged over the frequency. The significant rays contributing to the scattered field are depicted in Fig. 1.35.

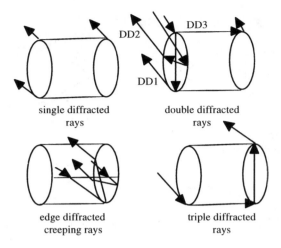

single diffracted
rays

double diffracted
rays

edge diffracted
creeping rays

triple diffracted
rays

Fig. 1.35. Rays contributing to the RCS calculation

Figure 1.36 exhibits the impulse response derived by using the integral equation method, for both polarizations of the incident field, designated by *para* (electric field in the plane of incidence) and *ortho* (electric field orthogonal to the plane of incidence). We note that the positions of the different peaks do indeed correspond to the lengths of the rays as predicted by GTD with the three highest peaks corresponding to the three diffracted rays. The doubly diffracted rays on the forward face (DD1 and DD2) produce peaks very close to those of the diffracted rays. Only DD2 appears in the *para* polarization, and it is located to the right of the last peak of the simple diffraction. The doubly diffracted rays, DD2 and DD3, are only important in the *para* polarization, because the diffraction coefficient vanishes at grazing incidence for the case of the other polarization.

There are two different types of diffracted creeping rays. One of these is symmetric with respect to the incident plane, while the two (only one is represented on the Fig. 1.35) other rays belonging to the second group are symmetric with respect to each other. The path lengths of these two types of rays are nearly the same so that their impulse response cannot separate them. These rays do not play a role in the *para* polarization because the incident electric field is in the tangent plane to the object at the attachment points of the creeping rays and, consequently, only the electric creeping rays, which are attenuated, are excited. In contrast, this field is normal to the object in the *ortho* polarization, and the diffracted creeping rays produce a contribution at the rear of the object. The last significant peak corresponds to the triply-diffracted ray, which is only of some significance in the *para* polarization. It is useful to note that, by

CYLINDER 20 x 30 CM

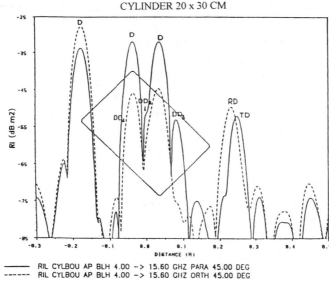

———— RIL CYLBOU AP BLH 4.00 -> 15.60 GHZ PARA 45.00 DEG
------ RIL CYLBOU AP BLH 4.00 -> 15.60 GHZ ORTH 45.00 DEG

Fig. 1.36. Impulse response of a finite circular cylinder

taking into account only the simply-diffracted rays, it is possible to obtain a rough estimate of the RCS. A more accurate prediction requires the inclusion of the higher-order effects.

We point out that in the above calculation we have only made sure that the amplitudes of the simply-diffracted rays, calculated in first approximation at the central frequency of 10 GHz, were close to the amplitudes extracted from the impulse response. The direct calculation of the amplitudes of the other rays was avoided because of its cumbersome nature.

We see from this example how the GTD can be helpful in interpreting the impulse response of the RCS of a scatterer. Conversely, by processing the intermediate frequency results derived from an integral equation code we can understand which rays make significant contributions to the RCS. The GTD can then be used to compute the RCS at higher frequencies. This leads us to conclude that these two methods are complementary.

1.8 Conclusions

The GTD is a physically intuitive method providing an accurate evaluation of the RCS of objects at high frequencies. In this chapter, we have reviewed all of the necessary techniques for the effective calculation of the diffracted field via the GTD. However, we have only enunciated the postulates upon which GTD is based, and have reviewed the rules of calculation of the different rays without mathematical justifications. Also, we did not discuss the general method of substitution in the zones where difficulties occur. The major part of this book will be devoted to the presentation of the mathematical foundations of the GTD and to the establishment of results whose domain of validity is more general. We are going to present, in Chap. 2, the asymptotic expansions which provide a justification of the GTD. Their domain of validity is the same as that of the GTD, since they apply in the zones of field of rays. In Chap. 3, we will see how to proceed in order to evaluate the field where these expansions are not valid.

References

[AC]   N. C. Albertsen and P. L. Christiansen, "Hybrid diffraction coefficients for first and second order discontinuities of two-dimensional scatterers," *SIAM J. Appl. Math*, **34**, 398-414, 1978.

[Be2] J. M. Bernard, "Diffraction by a metallic wedge covered with a dielectric material," *J. Wave Motion*, **9**, 543-561, 1987.

[Be2] J. M. Bernard, "On the diffraction of a an electromagnetic skew incident field by a non perfectly conducting wedge," *Annales Telecomm.*, **45**(1-2), 30-39, (Erratum 9-10, p.577), 1990.

[BF]   O. M. Bucci and G. Franceschetti, "Electromagnetic scattering by an half plane with two face impedances," *Radio Sci.*, **11**(1), 49-59, Jan. 1976.

[Bo]   D. Bouche, "GTD et réciprocité," *Annales des Télécomm*, **7-8**, 382-387, 1991.

[BS]   J. J. Bowman, T. B. A. Senior, and P.L. Uslenghi, "Acoustic and electromagnetic scattering by simple shapes," *Hemisphere*, 1987.

[Bu]   O. Buyakdura, "Radiation from sources near the edge of a perfectly conducting wedge," Thèse P.H.D., Université de l'Ohio, 1984.

[De]   G. A. Deschamps, "Ray techniques in electromagnetics," *Proc. IEEE*, **60**, 1022-1035, 1972.

[Fe1] L. B. Felsen, "Backscattering from wide-angle and narrow angle cones," *J. Appl. Physics*, **26**, 138-151, 1955.

[Fe2] L. B. Felsen, "Plane wave scattering by small angle cones," *IRE Trans. Ant. Prop.* **5**, 121-129, 1957.

[FK]   W. Franz and K. Klante, "Diffraction by surfaces of variable curvature," *IRE Trans. Ant. Prop.*, **AP-7**, S68-S70, Dec. 1959.

[Fr]   B. Friedman, *Principles of Applied Mathematics*, Dover.

[Go]   Gorianov, "An asymptotic solution of the diffraction of a plane electromagnetic wave by a conducting cylinder," *Radio Eng. Elec. Phys.* **3**(5), 23-39, 1958.

[Ha1] T. B. Hansen, "Diffraction of electromagnetic waves by corners on perfect conductors," Ph. D. Dissertation, The Technical University of Denmark, April 1991.

[Ha2] T. B. Hansen, "Corner diffraction coefficients for the quarter plane," *IEEE Trans. Ant. Prop.*, **AP-39**, 976-984 , July 1991.

[Hi]   K. C. Hill, "A UTD solution to the scattering by a vertex of a perfectly conducting plane angular sector," Ph.D Dissertation, Ohio State University, 1990.

[Ho]   S. Hong, "Asymptotic theory of electromagnetic and acoustic diffraction by smooth convex surfaces of variable curvature," *J. Math. Phys.*, **8**(6), 1223-1232, 1967.

[IM]   L. P. Ivrissimtzis and R. J. Marhefka, "A uniform ray approximation of the scattering by polyhedral structures including high order terms," *IEEE Trans. Ant. Prop.*, **AP-10**, 1150- 1160, Nov. 1992.

[Iv]   V. I. Ivanov, "Diffraction of short plane electromagnetic waves with oblique incidence on a smooth convex cylinder," Radiotechnika et Electronika, **5**(3), 524-528, 1960.

[Ja]   G. L.James, *Geometrical Theory of Diffraction for Electromagnetic Waves*, 3rd Ed., Peter Peregrinus, London, 1986.

[K1]   J. B. Keller, "A geometrical theory of diffraction," in *Calculus of Variations and its Applications*, Mc Graw Hill, New York, 1958.

[K2]   J. B. Keller, "Geometrical theory of diffraction," *J. Opt. Soc. Amer.*, **52**, 116-130, 1962.

[KA]   J. B. Keller and D. S. Ahluwalia, "Diffraction by a curved wire," *SIAM J. Appl. Math.*, **20**(3), 390-405, 1971.

[KK]   L. Kaminetzky and J. B. Keller, "Diffraction coefficients for higher order edges and vertices," *SIAM J. Appl. Math.*, **22**(1), 109-134, Jan. 1972.

[KL]   J. B. Keller and B. R. Levy, "Decay exponents and diffraction coefficients for surface waves on surfaces of nonconstant curvature," *IRE Trans. Ant. Prop.*, **AP-7**, S52-S61, Dec. 1959

[Ma]   G. D. Maliuzhinets, "Excitation, reflection and emission of surface waves from a wedge with given face impedances," *Sov. Phys. Dokl.* **3**, 752, 1958.

[Mi]   A. Michaeli, "Elimination of infinities in equivalent edge currents," *IEEE Trans. Ant. Prop.*, **AP-34**, 912-918, July 1986 and 1034-1037, Aug. 1986.

[Mo]   F. Molinet, "Geometrical theory of diffraction," *IEEE APS Newsletter*, part I, 6-17, Aug. 87; part II, 5-16, Oct. 1987.

[Pa]   P. H. Pathak, "Techniques for high frequency problems," in *Antenna Hand-book, Theory, Application and Design*, Y. T. Lo and S. W. Lee, Eds., Van Nostrand Rheinhold, 1988.

[SD]   T. Shamansky, A. Dominek, and L. Peters, "Electromagnetic scattering by a straight thin wire," *IEEE Trans. Ant. Prop.* **AP-37**, 1019-1025, 1989.

[Se1]   T. B. A. Senior, "Diffraction by an imperfectly conducting half-plane at oblique incidence," *Appl. Sci. Res*, **Sec B**(8), 45-61,1960.

[Se2]   T. B. A Senior, "The diffraction matrix for a discontinuity in curvature," *IEEE Trans Ant Prop.*, **AP-20**, 326-333, 1972.

[Se3]   T. B. A. Senior, "Solution of a class of imperfectly wedge problems for skew incidence," *Radio Sci., ***21**(2), 185-191, March-April 1986.

[Sh]   V. P. Shmyshlaev, "Diffraction of waves by conical surfaces at high frequencies," *Wave Motion* **12**, 329-339, 1992.

[St]   J. A. Stratton, *Electromagnetic Theory* , New York, McGraw Hill, 1941, p. 527.

[Stru] D. J. Struik, *Lectures on Classical Differential Geometry*, Dover, 1986.

[SV]   T. B. A. Senior and J. Volakis, "Scattering by an impedance right-angled wedge," *IEEE Ant Prop.*, **AP-34**, 681-689, May 1986.

[Sy, V] "An approximate skew incidence diffraction coefficient for an impedance wedge", Electromagnetics, vol 12, n° 1, 33-55, 1992

[TP]   K. D Trott., P. H. Pathak, and F. A. Molinet, "A UTD type analysis of a plane wave scattering by a fully illuminated perfectly conducting cone," *IEEE Trans. Ant. Prop.*, **AP-38**(8), 1150-1160, Aug. 1990.

[U]   P. Y. Ufimtsev, "Diffraction of plane electromagnetic waves by a thin cylindrical conductor," *Radio Eng. Elec. Phys.*, **7**, 241-249, 1962.

[Va]   V. G. Vaccaro, "The generalized reflection method in electromagnetism," AEU Band 34, Heft 12, 493-500, 1980.

[Vo]   D. R. Voltmer, "Diffraction by doubly curved convex surfaces," Ph.D. dissertation, Ohio State University, 1970.

[VS]   J. Volakis and T. B. A. Senior, "A simple expression for a function occurring in diffraction theory," *IEEE-AP*, **33**(11), 678-680.

[W]   V. H. Weston, "The effect of a discontinuity in curvature in high frequency scattering," *IRE Trans.Ant. Prop.*, **AP-10**, 775-780, Nov. 1962.

2. Search for Solutions in the Form of Asymptotic Expansions

The geometrical theory of diffraction, or GTD, was originally developed by starting from the general concepts presented in Chap. 1, viz., the localization principle; the generalized Fermat's principle; linear phase variation along a ray; power conservation in a tube of rays; and, polarization conservation. The GTD relies upon the known asymptotic solutions of canonical problems and these solutions play a two-fold role – they help validate the enunciated principles and enable us to determine the diffraction coefficients as well. We saw in Chap. 1 that, in a majority of cases, this approach not only provides useful tools for calculating the fields diffracted by an object, but also helps us to physically interpret the results in terms of rays.

However, the GTD approach, which attempts to generalize the results derived for special cases, has the disadvantage of requiring some intuitive guesswork on the part of the user, and this can sometimes lead to an error. In this chapter and the following two chapters, we will present an alternate approach, which is more systematic than the GTD, in that it derives the solutions of the Maxwell's equations in the form of asymptotic expansions. We will see that with such a method it is possible to recover, in a more deductive manner, all of the principles of the GTD enunciated in Chap. 1, except for the localization principle. In Sect. 2.1, we will provide a general survey of the perturbation methods and their use in the calculation of the field diffracted by an object. The above section not only helps us to understand the essential ideas but also to have a first exposure to the various notions developed in the rest of this book, viz., asymptotic expansions (Sect. 2.1.1); asymptotic series representing the ray field (Sects. 2.1.2 and 2.1.3); boundary layers (Sect. 2.1.4); and, uniform solutions (Sect. 2.1.5). Representative calculations and results for the conventional fields, i.e., the fields to which the conventional GTD analysis is applicable, are presented in the second part of this chapter. The methods of boundary layers and uniform solutions are developed in Chaps. 3 and 5, respectively.

2.1 Perturbation methods as applied to diffraction problems

2.1.1 Basic concepts of asymptotic expansion

Asymptotic expansions find frequent use in the solution of differential equations of mathematical physics. An application of the perturbation methods to these equations leads, in general, to a solution expressed as a series expansion in terms of integral powers of a small parameter η. The series in question satisfies the equation

$$f(\eta) - \sum_{n=0}^{N} a_n \eta^n = o(\eta^N), \tag{2.1}$$

where $f(0)$ is the solution of the unperturbed problem and o is the Landau symbol *small oh*. By definition, $o(u)$ tends toward 0 faster than u. The function $f(\eta)$, and the quantity $o(\eta^N)$ both depend upon the position of the point r. The asymptotic expansion method has been developed primarily by non-mathematicians who have not always defined, precisely, the manner in which the quantity $o(\eta^N)$ tends toward 0. In general, the convergence toward 0 is pointwise, and the asymptotic expansion is not uniform as a function of r.

A representation of the type in (2.1), which is limited to a finite number of terms, is called an *asymptotic expansion*. As η tends toward 0, the series in (2.1) converges to the exact solution of the problem, and the series represents an approximate solution for a small but non-zero η. From (2.1), it can be shown that for a given value of η, there exists an optimal value of N for which the difference between the exact solution and its asymptotic expansion is the smallest. Unfortunately, there does not exist a general rule by which we can predict the value of this optimum and a certain amount of experience is needed to determine the desired number of terms to be retained in the asymptotic expansion.

It should be noted that, in the practical sense, such an expansion is useful only if it is truncated to a finite number of terms and, consequently, we need not concern ourselves with its behavior as this number becomes infinite. This is useful to know, since there exists a distinct possibility that the infinite series may, in fact, be divergent.

Next, we introduce the important concepts of the Luneberg-Kline [Lu, Kl] expansion and its relationship to the geometrical optics. For the present, we will only concern ourselves with the basic principles of the Luneberg-Kline method, and defer the presentation of the detailed results until Sect. 2.2.1.

2.1.2 Luneberg-Kline Series and the Geometrical Optics

The Luneberg-Kline series is derived by applying the perturbation method to Maxwell's equations while regarding $1/k$ as a small parameter. The Luneberg-Kline series takes the form

$$U(r) = e^{ikS(r)} \sum_{n=0}^{N} (ik)^{-n} u_n(r) + o(k^{-N}), \tag{2.2}$$

where $U(r)$ denotes either the electric or the magnetic field at the observation point r, and $S(r)$ is their phase, which is real. The function $S(r)$ satisfies the eikonal equations of geometrical optics and the vectorial amplitudes $u_n(r)$ are solutions of the transport equations, that are coupled via their right-hand side terms in the expansion as a system of differential equations along a ray. The eikonal equations enable us to recover the laws of phase propagation. The zero-order transport equations lead to the law of power conservation in a tube of rays. Thus, the Luneberg-Kline series can be regarded as the foundation of geometrical optics. Its first term, viz., the approximation of the field to the first order with respect to $1/k$, represents the result given by the geometrical optics. The higher-order terms in the expansion serve two important purposes. First, if for a given frequency the second term is greater than the first, this indicates that the GO is not likely to be a good approximation of the solution; second, the inclusion of the higher-order terms can, in some cases, enable us to improve the accuracy of the GO results.

When an incident field of the form given in (2.2) impinges upon an object whose external surface is not discontinuous, by enforcing the boundary conditions that must be satisfied on the illuminated part of the surface we obtain a scattered field which has the same form as that of the incident field. The scattered field is the field due to the presence of the object which, when added to the incident field, yields the total field. The scattered field comprises not only the GO reflected field but also the field that annihilates the incident field in the shadow region. In the lit region, the zero-order ($n = 0$) term in the scattered field yields the GO reflected field, with a reflection coefficient identical to the one extracted by using the tangent-plane approximation, which is thus justified. The contribution of the $O(1)$ term to the reflection coefficient depends upon the curvatures of the surface as well as of the incident wavefront at the point of reflection. Similarly, the higher-order contributions also depend upon the higher-order derivatives of the surface and the wavefront.

The example of the Luneberg-Kline series, given above, serves as a good illustration of the contribution of the perturbation methods. In particular, it enables us to recover the GO result as the first term of the asymptotic expansion of the solution in terms of the powers of $1/k$. In addition, it also makes it possible for us to derive, in a deductive manner, the principles of GO; to delineate its domain of validity; and, to provide the correction terms that improve its predictions.

However, this series represents the field correctly only in the domain of validity of the GO. For this reason, we will now introduce a more general series representation that incorporates the diffraction phenomena as well.

2.1.3 Generalized Luneberg-Kline series and diffracted rays

The Luneberg Kline series is nothing but a particular case of an asymptotic series. In fact, in perturbation methods, rather than using an expansion in powers of $1/k$ as in (2.1), it is possible to represent f by means of an expansion in terms of a more general sequence of real and positive functions $\nu_n(\eta)$, of the small parameter η, where the latter is constrained to satisfy the condition

$$\nu_n(\eta) = o(\nu_{n-1}(\eta)). \tag{2.3}$$

Such a sequence is called an *asymptotic sequence* and it is possible to generalize the series (2.1) in the form of (2.3) as follows. An asymptotic expansion of a given function $f(\eta)$ in terms of a sequence of functions $\nu_n(\eta)$ takes the form

$$f(\eta) = \sum_{n=0}^{N} a_n \nu_n(\eta) + O(\nu_N(\eta)), \tag{2.4}$$

where the coefficients a_n are independent of $\nu_n(\eta)$, and are extracted by using the following limiting procedure

$$a_0 = \frac{\lim}{\eta \to 0} \frac{f(\eta)}{v_0(\eta)} \qquad a_p = \frac{\lim}{\eta \to 0} \frac{f(\eta) - \sum\limits_{n=0}^{p-1} a_n v_n(\eta)}{v_p(\eta)}. \tag{2.5}$$

The above definitions for the coefficients a_0 and a_n are meaningful only if there exist limits as defined in (2.5), which implies that it is necessary to impose certain restrictions on the choice of the asymptotic sequence $\nu_n(\eta)$. It is easy to see that, if b_n denote the non-vanishing coefficients in (2.5), the sequence $\nu_n(\eta)$ would be compatible with the function $f(\eta)$ if it contains the sequence $\mu_n(\eta)$ defined as follows, where Ord is an abbreviation for order.

$$Ord \,[\, \mu_0(\eta)] = Ord \,[f(\eta)] \qquad Ord[\, \mu_n(\eta)] = Ord \,[f(\eta) - \sum_{n=0}^{N} b_n v_n(\eta)]. \tag{2.6}$$

In practice, the function $f(\eta)$ is not known when applying the perturbation methods, because it is, in fact, just the solution we are seeking. However, since the solution $f(0)$ of the unperturbed equation is known, it is always possible to set $\nu_0(\eta) = 1$, which, in turn, leads to $a_0 = f(0)$. The determination of the other terms of the series requires supplementary information, which can be obtained by starting from the asymptotic solutions of the so-called canonical diffractions problems, e.g., a half-plane, wedge or a sphere, for which the solution $f(\eta)$ is known. A study of these solutions shows that there exists a one-to-one correspondence between the diffraction phenomena and the compatible asymptotic sequence defining the asymptotic expansion of the diffracted field associated with this phenomena. For instance, it is found that the asymptotic sequence associated with the reflected field is k^{-n}, where n is an integer, while the one describing the diffracted field by an edge is $k^{-n-1/2}$. The asymptotic sequences corresponding to other diffraction phenomena have also been identified by means of indirect approaches such as a perturbation solution [HO] of the integral equation satisfied by the surface current, or the boundary layer method as described in Chap. 3. The table below presents, for the 3-dimension case, the correspondence between the primary diffraction phenomena associated with surface singularities and their asymptotic sequences.

Diffraction phenomena	Asymptotic sequence $\nu_n(1/k)$ n = 0, 1, 2
reflection on a smooth surface	k^{-n}
diffraction on an edge	$k^{-n-1/2}$
diffraction through a discontinuity line of the curvature	$k^{-n-3/2}$
diffraction by a tip or a corner	k^{-n-1}

Just as we associate the terms in an asymptotic expansion with a class of rays in GO, we can also generalize the asymptotic sequence k^{-n} and write the diffracted field in the form

$$U^d(r) = \exp(ikS^d(r)) \sum_{n=0}^{N} \nu_n(1/k)u_n^d, \qquad (2.7)$$

where $\nu_n(1/k)$ is an asymptotic sequence satisfying (2.3) with $\eta = 1/k$, and $S^d(r)$ is a phase function which is real.

If expression (2.7) is to represent the diffracted field, it is necessary that $\nu_n(1/k)$ be a sequence compatible with the solution of the diffraction problem to be solved. Since the latter is not known, it is necessary to resort to an *Ansatz*. In choosing an *Ansatz* we are guided by the nature of the singularities occurring on the diffracting surface according to the prescription given in the table above.

The series (2.7) is not the most general form encountered in diffraction problems. The exact solutions of the circular cylinder and of the sphere show that the asymptotic series representing the diffracted field in the deep shadow region cannot be represented by using of a purely real phase function in (2.7). This is because the wave in the shadow region attenuates exponentially in a direction orthogonal to the creeping ray, and, hence, it has the character of an inhomogeneous plane wave (see Sect. 4.2, for the definition of these waves), i.e., its wave number is complex. To represent such a wave we can either extend (2.7) so that it represents a complex wave with phase and amplitude function that are both complex (see Appendix *Complex Rays*), or retain the real wave representation and simply augment the phase factor $\exp(ikS^d(r))$ by an exponential attenuation factor. When adopting the latter point of view, the solution of the cylinder problem in the deep shadow zone, far from the surface ($r \to \infty$), suggests the following form for the total field

$$U(r) = \exp(ikS(r) + ik^{1/3}\varphi(r)) \sum_{n=0}^{N} (ik)^{-n/3} u_n(r) + o(k^{-N/3}), \qquad (2.8)$$

where φ is a complex phase. The above type of expansion was first postulated by Friedlander and Keller [FK].

After substituting the expressions in either (2.7) or (2.8) in Maxwell's equations, and ordering the terms according to the powers $1/k$, we can demonstrate that the eikonal $S(r)$ satisfies the eikonal equation of geometrical optics. In addition, we find that, in a homogeneous medium, the equiphase surfaces of (2.8) are orthogonal to the surfaces of equal amplitudes given by $Im\{\varphi(r)\} = C$. The latter are consequently generated by rays which are straight lines and orthogonal to the surfaces $S(r) = C$. This result also implies that, far from the diffracting object, the rays are phase paths according to the terminology introduced by Felsen [Fe].

The amplitudes u^d_n satisfy a set of transport equations that are distinct for each diffraction phenomena, and this property is a consequence of (2.3). After integration along the characteristic curves (or rays) of the eikonal equation, the transport equations reduce to a system of coupled linear differential equations. In particular, the first term of the series satisfies the same transport equation as that obtained by using the Luneberg-Kline series, which implies that power is conserved in a tube of diffracted rays. All of these aspects will be discussed further in Sect. 2.2.

The above characteristics of the diffracted rays lead us to conclude that the diffracted rays satisfy the familiar laws of geometrical optics, namely the linear variation of the phase along a ray and the conservation of power in a tube of rays. We also observe that just as the Luneberg-Kline series is the basis of GO, the more general series, viz., (2.7) and (2.8), are the foundations of the geometric theory of diffraction, the GTD. Thus the asymptotic expansions enable us to demonstrate the laws of the GTD postulated in Chap. 1, where we saw that once we know the field at some point along the ray, then we can determine, using these laws, the field everywhere else on the ray. Thus, it only remains for us to calculate the field at some point on each ray in order to determine the diffracted field at an arbitrary observation point.

To determine the dyadic diffraction coefficients we used in Chap. 1, we use the method of canonical problems which proceeds as follows. Invoking the localization principle, we replace the object near the diffracting singularity by a simpler object, which is exactly treatable by the method of separation of variables.

It is also possible to calculate the diffraction coefficients by using the boundary layer method, which will be explained in detail in Chap. 3. Not only is this method more general than the previous one, but it provides a mathematical justification of the method of canonical problems as well.

Another important application of the boundary layer method is to provide solutions in zones where the ray methods fail. The essential physical ideas of the boundary layer method will be presented next.

2.1.4. Boundary layer method

The domain of validity of the expansions (2.7) or (2.8) is limited to the regions of space surrounding the diffracting object where the unperturbed solution is regular. The solution comprises the geometrical optics field and its perturbation, that includes the terms neglected in the GO solution, and the perturbation is given explicitly in Sect. 2.2.1. The domains where the regularity condition is not satisfied are the shadow boundaries and the caustics of the characteristic curves (or rays) associated with each diffraction phenomena. We have seen in Sect. 1.6 that, in these regions, the GTD generates infinite or discontinuous results. Obviously these results are not physical, and they can be traced to an incorrect choice of the asymptotic expansion. In the terminology of fluid mechanics, these regions are referred to as the *boundary layers*.

Generally speaking, a boundary layer is a thin layer inside which a function undergoes extremely rapid variation. The thickness of the layer and the rate of variation of the function within it depend upon a parameter which, in our case, is the wave number k. As this parameter tends to a limiting value ($k \rightarrow \infty$ in the present case), the thickness of the layer tends to zero and the function describing the layer tends to a limiting form. For instance, the boundary layer of the caustic around the edge of a wedge with plane faces is a circular cylinder, whose axis is the edge 0 and whose radius has an $O(1/k)$ behavior. This implies that when k tends to infinity, the radius of this region tends to zero. Consequently, a linear caustic can be seen as the limit of a circularly cylindrical region. Depending on k, the solution reaches a value inside this region which can be large; nonetheless it is finite when k itself is finite, and it tends to infinity as k tends to infinity. In a similar manner, a shadow boundary defines a boundary layer inside which the amplitude of the field varies rapidly in a direction orthogonal to the surface delineating the shadow zone. Some examples of boundary layers are shown on Figs. 2.1a, b, c below.

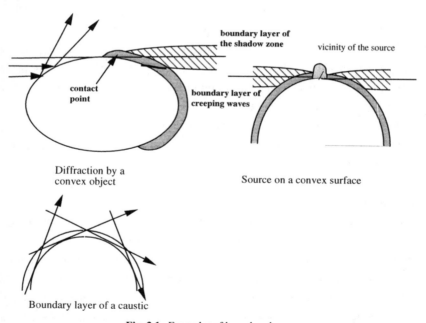

Fig. 2.1. Examples of boundary layers

We can see from the above figures that, in certain regions, two or more curves or surfaces can intersect and form multiple boundary layers.

The method of boundary layers is also called the method of matched expansions. This method is well known in fluid mechanics and was applied, for the first time to a diffraction problem, by Buchal and Keller [BK]. It involves the determination of the dyadic diffraction coefficient in the expansion exterior to the boundary layer, and this is accomplished by matching this expansion to a similar one for the region interior to the layer, in the intermediate region where both expansions are simultaneously valid. The determination of the interior expansion is achieved in two steps, viz., a stretching of the coordinates, and using a particular form for the solution based on an *Ansatz*.

Physically, the stretching takes into account of the rapid variation in the interior of the boundary layer. Mathematically, it enables us to transform the Maxwell's equations in the vicinity of the boundary layer in a way such that a perturbation method can be employed on the transformed equation, to yield a perturbation term which is regular with respect to the small parameter $1/k$. The coordinates are stretched in a direction orthogonal to the limiting surface S, to which the boundary layer tends, as $k\rightarrow\infty$. Let (X_1, X_2, X_3) be a curvilinear coordinate system in the boundary layer with the coordinate X_3 orthogonal to S (Fig. 2.2).

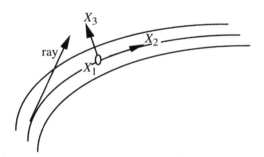

Fig. 2.2. Coordinate system in the boundary layer

The stretching transformation reads: $X'_3 = k^\alpha X_3$, with $\alpha > 0$. We will see, in Sect. 2.2.1, that we recover the GO results simply by neglecting a perturbation term of the form $k^{-2}\nabla^2 v$ in the wave equations. In order to determine the exponent α, we apply the stretching condition which consist in choosing α in a way such that the perturbation, which is singular when expressed in the original coordinates, is split up into two parts. The first of these is a term whose order in k is less than or equal to the highest-order term of the unperturbed equation, while the second is a regular perturbation term. As alternatives, we can either derive the stretching condition from the asymptotic solution of a canonical problem, or use a simple heuristic criterion which is based upon the notion that the boundary layer is the region where the rays lose their individuality and

the phase difference between two rays in the layer becomes $O(1)$. When applied correctly, all of these three methods yield the same results for the field in the transition
regions, as we will show via some examples in Chap. 3.

Once the exponent in question has been determined, the next step is to select an
Ansatz for the representation of the solution. Typically, the form of the expression
resembles either (2.7) or (2.8), although the amplitudes no longer depend on X_3, but on
the stretched coordinate X'_3 instead. The parameters of the representation are
determined by requiring that it satisfies the Maxwell's equations, the boundary
conditions, and matches the exterior expansion. The implementation of the procedure
just outlined requires some experience and it would be useful to illustrate its
applicability via some examples to be given in Chap. 3. The boundary layer method
provides solutions that are valid in all space, but the formulas obtained for the fields in
the boundary layers differ from those derived for outside these layers. From a practical
point of view, it is obviously more desirable to obtain field representations that are
simultaneously valid inside and outside the boundary layers and such results are
referred to as *uniform* in the literature. In Sect. 2.1.5 we will present the methods for
deriving the uniform solution. However, we will defer until Chap. 5 the presentation of
the precise definition of uniformity and a description of most of the uniform results that
have been derived.

2.1.5 Uniform asymptotic solution

The derivation of a uniform asymptotic solution, which is valid for all observation
points, is of great practical importance. We have pointed out that the matching of the
asymptotic expansions corresponding to the interior and exterior regions of a boundary
layer enables us to determine the diffraction coefficients. It is important to realize that,
although by following this procedure we identify the terms which have a similar origin,
their numerical values are not necessarily close to each other. Strictly speaking, the
method of matching the asymptotic expansions can be implemented only up to the
zero-order term in the domain of validity that is common to both expansions. In effect,
the phases associated with the terms in the exterior expansion satisfy the eikonal
equations exactly, while those related to the interior expansion do so only approximately, although, with an appropriate choice of the coordinates, it is possible to achieve
uniformity of the two phase terms. In principle, the coordinate surfaces must be based
on those employed for the exterior expansion, and the rays of the exterior expansion
play the role of defining these coordinates. By expressing the phase terms in the
boundary layer with respect to the coordinates attached to these rays, it is possible to

achieve the uniformity of the phases; however, in general, the amplitudes still differ in the dominant terms. Inside the boundary layer, the amplitude can be expressed by means of a special function. It is possible to achieve uniformity by appropriately modifying the argument of this function and multiplying it by a correction factor that tends to unity on the surface of the boundary. This procedure has been employed by Ivanov [Iv], who has constructed in this manner a uniform solution between the boundary layer of any convex cylinder and the exterior region. A similar procedure has been successfully applied to the case of a 3-D wedge with curved faces by Kouyoumjian and Pathak [KP], who have used the uniform asymptotic solution of the canonical problem for the wedge with plane faces as the starting point. They were able to achieve uniformity through the transition region by replacing the phase function and the divergence factor of the diffracted wave by their exact expressions. Also, they modified the argument of the transition function in a manner such that the total field, including the incident, reflected and diffracted fields, were all continuous. The uniform solutions obtained with any of the methods just described are not unique. In general, the differences appear in the higher-order terms in $1/k$, which, however, are negligible at high frequencies. These differences may be attributable to the approximations introduced in the phase terms while expressing the coordinates of the boundary layer as functions of those attached to the rays of the exterior expansion. They might also be caused by the conditions imposed on the amplitude, such as the continuity of the amplitude function, or the continuity of both this function and its first derivative. When the uniform solution of a diffraction problem is constructed from that of the associated canonical problem, it depends upon the type of uniform solution utilized in the canonical problem. For the case of the wedge, two uniform solutions have been devised by starting from two different forms of the uniform expansion of the radiation integral whose integrand has a pole near the stationary point of the phase. The first of these, called the Uniform Theory of Diffraction, or UTD, has been introduced by Kouyoumjian and Pathak [KP] and is based on the Pauli-Clemmow expansion. The second approach, which is referred to in the literature as the Uniform Asymptotic Theory, has been developed, among others, by Lee and Deschamps [LD], who have based it on the Oberhettinger and Van der Waerden type of expansions. It should be apparent to the reader after going through the material in this section that the construction of the uniform solution is an intricate procedure, which is not yet totally formalized. Nonetheless, for the benefit of the reader, we will describe the two most frequently employed uniform solutions in Chap. 5.

In conclusion, we have sketched in this section the essential ideas upon which the derivation of the solutions in the form of an asymptotic expansion is based. We will now discuss a number of examples that implement some of these ideas.

2.2 Ray fields

2.2.1 Geometrical Optics (GO)

Geometrical optics is perhaps the one asymptotic method that predates any of the others and, interestingly, it has been in existence even before being identified as an asymptotic method. To introduce the concept of GO and R fields, we first consider the scalar Helmholtz equation

$$(\nabla^2 + k^2)u = 0,$$

and look for the solution of this equation in free space such that it satisfies the Sommerfeld radiation condition and assumes some prescribed value u_0 on the edge G of an open set W, which is assumed to be bounded. Our problem, then, is to solve the problem stated in the following using a perturbational approach

$$(\nabla^2 + k^2)u = 0 \text{ in } \mathbb{R}^3-\Omega,$$
$$u = u_0 \text{ on } \Gamma,$$

$$u = O(1/r) \text{ and } \frac{\partial u}{\partial r} - iku = o(1/r), \text{ as } r \to \infty.$$

The existence and uniqueness of the solution to the problem, as stated above, can be proven mathematically. The role of the perturbation method, then, is to provide an explicit expression for the solution.

To derive a solution using the perturbational approach, we first divide the Helmholtz equation by k^2, to get

$$\left(\frac{1}{k^2}\nabla^2 + 1\right)u = 0.$$

The term $(1/k^2)\,\nabla^2 u$ could be viewed as a perturbation, but the solution of the unperturbed equation could then simply be the trivial solution $u = 0$. Consequently, to avoid this pitfall, we perform a transformation which leads to a form of equation to which the perturbation method can be applied.

We know that the solution of P should tend, in some sense, to that of geometrical optics when the wave vector k tends toward ∞. We seek, then, a transformation that results in an equation whose unperturbed solution is the GO. The main idea is to write the solution in the form of a product of a rapidly-varying phase and a slowly-varying amplitude. In 1911, Sommerfeld and Runge introduced the following transformation

$$U = \exp(ikS(r))v(r),$$

where S is a real phase function and v a complex amplitude function. The use of this transformation converts the wave equation into a new form as follows

$$(1 - \vec{\nabla}S^2)v + \frac{i}{k}\,(\nabla^2 S + 2\vec{\nabla}S \cdot \vec{\nabla}v) + \frac{1}{k^2}\nabla^2 v = 0, \tag{2.9}$$

where the different terms are ordered according to the parameter $1/k$ which is assumed to be small. Now, if we neglect the last term, we are led to the eikonal equation

$$\vec{\nabla}\,S^2 = 1, \tag{2.10}$$

and the transport equation

$$v\nabla^2 S + 2\,\vec{\nabla}\,S \cdot \vec{\nabla}\,v = 0. \tag{2.11}$$

We will now show that these equations contain all of the laws of the GO for the scalar waves and they enable us to derive an approximation to the solution in certain regions of space. To this end, we will first consider the eikonal equation in the following section.

2.2.1.1 Resolution of the eikonal equation

The eikonal is a first-order partial differential equation, the so-called Hamilton-Jacobi equation, which has the general form $F(x_i, p_i) = 0$, where $p_i = \partial S/\partial x_i$. An equation of this type is usually solved by using the method of the characteristics. The characteristics curves are called rays and coincide with the GO rays as introduced in Chap. 1. In a homogeneous medium, the parametric equations of the characteristic curves read [CH],

$$\frac{dx_i}{ds} = p_i, \qquad\qquad \frac{dp_i}{ds} = 0, \qquad\qquad \frac{dS}{ds} = 1.$$

The first equation shows that the tangent to a ray is directed along the gradient of the phase, whereas the second tells us that this tangent is constant along a ray. The rays are, then, straight lines directed along the gradient of the phase. The third equation shows that S is, except for an additive constant, the abscissa along the ray.

An alternate way of viewing the problem at hand is as follows. We begin by defining the rays as the integral curves of the gradient of the phase and assume, *a priori*, that the rays are not straight lines. Next, we consider two points A and B, located on the ray path. The variation of the phase along the ray is given by

$$S(A) - S(B) = \int_A^B \vec{\nabla} S \cdot \vec{ds} \, .$$

It is equal to the length of the ray $l(A,B)$, since $\vec{\nabla} S$ and $\vec{ds}$ are collinear, and that if the norm of $\vec{\nabla} S$ is equal to 1, then $\vec{\nabla} S \cdot \vec{ds} = \| \vec{ds} \|$. Now consider the linear segment AB. The variation $S(A) - S(B)$ of S is less than or equal to the length of the line segment AB, since $\vec{\nabla} S \cdot \vec{ds} \leq \| \vec{ds} \|$. Accordingly, $l(A,B)$ must be equal to $|AB|$, which is possible only if the ray is confined within the segment AB. Thus, in a homogeneous medium, the rays are straight lines. In Chap. 1, this result was obtained by invoking the Fermat's principle. The phase variation $S(A) - S(B)$ of S is simply the length $|AB|$. We thus recover, by solving the eikonal equation, the principle of the linear variation of the phase as postulated in Chap. 1.

The eikonal equation enables us to calculate the eikonal difference between two points located on the same ray. In order to calculate the eikonal S, it is necessary to know it on the *initial* surface Γ. This eikonal on Γ is obtained by starting from the initial data u_0, that are assumed to be expressed as the product of a phase function $\exp(ikS_0)$ and a slowly varying amplitude function. It is clear that this assumption is restrictive and that it does not let us treat an arbitrary field. The projection on Γ of the unit vector associated with the ray is equal to the surface gradient of S_0. We must distinguish the following three cases according to the value of the norm of this gradient (see Fig. 2.3).

(i) If the above norm is less than 1 and equal to $sin(\theta)$, the ray points in a direction whose projection on Γ is along the surface gradient of S_0, and its angle with the normal to the surface is θ. There exist two directions that are possible solutions and we choose the outward direction, i.e., the one that points toward the exterior of Γ, and construct the solution by following the ray. The solution of the eikonal equations is then unique in the neighborhood of Γ .

(ii) If the norm is equal to 1 on Γ, the rays are tangent to Γ, which then comprises a portion of the envelope of the rays, called the caustic surface. There exist two solutions on the convex side of the surface and none on the other side.

(iii) Finally, if the norm is greater than 1, the eikonal equations have no solution with a real S. It is possible to construct a solution in terms of rays by interpreting θ as a complex angle, which leads to complex values for S and also of the coordinates. We will return to this topic in the Appendix *Complex Rays*.

In the rest of this chapter we will see that, in the diffraction problems of homogeneous plane waves by objects, the phase of the diffracted field is equal to that of the incident field on the object. Consequently, the norm of the surface gradient is less than 1 at every point where the incident field is not grazing. On the light-shadow separatrix, when the field is grazing, the surface gradient has its norm equal to 1. Thus, the condition (i) exists outside the separatrix, while (ii) is satisfied on the separatrix itself.

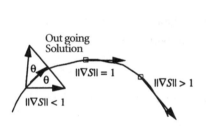

Fig. 2.3. Eikonal on a surface

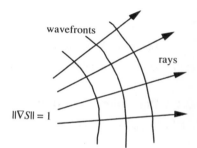

Fig. 2.4. Rays and wavefronts

Let us now introduce some important definitions. The rays in the direction of the gradient of S constitute a two-parameter family of straight lines labeled here as a and b (see Fig. 2.4). In mathematical language, such a family is called a congruence of straight lines. The surfaces orthogonal to this congruence, i.e., to the gradient of S, are $S = constant\ surfaces$, and they are referred to as the wavefronts. The wavefronts constitute a one-parameter family, which is simply the eikonal S characterizing the wavefront. Let the reference wavefront, which corresponds to $S = 0$ be termed W_0, and let (a, b) denote some surface coordinates on W_0. Then, S can be identified with the curvilinear abscissa s along a ray as measured from the intersection of this ray with W_0. The coordinates (a, b, s) are called the ray coordinates, and are shown in Fig. 2.5. The ray coordinates are regular in the neighborhood of Γ for the case (i). They become singular on the envelope of the rays, viz., the caustic. The caustic is the boundary of a region of space where there exists the solution of the eikonal equation corresponding to the initial conditions. In the shadow region of the caustic, none of the rays can penetrate. Evidently there exists a solution of the problem *(P)* on the shadow side of the caustic, though GO is unable to describe it, simply because it restricts itself only to real values of *S*. The caustic, defined as the envelope of the rays, is also the surface of the centers of the wavefronts. We will study its geometry in Sect. 6.2.

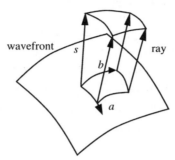

Fig. 2.5. Ray coordinates

Having completed the discussion of the solution of the eikonal equations, we now move on in the next section to study of the transport equation (2.11).

2.2.1.2 Resolution of the transport equation

As a first step toward the solution of the transport equation (3), we multiply it by v to get

$$\nabla \cdot (v^2 \vec{\nabla} S) = 0. \tag{2.12}$$

From this equation it follows that $v^2 \vec{\nabla} S$ has a vanishing divergence and consequently its flux in a tube of force lines is conserved. We apply this result to a tube of rays (see Fig. 2.5). The flux on the walls formed by the rays vanishes, since $v^2 \vec{\nabla} S$ is parallel to the rays. The flux on the walls comprising of the wavefronts W_1 and W_2 is given by the following expression

$$\int_{W_1} v^2 ds' - \int_{W_2} v^2 ds', \tag{2.13}$$

which leads to the conservation of the flux of the square of field amplitude. In particular, the square of the modules of the field, which is nothing but the power flux in a tube of rays, is conserved. We recover, from the above, the second postulate of

Geometrical Optics. We might note that this result can also be obtained by multiplying (2.11) with the conjugate of v.

The transport equation is also solved by using the ray coordinates. When expressed in terms of these coordinates, it becomes a simple differential equation along a ray. The wavefront W_0, which is the origin of the eikonal, is defined by the lines of curvature. Finally, we choose the surface coordinates as the parameters (a,b) characterizing a ray. In this coordinate system, the term $2\vec{\nabla} S \cdot \vec{\nabla} v$ of the transport equation takes the simple form $2\, dv/ds$.

On the other hand, by applying simple concepts of differential geometry, the Laplacian of S in the ray coordinate can be written [BS] as

$$\nabla^2 S = \frac{1}{\rho_1 + s} + \frac{1}{\rho_2 + s}, \tag{2.14}$$

where ρ_1 and ρ_2 are the radii of curvature of W_0 at the intersection of the ray and W_0. Equation (2.11) can be rewritten

$$2\frac{dv}{ds} + \left(\frac{1}{\rho_1 + s} + \frac{1}{\rho_2 + s}\right) v = 0. \tag{2.15a}$$

Hence

$$\frac{v(s)}{v(0)} = \sqrt{\frac{\rho_1 \rho_2}{(\rho_1 + s)(\rho_2 + s)}}. \tag{2.15b}$$

It is convenient to rewrite the previous equations in another form by introducing the Jacobian J of the ray coordinate.

$$\frac{J(s)}{J(0)} = \frac{(\rho_1 + s)(\rho_2 + s)}{\rho_1 \rho_2},$$

$$2\frac{dv}{v} + \frac{dJ}{J} = 0 \tag{2.16a}$$

$$\frac{v(s)}{v(0)} = \sqrt{\frac{J(0)}{J(s)}}. \tag{2.16b}$$

As we saw in Chap. 1, these equations are a direct consequence of the power conservation in a tube of rays as expressed by (2.13), but we will need the form of the transport equation given in (2.15a) and (2.16a), in the next section. Equations (2.15) and (2.16) enable us to calculate the amplitude of the field along a ray, once it is known at some point along this ray.

In conclusion, the eikonal and transport equations contain all of the laws of GO. The eikonal equation shows that the rays, identified as the characteristics of this equation, are straight lines in a homogeneous medium. The eikonal equation enables us to calculate the phase, starting from the initial phase on G, while the transport equation enables the calculation of the amplitude, starting from the initial amplitude on G.

2.2.1.3 Limit of validity and improvement of the solution provided by geometrical optics

We begin the discussion in this section with the remark that the previous formulas do not provide a good approximation to the desired ray solution everywhere, because, first of all, (2.15a) and (2.15b) yield a nonphysical result that diverges either when $s = -\rho_1$ or $s = -\rho_2$, i.e., on the caustic. Second, it is only when ρ_1 and ρ_2 are finite that the solution behaves as $1/r$ and satisfies the equation $\partial u/\partial r - iku = o(1/r)$, and thus the radiation condition, as r and consequently $s \to \infty$. However, if one of these radii of curvature is infinite, v, as given by (2.15b), decreases as $r^{-1/2}$. In this case, the GO solution is no longer valid and we have a caustic at infinity.

It should not come as a total surprise that geometrical optics leads to incorrect results under certain circumstances, because in order to establish the laws of GO, we have neglected the perturbation $(1/k^2) \nabla^2 v$ in (2.9). It is a singular perturbation; hence, we should expect some difficulties if the perturbation term is neglected. Thus, geometrical optics is valid if, and only if, this term is effectively negligible. In particular, geometrical optics is not valid in the regions where the derivatives of the field are large, i.e., the regions where the amplitude exhibits a substantial variation over distances on the order of a wavelength, which is precisely the case near the caustics. More precisely v, as given by GO, is proportional to $s^{-1/2}$ in the vicinity of the caustic where s is the distance to the caustic along the ray, or to $n^{-1/4}$ if n is the distance normal to the caustic. The term $(1/k^2) \nabla^2 v$ will then be proportional to $(kn)^{-2}$ and, consequently, it cannot be neglected near the caustic. More generally, as in all of the problems that can be classified as singular perturbations, we can distinguish a zone, called the ray field zone, where the term $(1/k^2) \nabla^2 v$ is small, and we can also distinguish the boundary layers, where it becomes of the same order as the other terms in (2.9). Outside these layers, the third term in (2.9) is small, and it can be treated as a perturbation. Since it involves only the amplitude and not the phase, the perturbation modifies only the function v, which can be written as a series in terms of the powers of $1/k$ as follows

$$v(r) = \sum_{n=0}^{N} (ik)^{-n} v_n(r) + O(v_N),$$ (2.17)

and is recognized as the Luneberg-Kline series. Inserting (2.17) into (2.9), and ordering the terms with respect to the power of $1/k$, we obtain a series in terms of the powers of $1/k$, each of whose coefficients must vanish separately. The zero-order (i.e., k^0) term, leads to the eikonal equations (2.10), which remain unchanged. The first-order term associated with $1/k$ yields the transport equation (2.11), which again remains unchanged. In contrast, the higher-order terms lead to transport equations but this time with a second member, viz.,

$$v_n \nabla^2 S + 2 \vec{\nabla} S \cdot \vec{\nabla} v_n = -\nabla^2 v_{n-1}.$$ (2.18)

These equations can be solved by iteration if the first term v_0, extracted from GO, is known. They reduce, in effect, to the differential equation along the rays

$$2 \frac{dv_n}{ds} + \frac{v_n}{J} \frac{dJ}{ds} = \frac{2}{\sqrt{J}} \frac{d}{ds} (\sqrt{J} v_n) = -\nabla^2 v_{n-1},$$ (2.19)

and can be integrated immediately to give

$$v_n(s) = \sqrt{\frac{J(0)}{J(s)}} v_n(0) - \frac{1}{2\sqrt{J(s)}} \int_0^s \sqrt{J(t)} \nabla^2 v_{n-1}(t) dt .$$ (2.20)

This procedure shows that geometrical optics is the first term of the Luneberg-Kline perturbation expansion (or of the asymptotic sequence) of the solution with respect to the powers $1/k$. We should note, however, that the laws of GO do not apply to the higher-order terms as there is no power conservation in a tube of rays because of the right hand side in (2.20).

We will see in the next section that the above results, which are valid for the case of the scalar wave equation, can be extended without difficulty to Maxwell's equations.

2.2.1.4 Case of Maxwell's equations

To extend the solution procedure given above to the case of Maxwell's equations, we begin by replacing the scalar wave equation by the vectorial Helmholtz equation, viz.,

$$(\nabla^2 + k^2)\, \vec{E} = 0 \quad \text{in} \quad \mathbb{R}^3 - \Omega, \tag{2.21}$$

where $\vec{E}$ denotes the electric field outside the diffracting object Ω and it is assumed that the medium is homogeneous and isotropic. Since the divergence of $\vec{E}$ must vanish, we have

$$\nabla \cdot \vec{E} = 0. \tag{2.22}$$

In addition, the tangential electric field on the boundary $\partial\Omega$ must satisfy a prescribed value and the Silver-Müller radiation condition at infinity.

Introducing the same transformation as we used for the scalar wave equation, viz.,

$$\vec{E} = \exp(ikS(r))\, \vec{e}\,(r), \tag{2.23}$$

and substituting (2.23) in (2.21) and (2.22), we get

$$(1 - \vec{\nabla} S^2)\, \vec{e} + \frac{i}{k}(\nabla^2 S + 2\vec{\nabla} S \cdot \vec{\nabla})\, \vec{e} + \frac{1}{k^2}\nabla^2\, \vec{e} = 0, \tag{2.24}$$

and

$$\vec{\nabla} S \cdot \vec{e} + \frac{1}{k}\nabla \cdot \vec{e} = 0. \tag{2.25}$$

The above equations apply to the magnetic field $\vec{H}$ as well. If we now neglect the third term in (2.24) and the second in (2.25), then (2.24) becomes identical to (2.9), and we conclude that S satisfies the eikonal equation in this case in the same manner as it did for the scalar wave equation.

The transport equation is the same as (2.11) for each component of the field. In particular, the flux of the modules squared of the electric and magnetic fields is conserved, i.e., the power flux conservation condition is satisfied.

Equation (2.25) provides some information on the polarization of the field as well, viz., that at first-order $\vec{E}$ and $\vec{H}$ are orthogonal to the ray. Finally, if we insert (2.23) in one of the Maxwell's equations, say for instance,

$$\vec{\nabla} \times \vec{E} = ikZ \vec{H}, \tag{2.26}$$

where Z is the impedance of the vacuum, we get

$$(\vec{\nabla}S) \times \vec{e} - Z\vec{h} - \frac{i}{k}\vec{\nabla} \times \vec{e} = 0, \tag{2.27}$$

The third term in (2.27) is of higher order than the other two terms, hence $Z\vec{h} \approx (\vec{\nabla}S) \times \vec{e}$. For the zero order, the fields $\vec{E}$ and $\vec{H}$ are orthogonal to each other, as well as to the direction of propagation of the ray, and the ratio of the modules of E/H is equal to the free-space impedance. The Poynting vector, derived from the first-order fields, is directed along the ray. This example shows that the local character of the GO solution is that of a plane wave.

Also, since the square of the amplitude of each field component is conserved in a ray tube, the polarization is also conserved along a ray. We observe, then, that if we neglect the higher-order terms in (2.27), we recover all the laws of Geometrical Optics from the Maxwell's equations. As we saw in Chap. 1, these laws enable us to determine the solution everywhere on the ray, once this solution is known at any one point on the ray. In our case, the initial data is the tangential electric field $\vec{E_t}$ on Γ. As in the scalar case, this data is assumed to be written in the form $\{\exp(ikS_0) \cdot \vec{E_t}\}$. The gradient of S_0 provides the direction of the ray, $\vec{E_t}$, and makes it possible for us to determine the transverse field at any point of Γ.

As in the case of the scalar wave equation, Geometrical Optics is valid only if the neglected terms are small. All of these terms involve the derivatives of the field, and the usual criterion for neglecting the higher-order terms is that the variation of the field over a distance of the order of a wavelength be relatively small.

Furthermore, by employing the previous perturbation method, we can identify the GO field as the first term of a perturbation series, shown below, which is an expansion in terms of the asymptotic sequence $(1/k)^n$.

$$\vec{E}(r) = \exp(ikS(r)) \sum_{n=0}^{N} (ik)^{-n} \, \vec{e_n}(r) + O(k^{-N}), \tag{2.28a}$$

$$\vec{H}(r) = \exp(ikS(r)) \sum_{n=0}^{N} (ik)^{-n} \, \vec{h_n}(r) + O(k^{-N}). \tag{2.28b}$$

Inserting the above expressions in Eqs. (2.24), (2.25), and (2.27), we obtain the set of equations

$$(\vec{\nabla} S)^2 = 1, \tag{2.29}$$

$$(\nabla^2 S + 2 \vec{\nabla} S \cdot \vec{\nabla}) \, \vec{e}_n = -\nabla^2 \vec{e}_{n-1}, \tag{2.30}$$

$$\vec{\nabla} S \cdot \vec{e}_n = -\nabla \vec{e}_{n-1}, \tag{2.31}$$

$$\vec{\nabla} S \times \vec{e}_n - Z\vec{h}_n = -\nabla \vec{e}_{n-1}. \tag{2.32}$$

Indeed, since $\vec{e}_{-1} = 0$, we can readily verify that the first term, viz., $\vec{e}_0$, is consistent with the results of Geometrical Optics.

Using (2.29), which is the eikonal equation, and (2.30), we can show that each component of the field satisfies the transport equation, which has a r.h.s. for the higher-order components, but not for $\vec{e}_0$. Equation (2.31) enables us to calculate the component directed along the direction of propagation. It vanishes for the order 0, but this is no longer true when the higher-order terms are taken into account. Also, (2.32) shows that the components of the fields $\vec{e}$ and $\vec{h}$ associated with higher-order terms are no longer orthogonal. As we saw for the scalar wave equation, only the dominant part of the field satisfies the laws of Geometrical Optics.

The Luneberg-Kline series provides an approximation to the solution when the electric or magnetic tangent field is given on the surface Γ. Next, we will investigate the problem of calculating the reflected field from an object illuminated by an incident ray field.

2.2.1.5 Field reflected by a regular object

Let us consider field $\vec{E}^i$ incident upon a regular object W, which is either perfectly conducting or whose surface satisfies an impedance boundary condition $\vec{E} - \hat{n}(\hat{n} \cdot \vec{E}) = Z\,\hat{n} \times \vec{H}$. We seek a solution $\vec{E}^r$ of Maxwell's equations in the exterior space $\mathbb{R}^3\_\Omega$. The field $\vec{E}^r$ must satisfy the Silver-Müller radiation condition, $(\vec{E}^r + \vec{E}^i)_{tan}$ must vanish on the surface of a perfect conductor, or must satisfy the appropriate surface impedance condition for the general case. The electric and magnetic fields, $\vec{E}^i$ and $\vec{H}^i$ are represented by their asymptotic expansions

$$\vec{E}^i \, (r) = \exp(ikS^i(r)) \sum_{n=0}^{N} (ik)^{-n} \, \vec{e}^i_n \, (r) + O(k^{-N}), \tag{2.33a}$$

$$\vec{H}^i \, (r) = \exp(ikS^i(r)) \sum_{n=0}^{N} (ik)^{-n} \, \vec{h}^i_n \, (r) + O(k^{-N}), \tag{2.33b}$$

where S^i satisfies the eikonal equation, and the amplitudes $\vec{e}^i{}_n$ and $\vec{h}^i{}_n$ are solutions to (2.30) through (2.32). It is evident from Fig. 2.6 that for the given incident wave, it is natural to divide up the surface of the object into lit and shadow regions, separated by a line called the separatrix (see Fig. 2.6), whose trace on $\partial\Omega$ is the light-shadow boundary C.

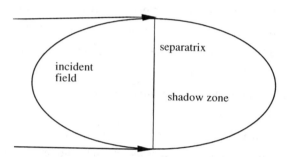

Fig. 2.6. Geometry of the separatrix

Next, we seek a representation of the reflected field in the form of a Luneberg-Kline series. When the object Ω is perfectly conducting, the cancellation condition imposed on total tangent field on $\partial\Omega$ provides the tangential reflected field on $\partial\Omega$, and the problem of finding the reflected field can be reduced to the one studied in the previous paragraph. The eikonal S^r of the reflected field is equal to the eikonal of the incident field on $\partial\Omega$. The surface gradient of the eikonal is simply the projection of the vector $\hat{i}$ of the incident ray on the tangent plane to the surface. The direction $\hat{r}$ of the reflected field is pointed outward from the surface and has the same projection on the tangent plane as $\hat{i}$. Thus, from the above, we obtain the relation

$$\hat{r} = \hat{i} - 2\,\hat{n}(\hat{n} \cdot \hat{i}), \tag{2.34}$$

in the lit region and recover the law of reflection, whereas

$$\hat{r} = \hat{i}, \tag{2.35}$$

in the shadow zone. Thus, the reflected field in the shadow region is negative of the incident field and the total field there vanishes. On C, $\hat{r}$ is tangent to $\partial\Omega$, and consequently we are on the caustic of the reflected field.

When the object satisfies the impedance condition $\vec{E}^t - \hat{n}(\hat{n} \cdot \vec{E}^t) = Z\,\hat{n} \times \vec{H}^t$, we replace $\vec{E}^i$, $\vec{E}^r$, $\vec{H}^i$, and $\vec{H}^r$ in this equation with their asymptotic expansions and order the resultant expression with respect to the powers of $1/k$. The eikonal of the reflected field is, as in the case of a perfect conductor, equal to the one of the incident fields, and (2.34) and (2.35) are valid. On the other hand, the first term of the reflected field takes the form $\vec{e}^r_0 = \vec{e}^i_0 \, \underline{R}$, in the lit zone, where $\underline{R}$ is the reflection dyad, which is derived by considering the canonical problem of the tangent plane defined in Chap. 1. Consequently, we recover the tangent plane approximation in the illuminated zone. For the higher-order terms this approximation is no longer valid, and, in particular, these terms depend on the local curvature of the object. However, the result still remains local, i.e., the field depends only upon the incident field and the geometry of the object near the point under consideration. In the shadow zone, $\vec{e}^r_0 = -\vec{e}^i_0$ for all orders, and the total field is zero.

Once the field is known on $\partial\Omega$, we can obtain the first term of the reflected field as in the past by applying the laws of GO, and by iteratively solving the differential transport equations along the rays.

The domain of validity of the solution, thus obtained, is similar to the one defined in the previous section, and, in particular, it is not valid on the caustics. We should note that if the object is convex, the caustic of the reflected field is located entirely within the interior of the object, and that it is tangent to the object on the shadow boundary. The first term of the Luneberg-Kline series remains bounded on the shadow boundary. In contrast, the second term, which is proportional to $\cos^{-3}\theta$, where θ is the incidence angle, becomes infinite on this boundary [BS].

In addition, the solution is also not valid in regions where the variation of the field is rapid and on the shadow boundaries through which the field and its derivatives exhibit discontinuities. It does not satisfy the radiation condition if one of the radii of curvature of the reflected wavefront vanishes. Furthermore, if the incident field is a plane wave, or more generally if it does not satisfy the radiation condition, neither does the reflected field which cancels the incident field in the shadow zone.

To summarize the discussion in this section, we observe that the Luneberg-Kline series provides an approximation to the solution of the diffraction problem of an incident field by an object. The first term of this approximation is the Geometrical Optics field in the lit region. It enables us to recover, in a deductive manner, the cancellation of the field in the shadow region, and the reflection law as well as the approximation of the tangent plane in the lit region.

As stated in Sect. 2.1, the Luneberg-Kline series, when applicable, makes it possible for us to recover the Geometrical Optics result as the first term of the asymptotic expansion of the solution of the problem in terms of the powers of $1/k$; to recover in a deductive manner all the principles of GO; and, to delineate the limits of validity as well as to suggest improvements of the GO approximation.

Next, we will turn our attention to the problem of calculating the field in the shadow zone, and we will do this by introducing the asymptotic expansions corresponding to the diffracted rays in this region.

2.2.2 Diffracted field by an edge

We saw in Sect. 2.1 that the asymptotic expansion representing the diffracted field by an edge takes the form (Eq. (2.7) in Sect. 2.1)

$$\vec{E}^d \ (r) = \exp(ikS^d(r)) \sum_{n=0}^{N} v_n(1/k) \ \vec{E}^d {}_n (r) + O(v_n(1/k)), \qquad (2.36)$$

with

$$v_n(1/k) = (1/k)^{-n-1/2}. \qquad (2.37)$$

The asymptotic sequence v_n is deduced from the canonical problem of the wedge. In the shadow region of the wedge, the diffracted field can, in fact, be written in the form of an expansion which resembles the expression in (2.36). Note that we can deduce the first term of the asymptotic series by the following simple method. For a two-dimensional problem, the field decreases as $1/\sqrt{r}$ when r is sufficiently large. The wedge does not have any characteristic dimensions and, consequently, the diffracted field can depend only upon the quantity kr which is dimensionless. Thus, we conclude that the first term of the asymptotic expansion is necessarily of the form $1/\sqrt{kr}$.

As in the previous section, we substitute the expression in (2.36) into the Maxwell's equations. The above expression is the product of a Luneberg-Kline series and the quantity $k^{-1/2}$. Consequently, we obtain the same eikonal and transport equations as for the Luneberg-Kline series, and (2.29) to (2.32) thus apply to the diffracted field. Also, we recover Keller's postulate, viz., that the rays diffracted by an edge behave similarly to the GO rays. In addition, this result is valid not only for the first term but for all of the terms of the asymptotic expansion of the diffracted field.

Let us now consider an object whose surface exhibits a discontinuity in the tangent plane that forms a sharp edge, which may either be straight or curved. Except for this discontinuity, the surface is assumed to be smooth with large radii of curvature in comparison to the wavelength. The incident field on the object not only generates a reflected field (r), as in the previous section, but also a field E^d (r) diffracted by the edge. Thus, the total field takes the form

$$\vec{E^s}\ (r) = \vec{E^r}\ (r) + \vec{E^d}\ (r). \tag{2.38}$$

All the diffracted rays emanate from the edge, because all of the rays originating from the regular part of the surface are already included in $\vec{E^r}$ (r). To determine the phase of the field at any point along the ray it is sufficient to know it just at the edge. The edge is a caustic of diffracted rays and the field is not a ray field in a small tube, which has a radius $O(1/k)$ and is centered at the edge. A rigorous derivation of the phase requires the use of a boundary layer method and the matching of the solution in the boundary layer in the vicinity of the edge with the ray solution given by (2.36). This procedure was explained in Sect. 2.1. and the details of the calculation can be found in Chap. 3. Following this technique, we arrive at a very simple matching principle that can be expressed as

$$S^d(r) = S^i(r). \tag{2.39}$$

The eikonal equation is the same for E^d (r) as it is for $\vec{E^r}$ (r); however, while the initial condition was imposed on a regular surface for the reflected field, it is now applied on the curve following the edge. It is interesting to note that this situation has also been studied in the literature, and the eikonal equation has been solved for this case in the usual manner, by using the method of the characteristics. However for a regular surface, the projection of the gradient of the eikonal on the surface is known, which leads to two possible alternatives. In the edge case, (2.39) only provides the projection of the gradient of the eikonal, i.e., of the unit vector $\hat{d}$ associated with a diffracted ray along the direction of the tangent to the edge.

$$\vec{\nabla} S^d(r) \cdot \hat{t} = \hat{d} \cdot \hat{t} = \vec{\nabla} S^i(r) \cdot \hat{t} = \hat{t} \cdot \hat{t} . \tag{2.40}$$

We thus find, once again, that the diffracted rays must lie on the surface of the Keller cone. It is useful to note that in this case we obtain a congruence, i.e., a two-parameter family of diffracted rays. The specificity of this congruence is such that one of the caustics reduces to a line.

As for the amplitude of the diffracted ray, it can not be determined by using the ray method. It is derived, instead, by matching the result in the boundary layer in the vicinity of the edge with the solution of Maxwell's equations obtained in this section.

2.2.3 Field diffracted by a line discontinuity

For the case of a line discontinuity, the asymptotic expansion of the diffracted field is the product of a function of k, as given in Table 2.1, and a series of the same form as the Luneberg-Kline series. Thus, for this case, the results are the same as those for an edge.

2.2.4 Diffracted field by a tip or a corner

The asymptotic expansion of the diffracted field is a product of $1/k$ with a series of the same form as the Luneberg-Kline series. The diffracted field is then a ray field, and the terms of the expansion satisfy equations (2.29) through (2.32). The eikonal is given at the point of diffraction and we can show that it is equal to that of the incident field. As in the past, this is achieved by matching the solution in a boundary layer in the vicinity of the tip, to the ray field solution in the exterior space. The only condition imposed on the gradient of the eikonal is that its norm be equal to 1; hence, the rays emanate in all directions outside the diffracting object. The amplitude of the diffracted field is also determined with the application of the boundary layer method.

2.2.5 Field in the shadow zone of a smooth object

As we saw in 2.2.1.5, Geometrical Optics predicts a vanishing field in the shadow zone of a smooth object. In effect, the asymptotic expansion of Luneberg-Kline series is not adapted to this zone where the field decreases as $\exp(-\alpha k^{1/3})$, i.e., the decay is faster than algebraic. We mentioned in Sect. 2.1 that the asymptotic expansion of the total field in the deep shadow zone of a smooth object is given by

$$U(r) = \exp(ikS(r) + ik^{1/3}\varphi(r)) \sum_{n=0}^{N}(ik)^{-n/3} u_n(r).$$

Let us insert this expansion into the wave equation and order the terms according to increasing powers of $k^{1/3}$. We then obtain, to $O(k^2)$, the result

$$(\vec{\nabla} S)^2 = 1, \tag{2.41}$$

which is the eikonal equation.

Let us consider a simple example, viz., the case of a convex cylindrical object. The standard GTD method assumes that the field in the shadow zone propagates along the creeping rays which follow the generatrix of the object and detach tangentially from it as shown in Fig. 2.7.

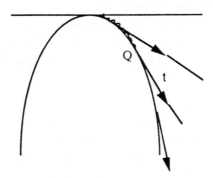

Fig. 2.7. Creeping rays

Let us now introduce the ray coordinates defined as follows. Let s be the curvilinear abscissa on the generatrix, and let l be the total length of the ray, namely the sum of s and the length t of the diffracted ray. This is the eikonal as obtained by applying the usual law of phase propagation along the creeping rays. A point M on the ray is then defined by

$$\vec{OM} = \vec{OQ} + (l - s)\hat{t} , \qquad (2.42)$$

where $\hat{t} = d\vec{OQ}/ds$ is the vector tangent to the cylinder, $\partial \vec{OM}/\partial s = -(l - s)(\hat{n}/\rho)$, and $\hat{n}$ is the normal vector exterior to the cylinder. We also have $\partial \vec{OM}/\partial l = \hat{t}$, which implies that the ray coordinates are orthogonal. Using these coordinates, we can calculate the gradient to find

$$(\vec{\nabla} l)^2 = 1. \qquad (2.43)$$

Thus, the total length l of the creeping ray is a solution of the eikonal equation. In order to complete the determination of the eikonal it is sufficient to know it at a particular point. Again, we use the phase matching principle, and apply it at the point C on the shadow boundary. The phase at that point is equal to that of the incident field, i.e., $S^i(C) = S^d(C)$. As we will see in Chap. 3, this result can be demonstrated by using the boundary layer techniques.

To $O(k^{5/3})$, we obtain the result

$$\vec{\nabla} S \cdot \vec{\nabla} \varphi = 0. \qquad (2.44)$$

The equiphase lines, i.e., S = constant, are then orthogonal to the lines φ = constant, which are simply the rays.

The higher-order terms lead us to a series of the transport equations. To a first order, the power is then conserved along a tube of creeping rays. We note that the surface of the object is a caustic for the creeping rays. Consequently, we must use a boundary layer technique to calculate the amplitude of these rays. We have thus deduced, from the asymptotic expansion (2.8) in Sect. 2.1, that, to first order, the creeping rays satisfy all of the laws of Geometrical Optics. To the next order, we obtain a series of transport equations; however, these equations are not the same as those for the diffracted field. For instance, for the term u_1 we get

$$\vec{\nabla} u_1 \cdot \vec{\nabla} S + u_1 \nabla^2 S = i\, u_0\, \vec{\nabla} \varphi^2. \tag{2.45}$$

Nevertheless, these transport equations enable us to calculate the field as they do for the other types of rays, by integration along the rays and by matching to the solution in the surface boundary layer. All of the previous results, obtained for the scalar wave equation, are easily extended to the Maxwell's equations by recognizing that each component of the fields must satisfy the transport equations just given above. The condition on the vanishing of the divergence provides, as it did for the GO rays, the information on polarization, which remains constant along a ray. Thus, we recover all of the laws of Geometrical Optics by following the procedure outlined above.

2.2.6 Conclusions

In conclusion, the method of asymptotic expansions not only enables us to understand the foundations of the ray techniques, but to recover the laws of Geometrical optics and those of the geometrical theory of diffraction as well. The latter includes the linear variation of the phase along the rays, power conservation in a tube of rays, and, the conservation of polarization. Additionally, the asymptotic expansion method also enables us to establish the laws of reflection. Finally, the diffraction laws, that were derived in Chap. 1 from an application of the generalized Fermat principle, are deduced here from the eikonal equations and from the equality principle of the phases of the diffracted and incident fields at the diffraction point. The demonstration of this principle, and that of the calculation of the amplitude on the diffracted rays, is accomplished by appealing to the boundary layer method, because the diffraction points are located in the zones where the previous procedures are not valid. We will address these questions in more detail in Chap. 3.

References

[BK]   R. N. Buchal and J. B. Keller, "Boundary layer problems in diffraction theory," *Comm. Pure Appl. Math.*, **13**, 85-114, 1960.

[BS]   J. J. Bowman, T. B. A. Senior, and P. L. Uslenghi, *Acoustic and Electromagnetic Scattering by Simple Shapes*, Hemisphere, 1987.

[CH]   R. Courant and D. Hilbert, *Partial Differential Equations*, Wiley, 1962.

[Fe]   L. B. Felsen, "Evanescent waves," *J. Opt. Soc. Amer.*, **66**, 751-760, 1976.

[FK]   F. C. Friedlander and J. B. Keller, "Asymptotic expansions of solutions of $(\Delta + k^2)U = 0$," *Comm. Pure Appl. Math.*, **8**, 387-394, 1955.

[Ho]   S. Hong, "Asymptotic theory of electromagnetic and acoustic diffraction by smooth convex surfaces of variable curvature," *J. Math. Phys.*, **8**(6), 1223-1232, 1967.

[Kl]   M. Kline, "An asymptotic solution of Maxwell's equations," *Comm. Pure Appl. Math.*, **4**, 225-262, 1951.

[Lu]   R. M. Luneberg, *Mathematical Theory of Optics*, Brown University Press, 1944.

3. The Boundary Layer Method

We have seen in Chap. 2, how a formal series representation, which is more general than the Luneberg-Kline series, can be used to describe the propagation of the diffracted rays. We have also seen that the formal series provides a description of the field only in regions where it is a ray field, and in the present chapter we will concern ourselves with the calculation of the field in the boundary layers. As explained in Sect. 2.1, this calculation entails the implementation of the following three steps:

(i)   Calculation of the field in the boundary layer, using the stretched coordinates. The field, thus calculated, satisfies the Maxwell's equations and the boundary conditions on the object.

(ii) Calculation of the field outside the boundary layer. This field depends upon the diffraction coefficient to be determined.

(iii) Matching to determine the diffraction coefficients.

We will begin by illustrating the first step by considering the most frequently encountered boundary layers. First, we will study the boundary layer for the creeping rays in the vicinity of two-dimensional (Sect. 3.1) and three-dimensional surfaces (Sect. 3.2). Following this we will discuss various other categories of boundary layers, viz, the boundary layers of the so-called whispering gallery modes (Sect. 3.3); neighborhood of a regular point of a caustic (Sect. 3.4); neighborhood of a light-shadow boundary (Sect. 3.5); and, neighborhood of the tip of a wedge (Sect. 3.6).

All these boundary layers are *simple*, in the sense that there is no overlap between the transition zones and the coordinate that is orthogonal to the direction along which the boundary layer is stretched. Later, we will treat the problem of the neighborhood of the contact point of a ray grazing on a smooth surface in (Sect. 3.7) and (Sect. 3.8), and will present some results for a surface which has an inflection point (Sect. 3.9).

The boundary layers discussed in Sect. 3.7 and Sect. 3.8 are more complex than the previous ones, because there is an overlap between two regions in these layers, which leads to dissimilar stretching of coordinates. Following the discussion of the subject of field computation in the inner and outer boundary layers, we will proceed to address the matching problem in Sect. 3.10. Next, we will treat the simple problem of matching the solution in the neighborhood of the contact point to the solution in the boundary layer close to the surface in Sect. 3.11 and this will enable us to calculate the field in the boundary layer. Following this, we will implement in Sect. 3.12, the matching of the boundary layer and the zone of the creeping rays, and this will, in turn, yield the diffraction coefficients D_h^p of Chap. 1. Similarly, we will see in Sect. 3.13, that the matching of the boundary layer in the neighborhood of the tip of the wedge to the zone of rays enables us to calculate the diffraction coefficient of a wedge. Finally, we will address the problem of the caustic in Sect. 3.14 and will conclude this chapter by considering the somewhat intricate problem of matching in the vicinity of the contact point in Sect. 3.15.

3.1 Boundary layers of creeping rays on a cylindrical surface (Fig. 3.1)

We begin by considering the case of cylindrical surfaces for which the Maxwell's equations reduce to scalar wave equations. It is useful to discuss this example not only because the calculations are much simpler than they are for the general case, but also because there is a great deal of commonality between the key steps in both. In order to familiarize the reader with the boundary layer method, we will present the details of all the steps of the calculations leading to the field computation in these layers.

To this end we will seek, as a first step, the solutions of the homogeneous problem i.e., modes (improper) of the problem and, as is typically the case with such homogeneous solutions, they will be defined within a multiplicative constant. These solutions will subsequently be used to build the solution, for the plane wave incidence case, in the vicinity of the surface of the object in the shadow region. The weight coefficient will be determined in Sect. 3.11 by enforcing the matching condition in the neighborhood of the contact point of the incident ray and the object.

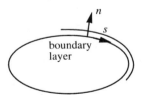

Fig. 3.1. (s, n) coordinates

3.1.1 Conditions satisfied by u

We seek a solution u of the wave equation: $(\nabla^2 + k^2)\, u = 0$ in the region outside a cylindrical object that satisfies a specified boundary condition on the surface of the cylinder.

In the shadow zone, the field u we are looking for is the total field, and, hence, the boundary condition is homogeneous. For a magnetically-polarized (TE) incident wave, and a cylinder described by an impedance boundary condition, u satisfies the equation

$$\frac{\partial u}{\partial n} + i\,k\,Zu = 0 \text{ on } \Gamma,$$

where Z is the relative impedance of the object with respect to that of the free space. Of course, the wave function u must also satisfy an appropriate condition at infinity, viz., the radiation condition. For a lossy medium $Im(k) > 0$, and this condition is expressed as $\lim\limits_{r \to +\infty} u(r) = 0$.

These conditions imply that $u = 0$, if we consider the usual space external to the cylinder, and solutions with finite energy. For the case of a finite cylinder, we consider the solution in a multi-sheeted space outside of the cylinder. For instance, when the cylinder is circular, it means that we extend the range of the angle θ from $[0, 2\pi]$ to $[-\infty, +\infty]$, realizing, of course, that the solution is no longer periodic in the extended range of θ. For the case of an arbitrary cylinder, extending θ over an infinite range is tantamount to looking for a solution in the form of a wave that circumnavigates the cylinder infinitely many times, without returning to its original value after each revolution. The solutions, thus obtained, decrease exponentially with θ, and their energy remains finite only in the interval $[a, +\infty]$. In practice, we admit these solutions for the circularly cylindrical case only in the shadow region, i.e., for $\theta > 0$. Likewise, the solutions derived for an arbitrary cylinder are also retained only in the shadow region, that is, in the region corresponding to positive curvilinear abscissas, with the origin of the abscissas taken on the shadow boundary.

3.1.2 Choice of the form of the solution

The solution is constructed in a manner such that it recovers the asymptotic expansion of the exact solution for the particular case of the circular cylinder (see Appendix 1). For a circular cylinder with a radius ρ, the rays in the shadow region propagate along the surface of the cylinder, and, consequently, the dominant term of the phase behavior is given by e^{iks}.

These rays attenuate as $\exp\{\alpha\,(k\rho)^{1/3}\theta\}$, where θ is the azimuthal angle of the observation point located on the circular cylinder and α is a complex constant. It is then natural to look for the phase behavior of the field that takes the form

$$ks + k^{1/3}\,\varphi(s),$$

where s is the curvilinear abscissa of the observation point on the arbitrary cylinder and $\varphi(s)$ is a complex phase function.

Returning to the circular cylinder, we note that the solution in the vicinity of the surface of the cylinder has the behavior of the Airy function given below

$$w_1\left(\xi - \left(\frac{2k^2}{\rho}\right)^{1/3} n\right),$$

where ξ is a constant which depends upon the impedance of the surface of the cylinder. For the circular cylinder, $n = r - \rho$ is the distance of the observation point to the surface of the cylinder.

For an arbitrary cylinder, it is then logical to introduce the distance as measured from the surface along the normal to the cylinder, and to stretch this coordinate. To this end, we first define

$$\nu_1 = k^{2/3} n,$$

and then recognize that we have several options that we can exercise in representing the behavior of the wave. The first of these is to write

$$w_1\left(\sum_{j=0}^{M} \beta_j(s, \nu_1) k^{-j/3}\right),$$

where M is an arbitrary integer, and to expand the phase in terms of the powers of $k^{-1/3}$ as follows [BB]

$$ks + k^{1/3} \varphi(s) + \sum_{j=0}^{M} \alpha_j(s, \nu_1) k^{-j/3}.$$

The second option [BK] is to expand the amplitude in terms of the powers of $k^{-1/3}$ and to write

$$\sum_{j=0}^{M} u_j(s, \nu_1) k^{-j/3}.$$

It is useful to point out that it is redundant to expand, simultaneously, the phase as well as the amplitude up to the order M, since it is possible to rewrite

$$\exp\left(i\sum_{j=1}^{M}\alpha_j(s,\nu_1)\,k^{-j/3}\right),$$

as a Taylor series, and to derive an expansion of the amplitude function in terms of $k^{-1/3}$.

Of the two choices listed above, the one that lends itself more conveniently to computation is the one given in the reference [BK]. In this approach, we begin by assuming that

$$u(s,\nu_1) \approx e^{i(ks+k^{1/3}\varphi(s))}\sum_{j=0}^{M}u_j(s,\nu_1)\,k^{-j/3}\,, \tag{D}$$

and then calculates the coefficients $u_j\,(s,\,\nu_1)$ such that they satisfy the wave equation subject to the boundary conditions mentioned above, viz., the impedance condition on the cylinder and the radiation condition at infinity.

In what follows, we will present in detail the approach developed in [BK] for the Neuman problem, i.e., for the case of the impedance $Z = 0$. The basic steps in this procedure entail the insertion of the assumed asymptotic expansion for u in the wave equation expressed in the (s, n) coordinates system, where, we might recall that s is the curvilinear coordinate on the cylinder, extended from 0 to $+\infty$, and n is the distance to the cylinder as measured along the normal, as well as in the boundary conditions on the surface of the cylinder.

3.1.3 Wave equation expressed with the coordinates (s, n)

For a smooth C^∞ convex cylinder, the system of coordinates (s, n) is C^∞ and orthogonal. In this system, the metric takes the form $dl^2 = h_s^2\,ds^2 + h_n^2 dn^2$. The Lamé coefficients h_s and h_n are given by $h_s = 1 + (n/\rho(s))$, where $\rho(s)$ is the curvature radius of the cylinder at the point with abscissa s; and, $h_n = 1$.

The Laplacian expressed with the coordinates s, n takes the form

$$\nabla^2 = \frac{1}{h_s h_n}\left\{\frac{\partial}{\partial s}\left(\frac{h_n}{h_s}\frac{\partial}{\partial s}\right) + \frac{\partial}{\partial n}\left(\frac{h_s}{h_n}\frac{\partial}{\partial n}\right)\right\},$$

and the Helmoltz equation becomes

$$\frac{1}{(1+n/\rho)}\left\{\frac{\partial}{\partial s}\left(1+\frac{n}{\rho}\right)^{-1}\frac{\partial u}{\partial s}+\frac{\partial}{\partial n}\left(1+\frac{n}{\rho}\right)\frac{\partial u}{\partial n}\right\}+k^2u=0 \ . \tag{3.1}$$

Next we introduce the stretched coordinate ν_1, and insert the asymptotic expansion (D) in (3.1). After ordering the resultant expression with respect to decreasing powers of $k^{1/3}$ and assuming that $\nu_1=O(1)$, we obtain

$$k^{4/3}\left[\frac{\partial^2 u_0}{\partial \nu_1^2}+\left(\frac{2\nu_1}{\rho}-2\varphi'(s)\right)u_0\right]+$$

$$+k\left[2i\frac{\partial u_0}{\partial s}+\frac{\partial^2 u_1}{\partial \nu_1^2}+\left(\frac{2\nu_1}{\rho}-2\varphi'(s)\right)u_1\right] \tag{3.2}$$

$$+ \ldots + k^{-j/3}\,(L_0\,u_j+L_1\,u_{j-1}+\ldots\,L_ju_0)+\ldots=0,$$

where the L_j's are differential operators. The two leading operators, viz., L_0 and L_1, are given simply by

$$L_0=\frac{\partial^2}{\partial \nu_1^2}+\left(\frac{2\nu_1}{\rho}-2\varphi'(s)\right), \tag{3.3}$$

and
$$L_1=2i\frac{\partial}{\partial s}. \tag{3.4}$$

From the wave equation, we can derive the following recursion relationships

$$L_0\,u_0=0, \tag{3.5}$$

$$L_0\,u_1+L_1\,u_0=0, \tag{3.6}$$

$$L_0\,u_j\ldots+L_j\,u_0=0 \ , \tag{3.7}$$

Our next step is to discuss the solution of the above equations.

3.1.4 Calculation of u_0

We begin with the first operator equation satisfied by u_0, i.e., (3.5). Defining a new variable ν via the equation

$$\nu=\left(\frac{2}{\rho(s)}\right)^{1/3}\nu_1, \tag{3.8a}$$

and setting

$$\xi = 2\varphi'(s) \left(\frac{\rho(s)}{2}\right)^{2/3},$$

(3.8b)

(3.5) can be reduced to

$$\frac{\partial^2 u_0}{\partial \nu^2} + (\nu - \xi)\, u_0 = 0,$$

(3.9)

which is the Airy equation in terms of the variable $\xi - \nu$.

Next, we impose the radiation condition by letting ν tend to $+\infty$. We can write the solution of the Airy equation as follows

$$u_0 = A(s)\, w_1\,(\xi - \nu) + B(s)\, w_2\,(\xi - \nu).$$

(3.10)

For a lossy medium, we impose the constraint that $u_0 \to 0$ as $\nu \to +\infty$. For such a medium, the wave number k has a small imaginary component which is positive and, consequently, the quantity $\xi - \nu$ is located below the negative real axis, i.e., argument is $\pi + \varepsilon$ where $\varepsilon > 0$. An examination of the functions w_1 and w_2 and, more precisely of their asymptotic expansions for large arguments, reveals that

$$w_2\,(t) \to +\infty \quad \text{when} \quad |t| \to +\infty \quad \text{with} \quad Arg\, t = \pi + \varepsilon$$

$$w_1\,(t) \to -0 \quad \text{when} \quad |t| \to +\infty \quad \text{with} \quad Arg\, t = \pi + \varepsilon$$

Thus, only the choice of function w_1 enables us to satisfy the radiation condition, and, hence, we must retain only the first term in expression (3.10). Thus we write u_0 as

$$u_0\,(s, \nu) = A(s)\, w_1\,(\xi - \nu).$$

(3.11)

Next, in terms of the coordinate variables (ν, s), the impedance condition becomes

$$\frac{\partial u}{\partial \nu} + i \left(\frac{k\rho}{2}\right)^{1/3} Zu = 0 \quad \text{for} \quad \nu = 0.$$

(3.12)

Finally, we introduce the notation

$$m = \left(\frac{k\rho}{2}\right)^{1/3},$$

(3.13)

in order to define a parameter mZ that plays an important role in our discussion. For instance, in order to recover the solution to the Neuman problem, which corresponds to $Z \rightarrow 0$, we will consider the case where mZ is $O(1)$. Although we could as well have made the choice of $Z = O(1)$, this would have forced us to consider the surface impedance problem as that of a perturbation of the Dirichlet problem, and the results would not be uniform in their behavior as Z is allowed to tend to 0 [BB]. Inserting the asymptotic expansion for u in the impedance condition, we obtain

$$\frac{\partial u_j}{\partial \nu} + imZu_j = 0 \quad \forall j \quad \text{for} \quad \nu = 0. \tag{3.14}$$

To first order $u_0 (s, \nu) = A(s) w_1 (\xi - \nu)$, and (3.14) becomes

$$w_1'(\xi) = imZw_1 (\xi). \tag{3.15}$$

The above equation has an infinite number of roots which we denote by $\xi_p (Z)$, where p is an integer. The roots are functions of s, since the impedance Z may depend upon s.

According to (3.8b), ξ is given by the expression

$$\xi = 2\varphi'(s) \left(\frac{\rho(s)}{2} \right)^{2/3},$$

which leads to

$$\varphi(s) = 2^{-1/3} \int_0^s \frac{\xi ds}{(\rho(s))^{2/3}} + \varphi(0) , \tag{3.16}$$

where $\varphi(0)$ is a constant which can be chosen equal to zero.

3.1.5 Calculation of $A(s)$

In this section we discuss the calculation of $A(s)$. To this end, we use (3.6), which reads $L_0 u_1 = -L_1 u_0$. The above equation has a solution only if the right-hand side, viz., $-L_1 u_0$ is orthogonal (in the sense of the scalar product in the L^2 space) to the solutions of the homogeneous equation. The above constraint provides us the necessary compatibility condition that enables us to calculate $A(s)$, as will be shown below.

3.1.6 Expression for the compatibility condition

Multiplying (3.6) by $w_1 (\xi - \nu)$, and integrating from $\nu = 0$ to $\nu = +\infty$, we get

$$\int_0^{+\infty} w_1(\xi - \nu) L_0 u_1 d\nu + \int_0^{+\infty} w_1(\xi - \nu) L_1 u_0 d\nu = 0 \ . \tag{3.17}$$

In view of the assumption that the medium has a finite loss, i.e., k has a positive imaginary component, the integrals can be shown to be convergent, since u_0 is proportional to $w_1 (\xi - \nu)$, $L_1 u_0$ is a linear combination of $w_1 (\xi - \nu)$ and $w_1'(\xi - \nu)$. Accordingly, the function u_1, which is a solution to the equation $L_0 u_1 = -L_1 u_0$, is a linear combination of $w_1 (\xi - \nu)$ and $w_1'(\xi - \nu)$ multiplied by powers of ν. Thus, we only have terms of the type $w_1^2 (\xi - \nu)$ and $w_1 (\xi - \nu) w_1'(\xi - \nu)$ in the integrand in (3.17).

All of these terms exhibit an exponential decay when the medium is lossy which, in turn, assures the convergence of these integrals. After integrating by parts we obtain

$$\int_0^{+\infty} w_1(\xi - \nu) L_0 u_1 d\nu = \int_0^{+\infty} (L_0 w_1(\xi - \nu)) u_1 d\nu + \left(\frac{2}{\rho}\right)^{2/3} \left[w_1(\xi - \nu) \frac{\partial u_1}{\partial \nu} - u_1 \frac{dw_1(\xi - \nu)}{d\nu} \right]_0^{+\infty} .$$
$$\tag{3.18a}$$

However, we can write

$$\left[w_1(\xi - \nu) \frac{\partial u_1}{\partial \nu} - u_1 \frac{dw_1(\xi - \nu)}{d\nu} \right]_0^{+\infty} =$$

$$\left[w_1(\xi - \nu) \left(\frac{\partial u_1}{\partial \nu} + imZu_1 \right) - \left(\frac{dw_1(\xi - \nu)}{d\nu} + imZw_1(\xi - \nu) \right) u_1 \right]_0^{+\infty} , \tag{3.18b}$$

and, for all $s \in \mathbb{R}$, we can express the boundary condition at $\nu = 0$ as

$$\frac{\partial u_1}{\partial \nu} + imZu_1 = 0, \tag{3.19a}$$

or alternatively as

$$\frac{dw_1(\xi - \nu)}{d\nu} + imZw_1(\xi - \nu) = 0. \tag{3.19b}$$

However, since the medium is lossy, we have

$$\lim_{\nu \to +\infty} w_1(\xi - \nu) = 0 \quad \text{and} \quad \lim_{\nu \to +\infty} w_1'(\xi - \nu) = 0,$$

and, consequently, the limit of all of the terms is zero as $\nu \to +\infty$. Hence, we have the result

$$\left[w_1(\xi - \nu)\frac{\partial u_1}{\partial \nu} - u_1 \frac{dw_1(\xi - \nu)}{d\nu} \right]_0^{+\infty} = 0. \tag{3.20}$$

On the other hand, according to (3.5), we have $L_0 w_1 (\xi - \nu) = 0$. Thus, we obtain

$$\int_0^{+\infty} w_1(\xi - \nu)L_0 u_1 d\nu = 0 \tag{3.21}$$

which leads to, using (3.17)

$$\int_0^{+\infty} w_1 L_1 u_0 d\nu = 0, \tag{3.22}$$

where $L_1 = 2i \, (\partial/\partial s)$. This is the compatibility condition, i.e., the orthogonality condition satisfied by w_1, the solution to the homogeneous equation satisfying the impedance condition and the radiation condition.

Next, we consider $L_1 u_0$ and we can write it as

$$L_1 u_0 = 2i \, \frac{d}{ds} A(s) w_1 (\xi - \nu) =$$

$$2i\left[A'(s)w_1(\xi - \nu) + \frac{1}{3}\frac{\rho'(s)}{\rho(s)} \nu A(s)w_1'(\xi - \nu) + \xi'(s)A(s)w_1'(\xi - \nu) \right], \tag{3.23}$$

and, consequently, we have

$$\int_0^{+\infty} w_1(\xi - v) L_1 u_0 dv =$$

$$2i\left\{ A'(s) \int_0^{+\infty} w_1^2(\xi - v) dv \right.$$

$$\left. + \frac{1}{3} \frac{\rho'(s)}{\rho(s)} A(s) \int_0^{+\infty} v w_1'(\xi - v) w_1(\xi - v) dv + \frac{1}{2} \xi'(s) A(s) w_1^2(\xi) \right\}. \quad (3.24)$$

After integrating by parts, and using the Airy equation, we obtain

$$\int_0^{+\infty} w_1^2(\xi - v) = -\left(w_1'^2(\xi) - \xi w_1^2(\xi) \right), \quad (3.25)$$

$$\int_0^{+\infty} v w_1(\xi - v) w_1'(\xi - v) dv = -\frac{1}{2}\left(w_1'^2(\xi) - \xi w_1^2(\xi) \right). \quad (3.26)$$

As previously discussed, the integrals above are convergent when the medium has a finite loss, in which event all of the integrands decay exponentially as the variable of integration becomes large. According to (3.22), (3.24), (3.25) and (3.26), we have

$$\frac{A'(s)}{A(s)} + \frac{1}{6}\frac{\rho'(s)}{\rho(s)} - \frac{1}{2}\xi'(s)\frac{w_1^2(\xi)}{w_1'^2(\xi) - \xi w_1^2(\xi)} = 0. \quad (3.27)$$

By setting $d(s) = w_1'^2(\xi) - \xi w_1^2(\xi)$, we can write for $\xi = \xi(s)$

$$d'(s) = -\xi'(s) \ w_1^2(x(s)) , \quad (3.28)$$

so that (3.27) becomes

$$\frac{A'(s)}{A(s)} + \frac{1}{6}\frac{\rho'(s)}{\rho(s)} + \frac{1}{2}\frac{d'(s)}{d(s)} = 0. \quad (3.29)$$

From the above, we finally obtain the desired result

$$\frac{A(s)}{A(0)} = \left(\frac{\rho(0)}{\rho(s)} \right)^{1/6} \left(\frac{d(0)}{d(s)} \right)^{1/2}, \quad (3.30)$$

which will be useful for deriving the representation for u, as will be seen below.

3.1.7 Final result for the first term u_0 of the expansion of u

Using the results derived in the last section, we write u as

$$u(s, \nu) \approx A(0) \left(\frac{\rho(0)}{\rho(s)}\right)^{1/6} \left(\frac{d(0)}{d(s)}\right)^{1/2} w_1\,(\xi-\nu) \exp\left(iks + ik^{1/3}\,2^{-1/3}\int_0^s \frac{\xi ds}{\rho(s)^{2/3}}\right).$$

$$(3.31)$$

The above expression is valid for $\nu = O(1)$, i.e., close to the surface of the scatterer.

As compared to the Neuman problem, treated in [BK], the present case is different in the following ways. First, the propagation constant satisfies the equation $w\,'_1\,(\xi) = imZ(s)\,w_1\,(\xi)$, instead of $w\,'_1\,(\xi) = 0$ for the Neuman problem. Second, the expression for w_1 has a multiplicative factor $(d(0)/d(s))^{1/2}$, which equals 1 if $mZ(s)$ is constant, i.e., if the product $\rho^{1/3}\,Z$ is constant. This particular case is often studied in the literature, despite the fact that it represents a somewhat peculiar situation.

In particular, the expression in (3.31) provides the field on the surface of the object for $\nu = 0$. In view of the relation $w\,'_1\,(\xi) = imZw_1\,(\xi)$, we can write

$$u(s, o) \approx A(0)\, w_1(\xi(0)) \left(\frac{\rho(0)}{\rho(s)}\right)^{1/6} \left(\frac{e(0)}{e(s)}\right)^{1/2} \exp\left(iks + \frac{ik^{1/3}}{2^{1/3}}\int_0^s \frac{\xi ds}{\rho(s)^{2/3}}\right), (3.32)$$

where $e(s) = \xi(s) + m^2 Z^2(s)$. The dependence of the form $(\rho(0)/\rho(s))^{1/6}$ is easily derived from the solution of the canonical problem of the perfectly conducting elliptical cylinder. However, the factor $(e(0)/e(s))^{1/6}$ can only be extracted by solving a more complicated problem, e.g., an elliptical cylinder with a surface impedance, for instance, which, to our knowledge, has not yet been treated. Thus, the method described herein enables us to extract new results as compared to the method of canonical problems, especially in the shadow zone. These solutions, that are constructed by determining the roots $\xi(s)$ as defined in (14), depend upon an arbitrary constant $A(0)$. However, as we will see in Sect. 3.11, by matching the solution in the transition zone it is possible to determine this coefficient $A(0)$.

Finally, we observe that the previous analysis provides a justification of the principle of locality, which, as we might recall, states that the field u depends only on the local properties of the scatterer.

The result given above has been expressed in the (s, n) coordinate system. We will now rewrite it in the ray coordinates that will enable us to match it to the outer solution in a more convenient way.

3.1.8 Fields expressed in terms of the ray coordinates (Fig. 3.2)

In the discussion we just presented in the last section, we have employed, for the sake of simplicity, the (s, n) coordinate system. The reader may recall, however, it was pointed out in Sect. 2.1 that, in principle, it is preferable to use the ray coordinates to express the fields. It turns out that the calculations with the latter coordinates follow along lines that are very similar to those carried out by the (s, n) coordinates. Consequently, we will restrict ourselves to presenting the principal results, primarily with the objective of illustrating the influence of the choice of coordinates on the expression of the fields.

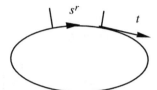

Fig. 3.2. Ray coordinates

Let us begin by recalling how the ray coordinates are defined for the case of the cylinder. In this coordinate system, the quantity s^r is the curvilinear abscissa up to the point of tangency of the ray, t is the length of the diffracted ray, and ℓ is the total length of the ray. The coordinates (s^r, ℓ) form an orthogonal system. In the boundary layer, close to the surface, $t \approx \sqrt{2\rho n}$. Previously, we had chosen a stretching factor $k^{2/3}$ for the coordinate n, and it is logical to use $k^{1/3}$ to stretch $t = \ell - s^r$, which is a proportional to the square root on n. The calculations performed by using these coordinates are direct but cumbersome. They have been carried out by Buslaev [Bu] and Ivanov [Iv] , for the case of constant mZ. The result for the first term of the expansion, generalized to mZ variable, can be expressed as

$$u \approx A \left(\frac{\rho(0)}{\rho(s^r)}\right)^{1/6} \left(\frac{d(0)}{d(s^r)}\right)^{1/2} w_1(\xi - Y)\, exp\left(ik\ell - i\frac{2}{3}Y^{3/2}\right)$$

$$exp\left(i\xi\sqrt{Y} + i\frac{k^{1/3}}{2^{1/3}}\int_0^{s^r} \xi\frac{ds}{\rho(s)^{2/3}}\right), \tag{3.33}$$

where Y is the stretched coordinate

$$Y = \left(m\frac{t}{\rho}\right)^2, \tag{3.34}$$

and ξ is taken at the point of detachment.

It can be shown (see Appendix 2 on differential geometry) that, in the boundary layer, we can write

$$
\left.
\begin{aligned}
Y &= v + O(k^{-1/3}) \\
\ell &= s + \frac{2}{3k} Y^{3/2} + O(k^{-4/3}) \\
\xi(s^r)\sqrt{Y} &= \frac{k^{1/3}}{2^{1/3}} \int_{s^r}^{s} \xi(s) \frac{ds}{\rho(s)^{2/3}} + O(k^{-1/3}) \\
\rho(s^r) &= \rho(s) + O(k^{-1/3}) \\
\xi(s^r) &= \xi(s) + O(k^{-1/3})
\end{aligned}
\right\} .
\tag{3.35}
$$

Inserting (3.35) into (3.33), we obtain

$$
u \approx A(0) \left(\frac{\rho(0)}{\rho(s)} \right)^{1/6} \left(\frac{d(0)}{d(s)} \right)^{1/2} w_1(\xi - v) \, \exp \left(iks + i \frac{k^{1/3}}{2^{1/3}} \int_0^s \xi \frac{ds}{\rho(s)^{2/3}} \right) + O(k^{-1/3}).
\tag{3.36}
$$

Note that the above expression is identical to that given in (3.31), as expected.

The advantage of (3.33) is that it is expressed directly in the same coordinates as the outer expansion, which facilitates the matching procedure tremendously. The drawback is that the calculations of the inner expansion leading to (3.33) are very tedious compared to those needed to derive (3.31). Historically, the ray coordinates were used until the beginning of the 60s. However, the Leningrad school, in particular Babitch and his collaborators, have more recently employed the (s, n) coordinates, which are easier to handle.

We will now discuss the general 3-D Maxwell's equations by using the (s, n) coordinates.

3.2 Boundary layers of creeping rays on a general surface

3.2.1 Introduction

In this section, we turn our attention to the boundary layer problem associated with creeping rays on a general surface. The creeping rays can be launched by a distant incident wave or by a source located on the surface of the object, as shown in Figs. 3.3 and 3.4, respectively.

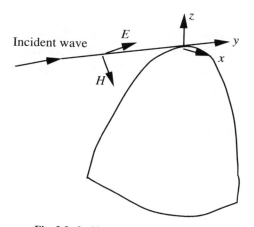

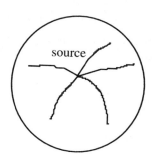

Fig. 3.3. Incident wave on a convex surface

Fig. 3.4. Source on a convex object

To address the boundary layer problem, we are going to follow exactly the same steps as we did for the case of the cylinder studied in Sect. 3.1. These are:

    (i) choosing an Ansatz

    (ii) expressing the equations in the (s, n) coordinate system

    (iii) introducing the Ansatz in the above equations

    (iv) solving the equations for each order.

Before going into the details of the procedure, we will make some preliminary observations pertinent to the method.

The generalization of the expansion (D) on the surface of the scatterer, given in the previous section, might be carried out by including a phase term $\exp(ikS)$, where S is an eikonal on the surface. However, the satisfaction of the Maxwell's equations by this field representation requires that the norm of the surface gradient be equal to 1. The eikonal equation on the surface is solved by using the method of characteristics, in a manner similar to that discussed in Chap. 2. Let us define the surface rays as the integral curves of the surface gradient of the phase. One can show, as in Chap. 2, that these curves minimize the path lengths between two points on the surface; hence, they are its geodesics. The phase term simply takes the form of $\exp(iks)$, where s is the abscissa along a geodesic. Thus, we find that even with this simple reasoning presented above, we recover the GTD postulate that the surface rays are the geodesics of the surface.

Following these introductory remarks, we will now proceed below with the task of deriving the solution.

3.2.2 Equations and boundary conditions

We seek a solution of the Maxwell's equations, satisfying the impedance condition $\vec{E} - (\vec{n} \cdot \vec{E}) = Z_a \vec{n} \times \vec{H}$ on the surface, where the complex number Z_a is the absolute impedance, and the Silver-Müller radiation condition at infinity. With the $\exp(-i\omega t)$ time convention, the Maxwell's equations read

$$\begin{cases} \nabla \times \vec{E} = i\omega\mu \vec{H} \\ \nabla \times \vec{H} = -i\omega\varepsilon \vec{E} \end{cases},$$

or, alternatively,

$$\begin{cases} \nabla \times \sqrt{\varepsilon} \vec{E} = i\omega\sqrt{\mu\varepsilon}\sqrt{\mu} \vec{H} \\ \nabla \times \sqrt{\mu} \vec{H} = -i\omega\sqrt{\mu\varepsilon}\sqrt{\varepsilon} \vec{E} \end{cases},$$

and the impedance condition takes the form

$$\sqrt{\varepsilon} \vec{E} - (\vec{n} \sqrt{\varepsilon} \vec{E}) \hat{n} = Z \hat{n} \times \sqrt{\mu} \vec{H},$$

where Z is the relative impedance, i.e., the absolute impedance normalized to the impedance of free space. Next, following Fock [F], we introduce scaled field quantities $\sqrt{\varepsilon} \vec{E}$ and $\sqrt{\mu} \vec{H}$ and rewrite Maxwell's equations as

$$\begin{cases} \nabla \times \vec{E} = ik \vec{H} \\ \nabla \times \vec{H} = -ik \vec{E} \qquad \text{in } \Omega', \\ \vec{E} - (\hat{n} \cdot \vec{E})\hat{n} = Z\hat{n} \times \vec{H} \quad \text{on } \Gamma \end{cases} \qquad (3.37)$$

where, for convenience, we have denoted $\sqrt{\varepsilon} \vec{E}$ and $\sqrt{\mu} \vec{H}$ simply by $\vec{E}$ and $\vec{H}$ in (3.37), and have written the relationships $\omega \sqrt{\mu\varepsilon} = \omega/c = k$, $Z_a = Z Z_0 = Z\sqrt{\mu/\varepsilon}$. Furthermore, $\vec{E}$ and $\vec{H}$ must satisfy the Silver-Müller radiation condition, which reads

$$\vec{E} \text{ and } \vec{H} = O\left(\frac{1}{r}\right)$$

$$\vec{E} + \hat{r} \times \vec{H} \text{ and } \vec{H} - \hat{r} \times \vec{E} = o\left(\frac{1}{r}\right) \quad \text{as } r \to +\infty,$$

where r is the distance between the observation point and the object, and $\hat{r}$ the unit vector along the direction of observation.

3.2.3 Form of the asymptotic expansion

Let us now turn to the derivation of the asymptotic representation. The form of the asymptotic expansion is suggested by Zauderer [Z] for the scalar wave equation. Zauderer introduces a system of geodesic coordinates as shown in Fig. 3.5.

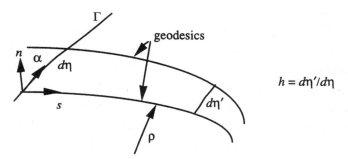

Fig. 3.5. Geodesic coordinate system of Zauderer

In this system of geodesic coordinates, s denotes the curvilinear abscissa along an arbitrary geodesic; α denotes the curvilinear abscissa along the orthogonal curve on Γ, which is the coordinate axis of the geodesic coordinates considered; and, n denotes the distance to the surface along the normal. The asymptotic expansion of the solution is written in the form

$$\exp\left(iks + ik^{1/3}\varphi(s,\,\alpha)\right)\left(u_0 + u_1\,k^{-1/3} + \ldots + u_j\,k^{-j/3}\right),$$

where the first bracket of the product represents the phase factor, while the second accounts for the amplitude behavior.

In above, $\varphi(s,\,\alpha)$ is a phase function which is complex, with its imaginary part describing the attenuation of the creeping rays, and u_i are the coefficients of the amplitude expansion with respect to the powers $-1/3$ of the wave number k. The u_i are complex-valued functions of the stretched coordinate $\nu_1 = k^{2/3}n$ and of the coordinate α, asymptotically ν_1 is of $O(1)$, and n is $O(k^{-2/3})$.

These estimations of the orders, that we just mentioned above, can be obtained from the solution of the canonical problem of the diffraction of an electromagnetic wave by a cylinder or a sphere. However, we assume that they continue to apply to the case of an arbitrary object.

To solve the Maxwell's equations, we propose the same type of expansions for the electric and magnetic fields, $\vec{E}$ and $\vec{H}$. Thus, we set

$$\begin{cases} \vec{E} = exp(iks + ik^{1/3}\varphi(s,\,\alpha))(\vec{E_0} + \vec{E_1}\,k^{-1/3} + \vec{E_j}\,k^{-j/3}...) \\ \vec{H} = exp(iks + ik^{1/3}\varphi(s,\,\alpha))(\vec{H_0} + \vec{H_1}\,k^{-1/3} + \vec{H_j}\,k^{-j/3}...) \end{cases} . \tag{3.38}$$

In the above, the fields $\vec{E_j}$ and $\vec{H_j}$ are vectorial functions of the variables α, ν_1 and s.

The calculation of $\varphi(s,\,\alpha)$, and of the fields $\vec{E_j}$ and $\vec{H_j}$, are performed by inserting the expansions (3.38) in the Maxwell's equations and by taking into account the impedance condition given in (1). Following this procedure, we can derive, for each order, the relevant equations and boundary conditions which, when combined with the radiation condition, enable us to determine the fields $\vec{E_j}$ and $\vec{H_j}$. In practice, we typically restrict ourselves to deriving the first order terms, that is to say to the calculation of $\vec{E_0}$ and $\vec{H_0}$. Using this procedure, Maxwell's equations and the boundary conditions will be satisfied only up to a certain order and, consequently, our solutions will be an approximation to the scattering problem at high frequencies. We should point out that we have not provided an upper bound for the terms we have neglected.

3.2.4   Derivation of the solution of Maxwell's equations, expressed in the coordinate system (s, α, n)

The coordinates (s, α, n) are neither orthogonal nor are they normalized. In order to express the differential operator $\nabla\times$, with these coordinates, we have to introduce (see for instance [Go]) the covariant and contravariant components of the vectors $\vec{E}$ and $\vec{H}$ that are under consideration. Denoting as $(V^s,\,V^\alpha,\,V^n)$ the contravariant components of $\vec{E}$ or $\vec{H}$ and as $(V_\alpha,\,V_s,\,V_n)$ those of the associated linear form (covariant), the expression for the curl operator in the coordinate system (s, α, n) becomes (see[Go])

$$\begin{cases} \nabla \times V^s = \dfrac{1}{\sqrt{g}}(\partial_\alpha V_n - \partial_n V_\alpha) \\[2mm] \nabla \times V^\alpha = \dfrac{1}{\sqrt{g}}(\partial_n V_s - \partial_s V_n)\,, \\[2mm] \nabla \times V^n = \dfrac{1}{\sqrt{g}}(\partial_s V_\alpha - \partial_\alpha V_s) \end{cases} \tag{3.39}$$

where
$$\begin{pmatrix} V_s \\ V_\alpha \\ V_n \end{pmatrix} = g_{ij} \begin{pmatrix} V^s \\ V^\alpha \\ V^n \end{pmatrix}.$$

g_{ij} is the matrix characterizing the metric of the coordinate system. Neglecting the terms $O(n^2)$, we can write g_{ij} in the form (see Appendix 2),

$$g_{ij} = \begin{pmatrix} 1+2b_{ss}n & 2b_{s\alpha}n & 0 \\ 2b_{s\alpha}n & g_{\alpha\alpha}+2b_{\alpha\alpha}n & 0 \\ 0 & 0 & 1 \end{pmatrix} + O(n^2),$$

where b_{ss}, $b_{s\alpha}$, $b_{\alpha\alpha}$ are the negatives of the coefficients of the second quadratic form of the surface in the (s, α) coordinates*. Since g is the determinant of this matrix, the components V_s, V_α and V_n can be written as

$$V_s = (1 + 2b_{ss}n)V^s + 2b_{s\alpha}n \, V^\alpha + O(n^2),$$

$$V_\alpha = (g_{\alpha\alpha} + 2b_{\alpha\alpha}n)V^\alpha + 2b_{s\alpha} \, V^s + O(n^2),$$

$$V_n = V^n + O(n^2). \tag{3.40}$$

However, as shown in Appendix 2, the following relationships hold: $b_{ss} = 1/\rho$, $b_{s\alpha} = -h\tau$, $b_{\alpha\alpha} = h^2/\rho_t$, $g_{\alpha\alpha} = h^2$, where ρ and τ are, respectively, the radius of curvature, and the torsion of the geodesic, and ρ_t is the normal radius of curvature of the surface in the direction orthogonal to the geodesic. The parameter h measures the narrowing (or broadening) of an infinitesimal geodesic pencil between a point with coordinates (s, α) and the axis of the α's. We restrict our consideration to the terms of $O(n)$ and neglect those that are $O(n^2)$ because of the reason explained below. After stretching, the latter terms are found to be of $O(k^{-4/3})$, since n is of $O(k^{-2/3})$. Finally, when we multiply these terms by k, in accordance with Maxwell's equations, they become of order $k^{-1/3}$.

The terms that have the behavior $k^{-1/3}$ are neglected in this analysis, and, hence, we will only retain the terms k^0, i.e., the terms of $O(1)$ in the expansion obtained by inserting (3.38) in Maxwell's equations.

The first Maxwell's equation $\nabla \times \vec{E} = ik \, \vec{H}$ can be written as three scalar equations, each of which yield four equations, if we sort out the terms that are of order k, $k^{2/3}$, $k^{1/3}$, and k^0. Thus, altogether a total of twelve equations are derived from the first Maxwell's equation. Likewise, the second Maxwell's equation, viz., $\nabla \times \vec{H} = -ik$

---

*We introduce the negative of the coefficients rather than the coefficients themselves so as to have + signs throughout the expansion of g_{ij}.

$\vec{E}$ provides twelve more equations — the dual set — which we could obtain from the first set of twelve equations by replacing $\vec{E}$ with $\vec{H}$, and $\vec{H}$ with $-\vec{E}$. The following is a catalog of the first set of equations.

Order k

$$H_0^s = 0. \tag{3.41}$$

$$E_0^n = -h\, H_0^\alpha. \tag{3.42}$$

$$H_0^n = h\, E_0^\alpha. \tag{3.43}$$

Order $k^{2/3}$

$$H_1^s = i\,h\, \frac{\partial E_0^\alpha}{\partial \nu_1}. \tag{3.44}$$

$$E_1^n = -h\, H_1^\alpha. \tag{3.45}$$

$$H_1^n = h\, E_1^\alpha. \tag{3.46}$$

Order $k^{1/3}$

$$H_2^s = i\,h\, \frac{\partial E_1^\alpha}{\partial \nu_1} - \frac{\partial \varphi}{\partial \alpha} - H_0^\alpha. \tag{3.47}$$

$$h\, H_2^\alpha + E_2^n = h\left(\varphi_s' H_0^\alpha - \nu_1\left(\frac{1}{\rho} + \frac{1}{\rho_t} \right)H_0^\alpha - \frac{\partial^2 H_0^\alpha}{\partial \nu_1^2} \right). \tag{3.48}$$

$$h\, E_2^\alpha - H_2^n = h\left(-\varphi_s' E_0^\alpha + \nu_1\left(\frac{1}{\rho} - \frac{1}{\rho_t} \right)E_0^\alpha \right). \tag{3.49}$$

Order k^0

$$H_3^s = i\,h\, \frac{\partial E_2^\alpha}{\partial \nu_1} - \frac{\partial \varphi}{\partial \alpha}\, H_1^\alpha + 2\, \frac{ih}{\rho_t}\, E_0^\alpha - ih\left(\frac{1}{\rho} - \frac{1}{\rho_t} \right)\nu_1\, \frac{\partial E_0^\alpha}{\partial \nu_1}. \tag{3.50}$$

$$h\, H_3^\alpha + E_3^n = h\left(\varphi_s' H_1^\alpha - \nu_1\left(\frac{1}{\rho} + \frac{1}{\rho_t} \right)H_1^\alpha - \frac{\partial^2 H_1^\alpha}{\partial \nu_1^2} \right) + \tag{3.51}$$

$$+\, 2ih\tau\, E_0^\alpha + 2i\nu_1\, h\tau\, \frac{\partial E_0^\alpha}{\partial \nu_1} - i\frac{\partial}{\partial s}\, (h\, H_0^\alpha) + i\frac{\partial \varphi}{\partial \alpha}\, \frac{\partial E_0^\alpha}{\partial \nu_1}.$$

$$h\, E_3^\alpha - H_3^n = h\left(-\varphi_s' E_1^\alpha + \nu_1\left(\frac{1}{\rho} - \frac{1}{\rho_t} \right)E_1^\alpha \right) \tag{3.52}$$

$$-\, 2i\nu_1\, h\tau\, \frac{\partial H_0^\alpha}{\partial \nu_1} + \frac{i}{h}\frac{\partial}{\partial s}\, (h^2\, E_0^\alpha) - i\frac{\partial \varphi}{\partial \alpha}\, \frac{\partial H_0^\alpha}{\partial \nu_1}.$$

The set of equations must be complemented by its *dual*, i.e. the equations obtained from the equations above, by replacing E with H and H with $-E$. We will speak of dual equations.

We will now interpret the equations associated with the different orders and will use them to calculate the functions $\vec{E}_0$ and $\vec{H}_0$, by taking into account the boundary and radiation condition.

3.2.5   Interpretation of the equation associated with the first three orders $(k, k^{2/3}, k^{1/3})$

The examination of the set of Eqs. (3.41) through (3.49) in the last section enables us to draw the following conclusions.

3.2.5.1 Equations of order k

The equations of order k, i.e., (3.41) through (3.43), and their dual equations show that H_0^n is a function of E_0^α, and E_0^n a function of H_0^α. They also show that the components with the superscript s, namely those in the direction of propagation of the creeping mode we are looking for, vanish for the zeroth order.

One can provide the following interpretation of the above result which is consistent with the Geometrical Theory of Diffraction. There are two types of creeping rays that are solutions of Maxwell's Equations. The first of these is called the electric creeping ray, because, for the zero order, the electric and magnetic fields, $\vec{E}$ and $\vec{H}$, are expressed as functions of the α-component of the electric field E_0^α. For the zero order, the components of this *electric* creeping ray are

$$E_0^\alpha,$$

$$H_0^n = h \ E_0^\alpha, \qquad\qquad (3.43)$$

$$H_0^s = E_0^s = 0, \qquad\qquad (3.41)$$

$$E_0^n = H_0^\alpha = 0.$$

At this point, it is useful to remember that the frame $(\vec{e}_s, \vec{e}_\alpha, \vec{e}_n)$ is not normalized. Thus, the Eq. (3.42) simply implies that H_0^n and E_0^α have the same length. The two dominant components (i.e., the highest order with respect to k) of the *electric* creeping ray are E_0^α and H_0^n; hence, the dominant order components the *electric* creeping ray propagate as plane waves (see Fig. 3.6). Similarly, for the zero and first order components of the *magnetic* creeping ray, the $\vec{E}$ and $\vec{H}$ fields are expressible in terms of

H_0^α and H_1^α. For the zero order, the components of such a creeping ray are H_0^α,

$$H_0^n = -h\,H_0^\alpha,\qquad\qquad\text{(3.42) or its dual equation}$$

$$H_0^s = E_0^s = 0,\qquad\qquad\text{(3.41) and its dual equation}$$

$$H_0^n = E_0^\alpha = 0.$$

Once again, the equation (3.43) simply means that E_0^n and H_0^α have equal lengths. The two dominant components of the *magnetic* creeping ray are H_0^α and E_0^n, and, to a first approximation, the magnetic creeping ray propagates like a plane wave (see Fig. 3.6).

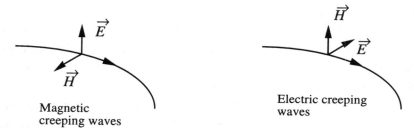

Fig. 3.6. The zero order components of the magnetic and electric creeping waves

We see from the above that the dominant order components of both the electric and magnetic creeping rays propagate as plane waves. The electric creeping ray is polarized with its electric field vector parallel to the surface, while correspondingly, the magnetic creeping ray has its magnetic field vector parallel to the surface.

3.2.5.2 Equations of order $k^{2/3}$

Let us now turn to the next higher-order equations, i.e., $O(k^{2/3})$, listed in Sect. 3.2.4. Equations (3.44) through (3.46) show that to a first order (i.e., $k^{-1/3}$), and in contrast to plane wave fields, there are now non-zero components in the direction of propagation. More precisely, the electric (magnetic) creeping ray has a first order component of its magnetic (electric) field, which does not vanish in the direction of propagation. In fact

$$H_1^s = i\,h\,\frac{\partial E_0^\alpha}{\partial v_1} \neq 0 \qquad\text{for the \textit{electric} creeping ray}\qquad(3.53)$$

and

$$E_1^s = -i\,h\,\frac{\partial H_0^\alpha}{\partial v_1} \neq 0 \qquad\text{for the \textit{magnetic} creeping ray}\qquad(3.54)$$

Once again, the coefficient h in the above equations is nothing but the norm of the vector $\vec{e}_\alpha$.

Equations (3.45), (3.46) are the counterparts of (3.42), (3.43), i.e. they express E_1^n and H_1^n as functions of H_1^α and E_1^α, respectively. To zero and first orders, all the components of $\vec{E}$ and $\vec{H}$ are expressible in terms of E_0^α, H_0^α, E_1^α, and H_1^α. It is therefore natural to choose E^α and H^α as the potentials in terms of which we express the rest of the field components. We should remark that we could have just as well chosen the normal components of the fields $\vec{E}$ and $\vec{H}$ for this purpose, and the results would have been equivalent. We also observe that E_0^α and H_0^α are yet to be obtained and we accomplish this by considering the equations of the next order.

3.2.5.3 Equations of $O(k^{1/3})$

We return once again to Sect. 3.2.4, and retrieve the equations corresponding to $O(k^{1/3})$. We note that Eqs. (3.47) and its dual yield H_2^s and E_2^s as functions of E_0^α and H_0^α, respectively, and of the derivatives of the first order components. Equations (3.48), (3.49), and their duals can be written in terms of $h\,H_2^\alpha + E_2^n$ and $h\,E_2^n - H_2^\alpha$ as expressed as functions of H_0^α and E_0^α. They are the counterparts of the two sets of equations, we have learned previously, viz.,

$$h\,H_0^\alpha + E_0^n = h\,E_0^\alpha - H_0^n = 0 \qquad (3.42, 3.43)$$

and

$$h\,H_1^\alpha + E_1^n = h\,E_1^\alpha - H_1^n = 0 \qquad (3.45, 3.46)$$

However, the Eqs. (3.48) through (3.49) are somewhat different from the equations of orders k and $k^{2/3}$, viz., now we have the presence of a non-zero right hand side that entails the derivatives of H_0^α or E_0^α.

Using (3.48) and the dual equation of (3.49), which lead to two different formulas for $h\,H_2^\alpha + E_2^n$, we derive a compatibility condition, which after some manipulation reduces to the equation below

$$\frac{\partial^2 H_0^\alpha}{\partial v_1^2} + 2\left(\frac{v_1}{\rho} - \varphi_s'\right) H_0^\alpha = 0 . \qquad (3.55)$$

Likewise, using (3.49) and the dual equation of (3.48), which lead to two different formulas for $h\ E_2^\alpha - H_2^n$, we obtain the second compatibility condition

$$\frac{\partial^2 E_0^\alpha}{\partial v_1^2} + 2\left(\frac{v_1}{\rho} - \varphi_s'\right) E_0^\alpha = 0. \tag{3.56}$$

Equations (3.55) and (3.56) are similar to Eq. (3.5) in Sect. 3.1 obtained for the case of the cylinder. The dependence of E_0^α and H_0^α, on the coordinate n is then determined, as in Sect. 3.1, by the function w_1, as in the case of the cylinder.

In order to solve the Eqs. (3.55) and (3.56), and to determine the dependence of the solution on n, we have to take into account the boundary conditions on the surface. This will be carried out in the following section, and will enable us to calculate the function $\varphi = \varphi(s,\ \alpha)$ appearing in (3.38). Finally, the remaining task of calculating the variation of the amplitude with respect to s, will be discussed in Sect. 3.2.7, by using the Eqs. (3.50) through (3.52).

Before concluding this section, we point out that (3.48), (3.49), and their dual equations enable us to derive an expression for E_2^n and H_2^n as functions of (H_0^α , H_2^α) and (E_0^α , E_2^α), respectively. For the second order, as the previous ones, all the components of $\vec{E}$ and $\vec{H}$ can be expressed in terms of the α-components of the $\vec{E}$ and $\vec{H}$ fields only.

3.2.6 Boundary conditions and the determination of φ
We will now turn our attention below to the boundary conditions to be satisfied by the various field components. We begin with the case of a perfect conductor, and express the boundary conditions on the surface of the object.

3.2.6.1 Perfect conductor
The electric field tangent to the surface of a perfect conductor must vanish; consequently, we have

$$E^\alpha = 0, \tag{3.57}$$

and

$$E^s = 0 \quad \text{for} \quad v_1 = 0. \tag{3.58}$$

Expanding (3.57) and (3.58) according to the powers of $k^{-1/3}$, we obtain

$$E_0^\alpha = \dots E_j^\alpha = 0 \quad \text{for} \quad v_1 = 0, \tag{3.59}$$

and

$$E_0^s = \dots E_j^s = 0 \quad \text{for} \quad v_1 = 0. \tag{3.60}$$

We note that $E_0^s = 0$ is satisfied for all ν_1; thus, the first condition given by (3.60) is applicable to E_1^s.

3.2.6.1.1 Boundary conditions on E_0^α and H_0^α

The boundary condition on E_0^α follows straightforwardly from (3.59). The one for H_0^α can be extracted from (3.60) after expressing E_1^s as

$$E_1^s = - ih \, \frac{\partial H_0^\alpha}{\partial \nu_1},$$

in accordance with the sdual equation of (3.44). The fields E_0^α and H_0^α satisfy the following boundary conditions

$$E_0^\alpha = 0 \quad \text{for} \quad \nu_1 = 0, \tag{3.61}$$

and

$$\frac{\partial H_0^\alpha}{\partial \nu_1} = 0 \quad \text{for} \quad \nu_1 = 0. \tag{3.62}$$

Thus we see that for the case of a perfect conductor, the electric creeping ray must satisfy Dirichlet and magnetic Neumann boundary conditions, respectively.

3.2.6.1.2 Calculation of φ

In order to solve (3.55) and (3.56), we follow [BK] and introduce the variable ν

$$\nu = \left(\frac{2}{\rho(s)} \right)^{1/3} \nu_1, \tag{3.63}$$

and we set

$$\xi = 2\varphi_s' \left(\frac{\rho(s, \alpha)}{2} \right)^{2/3}, \tag{3.64}$$

where $\rho(s, \dagger\alpha)$ is the radius of curvature of the geodesic and is equal to b_{ss}^{-1} (see Fig. 3.5) .

Equations (3.55) and (3.56), rewritten with the variable ν, become

$$\frac{\partial^2 E_0^\alpha}{\partial \nu^2} + (\nu - \xi)E_0^\alpha = 0, \tag{3.65}$$

and

$$\frac{\partial^2 H_0^\alpha}{\partial \nu^2} + (\nu - \xi)H_0^\alpha = 0. \tag{3.66}$$

The field E_0^α satisfies (3.65), the radiation condition, and the Dirichlet boundary condition (3.61). Correspondingly, the field H_0^α satisfies (16), the radiation condition, and the Neumann boundary condition (3.62).

We can recognize that (3.65) and (3.66) are Airy equations; consequently, E_0^α and H_0^α satisfying these equations are Airy functions. Once again, as in the case of the cylinder, we select the *good* Airy function by invoking the radiation condition in the form of the *limit absorption principle*. More precisely, we look for a solution which, when the medium is lossy, i.e., *Im* $k > 0$, decays as $\nu \rightarrow +\infty$, or equivalently, as the observation point moves away from the surface. A general Airy function is a linear combination of w_1 and w_2, where w_1 $(\xi - \nu) \rightarrow 0$ as $\nu \rightarrow +\infty$ with *Im* $\nu > 0$, while $w_2(\xi - \nu) \rightarrow +\infty$ under the same conditions.

The principle of limit absorption forces us to choose the Airy function w_1 for the representation of the scattered fields, and we write

$$E_0^\alpha \ (s, \ \alpha, \ \nu) = h^{-1} \, A(s, \ \alpha) \, w_1(\xi_E - \nu), \tag{3.67}$$

and

$$H_0^\alpha \ (s, \ \alpha, \ \nu) = h^{-1} \, B(s, \ \alpha) \, w_1(\xi_M - \nu), \tag{3.68}$$

where $A(s, \ \alpha)$ and $B(s, \ \alpha)$ are functions yet to be determined. The factor h^{-1} has been introduced to simplify the calculations given in Sect. 3.2. 7.

Inserting (3.67) in the boundary conditions (3.61) we get

$$w_1(\xi_E) = 0. \tag{3.69}$$

In a similar manner, by using (3.68) and (3.62), we obtain

$$w_1'(\xi_M) = 0. \tag{3.70}$$

Accordingly ξ_E is a zero of w_1, ξ_M a zero of the derivative of w_1, and ξ is given by (3.64). Thus, we deduce two values for φ_s', viz.,

$$\varphi_s' = \xi_E / 2^{1/3} \, \rho^{2/3}, \tag{3.71}$$

and

$$\varphi_s' = \xi_M / 2^{1/3} \, \rho^{2/3}. \tag{3.72}$$

The two equations (3.71) and (3.72) are not compatible because the function w_1 does not have a double zero, hence, $\xi_E \neq \xi_M$. From the above discussion it follows that either $H_0^\alpha = 0$ everywhere, or that $E_0^\alpha = 0$ for all ν. In the first case, we obtain an electric creeping ray all of whose components can be calculated from E_0^α. It propagates with a phase constant given by

$$ks + k^{1/3} \varphi_E (s, \alpha)$$

where

$$\varphi_E (s, \alpha) = 2^{-1/3} \int \frac{\xi_E ds}{\rho^{2/3}}, \qquad (3.73)$$

with

$$w_1(\xi_E) = 0.$$

In the second case, we obtain a magnetic creeping ray all of whose components can be calculated from H_0^α. It propagates with a phase

$$ks + k^{1/3} \varphi_M (s, \alpha),$$

where

$$\varphi_M (s, \alpha) = 2^{-1/3} \int \frac{\xi_M ds}{\rho^{2/3}}, \qquad (3.74)$$

with

$$w_1' (\xi_M) = 0.$$

The electric and magnetic creeping rays propagate with different velocities and attenuation constants. This result is well-known in GTD, and is recognized, for instance, when solving the canonical problem of the sphere. The advantage of the asymptotic expansion method is that it leads us to this result in a more deductive manner.

We have just derived φ for the electric and magnetic creeping rays propagating on a perfect conductor. To find E_0^α and H_0^α, we will have to impose the boundary conditions on E_1^α and H_1^α, as shown below, and to work with the equation of order k^0 as discussed in Sect. 3.2.2.1.

3.2.6.1.3 Boundary conditions on E_1^α and H_1^α

Let us turn now to the boundary condition on E_1^α and H_1^α. From (3.59) we have for E_1^α and more generally for E_j^α,

$$E_1^\alpha = 0 \quad \text{for} \quad \nu_1 = 0. \qquad (3.75)$$

The boundary condition on H_1^α is obtained from $E_2^s = 0$ and from the dual of (3.47) for E_2^s that relates it to H_1^α. For a perfect conductor, we obtain

$$E_2^s = -\frac{\partial \varphi}{\partial \alpha} E_0^\alpha - i\,h\,\frac{\partial H_1^\alpha}{\partial \nu_1}. \tag{3.76}$$

Furthermore, for $\nu = 0$, $E_0^\alpha = 0$; consequently, the boundary condition simply reads

$$\frac{\partial H_1^\alpha}{\partial \nu_1} = 0, \tag{3.77}$$

which corresponds to a homogeneous Neumann condition.

Thus for the case of a perfect conductor, the quantities E_1^α and H_1^α satisfy homogeneous Dirichlet and Neumann conditions, respectively. Next, we proceed to discuss the general case of a surface described by an impedance boundary condition.

3.2.6.2 Impedance conditions

Condition associated with an isotropic impedance
The impedance boundary condition relates the tangential electric and magnetic fields as follows

$$\vec{E}_{tg} = Z(\vec{n} \times \vec{H}_{tg}), \tag{3.78}$$

where

$$\vec{E}_{tg} = E^s\,\vec{e}_s + E^\alpha\,\vec{e}_\alpha \quad \text{and} \quad \vec{H}_{tg} = H^s\,\vec{e}_s + H^\alpha\,\vec{e}_\alpha.$$

Taking into account the fact that, on the surface, the frame $(\vec{e}_s, \vec{e}_\alpha, \vec{e}_n)$ is orthogonal, but not normalized, we have

$$\vec{n} \times \vec{e}_s = 1/h\,\vec{e}_\alpha,$$

$$\vec{n} \times \vec{e}_\alpha = -1/h\,\vec{e}_s.$$

Next, we can write (3.78) in the form

$$E^s = -Z\,h\,H^\alpha, \tag{3.79}$$

and

$$h\,E^\alpha = Z\,H^s. \tag{3.80}$$

Finally, by using the fact that $E_0^s = H_0^s = 0$, we get

$$k^{-1/3} E_1^s ... + k^{-j/3} E_j^s ... = - Zh \, [H_0^\alpha + k^{-1/3} H_1^\alpha ... + k^{-j/3} H_j^\alpha],\qquad(3.81)$$

$$k^{-1/3} H_1^s ... + k^{-j/3} H_j^s ... = \frac{1}{Z} h \, [E_0^\alpha + k^{-1/3} E_1^\alpha ... + k^{-j/3} E_j^\alpha].\qquad(3.82)$$

Equation (3.82) can also be derived directly from (3.81) by introducing the changes

$$E \to H, \qquad\qquad H \to -E, \qquad\qquad Z \to 1/Z.$$

3.2.6.2.1 Boundary conditions for E_0^α and H_0^α

To determine the boundary condition for E_0^α and H_0^α, it is necessary to determine the order of the surface impedance Z. The assumption $Z = O(1)$ has been examined in [Bo] and it has been found that it leads to the Dirichlet condition for E_0^α and H_0^α. However, this result is not uniform with respect to Z, because, when $Z \to 0$, H_0^α continues to satisfy the Dirichlet boundary condition, while when $Z = 0$, i.e., for the perfectly conducting case, it satisfies the Neumann condition instead.

In Sect. 3.1 we pointed out this behavior of the boundary condition for the case of a cylinder whose surface satisfies an impedance boundary condition. In order to avoid these problems alluded to above, and to derive a uniform solution with respect to Z when $Z \to 0$, we assume an impedance behavior as follows

$$Z = k^{-1/3} Z_H, \qquad \text{in Eq. (3.81)}$$

$$Z = k^{1/3} Z_E, \qquad \text{in Eq. (3.82)}$$

where Z_H and Z_E are assumed to be of $O(1)$. This, in turn, implies that the impedance is small in (3.81) and large in (3.82).

By equating the coefficients of the powers $k^{-1/3}$, we obtain, to the first order

$$- Z_H h \, H_0^\alpha = E_1^s = - i h \frac{\partial H_0^\alpha}{\partial v_1}.$$

By introducing the variable v and the quantity m defined in Sect. 3.1, the above equation can be written in the form

$$\frac{\partial H_0^\alpha}{\partial v} + im \, Z H_0^\alpha = 0 \quad \text{for} \quad v = 0.\qquad(3.83)$$

Similarly for (3.82) we have, to the first order

$$\frac{\partial E_0^\alpha}{\partial v} + i\frac{m}{Z} \, E_0^\alpha = 0 \text{ for } v = 0, \tag{3.84}$$

which can be deduced from (3.83) by substituting E in place of H and $1/Z$ in place of Z.

As for the perfectly conducting case, the use of (3.67) and (3.68), in conjunction with the boundary conditions (3.83) and (3.84), enable us to calculate ξ_E and ξ_M, and, hence, φ_s'.

E_0^α is proportional to $w_1\,(\xi_E - v)$ where ξ_E satisfies the equation

$$w_1'(\xi_E) = i\frac{m}{Z} w_1\,(\xi_E). \tag{3.85}$$

Likewise, H_0^α is proportional to $w_1\,(\xi_M - v)$ where ξ_M satisfies

$$w_1'(\xi_M) = i\,mZ\,w_1\,(\xi_M). \tag{3.86}$$

As for the case of a perfect conductor, the quantities ξ_E and ξ_M cannot be equal except when $Z = 1$, because w_1 and w_1' do not vanish simultaneously.

For the perfect conductor we define two modes. The first of these is associated with the electric creeping ray. For this mode, $E_0^\alpha \neq 0$, but $H_0^\alpha = 0$ everywhere, which implies that (3.55) and (3.83) are satisfied.

Equations (3.64) and (3.85) enable us to calculate φ_E and the resulting expressions given below, are equivalent to (3.73) and (3.67) for the perfect conductor.

$$\varphi_E\,(s,\,\alpha) = 2^{-1/3} \int \xi_E \frac{ds}{\rho^{2/3}}, \tag{3.73'}$$

and

$$E_0^\alpha\,(s,\,\alpha,\,v) = \frac{A(s,\alpha)}{h} w_1\,(\xi_E - v). \tag{3.67'}$$

(The primed equation numbers simply refer to the unprimed ones.) However, this time, ξ_E depends upon Z and is determined from Eq. (3.85). The second mode for the perfectly conducting case is associated with a magnetic creeping ray, with $H_0^\alpha \neq 0$, $E_0^\alpha = 0$.

The equations are the same as those derived for a perfect conductor

$$H_0^\alpha \ (s, \alpha, v) = \frac{B(s, \alpha)}{h} w_1 \ (\xi_E - v),$$ (3.68')

and

$$\varphi_M \ (s, \alpha) = 2^{-1/3} \int \xi_M \ \frac{ds}{\rho^{2/3}} .$$ (3.74')

Here ξ_M is determined from Eq. (3.86). Equations (3.85) and (3.86) can be deduced from each other by replacing Z with $1/Z$ and vice-versa. We can verify, *a posteriori*, that the assumptions embodied in (3.81) and (3.82) are not contradictory. Equation (3.81) assumes that the impedance is small $(O(m^{-1}))$ for the magnetic creeping ray, whereas (3.82) implies that the impedance is large for the electric creeping ray. Consequently, we observe that these two assumptions are indeed uncoupled. Furthermore, they yield good results even outside their initial domain. In effect, if we let Z tend to infinity in (3.83) even though it should be small, in principle, we regain the Dirichlet boundary condition for H_0^α, which corresponds to the magnetic conducting case. We conclude, therefore, that the limit obtained is correct for $Z \to +\infty$. Similarly, if we let Z tend to 0 in (3.85), despite the fact that it is assumed to be large, we recover the Dirichlet boundary condition for E_0^α for the perfect electric conductor case. Again, the limit is correct as $Z \to 0$, and we conclude, therefore, that the solution for the impedance boundary condition case is general enough to include the conducting case.

3.2.6.2.2 Boundary conditions satisfied by E_1^α and H_1^α

To determine E_0^α and H_0^α we will also need the boundary conditions on E_1^α and H_1^α. To this end, we return to (3.81) and (3.82), and obtain a homogeneous boundary condition on H_1^α for the magnetic creeping ray by using the second-order terms. The condition reads

$$\frac{\partial H_1^\alpha}{\partial v} + im\, Z H_1^\alpha = 0 \quad \text{for} \quad v = 0.$$ (3.87)

The condition on E_1^α also obtained by using the second-order terms is

$$\frac{\partial E_1^\alpha}{\partial v} + i\, \frac{m}{Z}\, E_1^\alpha = -ih^{-1}\, \frac{\partial \varphi}{\partial \alpha}\left(\frac{\rho}{2}\right)^{1/3} H_0^\alpha \quad \text{for} \quad v = 0.$$ (3.88)

In the same way, we obtain for the electric creeping ray a homogeneous condition on E_1^α, viz.,

$$\frac{\partial E_1^\alpha}{\partial \nu} + i \frac{m}{Z} E_1^\alpha = 0 \text{ for } \nu = 0, \qquad (3.89)$$

and an inhomogeneous condition on H_1^α, which reads

$$\frac{\partial H_1^\alpha}{\partial \nu} + i \, mZ H_1^\alpha = - ih^{-1} \frac{\partial \varphi}{\partial \alpha} \left(\frac{\rho}{2} \right)^{1/3} E_0^\alpha \text{ for } \nu = 0. \qquad (3.90)$$

These boundary conditions will enable us to determine E_0 and H_0 as shown below.

3.2.7 Complete determination of E_0 and H_0

3.2.7.1 Equations of order k^0

Equations (3.41) to (3.43), (3.44) to (3.46), (3.47) to (3.49) in Sect. 3.2.4 and the associated boundary conditions on the surface Γ have enabled us to calculate $\varphi(s)$ and to demonstrate that the variation of the solution as a function of ν is described by an Airy function. We still need to calculate the factors $A(s, \alpha)$ and $B(s, \alpha)$ in order to determine the amplitude of the fields E_0^α and H_0^α on the surface. Toward this end, we use the previous boundary conditions and the equations for the zero order k^0. We find that (3.51) and the dual equation of (3.52) in Sect. 3.2.4 provide two different expressions for $h \, H_3^\alpha + E_3^n$, whereas (3.52) and the dual equation of (3.51) yield corresponding expressions for $h \, E_3^\alpha - H_3^n$.

From the above equations, we obtain two compatibility conditions and the expression of E_3^n and H_3^n as functions of E_3^α and H_3^α, E_0^α and H_0^α, and E_1^α and H_1^α. Likewise, Eqs. (3.50) and ist dual provide the expressions for E_3^s and H_3^s as functions of the same quantities, viz., E_3^α, H_3^α, etc.

Finally, we obtain all of the field components as functions of the components labeled with α. To simplify the notations, we define the quantities E_0 and H_0 via the equations $E_0 = h \, E_0^\alpha$, $H_0 = h \, H_0^\alpha$, where E_0 and H_0 represent the fields with respect to a unit vector $\hat{\alpha} = \vec{e}_\alpha / \| \vec{e}_\alpha \|$. We can also define the scaled versions of the fields E_1 and H_1 in a similar manner. The compatibility conditions read

$$-i \left[\frac{\partial^2 H_1}{\partial \nu_1^2} + 2 \left(\frac{\nu_1}{\rho} - \varphi_s' \right) H_1 \right] - 2\tau E_0 + \frac{2 \partial H_0}{\partial s} + h^{-1} \left(\frac{\partial h}{\partial s} \right) H_0 = 0, \qquad (3.91)$$

$$-i \left[\frac{\partial^2 E_1}{\partial \nu_1^2} + 2 \left(\frac{\nu_1}{\rho} - \varphi_s' \right) E_1 \right] + 2\tau H_0 + \frac{2 \partial E_0}{\partial s} + h^{-1} \left(\frac{\partial h}{\partial s} \right) E_0 = 0. \qquad (3.92)$$

After introducing a change of variables, we recognize the presence of

$$L_0 = \frac{\partial^2}{\partial v^2} + (v - \xi),$$

appearing in (3.91) and (3.92). We then rewrite these equations as

$$\left(\frac{2}{\rho}\right)^{2/3} i L_0 H_1 = \frac{2}{h^{1/2}} \frac{\partial}{\partial s}(h^{1/2} H_0) - 2\tau E_0, \tag{3.93}$$

$$\left(\frac{2}{\rho}\right)^{2/3} i L_0 E_1 = \frac{2}{h^{1/2}} \frac{\partial}{\partial s}(h^{1/2} E_0) + 2\tau H_0. \tag{3.94}$$

We remark here that the convention we adopted for the torsion, viz.,

$$d\hat{n}/ds = -1/\rho \ \hat{s} + \tau \hat{b},$$

where $(\hat{s}, \hat{n}', \hat{b}) = (\hat{s}, -\hat{n}, \hat{\alpha})$ is the Frenet system of the geodesic. The opposite convention is sometimes adopted in the literature and affects the sign of the last term in (3.93) and (3.94).

3.2.7.2 Resolution of the previous equations

For the magnetic creeping ray, $E_0 = 0$, and (3.93) becomes

$$\left(\frac{2}{\rho}\right)^{2/3} i L_0 H_1 = \frac{2}{h^{1/2}} \frac{\partial}{\partial s}(h^{1/2} H_0), \tag{3.95}$$

with the boundary conditions

$$\frac{\partial H_0}{\partial v} = 0 \quad (3.62), \quad \frac{\partial H_1}{\partial v} = 0 \quad (3.77) \quad \text{for the perfect conductor,}$$

$$\frac{\partial H_0}{\partial v} + imZH_0 = 0 \quad (3.83), \quad \frac{\partial H_1}{\partial v} + imZH_1 = 0 \quad (3.87) \quad \text{for the impedance case.}$$

The field H_0 is given by (3.68), which is now rewritten as

$$H_0 = B(s, \alpha) w_1 (\xi_M - v),$$

and the coefficient $B(s, \alpha)$ can be determined by using Eq. (3.95) by following the same procedure as used in Sect. 3.1 for the cylinder. Equation (3.95) is a Sturm-Liouville equation, which has a solution only if its right hand side is orthogonal to the solutions of the homogeneous equation. The orthogonality condition will provide us with an equation for $B(s, \alpha)$. In practice, we obtain this condition by multiplying (3.95) by the solution w_1 $(\xi_M - \nu)$ of the homogeneous equation and integrating from 0 to $+\infty$ with respect to the variable ν. We obtain, after performing the calculations similar to those of Sect. 3.1.6, the differential Eq. (3.96) given below, which is valid along a geodesic, where α is constant. In order to simplify (3.96) we omit the dependence of the functions B, h, ρ and d_M on α, and write the equation as

$$\frac{B'(s)}{B(s)} + \frac{1 h'(s)}{2 h(s)} + \frac{1}{6} \frac{\rho'(s)}{\rho(s)} + \frac{1}{2} \frac{d'_M(s)}{d_M(s)} = 0, \tag{3.96}$$

where $B(s)$, which is implicitly a function of α, is the amplitude of the creeping mode on a creeping ray.

Let us recall that $h(s)$ represents the broadening (or the narrowing) of the geodesic pencil with respect to the axes of α. It may also be identified as the ratio $d\eta'/d\eta$ in the GTD (see Fig. 3.5), where $d\eta'$ is compiled at the observation point, while $d\eta$ is calculated on the axes of α. Also, $\rho(s)$ is the radius of curvature of the geodesic along s (see Fig. 3.5). Finally, d_M is defined by the following equation

$$d_M(s) = w_1'^2(\xi_M) - \xi_M w_1^2(\xi_M)$$

where ξ_M satisfies Eq. (3.86).

For a perfect conductor both ξ_M and $d_M(s)$ are constants. For the impedance case, (3.96) implies

$$B(s) = B(s, \alpha) = B(\alpha)\, h^{-1/2}(s)\, \rho^{-1/6}(s)\, d^{-1/2}_M (s) \,. \tag{3.97}$$

Inserting (3.97) into (3.68), we obtain H_0, while (3.74) provides the phase $\varphi_M(s, \alpha)$. Inserting the above results in the asymptotic expansion (3.38) of Sect. 3.2.3, and neglecting terms of order $O(k^{-1/3})$, we obtain the following expression for the field expansion along a magnetic creeping ray

$$H(s, \alpha, \nu) \approx B(\alpha)\, h^{-1/2}(s)\, \rho^{-1/6}(s)\, d^{-1/2}_M (s)\, w_1 (\xi_M - \nu)$$

$$\exp\left(iks + i\frac{k^{1/3}}{2^{1/3}} \int_{p(\alpha)}^{s} \frac{\xi_M ds}{\rho^{2/3}}\right). \tag{3.98}$$

Similarly, for the electric creeping ray, we can obtain

$$E(s, \alpha, \nu) \approx A(\alpha) \, h^{-1/2}(s) \, \rho^{-1/6}(s) \, d_S^{-1/2}(s) \; w_1 \, (\xi_E - \nu)$$

$$\exp \left(iks + i \frac{k^{1/3}}{2^{1/3}} \int_{p(\alpha)}^{s} \frac{\xi_E ds}{\rho^{2/3}} \right), \tag{3.99}$$

where $d_E(s) = w_1^2(\xi_E) - \xi_E \, w_1^2(\xi_E)$.

To summarize, we can say that the field calculation for each creeping mode, whether H or E-type, entails the implementation of the following steps. We have to determine:

(i)   the root ξ of Eq. (3.86) or Eq. (3.85).

(ii)   the function $B(\alpha)$ or $A(\alpha)$ that yields the amplitudes of the H or E fields on each geodesic.

(iii) the function $p(\alpha)$, which is the lower bound of integration of the phase φ.

Each determination of ξ leads to a specific mode, labeled in Chap. 1 with its index p. For the case of the diffraction by an object, the functions A, B and p of the parameter α are determined in Sect. 3.11 by matching the creeping ray solutions that we just have determined, to the solution in the neighborhood of the light-shadow boundary. The solution in the shadow zone is then completely specified.

Equations (3.98) and (3.99) provide the fields due to a creeping ray in the vicinity of the surface. One extracts the surface fields by setting $\nu = 0$ in these formulas. The case of the electric creeping ray for the perfect conductor represents a special case, because (3.99) reduces to $E_0 = 0$ on the surface. In contrast, the tangential magnetic field does not vanish on a perfect conductor, and its calculation is detailed in Sect. 3.2.7.4. Before going into the details of the magnetic field, H_0, however, we will first provide a physical interpretation of the expressions in (3.98) and (3.99).

3.2.7.3 Physical interpretation of the results

3.2.7.3.1 Satisfaction of the impedance condition

The expressions in (3.98) and (3.99) have a somewhat unexpected character because the only zero-order component in these equations is along α. One might wonder how the impedance condition, which imposes the ratio of the orthogonal components of $\vec{E}_{tan}$, and $\vec{H}_{tan}$, can be satisfied. The answer to the above puzzle is that if we consider the terms that are $O(k^{-1/3})$, the corresponding fields do indeed have components along the direction of propagation. As an example, for the magnetic creeping ray, there exists a component E_S^1 along the direction of propagation, and, for the dominant order, the field E_s is equal to ZH_0. Similarly, for the electric creeping ray the field H_s to the dominant order, is given by $1/Z\ E_0$. These fields are $O(1)$ in the analysis, because Z is of order $k^{-1/3}$ and $k^{1/3}$, respectively, for the magnetic and electric creeping rays.

Figure 3.7 depicts the propagation of the creeping modes on the surface, each of which intrinsically satisfies the impedance condition. It is worthwhile noting that by taking into account the fields of $O(k^{-1/3})$ leads to an interpretation that is different from the one given in Fig. 3.6.

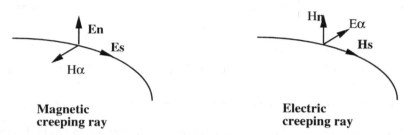

Magnetic creeping ray   **Electric creeping ray**

Fig. 3.7. Magnetic and electric creeping waves, with first-order components taken into account

3.2.7.3.2 Interpretation of the geometrical factors
The factor $h^{-1/2}(s)$ has a simple interpretation in the context of GTD. The width of a pencil of surface rays is proportional to $h(s)$, the power is conserved in a tube of surface rays, and this power is proportional to the product of $h(s)$ and the squared field; hence $E_0^2\ h = H_0^2\ h$ = constant.

An alternate interpretation, which is more global but is somewhat unconventional, can be provided by considering the thickness of the boundary layer, which is proportional to $k^{-2/3}\ \rho^{1/3}$. The creeping ray can be seen as a volumetric tube of rays with *width* proportional to h tangential to the surface, and *height* proportional to $k^{-2/3}$ $\rho^{1/3}$ along the normal. The conservation of the power in this tube of rays lead to the law $E_0^2\ h\rho^{1/3} = H_0^2\ h\rho^{1/3}$ = *constant,* and enables us to determine the factor $\rho^{-1/6}(s)$.

The factors $d_M^{-1/2}(s)$ and $d_E^{-1/2}(s)$ can be interpreted as indices of refraction. For ordinary rays, the amplitude is in fact proportional to $n^{-1/2}$, where n is the reflective index of the medium. We regain the factor $d_M^{-1/2}(2)$ by identifying d_M with an equivalent index.

Finally, the lower bound $p(\alpha)$ in the integrals appearing in (3.98) and (3.99) determine the point where the attenuation starts. Evidently, we expect this point to be at the intersection of the creeping ray and the light-shadow boundary.

Consequently, all of the factors appearing in the expressions (3.98) and (3.99) may be interpreted by recognizing the similarity of behavior between the creeping and space rays.

3.2.7.3.3 Particular case of a perfect conductor with $Z = 0$

The creeping ray propagating on the surface of a perfect conductor represents a singular case, because the electric field, as given by (3.99), vanishes on the surface. The zero-order magnetic field, with respect to k, on the surface of the conductor, also goes to zero and the first non-vanishing components are H_1^s and H_1, which are $O(k^{-1/3})$. H_1^s is given by

$$H_1^s = i \frac{\partial E_0}{\partial v_1} = i \left(\frac{2}{\rho}\right)^{1/3} \frac{\partial E_0}{\partial v} = -i \left(\frac{2}{\rho}\right)^{1/3} A w_1'(\xi_E - v), \qquad (3.100)$$

where $E_0 = A w_1 (\xi_E - v)$, and A is defined in (3.99). Similarly, H_1 satisfies (3.93), which leads to

$$\left(\frac{2}{\rho}\right)^{2/3} L_0 H_1 = 2i\tau E_0, \quad \text{with} \quad L_0 = \frac{\partial^2}{\partial v^2} - (\xi_E - v),$$

and the boundary conditions (3.62)

$$\frac{\partial H_1}{\partial v} = 0.$$

A particular solution of (3.93) satisfying the radiation condition is

$$2i\tau A \; w_1'(\xi_E - v) \left(\frac{\rho}{2}\right)^{2/3} .$$

The general solution of the homogeneous equation satisfying the radiation condition is $w_1(\xi_E - v)$. Thus, the general solution of (3.93) satisfying the radiation condition is $2i\tau A \, w \, w_1' \; (\xi_E - v)(\rho)^{2/3} + C \, w_1(\xi_E - v)$ where C is a constant which is determined by using the boundary condition (3.62). We can show that $C = 0$, and hence

$$H_1 = 2i\tau A \; w \, w_1' \; (\xi - v) \left(\frac{\rho}{2}\right)^{2/3} . \qquad (3.101)$$

Using the notation $m = (k\rho/2)^{1/3}$, we can write the $\vec{H}$ field as

$$\vec{H} = k^{-1/3} H_1 \hat{\alpha} + k^{-1/3} H_1^s \hat{s},$$

$$\vec{H} = -\frac{i}{m} A w_1'(\xi_E - v) \; [-(\tau\rho) \hat{\alpha} + \hat{s}]. \tag{3.102}$$

The magnetic field on the surface is directed along the vector $[-(\tau\rho) \hat{\alpha} + \hat{s}]$. For the particular case of the electric creeping mode on a circular cylinder, the geodesics follow a helical path. The product $\tau\rho$ is equal to $cot\beta$ where β is the angle of the helix with the axis of the cylinder (Fig. 3.8).

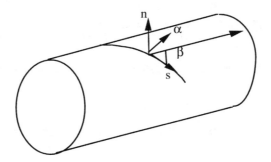

Fig. 3.8. Geodesics on a circular cylinder

For this particular case, $[-(\tau\rho) \hat{\alpha} + \hat{s}] = \hat{\theta}/sin\beta$ and, hence, we obtain the well-known result that the magnetic field is directed along the circumference. This follows from the fact that for TM polarization, and for a metallic cylinder illuminated at oblique incidence, the surface field is TM, i.e., directed along the circumference, and the surface currents flow along the generatrix of the cylinder.

3.2.8 Special case of the surface impedance given by Z = 1

The case of surface impedance $Z = 1$ is a special one, because the electric and magnetic creeping rays satisfy the same boundary condition, viz., $(\partial H_0/\partial n) + imH_0 = 0$ (resp. $E_0 = 0$). Thus, we can no longer deduce that $H_0 = 0$ or $E_0 = 0$ as in Sect. 3.2.6. For this particular case it can be shown that the electric and magnetic creeping rays are coupled.

Starting from Eqs. (3.65) and (3.66) of Sect. 3.2.6.1.2, and setting

$$K_0 = H_0 + iE_0 \quad \text{and} \quad J_0 = H_0 - iE_0,$$

$$K_1 = k^{-1/3}(H_1 + iE_1) \quad and \quad J_1 = k^{-1/3}(H_1 - iE_1),$$

We obtain $$L_0 K_0 = L_0 J_0 = 0. \tag{3.103}$$

Likewise, from (3.93) and (3.94) in 3.2.7.1, we have

$$L_0 K_1 = \frac{\tau\rho}{m} K_0 - i\frac{\rho}{h^{1/2}m}\frac{\partial}{\partial s}(h^{1/2}K_0), \tag{3.104}$$

$$L_0 J_1 = -\frac{\tau\rho}{m} J_0 - i\frac{\rho}{h^{1/2}m}\frac{\partial}{\partial s}(h^{1/2}J_0). \tag{3.105}$$

For $\nu = 0$, the boundary conditions are

$$\frac{\partial K_0}{\partial v} + imK_0 = 0, \tag{3.106}$$

$$\frac{\partial J_0}{\partial v} + imJ_0 = 0. \tag{3.107}$$

Using (3.88), (3.90), and the value of φ' given by (3.71), we obtain the boundary conditions on K_1 and J_1 which read

$$\frac{\partial K_1}{\partial v} + imK_1 = \frac{\partial\varphi}{\partial\alpha}\frac{1}{hm}\left(\frac{\rho}{2}\right)^{2/3}K_0, \tag{3.108}$$

$$\frac{\partial J_1}{\partial v} + imJ_1 = \frac{\partial\varphi}{\partial\alpha}\frac{1}{hm}\left(\frac{\rho}{2}\right)^{2/3}J_0. \tag{3.109}$$

Equations (3.103), (3.106), and (3.107) enable us to write

$$K_0 = K(s, \alpha)\, w_1\, (\xi - \nu) \quad and \quad J_0 = J(s, \alpha)\, w_1\, (\xi - \nu),$$

where ξ is a solution of $w_1'(\xi) = imw_1(\xi)$.

In order to calculate $K(s, \alpha)$ and $J(s, \alpha)$, we write, as in Sect. 3.1, the condition of orthogonality to the solution $w_1(\xi - \nu)$ of the homogeneous equation. This leads us to the following equation for determining K_0, viz.,

$$-\frac{\partial\varphi}{\partial\alpha}\frac{1}{hm}\left(\frac{\rho}{2}\right)^{2/3}\frac{w_1^2(\xi)}{w_1'^2(\xi)-\xi w_1^2(\xi)} = -\frac{\tau}{m}+\frac{i}{m}\frac{d}{ds}\ell n\,[K_0\, h^{1/2}\rho^{1/6}d^{1/2}]. \tag{3.110}$$

Similarly J_0 is determined from an equation which is derived from (3.110) simply by replacing $-\tau$ with τ in (3.110). Since $w_1'(\xi) = imw_1(\xi)$, the left hand side of (3.110) is of $O(1/m^2)$ compared to the right hand side term, and can be neglected.

Finally, after solving (3.110) with a zero left hand side, we obtain

$$K_0(s,\,\alpha) = K(\alpha)\,h^{-1/2}(s)\,\rho^{-1/6}(s)\,d^{-1/2}(s)\,\exp\left(-i\!\int \tau\,ds\right). \qquad (3.111)$$

Similarly, for $J_0\,(s,\,\alpha)$ we can derive a companion equation to (3.111) by changing i into $-i$. Equation (3.111) and its companion equation show, once again, that the factor $h^{-1/2}\rho^{-1/6}d^{-1/2}$ is associated with the variation of the amplitude. The factor exp $\left(-i\!\int \tau\,ds\right)$ corresponds to a rotation of the polarization. The angle P of polarization is given by $dP/ds = \tau$, using the notations of Fig. 3.9. For a positive torsion, the fields $\vec{E}$ and $\vec{H}$ move in the counterclockwise direction, and we recover the law for the ordinary rays. Before closing this discussion, we should point out that the sign difference between the result presented here and that given in Russian works [BB], for example. The difference arises from the fact that we have chosen an opposite sign convention in the definition of the torsion. This is done for the special case of the impedance $Z = 1$, or, more generally, when Z is close to 1.

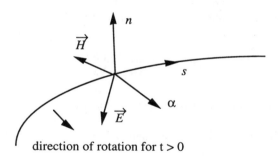

direction of rotation for t > 0

Fig. 3.9. Rotation of polarization along a ray with torsion

To summarize this section, we present the following observations

(i)   For an impedance Z sufficiently different from 1, the electric and magnetic creeping rays are uncoupled.

(ii)   For an impedance Z equal 1, the electric and magnetic modes satisfy the same boundary conditions and we extract one from the other by rotating $\vec{E}$ and $\vec{H}$ by $\pi/2$, the rotating axis being the tangent to the geodesic. The torsion of the geodesics thus induces a rotation of the fields $\vec{E}$ and $\vec{H}$ according to Rytov's law.

The above effect was also observed in a work by Lyalinov [L], who assumed that the impedance $Z = O(1)$, for all the cases studied. With this hypothesis, we can solve, to the leading order, the Dirichlet problem for both the electric and magnetic creeping rays, and we can derive both the coupling and the rotation for them. However, when Z is sufficiently different from 1, one of the two modes attenuates much more rapidly than the other and the dominant effect is no longer the rotation. For instance, for the perfectly conducting case, the electric creeping ray attenuates much more than does the magnetic creeping ray, and the creeping rays can be treated separately, as is the procedure in the GTD. Thus, the approximation $Z = O(1)$ is well-suited for describing of the physics of the problem when the relative impedance of the surface is close to 1. However, it yields non uniform results when the impedance deviates significantly from 1.

Finally, we should mention that one drawback of the method of impedance stretching is that the rotation effect in the creeping rays is only predicted for $Z = 1$ while, in reality, it also exists for Z in the neighborhood of 1 and it would have certainly been desirable to be able to recover the torsion effect over a broader range of impedance than it is possible with the impedance stretching method.

3.2.9 Conclusions

By using asymptotic expansion we have been able to describe the creeping rays on a convex object characterized by an impedance condition. This technique enables us to deduce directly the expression of the creeping rays on a general surface, albeit at a cost of extensive and cumbersome calculations. The asymptotic technique does not rely as much on the solution of the canonical problems as does the conventional GTD approach. This is because the solution to the canonical problems are only used to guess an Ansatz of the solution.

The asymptotic solution is only uniform when the impedance $Z \to 0$ or ∞, in the sense that we regain the case of the perfect electric or magnetic conductor, respectively, for the above limiting values of Z. However, to obtain this result we implicitly assumed that the impedance is either $O(k^{1/3})$ or $O(k^{-1/3})$. By contrast, the assumption that $Z = O(1)$ does not lead to uniform results.

We have shown that all of the components of the fields $\vec{E}$ and $\vec{H}$ can be expressed in terms of tangential components orthogonal to the geodesics of $\vec{E}$ and $\vec{H}$, that can be considered as potentials. For the assumed surface impedances, we have shown that there exists two types of creeping rays viz., electric and magnetic. The principal tangential component of the electric and magnetic rays are the electric and magnetic fields, orthogonal to the geodesics followed by the creeping rays. These two creeping rays propagate with different velocities, they have different attenuation characteristics, and are uncoupled when Z is different from 1.

The advantage of the asymptotic expansion method is that we can recover, deductively, the results previously obtained in the context of GTD in a heuristic manner. Close to the surface, the dependence of the field along the normal is given by an Airy function, and the propagation constants ξ satisfies the equation $w_1'(\xi) - mZw_1(\xi) = 0$ for the magnetic creeping ray, and $w_1'(\xi) - m/Zw_1(\xi) = 0$ for the electric one. The magnitude of the field is inversely proportional to the square root of the divergence factor of the geodesic pencil. We also draw attention to two effects that are difficult to obtain starting from the usual canonical problems, these being the $\rho^{-1/6}$ dependence on the radius of curvature ρ and the geometrical factor $d^{-1/2}$, which depends upon the impedance. All of these effects can be interpreted physically by considering the creeping ray as a tube of rays of thickness $k^{-2/3}\rho^{1/3}$, with an index that depends on the impedance. For $Z = 1$, the electric and magnetic creeping rays are coupled, and their propagation exhibits a rotation of the polarization similar to that dictated by Rytov's law as it applies to the case of conventional rays in geometrical optics.

The method of asymptotic expansions not only enables us to extract interesting and useful results, but it also does so in a more deductive manner than does the GTD approach. In this section, we have restricted ourselves to the calculation of the first term of the creeping wave representation. For further details we refer the reader to the literature, in particular to Andronov and Bouche [AB1,2,3] who have established the following complementary results.

The above authors discuss the calculation of the first order term, and, in particular, highlight the effect of the transverse curvature on the geodesic, its torsion, and the variation of the radius of curvature. The effect of the impedance variation on the propagation constant of the creeping waves is also discussed. Additionally, the authors evaluate the effect of a small radius of curvature on the propagation of the creeping waves and show that the attenuation of the magnetic creeping mode is weaker than for the case treated in this section, where the transverse radius of curvature is assumed to be of $O(1)$. This last result is helpful in understanding the phenomenon associated with traveling waves propagating on the surface of a slender body. Finally, the authors also investigate case of an anisotropic impedance and show that, for this case, the electric and magnetic creeping modes described in this section are replaced by mixed modes that have non-vanishing components along the binormal of both the electric and magnetic fields.

In addition to the case of $Z = 1$, discussed above, it would also be useful to study the case of an impedance close to unity because the rotation effect exists not only when $Z = 1$, but also when it is close to unity. One approach to handling the situation is to assume that the impedance is $O(1)$, the assumption being quite natural in this case. However, if we follow this approach, we lose the uniformity when the impedance is either small or large compared to 1.

Next, we are going to discuss the case of two boundary layers associated with whispering gallery modes (Sect. 3.4) and with the caustic (Sect. 3.5). Although these two types of boundary layers are associated with different physical phenomenon, viz., the boundary layer of the whispering gallery modes, and that of a caustic, they lead to calculations very similar to those presented in this section.

3.3 Boundary layer of the whispering gallery modes

To investigate the boundary layer of the whispering gallery mode (see Fig. 3.10) we seek the source-free solutions in the interior of a closed convex cylinder Ω, whose cross-section is bounded. These solutions must satisfy the equation

$$(\nabla^2 + k^2)\, u = 0 \quad \text{in } \Omega,$$

$-(\partial u/\partial n) + ik\, Zu = 0$ on $\partial\Omega$, where n denotes the outward normal.

Fig. 3.10. Whispering gallery mode

Specifically, we seek the solutions which have a form similar to that of creeping waves, i.e., described by the expansion in Eq. (D) of Sect. 3.1. However, in contrast to Sect. 3.1, where we sought the solutions corresponding to $n > 0$, we now search for those with $n < 0$. We note that the equations are exactly identical to those found in Sect. 3.1, and only the boundary condition on $\partial\Omega$ and the application of the radiation condition are different. Thus the solution of our problem can still be written as (see (3.31) in Sect. 3.1)

$$u \approx A(0) \left(\frac{\rho(0)}{\rho(s)}\right)^{1/6} \left(\frac{d(0)}{d(s)}\right)^{1/2} F(\xi - \nu) \exp\left(iks + i\frac{k^{1/3}}{2^{1/3}} \int_0^s \frac{\xi\, ds}{\rho(s)^{2/3}}\right), \quad (3.112)$$

where F is an Airy function. Next, we impose the condition that the solution must tend toward zero as $\nu \to -\infty$. The variable $\xi - \nu$ will then be above the positive real axis. The only Airy function which satisfies this condition is the Ai function; hence, we have

$$F(\xi - \nu) = Ai(\xi - \nu). \tag{3.113}$$

Inserting (3.112) into the expression for the boundary condition on the object and still assuming that mZ is of $O(1)$, we are led to the following equation for ξ

$$Ai'(\xi) + imZAi(\xi) = 0, \tag{3.114}$$

which is the counterpart of Eq. (3.14) in Sect. 3.1. It is useful to note that ξ is real and negative if either $Z = 0$ or $Z = \infty$, which is true for a perfect conductor, or if Z is purely imaginary.

The solution, thus obtained, propagates without attenuation, i.e., its phase remains pure imaginary. However, whenever Z has a non-vanishing real part, ξ becomes complex, the phase acquires a non-vanishing real part and the solution becomes evanescent. For large ξ and ν of $O(1)$ these solutions which are referred to as the whispering gallery modes, may be visualized as rays reflecting from the surface $\partial\Omega$. For this case we can, in effect, approximate the Airy function by its asymptotic expansion in the form of the sum of two exponential functions, which can be interpreted as rays. The case of the rays propagating without attenuation is the same as the one in which the attenuation coefficient on the wall has a modulus equal to 1. The attenuation of the whispering gallery modes is due to the reflection from the wall, or more specifically, from the material coating on the wall. Note that, unlike the conventional creeping rays, the whispering gallery mode attenuation is not due to the curvature of the surface.

The previous discussion extends to the three-dimensional case in exactly the same manner as it did for the creeping waves. One obtains similar formulas as (3.98) and (3.99) in Sect. 3.2, with ξ satisfying, according to the type of mode, Eq. (3.114) or the companion equation that is obtained by replacing Z with $1/Z$ in (3.114).

The whispering gallery modes occur, for example, in the calculation of the field radiated by a source located on a concave surface [BB], and that of the field diffracted by a wedge with concave faces.

In both of these problems, the amplitude $A(0)$ of these modes, which is chosen to be arbitrary in this section, is determined through the matching to the solution close to the source or to the tip. The whispering gallery modes propagating along closed geodesics give rise to eigenmodes of cavities, as explained in [BB]. The method described in the previous section enables us to calculate the frequencies of these modes [BB].

Finally, the ray interpretation of the gallery modes shows that these rays have an envelope, or a caustic, which is located on the wall of the object supporting this mode. We now discuss the general topic of field computation in the neighborhood of the caustics.

3.4 Neighborhood of a regular point of a caustic (Fig. 3.11)

The caustics occur in a number of different situations, as for instance in the study of the reflection of a plane wave by a concave surface, or, as we will see in Chap. 7, of the whispering gallery modes we have just discussed above. In three dimensions, the caustic, i.e., the envelope of the rays, is comprised of two sheets that are joined together along cusp edges. As in the two-dimensional case, we will restrict ourselves to regions that are sufficiently far from the cusp of the caustic.

Fig. 3.11. A portion of the caustic

The convex side of the caustic accommodates two rays, while there are none on its concave side (see Fig. 3.11). Since the rays are tangent to the caustic, the norm of the gradient of the phase on the caustic equals 1. Consequently, following the same argument as in Sect. 3.2, we can show that the integral curves of the gradient of the phase on the caustic are geodesics of the caustic. This result can also be found by regarding the caustic as the evolute of the wavefront [Da].

A simple canonical solution of the wave equation describing a circular caustic is: $u(r,\theta) = J_{ka}(kr)\, e^{ika\theta}$, where ka is large [BB]. If we replace the Bessel function by its asymptotic expansion of Debye, we find

$$u \approx \frac{1}{2}\,\sqrt{\frac{2}{\pi ka}}\left(\frac{a^2}{r^2 - a^2}\right)^{1/4}\left\{exp\left[ik\left(\sqrt{r^2 - a^2} - a\,arcos\frac{a}{r} + a\theta\right) - i\frac{\pi}{4}\right]\right.$$

$$\left. + exp\left[ik\left(-\sqrt{r^2 - a^2} + a\,arcos\frac{a}{r} + a\theta\right) + i\frac{\pi}{4}\right]\right\}\cdot\left(1 + O\!\left(\frac{1}{k}\right)\right).\qquad(3.115)$$

Equation (3.115) can be interpreted as the sum of two rays, the first term behaving like a ray which leaves the circle, which has a radius a, the second corresponding to a ray going toward the circle. The two rays are tangent to the circle which is, consequently, a caustic. We note that there is a $-\pi/2$ jump of the phase at the crossing of the caustic.

Expansion (3.115), which is in terms of the integer powers of k^{-1}, corresponds to the Luneberg-Kline expansion of the solution discussed in Chap. 2. It is not valid for $kr \approx ka$, that is to say close to the caustic, where it is necessary to resort to the Watson expansion. This expansion is valid when the difference between the argument and the index of the Bessel function is on the order of the 1/3 power of the argument. As in the past, introducing the variable $\nu = (r - a)\,(2/a)^{1/3}\,k^{2/3}$, we obtain

$$J_{ka}(kr) \approx -\frac{i}{m}\,Ai(-\nu).\tag{3.116}$$

In this case also, the canonical problem of the circular caustic suggests that we stretch the normal coordinate by the factor $k^{2/3}$, and once again this leads to the Airy function obtained in the previous section. This result is not too surprising because the whispering gallery modes exhibit caustics.

Let us apply other techniques to find again this $k^{2/3}$. In the $(s,\ n)$ coordinates used in Sect. 3.2, the Laplacian contains a factor $(1 + (n/\rho))^{-1}$, which for small n, will give rise to a term of the form $k^2(n/\rho)$, which behaves as $k^{2-\alpha}$, if α is the dilatation exponent of n. On the other hand, the term associated with d^2/dn^2 yields a term which is $k^{2\alpha}$, i.e., $O(1)$. From the arguments presented in Sect. 2.1, it follows that these two terms must be of the same order, implying that $\alpha = 2/3$. Finally, with simple calculations of differential geometry, it can be shown that the order of magnitude of the phase difference between the two rays passing through the same point is $kn^{3/2}$. Thus, this difference is of $O(1)$ for $n = O(k^{-2/3})$, and the thickness of the boundary layer is of $O(k^{-2/3})$. Note that the previous arguments regarding the stretching of the coordinates also apply to the case of the creeping waves and to the gallery modes. However, these arguments are particularly easy to follow for the case of the caustic.

The previous considerations suggest that we look for an Ansatz, which takes the following form for the two-dimensional case:

$$u(s,\ \nu) = \exp\,(iks) \sum_{j=0}^{M} u_j(s,\ v)k^{-j/3}.\tag{3.117}$$

Inserting this Ansatz into the wave equation and sorting out terms of different order, we can derive a set of equations that are very similar to the ones we obtained in Sect. 3.1. In particular, for the dominant order we have

$$\frac{\partial^2 u_0}{\partial v^2} + \nu u_0 = 0,\tag{3.118}$$

whose solution is $u_0 = A(s)\,F(-\nu)$, where F is an Airy function. This solution must be evanescent on the shadow side of the caustic, where there is no ray, i.e., where $\nu \to \infty$. This leads us to the following choice for the function Ai to express u_0 as

$$u_0 = A(s)Ai(-\nu). \tag{3.119}$$

In contrast to the cases studied in the previous sections, there is no boundary condition on the caustic; instead, this condition is now replaced by a matching condition to the outer solution, which will be detailed in Sect. 3.12.

In the discussion above, we have chosen to use the coordinate system (s, n). However, in order to facilitate the matching, it desirable to choose an alternate system, viz., (S, δ) where S is the average of the two eikonals of the two rays, and δ is the half-difference between these eikonals. Inside the boundary layer, $S = s + O(k^{-4/3})$ and $\delta = (2/3)\sqrt{2/\rho}\; n^{3/2} + O(k^{-4/3})$, so that we obtain, from (3.119), the first order solution by expressing s and ν as functions of S and δ.

Finally, the case of the caustic for the Maxwell's equations in three dimensions is treated in a similar manner. The integral curves of the tangent vector field are the geodesics of the caustic. This can be demonstrated in two different ways: (i) by noticing that $|\nabla S| = 1$ on the caustic and then using the same argument as that given in Sect. 3.2.1; or, (ii) by considering the caustic as an evolute of the wavefronts, and then using the properties of the evolute which have been studied in detail, especially by Darboux [Da] (see Chap. 6).

For the three-dimensional case, the Ansatz takes the form

$$\vec{E}(s,\, \alpha,\, v) = exp\,(iks)\sum_{j=0}^{M}\vec{E}_j(s,\, \alpha,\, v)k^{-j/3}\,, \tag{3.120a}$$

$$\vec{H}(s,\, \alpha,\, v) = exp\,(iks)\sum_{j=0}^{M}\vec{H}_j(s,\, \alpha,\, v)k^{-j/3}\,. \tag{3.120b}$$

As in Sect. 3.2, inserting (3.120a) and (3.120b) in the Maxwell's equations, yields a system of equations similar to the one given in this catalog of Eqs. (3.41) through (3.52) in Sect. 3.2. More precisely, we derive the equation of the caustic by suppressing the terms containing φ and its derivatives in this set of equations.

The components E_0 and H_0 of the electric and magnetic fields that are binormal to the geodesic, satisfy the equations

$$\frac{\partial^2 E_0}{\partial v^2} - \nu E_0 = 0, \qquad (3.121a)$$

$$\frac{\partial^2 H_0}{\partial v^2} - \nu H_0 = 0, \qquad (3.121b)$$

whose solutions are given by

$$E_0(s, \alpha, v) = E(s, \alpha) \, Ai(-v), \qquad (3.122a)$$

$$H_0(s, \alpha, v) = H(s, \alpha) \, Ai(-v) . \qquad (3.122b)$$

As in the 2D case, the boundary condition in Sect. 3.2 is again replaced by a matching condition to the outer solution. This matching aspect will be dealt with later in Sect. 3.14.

3.5 Neighborhood of the light-shadow boundary

Let us now treat the neighborhood of the light-shadow boundary, which is illustrated in Fig. 3.12. In order to simplify the calculations, we will restrict ourselves to the 2-D case.

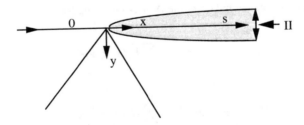

Fig. 3.12. Boundary layer of the shadow zone

Let us consider the boundary layer close to the shadow boundary of a wedge with curved faces and a straight edge illuminated by a scalar plane wave. In this case the boundary surface S of the layer is coincident with the surface constituting the shadow boundary. This is a plane surface passing through the edge O of the wedge (Fig. 3.12). This boundary layer intersects the boundary layer close to the edge and we will denote this as Domain-I. Similarly, we will refer to the domain of the boundary layer outside I as Domain-II.

In Domain-II, we choose the polar coordinates (ρ, φ) with the origin O, the coordinate normal to the plane S being $X_3 = \rho \varphi$ with $\rho = constant$. Then, the stretching transformation takes the form: $\varphi' = k^\alpha (\varphi - \varphi_0)$, where φ_0 is the angular position of the plane S.

For scalar waves, the solution of the diffraction problem satisfies the scalar Helmholtz equation $\nabla^2 u + k^2 u = 0$. Introducing the Sommerfeld and Runge transformation

$$u(\vec{r}) = e^{ikS(\vec{r})} \, v(\vec{r}),$$

(3.123)

the wave equation becomes

$$(1-|\vec{\nabla}S|^2) + \frac{i}{k}\left(\nabla^2 S + 2\vec{\nabla}S \cdot \frac{\vec{\nabla}v}{v}\right) + \frac{1}{k^2}\frac{\nabla^2 v}{v} = 0.$$

(3.124)

Outside the boundary layers I and II of Fig. 3.12, the term $1/k^2 \, (\nabla^2 v/v)$ is small in comparison to the first two terms of (3.124) and can be treated as a perturbation. In the boundary layer II, the third term reads

$$\frac{1}{k^2}\frac{\nabla^2 v}{v} = \frac{1}{k^2}\left(\frac{1}{\rho}\frac{\partial}{\partial\rho} + \frac{\partial^2}{\partial\rho^2} + \frac{k^{2\alpha}}{\rho^2}\frac{\partial^2}{\partial\varphi'^2}\right),$$

(3.125)

when expressed with the stretched coordinates. We note that by setting $\alpha = 1/2$ in the third term in the right hand side of (3.125), we can separate a term of $O(k^{-1})$ as the second term in (3.124). The initial perturbation term is thus decomposed into a term of $O(1/k)$, that must be combined with the terms of the same order in (3.124), and a term of $O(1/k^2)$ which is a regular perturbation term.

The stretching condition employed in the previous problem can also be obtained by starting from the asymptotic expansion of the wedge with planar faces. For this case, we find that in the boundary layer close to the edge the solution behaves as a Fresnel function $F(\sqrt{kL}(\varphi - \varphi_0))$, where L is a distance parameter. This leads also to the stretching condition: $\varphi' = k^{1/2}(\varphi - \varphi_0)$.

Finally, this stretching condition can be derived by using a heuristic criterion for distinguishing between the rays. In the vicinity of the shadow boundary, the phase difference between the geometrical optical ray and the diffracted ray is $k\rho^2(\varphi - \varphi_0)^2$. It becomes large outside a zone defined by $\rho(\varphi - \varphi_0) = O(k^{-1/2})$; hence the stretching condition is again given by $\varphi' = k^{1/2}(\varphi - \varphi_0)$.

Our next task is to seek an Ansatz in the vicinity of the shadow boundary. Let us choose, for instance, the phase of the diffracted ray, for the phase of the Ansatz: $S = \rho$, i.e., we let

$$u\,(\rho, \varphi) = e^{ik\rho}\, v(\rho, \varphi').$$

(3.126)

The next step is to insert (3.126) into (3.124), and order the resulting expression in accordance with the powers of k. For the highest order in k, we obtain a parabolic type of equation which reads

$$i\frac{v}{\rho} + 2i\frac{\partial v}{\partial \rho} + \frac{1}{\rho^2}\frac{\partial^2 v}{\partial \varphi'^2} = 0.$$

(3.127)

We then seek a particular solution of this equation in the form $v(\sqrt{\rho}\,\varphi')$, and let $\xi = \sqrt{\rho}\,\varphi'$, to get

$$iv + i\xi v' + v'' = 0.$$

(3.128)

A particular solution of this equation is

$$v = \exp(-i\xi^2/2)\, F\!\left(\frac{\xi}{\sqrt{2}}\right),$$

(3.129)

where F is the Fresnel function.

More generally, let us consider a shadow boundary due to a wedge or a smooth object. We choose the Cartesian coordinates to describe the shadow boundary, with the x axis directed along this boundary and the y axis being perpendicular to it. Let us stretch the coordinate y with a factor $k^{1/2}$ and choose as an Ansatz

$$u = e^{ikx}\, v(x, Y),$$

(3.130)

where $Y = k^{1/2}\, y$.

Next we insert (3.130) into (3.124), and order the terms in the resulting equation with respect to the powers of k, as we have done in the past. The highest order in k yields the equation

$$2i\,\frac{\partial v}{\partial x} + \frac{\partial^2 v}{\partial Y^2} = 0.$$

(3.131)

Which is again a parabolic type and can be solved in the same manner as a diffusion equation [Za1]. As a particular solution, it admits a function of the type

$$F\left(\frac{Y}{\sqrt{2x}}\right). \tag{3.132}$$

The above discussion is helpful in understanding, at least partially, the role of the Fresnel function in describing the field in the shadow zone. However, the solution to our problem is not complete until we enforce the boundary conditions that enable us to match this solution to the field in the outer region. Thus, at this stage we have only constructed one of several possible solutions, and to determine the desired solution uniquely, it is necessary to match the solution to the incident plane wave in the shadow zone, for large negative values of the argument, and to the diffracted field for large positive values of the same. The solution given in (3.132) tends toward a constant when $Y \to \infty$, and behaves as $1/\sqrt{Y}$ when $Y \to -\infty$. Thus, the solution assures the progressive disappearance of the incident field as we move deeper into the shadow zone where only the diffracted field, which decreases as $1/\sqrt{Y}$, is present. For a more detailed treatment of the shadow zone of the half-plane, we refer the reader to the work by Zauderer [Za2].

Let us now compare the two solutions (3.129) and (3.132). We will now show that these two solutions are in fact equivalent to each other for the case of the wedge. To this end, we express (3.129) in the Cartesian coordinate system, and write ρ as

$$\rho = x + \frac{1}{2}\frac{y^2}{x} + 0\left(\frac{y^4}{x^4}\right). \tag{3.133}$$

For $y = O(k^{-1/2})$

$$\rho = x + \frac{1}{2}\frac{y^2}{x} + O(k^{-2}). \tag{3.134}$$

On the other hand,

$$\xi = \sqrt{\rho}\,\varphi' = \frac{y}{\sqrt{x}} + O(k^{-3/2}), \tag{3.135}$$

so that

$$e^{ik\rho}\,e^{-i\xi^2/2}\,F\left(\frac{\xi}{\sqrt{2}}\right) = e^{ikx}\,F\left(\frac{Y}{\sqrt{2x}}\right) + O(k^{-1}). \tag{3.136}$$

From (3.136), we conclude that the two solutions are equal to each other to within $O(k^{-1})$. Instead of using the Cartesian coordinates, we could have used the optimal coordinates, i.e., those along the rays. For the case of the wedge, these coordinates are: S, one-half the sum of the eikonals of the two rays, viz., geometrical optics and diffracted; and δ, one-half the difference of these eikonals, called the detour parameter in the literature. For smooth objects, the optimal coordinates are those defined in Sect. 3.1.8. One can verify that the use of these coordinates in the boundary layer, leads to results equivalent to those just presented above in (3.129) through (3.136). In Chap. 5, we will again find a description of the field close to the shadow boundary of a wedge in which the Fresnel function and the detour parameter will appear.

In the following section, we will discuss the boundary layer in the vicinity of an edge of wedge with curved faces.

3.6 Boundary layer in the neighborhood of an edge of curved wedge (Fig. 3.13)

Let us first, as in the previous sections, determine the order of coordinate stretching in the boundary layer. In the neighborhood of an edge of a wedge, the rays diffract in all directions and the phase difference between the incident ray and the diffracted rays is $O(kr)$, where r denotes the distance to the tip of the wedge, excepting outside the neighborhood of the light-shadow boundary, where they are nearly the same. This phase difference will be of $O(1)$ in a neighborhood $O(1/k)$ of the tip. We can then expect a boundary layer of thickness $1/k$ and a stretching factor k of the coordinates. The above assertion can be validated by considering the canonical problem of the plane wedge.

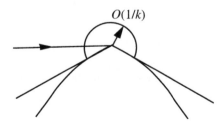

$O(1/k)$

Fig. 3.13. Diffraction by a curved wedge

Finally, the stretching factor k is natural in this problem of singular perturbation as the perturbation term ∇^2/k^2 becomes $O(1)$ in the boundary layer whose thickness is $1/k$. Thus, the three methods lead to the same stretching factor k. In this boundary layer, the Helmholtz equation becomes

$$(\nabla^2 + 1)\, u = 0. \tag{3.137}$$

Let us place the origin of the coordinates at the tip of the wedge. The equation of the face on the positive side $x > 0$ can be written as

$$y = f(x). \tag{3.138}$$

Using the stretched coordinates, viz., $X = kx$, $Y = ky$, it becomes

$$Y = f'(0)X + O(k^{-1}). \tag{3.139}$$

The boundary condition, which, for the sake of simplicity, is assumed to be of the Dirichlet type on the face of the wedge, takes the form

$$u(X, f'(0)X) + O(k^{-1}) = 0. \tag{3.140}$$

The first-order terms yield the result

$$u(X, f'(0)X) = 0, \quad \text{for} \quad X > 0. \tag{3.141}$$

Equation (3.141) can be interpreted to mean that we impose the boundary condition on the plane tangent to the face. This argument also holds true for the other face; hence to first order, the boundary conditions apply to the local tangent wedge.

The previous discussion can be easily extended to Maxwell's equations. Thus, we have justified the approximation that consists of replacing the edge by its local tangent wedge at each point. This approximation is only valid for the highest order in k. Borovikov [Bor] has calculated the next term in the expansion for the perfectly conducting case, and has thus derived the term that shows the effect of the curvature of the faces. For the special case in which the angle of the wedge equals π, the first term vanishes, the diffraction coefficient is of $O(k^{-1})$, and reduces to the diffraction coefficient of the curvature discontinuity given in Chap. 1.

Finally, Bernard [Be] has extended Borovikov's work to the study of the wedge described by an impedance condition. Bernard's solution also covers the case of the discontinuity in the curvature of a surface which is described by an impedance condition.

To conclude this section, we observe that we recover the same type of results as those derived in Chap. 2 for reflection from a curved surface, viz., that the GTD, in common with the geometrical optics, yields the first term of a perturbation series. All of the boundary layers treated in the previous sections were *simple* in the sense that there was no overlap of the transition zones. In the next section, we will study the propagation of creeping rays in the neighborhood of the contact point on a smooth surface.

3.7 Neighborhood of the contact point of a creeping ray on a smooth surface

In the neighborhood of the contact point of a creeping ray on a smooth surface, there is an overlap of two transition zones, the first is the neighborhood of the surface of the object, while the second is the light-shadow boundary, which was treated in Sect. 3.5. To discuss this case, it would be useful to consider the example of the cylinder shown in Fig. 3.14.

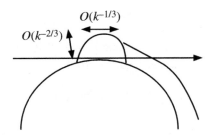

Fig. 3.14. Boundary layer of the point of contact

As in Sect. 3.1, once again we use the (s, n) coordinates to treat this case. The stretching factors, suggested by the canonical problem of the circular cylinder, are $k^{1/3}$ for s and $k^{2/3}$ for n. This result can also be derived by using the equalization method of the weights of the dominant terms. After stretching, the dominant terms turn out to be the following

$$2ik\frac{\partial u}{\partial s}, \, 2\, \frac{n}{\rho}\, k^2\, u, \frac{\partial^2 u}{\partial n^2}. \tag{3.142}$$

In order that all of the terms appearing in (3.142) above be of the same order, we stretch the coordinate s with the factor k^{α}, and the coordinate n with k^{β}. This leads to the following relation ship between α and β

$$1 + \alpha = 2 - \beta = 2\beta.$$

From the above it immediately follows that $\alpha = 1/3, \beta = 2/3$, which is the same result we obtained previously.

Finally, we can apply the criterion on the phase difference between the rays. sing the (s, n) coordinates, and the techniques of differential geometry, we can show that in the vicinity of C, the phase difference between the direct ray of optics and the creeping ray is given by

$$k\left(-\frac{s^3}{6\rho^2} + \frac{ns}{\rho} - \frac{2}{3}n\sqrt{\frac{2n}{\rho}}\right). \tag{3.143}$$

In order that all of these terms be $O(1)$, it is necessary that s be $O(k^{-1/3})$ and $n = O(k^{-2/3})$, respectively. This leads again to the stretching of s with the factor $k^{1/3}$, and n with the factor $k^{2/3}$.

We should point out that the present problem is a little different from the one we solved earlier in the shadow zone. Although the radiation condition is still imposed by requiring that $lim_{,r\to+\infty}\ u(r) = 0$ if the medium is lossy, it is applied now to the diffracted field denoted by u^d, and not to the total field as was done in the shadow zone.

We seek a solution for the diffracted field u^d that satisfies the following criteria:

(i)   the wave equation: $(\nabla^2 + k^2)\, u^d = 0$

(ii)   the *radiation condition* $lim_{,r\to+\infty}\ u^d(r) = 0$ if the medium is lossy

(iii) $(\partial(u^d + u^{inc})/\ \partial n) + ikZ(u^d + u^{inc}) = 0$ at the surface of the cylinder, where u^{inc} denotes the incident plane wave with unit amplitude i.e., $u^{inc} = e^{ikx}$.

Using the (s, n) coordinates, the field u_d may be written in the form suggested by [BK]

$$u^d = e^{iks} \sum_{j=0}^{M} u_j^d(\sigma, v)\, k^{-j/3}, \tag{3.144}$$

where v is the same coordinate as we employed in our analysis for the shadow zone, and it is given by

$$v = \left(\frac{2k^2}{\rho}\right)^{1/3} n = 2m^2\, \frac{n}{\rho},$$

where $\rho = \rho(s = 0)$, and $\sigma = m\ (s/\rho)$.

As in the case of the shadow zone, the insertion of the previous expansion in the wave equation provides the following set of recursion relationships for the sequence u_j^d

$$L_0 u_0^d = 0,$$

(3.145)

$$L_0 u_1^d + L_1 u_0^d = 0,$$

(3.146)

$$L_0 u_j^d \ldots + L_j u_0^d = 0.$$

(3.147)

The operators L_j are different from those in the shadow zone.

In particular, for the zero order, Eq. (3.145) leads to Fock's parabolic equation

$$\frac{\partial^2 u_0^d}{\partial v^2} + v u_0^d + i \, \frac{\partial u_0^d}{\partial \sigma} = 0.$$

(3.148)

If we Fourier transform $\sigma \to \xi$ this equation, with respect to σ we obtain once again just as in the shadow zone, the Airy equation in the transformed variable ξ satisfied by $\tilde{u}_0^d$

$$\frac{\partial^2 \tilde{u}_0^d}{\partial v^2} + (v - \xi) \, \tilde{u}_0^d = 0.$$

(3.149)

The above result was somewhat anticipated, since the Fourier transformation equivalently represents the solution as a superposition of exponential functions. Because we have represented the solution in the shadow zone in the form of an exponential, it is not surprising to find that the equation in the present case is similar to the one in the shadow zone. Of course, it is necessary to verify that the functions in question are indeed Fourier transformable, although this usually does not present a problem.

If, as in the case of the shadow zone, we now impose the condition that the solution $\tilde{u}_0^d$ satisfies the radiation condition, we are led to the expression

$$\tilde{u}_0^d = \alpha(\xi) \, w_1(\xi - v),$$

(3.150)

where $\alpha(\xi)$ is a function yet to be determined.

It remains now to express the boundary condition on the cylinder which, for $mZ = 0$, can be written in the form

$$\left(\frac{\partial u^d}{\partial v}\right) + imZu^d = -\left[\frac{\partial u^{inc}}{\partial v} + imZu^{inc}\right],$$ (3.151)

as in the case of the shadow zone. After inserting the following expansion

$$u^{inc} = e^{iks} \sum \frac{u_j^{inc}}{k^{j/3}},$$ (3.152)

in Eq. (3.151), we are led to uncoupled boundary conditions on u_j^d

$$\left(\frac{\partial u_j^d}{\partial v}\right) + imZu_j^d = -\left[\frac{\partial u_j^{inc}}{\partial v} + imZu_j^{inc}\right].$$ (3.153)

We now consider the particular case of $j = 0$ and we perform the Fourier transformation $\sigma \rightarrow \xi$ on the boundary condition to get

$$\frac{\partial \tilde{u}_0^d}{\partial v} + imZ\tilde{u}_0^d = -\left[\frac{\partial \tilde{u}_0^{inc}}{\partial v} + imZ\tilde{u}_0^{inc}\right].$$ (3.154)

One has now to calculate $\tilde{u}_0^{inc}$. This is achieved by expressing kx in the (s, n) coordinate system, and restricting ourselves to the terms of order $O(1)$, since higher order terms, i.e., those having $k^{-j/3}$, contribute only to the u_0^{inc} terms with $j \geq 1$.

Since the coordinates s and n are of $O(k^{-1/3})$ and $O(k^{-2/3})$, respectively, we retain only the terms of the type s, s^2, s^3 and sn in the expansion of x with respect to the powers of s, n; which lead to terms of the type k^α, where $\alpha \geq 0$. After performing a Taylor expansion (see Appendix 2) we find the expression

$$x = s - \frac{s^3}{6\rho^2} + n\frac{s}{\rho} + O(s^4, n^2, n\, s^2).$$ (3.155)

Using the coordinate system (σ, v), we have

$$kx = ks + i\,(\sigma v - \sigma^3/3 + O(k^{-1/3})).$$ (3.156)

This result can also be derived by expressing the eikonal equation in the (σ, v) coordinate system (see [BK]).

Using (3.156) we can write

$$u_0^{inc} = e^{i(\sigma v - \sigma^3/3)} . \tag{3.157}$$

It is demonstrated in [BK] that the result is valid not only for a plane wave, but also for any incident ray field whose phase gradient is tangent to the diffracting object at point C. Fourier transforming the expression in (3.157) and introducing the notation $v = \sqrt{\pi}\ Ai$ that was also employed in [BK], we get

$$\tilde{u}_0^{inc} = \frac{1}{\sqrt{\pi}}\ v(\xi - v). \tag{3.158}$$

The application of the boundary condition on the cylinder leads to the equation for α that we were seeking. It reads

$$\alpha = -\frac{1}{\sqrt{\pi}}\ \frac{v'(\xi) - imZv(\xi)}{w_1'(\xi) - imZw_1(\xi)} . \tag{3.159}$$

Finally, using the results derived above, we obtain the following expression of the total field u_0

$$u_0(\sigma, v) = u_0^{inc} + u_0^d = \frac{1}{\sqrt{\pi}} \int_{-\infty}^{+\infty} \left(v(\xi - v) + \frac{v'(\xi) - imZv(\xi)}{w_1'(\xi) - imZw_1(\xi)} w_1(\xi - v) \right) e^{i\sigma\xi} , \tag{3.160}$$

which is the well known Fock's *universal function* $V(\sigma, v, mZ)$. The convergence of the integral in (3.160) has been studied by Logan [Lo].

From (3.160), the value of the field on the surface of the cylinder may be written

$$u_0(\sigma, 0) = \frac{1}{\sqrt{\pi}} \int_{-\infty}^{+\infty} \frac{v(\xi)w_1'(\xi) - v'(\xi)w_1(\xi)}{w_1'(\xi) - imZw_1(\xi)} e^{i\sigma\xi} d\xi , \tag{3.161a}$$

$$u_0(\sigma, 0) = \frac{1}{\sqrt{\pi}} \int_{-\infty}^{+\infty} \frac{1}{w_1'(\xi) - imZw_1(\xi)} e^{i\sigma\xi} d\xi , \tag{3.161b}$$

since the Wronskian $W(v, w_1)$ appearing in (3.161a) is equal to 1.

We recognize in the expression (3.161b) for u the Fock current function $F_Z(\sigma)$. Thus the surface field is given by

$$u \approx e^{iks} F_Z(\sigma),$$

to the first order.

Having completed the discussion of the derivation of the expression for u in the neighborhood of the shadow boundary of the surface, we proceed next, in the following section, to the more complicated case of three dimensions involving Maxwell's equations.

3.8 Calculation of the field in the neighborhood of
a point of the shadow boundary

3.8.1 Calculation of the fields in the vicinity of the surface

Let us consider a plane wave with unit amplitude, propagating along the x-axis and impinging upon a 3-D object (see Fig. 3.15). The incident field may be written as $\vec{E}^{inc} = \exp(ikx)\,\hat{y}$.

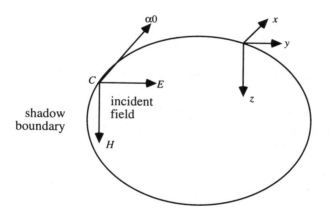

Fig. 3.15. Obstacle as seen from the direction of incidence

In the vicinity of point C of the shadow boundary, we introduce the coordinate system (s, α, n) and the stretched coordinate system (σ, α, ν) as employed in Sect. 3.2.

We seek representations of the diffracted fields $\vec{E}$ and $\vec{H}$ satisfying the Maxwell's equations, the impedance condition $(\vec{E}_{tg} - Z\,\vec{n} \times \vec{H}_{tg}) = -(\vec{H}_{tg}^{inc} - Z\,\vec{n} \times \vec{H}_{tg}^{inc})$, and the Silver-Müller radiation condition given in Sect. 3.2.2.

Our first step is to express $\vec{E}$ and $\vec{H}$ in the form

$$\vec{E} = \exp(iks)\, (\vec{E}_0\,(\sigma,\,\alpha,\,\nu) + k^{-1/3}\,\vec{E}_1\,(\sigma,\,\alpha,\,\nu) + ...), \tag{3.162}$$

and

$$\vec{H} = \exp(iks)\, (\vec{H}_0\,(\sigma,\,\alpha,\,\nu) + k^{-1/3}\,\vec{H}_1\,(\sigma,\,\alpha,\,\nu) + ...). \tag{3.163}$$

The insertion of (3.162) and (3.163) into the Maxwell's equations shows, as it did in Sect. 3.2, that it is possible to express all of the components of the fields in terms of the α components alone. Specifically, we obtain the Eqs. (3.41) through (3.43) for $O(k)$, and (3.44) through (3.46) of Sect. 3.2 for $O(k^{2/3})$, since the term $k^{1/3}$ of the phase, defined in Sect. 3.2, does not enter into these equations.

The compatibility conditions (3.55) and (3.56) of Sect. 3.2 are now replaced by two *parabolic* equations

$$\frac{\partial E_0}{\partial \nu^2} + \nu E_0 + i\,\frac{\partial E_0}{\partial \sigma} = 0, \tag{3.164}$$

and

$$\frac{\partial H_0}{\partial \nu^2} + \nu H_0 + i\,\frac{\partial H_0}{\partial \sigma} = 0. \tag{3.165}$$

E_0 and H_0 are the components along the normalized vector $\hat{\alpha}(s,\,\alpha) = \vec{e}_\alpha /\! \parallel \vec{e}_\alpha \parallel$, of $\vec{E}_0$ and $\vec{H}_0$.

Since the incident wave obviously satisfies the Maxwell's wave equations, we can express all the components of the incident wave as functions of the components along $\hat{\alpha}$ alone. These are given by

$$\vec{E}^{inc} \cdot \hat{\alpha} = (\hat{y} \cdot \hat{\alpha})\, \exp(ikx), \tag{3.166}$$

$$\vec{H}^{inc} \cdot \hat{\alpha} = (\hat{y} \cdot \hat{\alpha})\, \exp(ikx). \tag{3.167}$$

Next, we only look for the dominant terms, viz., E_0^{inc} and H_0^{inc} of the above components, and we work in the neighborhood of $O(k^{-1/3})$ of the shadow boundary, where we write

$$\hat{\alpha} = \hat{\alpha}_0 + O(k^{-1/3}). \tag{3.168}$$

In view of (3.168), we can then replace $\hat{\alpha}$ by $\hat{\alpha}_0$ in (3.166) and (3.167) without modifying the dominant terms of the incident field.

It remains to express x in terms of the coordinate system (s, α, n). To this end, we construct a Taylor expansion in terms of the powers of (s, α, n). Using the principles of differential geometry we can derive

$$x = s + (s + d(\alpha)) \, \frac{n}{\rho} - \frac{(s + d(\alpha))^3}{6\rho^2} + O(n^2, s^2 n, s^4), \qquad (3.169)$$

where $d(\alpha)$ is the distance between the shadow boundary C on the object and the α axes (see Appendix 2). Let us introduce the coordinate $s' = s + d(\alpha)$, which is tantamount to assuming that the origin of s' is the shadow boundary. With this choice, the phase term kx takes the form

$$kx = ks + ks' \, \frac{n}{\rho} - k \frac{s'^3}{6\rho^2} + kO(n^2, s'^2 n, s'^4). \qquad (3.170)$$

Taking the stretching of the coordinates into account, we find that the term $kO(n^2, s'^2 n, s'^4)$ is $O(k^{-1/3})$ and, consequently, it can be neglected in the calculation of E_0^{inc} and H_0^{inc}.

Inserting (3.168) and (3.170) in (3.166) and (3.167), and introducing, as in the case of the cylinder, the stretched coordinates $\sigma = m \; (s'/\rho)$ and ν, we obtain

$$\vec{E}^{inc} \cdot \hat{\alpha} = exp(iks) \, (\hat{y} \cdot \hat{\alpha}_0) \; exp \, i(\sigma\nu - \sigma^3/3) + O(k^{-1/3}), \qquad (3.171)$$

$$\vec{H}^{inc} \cdot \hat{\alpha} = exp(iks) \, (\hat{z} \cdot \hat{\alpha}_0) \; exp \, i(\sigma\nu - \sigma^3/3) + O(k^{-1/3}), \qquad (3.172)$$

and

$$E_0^{inc} \, (\sigma, \nu) = (\hat{y} \cdot \hat{\alpha}_0) \; exp \, i(\sigma\nu - \sigma^3/3), \qquad (3.173)$$

$$H_0^{inc} \, (\sigma, \nu) = (\hat{z} \cdot \hat{\alpha}_0) \; exp \, i(\sigma\nu - \sigma^3/3). \qquad (3.174)$$

Note that E_0^{inc} and H_0^{inc} are functions of $\sigma = (s + d(\alpha))m/\rho$, and this remark applies to the diffracted fields also. This result is expected since $s + d(\alpha)$ is the curvilinear abscissa as measured from the shadow boundary, and, consequently, it is the pertinent quantity to use to calculate the attenuation due to the propagation on the object (see Appendix 2).

The diffracted fields are calculated by following the same approach as employed for the case of the cylinder, which was discussed in the previous section. Noting that E_0^{inc} and H_0^{inc} have the same form as u_0^{inc} (see (3.157) in Sect. 3.7), we write the following expressions for the total field along $\hat{\alpha}$

$$H_0 = (\hat{y} \cdot \hat{\alpha}_0) \ V(\sigma, \nu, mZ), \tag{3.175}$$

and

$$E_0 = (\hat{z} \cdot \hat{\alpha}_0) \ V(\sigma, \nu, m/Z). \tag{3.176}$$

Our next task is to calculate the surface fields using the results we derived above. We will use these results in Chap. 7, where we will need to know the surface field.

3.8.2 Surface field

We turn our attention in this section to the computation of the surface field on the surface of the object, i.e., for $\nu = 0$. The Eqs. (3.175) and (3.176), which give the total field along $\hat{\alpha}$, yield

$$E_0 = e^{iks} \ (\hat{y} \cdot \hat{\alpha}_0) \ F_{1/Z}(\sigma), \tag{3.177}$$

and

$$H_0 = e^{iks} \ (\hat{z} \cdot \hat{\alpha}_0) \ F_Z(\sigma), \tag{3.178}$$

where $F_{1/Z}$ and F_Z are the Fock functions. (See Appendix 5.)

Let us now calculate the total field along $\hat{s}$. The fields E_0^s and H_0^s vanish along $\hat{s}$, and, as in Sect. 3.2, the terms of highest order entering into the expression of the fields E^s and H^s are E_1^s and H_1^s. For the dominant order, E^s and H^s are equal, respectively, to $-ZH_0^t$ for the magnetic creeping ray and $(1/Z)E_0^t$ for the electric creeping ray. The physical interpretation of these fields is the same as given in Sect. 3.2.

For the dominant order terms, the fields tangent to the surface are then given by

$$\vec{E} = e^{iks} \{ (\hat{y} \cdot \hat{\alpha}_0) \ F_{1/z}(\sigma) \ \hat{\alpha} - Z(\hat{z} \cdot \hat{\alpha}_0) \ F_z(\sigma) \ \hat{s} \}, \tag{3.179}$$

$$\vec{H} = e^{iks} \{ (\hat{z} \cdot \hat{\alpha}_0) \ F_z(\sigma) \ \hat{\alpha} + \frac{1}{Z} \ (\hat{y} \cdot \hat{\alpha}_0) \ F_{1/z}(\sigma) \ \hat{s} \}. \tag{3.180}$$

The case $Z = 0$ is singular because (3.179) reduces then to $\vec{E} = 0$. It becomes necessary, as in Sect. 3.2.7.4, to calculate the magnetic field on the surface, whose dominant component is of $O(k^{-1/3})$.

The dominant component of the magnetic field along s induced by E_0^{inc} is given by

$$H^s = \exp(iks) \, \frac{i}{m} \, \frac{\partial E_0}{\partial v} . \tag{3.181}$$

Substituting E_0 into (3.181) we obtain the following expression for the surface field $H^s(s, \sigma)$

$$H^s(s, \sigma) = \exp(iks) \, \frac{i}{m} \, (\hat{y} \cdot \hat{\alpha}_0) \, \frac{1}{\sqrt{\pi}} \int_{-\infty}^{+\infty} \frac{e^{i\sigma\xi}}{w_1(\xi)} \, d\xi \tag{3.182}$$

Recognizing the Fock electric function $f(\xi)$ in (3.182) (see Appendix 5), we can write

$$H^s(s, \sigma) = \exp(iks) \, \frac{i}{m} \, (\hat{y} \cdot \hat{\alpha}_0) \, f(\sigma). \tag{3.183}$$

The dominant component of the magnetic field along α can be obtained from an equation similar to (3.93) in Sect. 3.2.7.1, and it is given by

$$\left(\frac{2}{\rho}\right)^{2/3} L_0 \tilde{H}_1 = 2i \, \tau \tilde{E}_0 , \tag{3.184}$$

where ~ denotes the Fourier transform, and the value of H^{inc}.

The procedure that we follow is similar to that of Sect. 3.2.7.1, and for the dominant component of the magnetic surface field generated by E_0^{inc} we get

$$\vec{H}(s, \sigma) = \exp(iks) \, (\hat{y} \cdot \hat{\alpha}_0) \, [-\tau\rho\hat{\alpha} + \hat{s}] \, \frac{i}{m} f(\sigma), \tag{3.185}$$

to which we must obviously add the field $\exp(iks) \, (\hat{z} \cdot \hat{\alpha}_0) \, g(\sigma) \, \hat{\alpha}$, generated by H_0^{inc}, to complete the solution. Examining it we find that the boundary layer method recovers the Fock solutions [F], viz., that the surface magnetic field on a perfect conductor in the vicinity of the boundary is the sum of two terms: (i) magnetic part described by the function g; and (ii) an electric part described by the function f. We will see in Sect. 3.11 how this Fock field generates the creeping rays described in Sect. 3.2.

We have discussed the case of boundary layers for a scatterer with concave and convex parts of a surface that support whispering gallery modes and creeping ray propagation, respectively. In the next section, we will go on to follow Popov's works [Po] to study the neighborhood of an inflection point.

3.9 Whispering gallery modes incident upon an inflection point (Fig. 3.16)

In this section we consider the case of a whispering gallery mode incident upon a point of inflection. Unfortunately, unlike in the previous section, we do not have access to a canonical problem that would suggest the correct stretching of the coordinates for this case. However, the criterion enunciated in Chap. 2 still is applicable and we can proceed, as in Sect. 3.8, to introduce an Ansatz and represent u as

$$u = \exp(iks) \sum_{j=0}^{M} u_j(s, n)$$

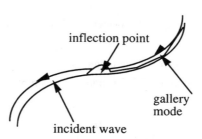

Fig. 3.16. Region of the point of inflection

For the dominant terms with respect to k, we obtain

$$2ik\frac{\partial u}{\partial s}, \; 2ns\,c\;k^2 u, \; \frac{\partial^2 u}{\partial n^2}, \tag{3.186}$$

where c is the derivative of the curvature $K(s)$, i.e.,

$$K(s) = -cs + O(s^2), \tag{3.187}$$

with $c > 0$. This implies that the whispering gallery modes correspond to $s < 0$ while the creeping wave modes correspond to $s > 0$.

Let us stretch the coordinate s with the factor k^a and the coordinate n with k^b. We then obtain the following orders for the different terms of (3.186)

$$k^{1+a}, \; k^{2-a-b}, \; k^{2b}, \tag{3.188}$$

which lead us to the relationship between a and b given below, viz.,

$$1 + a = 2 - a - b = 2b. \qquad (3.189)$$

From (3.189), we get

$$a = \frac{1}{5}, b = \frac{3}{5}. \qquad (3.190)$$

Next, following Popov [Po], we define the dimensionless or scaled coordinates via the following equations

$$\sigma = (k\,b^2)^{1/5}\,s, \qquad (3.191a)$$

$$\nu = (k^3\,c)^{1/5}\,n, \qquad (3.191b)$$

and obtain the following parabolic equation for the principal term u_0

$$2\,k^{6/5}\,c^{2/5}\left(i\,\frac{\partial u_0}{\partial \sigma} + \sigma\nu\,u_0 + \frac{1}{2}\,\frac{\partial^2 u_0}{\partial \nu^2}\right) = 0. \qquad (3.192)$$

To determine u_0 we must complement (3.192) with the following boundary conditions:

(i)   matching to the whispering gallery modes for $\sigma \to -\infty$

(ii) matching to the creeping wave modes for $\sigma \to +\infty$

(iii) enforcing the boundary condition $u_0(\sigma, 0) = 0$ for the perfectly conducting case and TM waves .

The solution to (3.192) has been studied by Popov [Po] from the points of view of existence and uniqueness, and has also presented numerical calculations for a few cases.

In conclusion, the previous example illustrates the flexibility and the versatility of the boundary layer method which does not depend upon the availability of associated canonical problems, and seeks instead a direct solution to the problem.

All the solutions presented in the previous sections, except for those given in Sects. 3.7 and 3.8, are particular solutions, satisfying the Maxwell's equations in certain regions of space, the absorption principle, and the boundary conditions on the object. In general, these solutions include constants that are determined by using the matching principles to be elaborated on in the next section.

3.10 Matching principle

In this section, we address the remaining but important problem of matching the solutions we have derived in the last two sections for the regions in the interior and exterior of the boundary layer. Although several different matching principles have been enunciated by different authors, to our knowledge none has been demonstrated rigorously. The underlying concept of the matching procedure is as follows. The coordinate normal to the boundary layer is a stretched coordinate. Typically there exists a zone in which this stretched coordinate is large, while simultaneously, the conventional coordinate is small. In this zone, it is possible to use the asymptotic expansion for large values of the stretched coordinate for the inner expansion, and an asymptotic expansion of the outer solution valid for small values of the conventional coordinate. It is natural to assume that these two expansions will have a region of overlap; however, this is not always true.

In addition, the above idea is somewhat vague as it does not clearly define what is meant by *small* or *large*. A very simple approach to implementing the matching procedure has been proposed by Van Dyke [VD] and it entails the following steps:

(i)   Calculation of the asymptotic expansion of the inner expansion for large values of the stretched coordinate up to the order m.   The result thus obtained is referred to as the *outer limit of the inner solution.*

(ii)   Calculation of the asymptotic expansion of the outer expansion for small arguments of the conventional coordinate, followed by the replacement of the conventional coordinate by its representation in terms of the stretched coordinate, and truncation of the resulting expansion to the order m, that is referred to as the *inner limit of the outer solution.*

(iii) Matching of the inner and outer solutions thus obtained.

In the Van Dyke approach, the inner and outer expressions are calculated by using the stretched coordinates. Evidently, it is possible to express both of the above quantities in the conventional coordinates as well.

Yet another matching principle, which is in fact more general in nature, has been proposed by Lagerstrom and Kaplun [KL]. We recall that the matching procedure proposed by Van Dyke is applicable only when an overlapping zone exists in which the inner and outer expansions are simultaneously valid. It becomes necessary to resort to the Lagerstrom and Kaplun approach when a gap exists between the above two expressions.

However, more often than not, the Van Dyke approach is found to be adequate, although we must obviously verify that there exists a domain of validity that is common to both expansions. To illustrate the various possibilities, we will sometimes work with the (s, n) coordinates, and other times with the ray coordinates. In all of the examples, we will assume that the incident field is a plane wave propagating along the x-axis. For the sake of simplicity, we will initially present the two-dimensional case in detail, and then provide some indication as to how the three-dimensional case can be handled.

In the two-dimensional case, we will find it convenient to choose the origin whose neighborhood is of interest to us, viz., the contact point of the object and the grazing ray in Sects. 3.11, 3.12 and 3.15, and the edge of the wedge in Sect. 3.14. The incident field is normalized to unity at the origin.

3.11 Matching the solution expressed in the form of a creeping ray at the contact point

To illustrate the matching of the creeping ray solution at the contact point, we will restrict ourselves to the 2-D case and the terms that are $O(1)$.

In the vicinity of the contact point, let the stretched variable σ tends toward $+_-$ in the inner expansion, i.e., in the solution given in (3.160) of Sect. 3.7. By using the properties of the Fock functions given in a paper by Logan [Lo], we can write the inner expression as

$$u \approx e^{iks}\, V(\sigma, \nu, Z) \approx e^{iks} \sum_{p} \frac{2i\sqrt{\pi}}{(\xi^p + m^2 Z^2) w_1^2(\xi^p)}\, exp(i\xi^p \sigma)\, w_1(\xi^p - \nu).\text{(3.193)}$$

Next, we consider the outer expansion given in (3.36) in Sect. 3.1.8, which reads

$$u \approx \sum A^p(0)\left(\frac{\rho(0)}{\rho(s)}\right)^{1/6}\left(\frac{d(0)}{d(s)}\right)^{1/2} exp\left(iks + i\frac{k^{1/3}}{2^{1/3}}\int_0^s \frac{\xi^p(s)}{\rho(s)^{2/3}}\, ds\right) w_1(\xi_p - \nu),$$

$$(3.194)$$

where $A^p(0)$ is as yet undetermined. For s small, (3.194) becomes

$$u \approx \sum A^p(0)\, e^{iks}\, exp\left(i\frac{k^{1/3}\xi^p(0)s}{2^2 \rho^{2/3}(0)}\right),$$

$$(3.195)$$

which can be rewritten in terms of the inner variable σ as

$$u \approx \sum A^p(0)\, e^{iks}\, exp(i\xi\sigma).$$

$$(3.196)$$

A comparison of (3.193), the outer limit of the inner expansion, with (3.196), the inner limit of the outer expansion, yields the following expression for the coefficients $A^p(0)$

$$A^p(0) = \frac{2i\sqrt{\pi}}{(\xi^p + m^2 Z^2) w_1^2(\xi^p)} = -\frac{2i\sqrt{\pi}}{d^p(0)}. \tag{3.197}$$

All the quantities in (3.197) are calculated at the contact point with curvilinear abscissa equal to 0.

We have illustrated above how a direct application of the matching procedure enables us to extract the constants $A^p(0)$. At this point some remarks on this procedure are in order. We note, first of all, that (3.193) is valid as long as σ is *large*, i.e., $\sigma = O(k^\varepsilon)$ with $\varepsilon > 0$. We also observe that (3.195) represents the first term of the expansion of (3.194) in terms of the powers of s, which is assumed to be small. Since the terms $O(s)$ and $O(k^{1/3} s^2)$ have been assumed to be small and have been neglected in (3.195), it is valid only if

$$s = O(k^{-\eta}) \text{ and } s = O(k^{-1/6-\eta'}) \text{ with } \eta \text{ and } \eta' > 0,$$

the latter being the most restrictive condition.

The range of σ for which both (3.193) and (3.195) are simultaneously valid is given by

$$O(k^\varepsilon) < \sigma < O(k^{1/6-\eta'}),$$

which requires that the condition $\varepsilon + \eta' < 1/6$ be satisfied.

We might recall that the results of Sect. 3.7 were valid not only for a plane wave but also for a ray field whose gradient of the eikonal is tangent to the object, provided that we restrict ourselves to retaining only the dominant order term. It is useful to point out that the same remark applies to the results presented in this section as well.

Finally, the extension of the present matching procedure to the 3-D case is straightforward. Both the electric and magnetic creeping wave modes are excited in this case, and the matching enables us to calculate their amplitudes and to show that the creeping rays begin to attenuate as they propagate away from the shadow boundary. Explicit formulas similar to (3.197) will be given in the next section for the 3-D case.

3.12 Matching of the solution in the boundary layer in the vicinity of the surface to the solution in the form of creeping rays and determination of the solution in the shadow zone

The matching of the solution in the boundary layer can be carried out by using the coordinates (s, n), and the procedure prescribed by Babich and Kirpicnikova [BK]. However, it is more convenient to use the ray coordinate and to start with formula (3.33) of Sect. 3.1 to describe the inner solution. Using the asymptotic expansion of w_1 for large negative values of the argument in (3.33) we get

$$w_1(\xi - Y) \approx Y^{-1/4} \exp\left(i\frac{2}{3}Y^{3/2} - i\xi\sqrt{Y} + i\frac{\pi}{4}\right), \qquad (3.198)$$

$$u \approx A \left(\frac{\rho(0)}{\rho(s)}\right)^{1/6}\left(\frac{d(0)}{d(s)}\right)^{1/2} Y^{-1/4} \exp\left(i\frac{\pi}{4}\right)\exp\left(ik\ell + i\frac{k^{1/3}}{2^{1/3}}\int_0^s \xi\frac{ds}{\rho(s)^{2/3}}\right), \qquad (3.199)$$

where, for convenience we have omitted index r in s^r, and the index of the creeping mode. Using the relation $Y^{1/4} = m^{1/2}(t^{1/2}/\rho^{1/2})$, the above equation can be written as

$$u \approx A \left(\frac{2}{k}\right)^{1/6} \frac{(\rho(0)\rho(s))^{1/6}}{\sqrt{t}} \left(\frac{d(0)}{d(s)}\right)^{1/2} \exp\left(i\frac{\pi}{4}\right)\exp\left(ik\ell + i\frac{k^{1/3}}{2^{1/3}}\int_0^s \frac{\xi ds}{\rho(s)^{2/3}}\right). \qquad (3.200)$$

Equation (3.200) is the outer limit of the inner expansion which has been derived directly by using the ray coordinates adapted to the outer expansion. In passing, we note that the above result gives another reason for choosing the Airy function w_1. The function w_1 corresponds to the outgoing rays, while the function w_2 corresponds to incoming rays. Therefore, we have to choose w_1 to match the outer expansion.

Next we turn our attention to the outer expansion. The Friedlander-Keller outer expansion, presented in Sect. 2.5 reads

$$u \approx \frac{u_0(s)}{\sqrt{t}} \exp\left(ik\ell + ik^{1/3}\varphi(s)\right). \qquad (3.201)$$

It is not necessary to determine the limit of this expansion for the purpose of matching because we can directly extract the following relationship for $\varphi(s)$ and $u_0(s)$ by comparing (or identifying) (3.200) and (3.201).

$$\varphi(s) = \frac{1}{2^{1/3}} \int_0^s \frac{\xi ds}{\rho(s)^{2/3}},\tag{3.202}$$

$$u_0(s) = A \left(\frac{2}{k}\right)^{1/6} (\rho(0)\rho(s))^{1/6} \left(\frac{d(0)}{d(s)}\right)^{1/2} exp\left(i\frac{\pi}{4}\right).\tag{3.203}$$

Next, we rewrite (3.202) and (3.203) by using the usual notations of GTD, i.e., in terms of the Airy function Ai.

The parameter ξ satisfies (see (3.14) in Sect. 3.1).

$$w_1'(\xi) = imZw_1(\xi).\tag{3.204}$$

However, since $w_1(\xi) = 2\sqrt{\pi}\ e^{i\pi/6} Ai\ (\xi e^{2i\pi/3})$, we have

$$\xi = e^{i\pi/3}q,\tag{3.205}$$

where q satisfies the equation

$$A'i\ (-q) - mZ\ e^{-i\pi/6} Ai\ (-q) = 0,$$

which was given earlier in Sect. 1.5 as Eq. (1.117). Next we manipulate the above equations as follows. We replace w_1 and w_1' by their expressions in terms of Ai; substitute ξ in (3.202) by its expression in (3.205), and insert the expression (3.197) of Sect. 3.11 for A in (3.203). After some rearrangement, we obtain the following expression for u

$$u(s,\ t) \approx \frac{(2\pi kt)^{-1/2}(m(0)m(s))^{1/2}\,e^{i\pi/12}}{(q(0)Ai^2(q(0)) + A'i^2(q(0))^{1/2}(q(s)A'i^2(q(s)) + Ai^2(q(s))^{1/2}))}$$

$$exp\left(ik\ell + i\frac{k^{1/3}}{2^{1/3}}\,e^{i\pi/3} \int_0^s \frac{q(s)ds}{\rho(s)^{2/3}}\right),\tag{3.206}$$

which can be rewritten in its final form that is useful for GTD applications. It reads

$$u(s,\ t) \approx \frac{D_h(0)D_h(s)}{\sqrt{t}} \exp\left(ik(s+t) + i\int_0^s \alpha_h(s)\ ds\right),\tag{3.207}$$

where α_h and D_h are given, respectively, by (1.115) and (1.120) of Sect. 1.5. We may readily verify that (3.207) is identical to the GTD result we derived in Chap. 1 using the method of canonical problems.

In passing we note the advantages of the optimal coordinates, viz., that the matching is achieved more easily and its validity is guaranteed, if these coordinates are used, since the solution in the ray format appears simply as the asymptotic expansion for large Y of the solution in the boundary layer.

To summarize this section, we have recovered, in a deductive manner, the GTD results for the 2-D object via the method of boundary layers. We now list the major steps involved in the boundary layer method:

(i)   constructing a solution in the neighborhood of the surface (Sect. 3.1)

(ii)   deriving a solution close to the contact point (Sect. 3.8)

(iii) matching the two solutions in the neighborhoods of the contact point and the surface (Sect. 3.11)

(iv)   matching of solution in the neighborhood of the surface and the external space (Sect. 3.12).

Next, without going through the details of the calculations, we will apply to 3-D objects the approach we employed earlier for the 2-D for 3-D objects. We now deal with the Maxwell equation for the 3-D object, using the same method as for the 2-D object. The solutions close to the surface and of the contact point are the same as those given in Sects. 3.2 and 3.9. The matching is performed by following the same technique used for the 2-D object. The matching between the contact point and the boundary layer in the vicinity of the surface not only enables us to identify the function $p(\alpha)$ in formulas (3.98) and (3.99) of Sect. 3.2.7.2 for the creeping waves as the distance between the shadow zone and the α axis, but to also determine the amplitudes $A(\alpha)$ and $B(\alpha)$ of the electric and magnetic creeping rays (see (3.98) and (3.99) of Sect. 3.2). The expressions for $A(\alpha)$ and $B(\alpha)$ take the form

$$A(\alpha) = -\frac{2i\sqrt{\pi}}{(d_{E(0)})^{1/2}} \; (\vec{E}^i \cdot \hat{\alpha})\rho^{1/6}(0), \tag{3.208}$$

$$B(\alpha) = -\frac{2i\sqrt{\pi}}{(d_{H(0)})^{1/2}} \; (\vec{H}^i \cdot \hat{\alpha}) \, \rho^{1/6}(0), \tag{3.209}$$

where we have again omitted the index p of the creeping mode, and have used the relationships

$$w_1'(\xi_E) = i\, \frac{m}{Z}\, w_1(\xi_E), \tag{3.210}$$

and

$$w_1'(\xi_H) = im\, Zw_1(\xi_H). \tag{3.211}$$

In (3.208) $\hat{\alpha}$ is the unit vector at the contact point C, residing in the tangent plane and is orthogonal to the ray direction. It is also the binormal vector to the geodesic whose tangent vector is parallel to the incident ray at point C. Finally, $\vec{H}^i \cdot \hat{\alpha} = -\vec{E}^i \cdot \hat{n}$, where $\hat{n}$ is the normal to the surface at point C.

Let us rewrite (3.98) and (3.99) of Sect. 3.2 using the ray coordinates and replace w_1 by its asymptotic expansion in (3.198). Again, omitting the index r of s^r as well as the index of the mode, the outer limit of the inner expansion can be written as

$$\vec{E} \approx -\left(\frac{2}{k}\right)^{1/6} \frac{(\rho(0)\rho(s))^{1/6}}{t^{1/2}}\, h^{-1/2}(s)$$

$$\left\{ d_E^{-1/2}(s)\, A(\alpha)\, exp\left(i\frac{k^{1/3}}{2^{1/3}} \int_0^s \frac{\xi_E(s)ds}{\rho(s)^{2/3}}\right) (\vec{E}_i \cdot \hat{b}(0))\, \hat{b}(s) \right.$$

$$\left. + d_H^{-1/2}(s)\, B(\alpha)\, exp\left(i\frac{k^{1/3}}{2^{1/3}} \int_0^s \frac{\xi_H(s)ds}{\rho(s)^{2/3}}\right) (\vec{E}_i \cdot \hat{n}(0))\, \hat{n}(s) \right\}. \tag{3.212}$$

As previously, by replacing $A(\alpha)$ and $B(\alpha)$ by their expressions given in (3.208) and (3.209), and ξ_E and ξ_H by their corresponding expressions, we obtain, from (3.212), an explicit formula for $\vec{E}$. It reads

$$\vec{E} \approx \frac{D_s(0)D_s(s)}{\sqrt{ht}}\, (\vec{E}_i \cdot \hat{b}(0))\, \hat{b}(s) \left(exp(ik(s+t) + i\int_0^s \alpha_s(s)\, ds \right)$$

$$+ \frac{D_h(0)D_h(s)}{\sqrt{ht}}\, (\vec{E}_i \cdot \hat{n}(0))\, \hat{n}(s) \left(exp(ik(s+t) + i\int_0^s \alpha_n(s)\, ds \right). \tag{3.213}$$

It may be verified that the coefficients D_s, D_h, α_s and α_h appearing in (3.213), are given by the formulas in Sect. 1.5. Consequently, the outer limit of the inner solution does not contain any undetermined constants or arbitrary functions.

Turning now to the outer expansion, it can be written in the following form

$$\vec{E}(s, t) \approx \vec{E}_0(s) \sqrt{\frac{\rho_d}{(\rho_d + t)t}} \, \exp(ik(s + t) + ik^{1/3}\varphi(s)), \qquad (3.214)$$

where $\vec{E}_0(s)$ and $\varphi(s)$ are arbitrary functions.

The inner limit of the expansion in (3.214) is

$$\vec{E}(s, t) \approx \frac{\vec{E}_0(s)}{\sqrt{t}} \exp(ik(s + t) + ik^{1/3}\,\varphi(s)), \qquad (3.215)$$

since t is assumed to behave as $O(k^{-1/3})$.

A comparison of (3.213) with (3.215) enables us to determine the functions appearing in the outer expansion, including the terms $\varphi(s)$ and $\vec{E}_0(s)$ as well.

The first and second terms of (3.213) correspond to the electric creeping ray and the second to the magnetic creeping ray, respectively. We have

$$k^{1/3}\,\varphi(s) = \int_0^s \alpha_s(s)\,ds, \text{ for the electric creeping ray,} \qquad (3.216)$$

$$k^{1/3}\,\varphi(s) = \int_0^s \alpha_h(s)\,ds, \text{ for the magnetic creeping ray.} \qquad (3.217)$$

Finally,

$$\vec{E}_0(s) = \frac{D_s(0)D_s(s)}{\sqrt{h}}\,(\vec{E}_i \cdot \hat{b}(0))\,\hat{b}(s), \text{ for the electric creeping ray,} \quad (3.218)$$

$$\vec{E}_0(s) = \frac{D_h(0)D_h(s)}{\sqrt{h}}\,(\vec{E}_i \cdot \hat{n}(0))\,\hat{n}(s), \text{ for the magnetic creeping ray,} \quad (3.219)$$

where h is identical to the quantity $(d\eta(Q)/d\eta(Q'))$ of the GTD.

If we replace $\varphi(s)$ and $\vec{E}_0(s)$ in (3.214) by their values given in (3.216) through (3.219), we recover the GTD results given in Chap. 1.

In summary, the boundary layer method has enabled us to demonstrate the GTD formulas and to provide the field in the shadow zone of a smooth object, albeit at the expense of some cumbersome calculations, especially in three dimensions. In addition, in contrast to the GTD, the boundary layer approach has the advantage of providing explicit results near to the surface of the object, which is a caustic of the diffracted rays.

We have seen in this section how we can recover the results for the creeping ray propagation by using the boundary layer method. In the next section, we will discuss the application of the boundary layer method to the problem of edge diffraction.

3.13 Matching of the boundary layer in the neighborhood of the edge of a wedge

We have seen in Sect. 3.6 that, at a distance $r = O(1/k)$ from the edge of a three-dimensional wedge, the dominant order fields is identical to the exactly known field in the vicinity of a locally-tangent wedge. Outside of the neighborhood of the light-shadow boundary of the incident and reflected fields, and the neighborhood of the tangents to the faces, the field is expressed, for large kr, in terms of the wedge diffraction coefficient $D(\theta)$, where θ is the polar angle. For an incident field e^{ikx}, or more generally, for one whose first term of its Luneberg-Kline series is e^{ikx}, the diffracted field u^d reads, for large kr

$$u^d(r,\ \theta) = \frac{e^{ikr}}{\sqrt{r}} D(\theta) + O((kr)^{-3/2}). \qquad (3.220)$$

Equation (3.220) is the limit of the inner solution as kr becomes large, and, under that condition, the solution also represents a ray field. According to Sect. 2.2.2., the outer solution takes the form of (3.220) but contains a function $D'(\theta)$ which is as yet undetermined. The application of the matching simply leads to the equality of D and D'.

Let us now treat the 3-D acoustic case with the Neuman condition. As a first step we introduce, in the vicinity of the point 0 under consideration, a cylindrical coordinate system $(r,\ \theta,\ z)$ whose origin is located at 0. In the boundary layer, the dominant order field is identical to the one obtained from the diffraction of an obliquely incident plane wave on an infinite wedge.

The outer limit of the inner expansion reads

$$u^d(M) \approx \frac{e^{ikr\sin\beta}}{\sqrt{r}} \frac{D(\theta)}{\sqrt{\sin\beta}} \exp{(ik\cos\beta z)}\, u^i(0). \qquad (3.221)$$

Next, we introduce the following ray coordinates in the boundary layer. We let P be the point emitting a diffracted ray at point M, z_0 its ordinate, and s as the distance PM. Using the above conventions we can write $z = z_0 + s \cos\beta$ and $r = s \sin\beta$. Using these new coordinates, and recognizing that $\exp{(ikz_0 \cos\beta)}\, u^i(0)$ is simply the incident field at point P, we can rewrite (3.221) as

$$u^d(M) \approx \frac{e^{iks}}{\sqrt{s}} \frac{D(\theta)}{\sin\beta}\, u^i(P). \qquad (3.222)$$

Since the rays emanate from the edge, the outer solution can be written as

$$u^d(M) = \frac{e^{iks'}}{\sqrt{s'}} \sqrt{\frac{\rho_d}{\rho_d + s}} \ D'(\theta)u^i(P'),$$ (3.223)

where s' is defined similar to s, but which can subtend an angle that can differ from β.

If we now approach the inner limit of (3.223), the factor $(\rho_d/\rho_d + s)$ in front of $D'(\theta)$ tends to 1 and consequently can be omitted. Thus, we can rewrite (3.223) as

$$u^d(M) \approx \frac{e^{iks'}}{\sqrt{s'}} \ D'(\theta)u^i(P').$$ (3.224)

If we now match (3.222) and (3.224), we in effect impose the equality of the coordinates s and s' and of P and P'. This, in turn, turns out to be equivalent to enforcing the Keller law and also helps us to derive the relationship $D'(\theta) = D(\theta)/\sin \beta$. The results above serve to demonstrate that, to first order, the diffracted field can be derived by replacing the edge by its local tangent wedge.

3.14 The case of caustics

For the cases studied in Sects. 3.12 and 3.13, we find that the inner solution inside the boundary layer was completely determined. By passing to the outer limit of this solution, and subsequently applying the principle of matching of the inner and outer solutions, we were able to identify as well as determine the arbitrary functions that appeared in the representation of the solutions.

In the case of the caustic which we are considering in this section, the solution outside the boundary layer is known and our task is to determine the field inside the boundary layer.

The dominant order component of the inner solution is given by Eq. (3.119) of Sect. 3.4 and it reads

$$u(s, \nu) = e^{iks} A(s)Ai(-\nu) + O(k^{-1/3}).$$ (3.225)

To go to the outer limit of (3.225) we replace Ai by its asymptotic expansion for large negative values of ν. This leads us to the following outer expansion of the inner solution

$$u(s, \nu) \approx \frac{A(s)}{2\sqrt{\pi}} e^{iks} \ \nu^{-1/4} \ e^{i\,\pi/4} [e^{-i2/3\nu^{3/2}} - i \ e^{i2/3\nu^{3/2}}].$$ (3.226)

The dominant order term of the outer solution is just the Geometrical Optic field. We will denote with the index i the ray which approaches the caustic and with 0 the one that recedes from it. In addition, we denote by s_i and s_0 the abscissa of the contact point on the caustic, and by t_i and t_0 the lengths of the i and 0 rays. The outer solution then can be written

$$u = B(s_i) \frac{e^{ik(s_i - t_i)}}{\sqrt{t_i}} + dB(s_0) \frac{e^{ik(s_0 - t_0)}}{\sqrt{t_0}}, \tag{3.227}$$

where d is the phase-shift that the ray undergoes when passing through the caustic.

By using differential geometric relationships, we can show that (see Appendix 2 for details) if ν is of order 1, then we can write

$$\frac{1}{2}\big((s_i - t_i) + (s_0 + t_0)\big) = s + O(k^{-4/3}), \tag{3.228}$$

where s is the abscissa of the observation point, and

$$\frac{1}{2}\big((s_0 - t_0) + (s_i + t_i)\big) = \frac{2}{3}v^{2/3} + O(k^{-4/3}). \tag{3.229}$$

Since, the difference term $s_i - s_0$ is of $O(k^{-1/3})$ inside the boundary layer, we have

$$\left.\begin{array}{l} B(s_i) = B(s) + O(k^{-1/3}) \\ B(s_0) = B(s) + O(k^{-1/3)} \end{array}\right\}. \tag{3.330}$$

Finally, still assuming that ν is of $O(1)$ we can write

$$\left.\begin{array}{l} t_i = 2^{1/3} \rho^{2/3} k^{-1/3} v^{1/2} + O(k^{-1/3}) \\ t_0 = 2^{1/3} \rho^{2/3} k^{-1/3} v^{1/2} + O(k^{-1/3}) \end{array}\right\}. \tag{3.231}$$

Inserting approximations (3.228), (3.229), (3.230) and (3.231) in (3.227), we obtain

$$u(s, \nu) \approx B(s) \, 2^{-1/6} \rho^{-1/3} k^{1/6} \nu^{-1/4} e^{iks} \, [e^{-i2/3 \nu^{3/2}} + d \, e^{i2/3 \nu^{3/2}}], \tag{3.232}$$

which is the desired inner expansion of the outer solution.

A comparison of (3.226) and (3.232) enables us to determine $A(s)$ and d. The result is

$$A(s) = 2\sqrt{\pi}\; e^{-i\pi/4} \left(\frac{k}{2}\right)^{1/6} \rho^{-1/3}\, B(s),$$

(3.233)

$$d = -i.$$

(3.234)

Equation (3.233) provides us the expression for the amplitude $A(s)$ and enables us to determine the field in the boundary layer, while (3.234) tells us that the ray undergoes a phase-shift of $-(\pi/2)$ when passing through the caustic. It may be interesting to note that we could have chosen $B(s_i)$ instead of $B(s)$ in (3.233).

The formalism presented above also applies to three dimensions. In the boundary layer, the binormal component of the electric field is given by (3.121) of Sec. 3.4. It can be expressed as

$$E^{\alpha}(s, \alpha, \nu) \approx E^{\alpha}(s, \alpha)\, Ai(-\nu).$$

(3.235)

The normal component of the electric field, which, in accordance with the notation of Sect. 3.2, is equal to the binormal component of $\vec{H}$, can be written

$$E^{n} \approx E^{n}(s, \alpha)\, Ai(-\nu).$$

(3.236)

As for the outer solution, it is given by Geometrical Optics as in the scalar case. Let us denote by $\hat{b}_0$ and $\hat{n}_0$ the binormal and normal vectors to the geodesic at point 0, which is the contact point for the outgoing ray, and by $\hat{b}_i$ and $\hat{n}_i$, the corresponding vectors for the contact point I of the incoming ray. We can then write the outer solution as

$$\vec{E} = (E'^{\alpha}(s_i, \alpha_i)\, \hat{b}_i + E'^{n}(s_i, \alpha_i)\, \hat{n}_i)\, \sqrt{\frac{\rho_i}{t_i(\rho_i + t_i)}}\; e^{ik(s_i + t_i)} +$$

$$d(\,E'^{n}(s_0, \alpha_0)\, \hat{b}_0 + E'^{n}(s_0, \alpha_0)\, \hat{n}_0)\, \sqrt{\frac{\rho_0}{t_0(\rho_0 + t_0)}}\; e^{ik(s_0 + t_0)}.$$

(3.237)

For the general case, I and 0 may be located on distinct geodesics. However, the important point is that I and 0 are only separated by a distance of order $k^{-1/3}$, and this enables us to replace $E'^{\alpha}(s_i, \alpha_i)$, $E'^{n}(s_i, \alpha_i)$, $\hat{b}_i$, $\hat{n}_i$ by quantities without the index, i (and similarly for the index 0) calculated at the point (s, α).

In addition, we can write

$$\sqrt{\frac{\rho_i}{(\rho_i + t_i)}} = 1 + O(k^{-1/3}). \qquad (3.238)$$

Finally, we can show that the geometrical equalities (3.228), (3.229), (3.231) are still valid in 3D (see [BB] for instance). Thus, we can pass to the inner limit of the outer solution (3.237), as in the 2-D case and obtain

$$\approx (E'^{\alpha}(s, \alpha)\ \hat{b} + E'^{n}(s, \alpha)\ \hat{n})\ 2^{-1/6}\rho^{-1/3}k^{1/6}\ \nu^{-1/4}\ e^{iks}\ [e^{-i2/3\nu^{3/2}} - i\ e^{i2/3\nu^{3/2}}]. \qquad (3.239)$$

Finally, passing to the outer limit in the inner solution given by (3.235) and (3.236), and comparing the inner and outer solutions in the same manner in the 2-D case, we get

$$E^{\alpha}(s, \alpha) = 2\sqrt{\pi}\ e^{-i\pi/4}\left(\frac{k}{2}\right)^{1/6}\rho^{-1/3}\ E'^{\alpha}(s, \alpha), \qquad (3.240a)$$

$$E^{n}(s, \alpha) = 2\sqrt{\pi}\ e^{-i\pi/4}\left(\frac{k}{2}\right)^{1/6}\rho^{-1/3}\ E'^{n}(s, \alpha), \qquad (3.240b)$$

which enable us to determine the amplitude of both of the components of the electric field inside the boundary layer. In addition, we find that $d = e^{-i\pi/2}$, and this directly provides the value of the phase shift as the ray crosses the caustic. Thus, we conclude that for the caustic case, the matching procedure helps us determine, in its entirety, the solution inside the boundary layer from the solution of geometrical optics.

3.15 Matching in the neighborhood of the contact point (Fig. 3.17)

In Sect. 3.11 we considered the problem of matching between the neighborhood of the contact point and the boundary layer close to the surface. We now study a more complicated situation that involves the matching of the solutions in the neighborhood of the contact point to the boundary layer located close to the light-shadow boundary. This problem has been studied by Brown [Br], and we will follow his approach, though simplifying it somewhat. We restrict ourselves to the case of the cylinder. The present case is different from the ones we have studied previously because the two boundary layers overlap close to the contact point C. The first one of these is a classical boundary layer associated with the light-shadow boundaries and has been discussed in Sect. 3.5. The second, which has an angular width of $(k\rho)^{-1/3}$, is specific to the smooth objects.

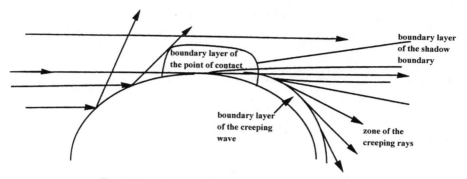

Fig. 3.17. Boundary layer diffraction by a smooth obstacle

The present problem can be analyzed either by following the investigation of the canonical problem of the circular cylinder treated by Pathak [Pa], or by considering the phase difference δ between the direct geometrical optics ray, which follows a direct path through the object, and the creeping ray (see Appendix 2).

The above phase difference is equal to $k(s + t - x)$ where s, t, and x are, respectively, the curvilinear abscissa of the detachment point of the creeping ray, the length of the space ray, and the abscissa of the observation point. To first order

$$\delta \approx k\left(\rho\frac{\theta^3}{6} + t\frac{\theta^2}{2}\right), \tag{3.241}$$

where θ denotes the angle of the space ray with respect to the x-axis.

The phase δ is of $O(1)$ if $k\rho\theta^3$ or if $kt\theta^2 = O(1)$. Thus θ has to satisfy one of the two conditions (3.242) or (3.243)

$$\theta = O((k\rho)^{-1/3}), \tag{3.242}$$

$$t\theta^2 = O(k^{-1}). \tag{3.243}$$

The condition (3.242) is specifically applicable to the smooth object problem whereas condition (3.243) is pertinent to the shadow boundary problem treated in Sect. 3.5. We refer the reader to Zworski's thesis [ZW], for a more rigorous justification of these results starting from techniques of microlocal analysis.

Let us now express the Helmholtz equation in terms of the ray coordinates, $\ell = s + t$, s, that are orthogonal, and the Helmholtz equation in this system becomes

$$\nabla^2 u + k^2 u = \frac{1}{t}\frac{\partial}{\partial \ell}\left[t\frac{\partial u}{\partial \ell}\right] + \frac{\rho}{t}\frac{\partial}{\partial s}\left[\frac{\rho}{t}\frac{\partial u}{\partial s}\right] + k^2 u = 0. \tag{3.244}$$

As for the total field, we choose the following Ansatz

$$u = e^{ik\ell} v, \tag{3.245}$$

and insert the last expression in (3.244) to get

$$ik + \left[2\frac{\partial v}{\partial \ell} + \frac{1}{t}v\right]\frac{1}{t}\frac{\partial}{\partial \ell}\left[t\frac{\partial v}{\partial \ell}\right] + \frac{\rho}{t}\frac{\partial}{\partial s}\left[\frac{\rho}{t}\frac{\partial v}{\partial s}\right] = 0, \tag{3.246}$$

where we have set $t = \ell - s$.

Let us stretch the coordinate s with a factor $k^{1/2}$ by setting $\sigma' = k^{1/2} s$. Just as the first term, the third term of (3.246) also becomes of $O(k)$.

Writing v in the form

$$v = v_0 + O(k^{-1}). \tag{3.247}$$

We obtain, to within the dominant order,

$$2i\frac{\partial v_0}{\partial \ell} + \frac{i}{\ell}v_0 + \frac{\rho^2}{\ell^2}\frac{\partial^2 v_0}{\partial \sigma'^2} = 0. \tag{3.248}$$

Equation (3.248) is a parabolic equation similar to (3.127) of Sect. 3.5. In fact, the two are identical if the variable s and the angle θ of the diffracted ray are interchanged.

Now, if we no longer stretch s with the factor $k^{1/2}$, but with the factor $k^{1/3}$, as in Sect. 3.7, then only the first term is of $O(k)$, and we find to within the dominant order the relation

$$2\frac{\partial v}{\partial \ell} + \frac{1}{\ell}v = 0, \tag{3.249}$$

which yields

$$v = A(\sigma)\,\ell^{-1/2}, \tag{3.250}$$

where $\sigma = ms/\rho$ was defined in Sect. 3.7.

This example illustrates fairly well how the choice of the thickness of the zones where we seek the solution and, consequently, the choice of the stretching factors of the coordinates lead to different forms of equations and their solutions. In the present case, the solution is described by (3.250) in a zone of angular thickness $k^{-1/3}$ around the shadow boundary. However, this solution is not valid in a small region of the shadow boundary where s is stretched by the factor $k^{1/2}$ and the behavior of the solution is described by (3.248). Equation (3.248) admits particular solutions that are functions of only the variable $\sigma \sqrt{\ell}$. One can then expect a boundary layer described by a parabola and located within the previous boundary layer. Finally, we have seen in Sect. 3.7 that there exists a boundary layer with a normal height of $O(k^{-2/3})$ and a width of $O(k^{-1/3})$ in the vicinity of the contact point.

The solution in such a boundary layer was obtained in Sect. 3.7 by using the (s, n) coordinates although it is also possible to work with the ray coordinates. In the latter case, the coordinate t is proportional to the square root of the coordinate n and the stretching coefficient to be used on the coordinates ℓ and s is $k^{1/3}$.

Let us define a new set of stretched coordinates $L = (m/\rho) \ell$ and $S = (m/\rho) s$ where $m = (k\rho/2)^{1/3}$ and ρ is the radius of curvature of the cylinder at point C. Inserting these stretched coordinates in (3.246), we obtain the following equation for the dominant order $k^{4/3}$

$$4i \frac{\partial v_0}{\partial L} + 2i \frac{v_0}{L-S} + \frac{\partial}{\partial S}\left[\frac{1}{L-S} \frac{\partial v_0}{\partial S} \right] = 0. \tag{3.251}$$

This equation has been obtained by Fock and Weinstein, as well as Malhiuzinets, for the case of the circular cylinder [F]. Following these authors, we introduce the new variable

$$Y = (L - S)^2 = m^2 \frac{t^2}{\rho^2}, \tag{3.252}$$

and seek an unknown function v_0 in the form

$$v_0 = \exp\left(-\frac{2}{3} iY^{3/2} \right) w. \tag{3.253}$$

Implementing these changes in Eq. (3.251), we obtain

$$\frac{\partial^2 w}{\partial Y^2} + Yw + i \frac{\partial w}{\partial L} = 0. \tag{3.254}$$

Comparing the above equation with (3.148) of Sect. 3.7 note that it is the same equation, provided we identify L with σ, and Y with ν. The result is not unexpected, since Y is nothing but the coordinate y defined with (3.34) of Sect. 3.1.8, and we have seen from (3.35) of Sect. 3.1.8 that

$$Y = \nu + O(k^{-1/3}). \tag{3.255}$$

Similarly, again according to (3.35) of Sect. 3.1.8

$$\ell = s + \frac{2}{3k} Y^{3/2} + O(k^{-4/3}). \tag{3.256}$$

We find that the second term of (3.256) is of $O(k^{-2/3})$, relative to the first, since inside the boundary layer $Y = O(1)$, $s = O(k^{-1/3})$. The replacement of L by σ in Eq. (3.254) does not modify the result up to the dominant order. It is possible to solve equation (3.254), using the limit absorption principle and the impedance condition on the object in the same manner as in Sect. 3.7. A calculation similar to that performed in Sect. 3.7 shows that x can be written

$$x = \ell - \frac{\ell^3}{6\rho^2} + \frac{st^2}{2\rho^2} + \frac{t^3}{6\rho^2} + O(\ell^4, t^4). \tag{3.257}$$

Introducing the stretched coordinates, this equation takes the following form inside the boundary layer, where the relations $\ell = O(k^{-1/3})$, $t = O(k^{-1/3})$ are satisfied

$$kx = k\ell - \frac{L^3}{3} + LY - \frac{2}{3} Y^{3/2} + O(k^{-1/3}). \tag{3.258}$$

Finally the incident field becomes

$$\exp{(ikx)} = \exp\left(ik\ell - \frac{L^3}{3} + LY - \frac{2}{3} Y^{3/2}\right) + O(k^{-1/3}). \tag{3.259}$$

One recognizes that the incident field has the same form as that given in Sect. 3.7, multiplied by the factor $\exp{(-i\ (2/3)\ Y^{3/2})}$. Again, as in Sect. 3.7, by using the Fourier transformation, we can derive the following expression for the total field

$$u(l, L, Y) = \exp{(ik\ell)} \exp\left(-\frac{2}{3} iY^{3/2}\right) V(L, Y, mZ) + O(k^{-1/3}), \tag{3.260}$$

where V is the Fock function introduced in Sect. 3.7.

The neighborhood of the point C is then the confluence of all the boundary layers, viz., the boundary layer of the creeping rays and that of the shadow boundary, as shown in Fig. 3.17. We are now going to exploit this explicit solution in the vicinity of C in order to precisely determine the solution in the two overlapping boundary layers in the neighborhood of the shadow boundary. As in the previous examples, it is faster to use the ray coordinates in order to perform the matching and to start with (3.260) as the inner expansion. To this end, let us first consider the outer limit of (3.260). More precisely, we consider the limit of the function V as $Y \to +\infty$ and $L \to +\infty$.

Logan [Lo] has studied this case and from the Eqs. (15)-(16) of his work we have

$$V^{-1/4} \exp\left(i \frac{2}{3} Y^{3/2}\right) P\left(L - \sqrt{Y}, mZ\right) + G(L, Y), \qquad (3.261)$$

with

$$L = m \frac{(s+t)}{\rho} \sqrt{Y} = m \frac{t}{\rho}. \qquad (3.262)$$

Hence

$$L - \sqrt{Y} = m \frac{s}{\rho} = \sigma, \qquad (3.263)$$

$$Y^{-1/4} = \sqrt{\frac{\rho}{m}} \frac{1}{\sqrt{t}}, \qquad (3.264)$$

where P is the modified Pekeris-Fock function given by

$$P(z, q) = -\frac{e^{i\pi/4}}{\sqrt{\pi}} \int_{-\infty}^{+\infty} \exp(izt) \frac{v'(t) - qv(t)}{w_1'(t) - qw_1(t)} dt - \frac{e^{i\pi/4}}{2\sqrt{\pi}z}. \qquad (3.265)$$

Finally, the first term in (3.261) can be written as

$$\exp\left(i \frac{2}{3} Y^{3/2}\right) \sqrt{\frac{\rho}{m}} \frac{1}{\sqrt{t}} P(\sigma, mZ). \qquad (3.266)$$

Let us now turn to the second term in (3.261) which takes two different forms according to the sign of $L - \sqrt{Y}$. If $L - \sqrt{Y} < 0$, which corresponds to a ray (pseudo) detaching tangentially from the object in the lit zone, G can be written as

$$G(L, Y) = \exp\left(iLY - i\frac{L^3}{3}\right) - \exp\left(-i\frac{2}{3} Y^{3/2}\right) K(-Y^{1/4}(L - \sqrt{Y})). \qquad (3.267)$$

The first term in (3.267) simply corresponds to the incident field. In effect, using (3.259) we can write

$$\exp(ik\ell) \exp\left(iLY - i\frac{L^3}{3}\right) - \exp\left(-\frac{2}{3} iY^{3/2}\right) = \exp(ikx) + O(k^{-1/3}).$$

In the second term of (3.267), K is a Fresnel modified function given by

$$K(\tau) = \exp\left(-i\tau^2 - i\frac{\pi}{4}\right)\frac{1}{\sqrt{\pi}}\int_{\tau}^{\infty} exp(it^2)dt \,, \tag{3.268}$$

$$Y^{1/4}(L - \sqrt{Y}) = k^{1/2}\frac{s\sqrt{t}}{\rho} = \sigma'\frac{\sqrt{t}}{\rho} \,. \tag{3.269}$$

If, on the other hand, $L - \sqrt{Y} > 0$

$$G(L, Y) = \exp\left(i\frac{2}{3}Y^{3/2}\right)K(Y^{1/4}(L - \sqrt{Y})). \tag{3.270}$$

Using (3.260) through (3.270) we finally obtain the outer limit of the inner solution as follows

$$u - u^i \approx \sqrt{\frac{\rho}{m}}\frac{1}{\sqrt{t}}\,P(\sigma, mZ)\,e^{ik\ell} + K\left(-\sigma'\frac{\sqrt{t}}{\rho}\right)e^{ik\ell}, \tag{3.271}$$

in the lit zone, and

$$u \approx \sqrt{\frac{\rho}{m}}\frac{1}{\sqrt{t}}\,P(\sigma, mZ)\,e^{ik\ell} + K\left(\sigma'\frac{\sqrt{t}}{\rho}\right)e^{ik\ell}, \tag{3.272}$$

in the shadow zone.

We have seen at the beginning of this section (see (3.245) and (3.250)) that in the boundary layer with an angular width of $k^{-1/3}$, u can be written as

$$u \approx e^{ik\ell}A(\sigma)\,\ell^{-1/2}. \tag{3.273}$$

Since $\ell = t + s$, and $s = O(k^{-1/3})$, the above expression becomes

$$u \approx e^{ik\ell}A(\sigma)\,t^{-1/2}. \tag{3.274}$$

Finally, in the inner boundary layer close to the shadow boundary, we can write

$$u \approx e^{ik\ell}v_0, \tag{3.275}$$

where v_0 is a solution of (3.248).

Again, using $\ell = t + s$, and for $s = O(k^{-1/2})$, (3.248) can be written as

$$2i \frac{\partial v_0}{\partial t} + \frac{i}{t} v_0 + \frac{\rho^2}{t^2} \frac{\partial^2 v_0}{\partial \sigma'^2} = 0. \tag{3.276}$$

It is possible to verify that $K(-\sigma' (\sqrt{t} / \rho))$ and $K(\sigma' (\sqrt{t} / \rho))$ are solutions of (3.276).

Thus, the outer limit of the inner solution can be expressed as the sum of two terms: first, a solution in the boundary layer with angular width $k^{-1/3}$, described with a Fock-Pekeris function; and second, a solution in the boundary layer close of the shadow boundary, described by a Fresnel function. Equations (3.271) and (3.272) provide the solutions in the lit and shadow zones inside the two overlapping boundary layers. We note that the Fresnel contribution is dominant in the immediate neighborhood of the shadow boundary. Indeed it is of $O(k^{1/6})$ as compared to the Fock-Pekeris contribution. Consequently, in the immediate neighborhood of the shadow boundary, we need retain only the second term in (3.271) and (3.272).

For large values of σ', it is possible to replace K by its asymptotic expansion. Expression (3.272) then becomes

$$u(\sigma, l, t) \approx \sqrt{\frac{\rho}{m}} \frac{1}{\sqrt{t}} \hat{p} \ (s, \ mZ) \ e^{ik\ell}, \tag{3.277}$$

where $\hat{p}$ is the Fock-Pekeris function (see Appendix 5)

$$\hat{p} \ (z, q) = P \ (z, q) + \frac{e^{i\pi/4}}{2\sqrt{\pi z}}. \tag{3.278}$$

For large values of σ, this function can be written in the form of a series expansion in terms of creeping rays, and we recover the Friedlander-Keller expansion. It can be shown that we can obtain once again the same field in the shadow region as was calculated in Sect. 3.12; therefore, the results are entirely consistent.

In conclusion, the application of the matching procedure of the outer expansion of the solution in the vicinity of the contact point provides the solution in the vicinity of the shadow boundary. The solution is somewhat more involved in this case than in the previous examples, because of the overlapping of the boundary layers.

We could include many other examples that illustrate the application of the use of the boundary layer technique in diffraction studies. It is hoped that the overview presented herein will provide the reader sufficient information to apply this technique in various applications. It is also interesting to point out that, although we have restricted ourselves in this chapter to deriving separate solutions inside and outside the boundary layer, it is often more convenient to have access to *uniform* solutions that are valid simultaneously in both regions. The solutions we have established in this chapter will serve as the starting point to the derivation of these solutions. Before discussing these uniform solutions in Chap. 5, we will first present in Chap. 4 the Spectral Theory of the Diffraction, which has proven to be useful for the derivation of some of these uniform solutions.

References

[AB1]   I. Andronov and D. Bouche, "Calcul du second terme de la constante de propagation des ondes rampantes par une méthode de couche-limite," *Annales des Télécomm.*, 49, n° 3, pp 199-204, 1994.

[AB2]   I. Andronov and D. Bouche, "Etude des ondes rampantes sur un corps élancé," ibid. pp 205-210.

[AB3]   I. Andronov and D. Bouche, "Ondes rampantes sur un objet convexe décrit par une condition d'impédance anisotrope," ibid. pp 194-198.

[BK]   V. M Babich and N. Y. Kirpicnikova, "The boundary-layer method in diffraction problems," Springer-Verlag, Berlin, Heidelberg, New York, 1979.

[BB]   V. M Babich and V. S. Buldyrev, "Asymptotic methods in shortwave diffraction theory," Springer, 1990.

[Be]   J. M. Bernard, "Diffraction par un dièdre à faces courbes non parfaitement conducteur," *Revue technique Thomson*, **23**(2), 321-330, 1989.

[Bo]   D. Bouche, "Etude des ondes rampantes sur un corps convexe décrit par une condition d'impédance par une méthode de développement asymptotique," *Annales des Télécomm*, (47), 400-412, 1992.

[Bor]   V. A. Borovikov, "Diffraction by a wedge with curved faces," *Sov. Phys. Acoust.* **25**(6), Nov-Dec. 1979.

[Br]   W. P. Brown, "On the asymptotic behavior of electromagnetic fields scattered from convex cylinders near grazing incidence," *J. Math. Anal. and Appl.*, **15**, 355-385, 1966.

[Bu]   V. S. Buslaev, "Shortwave asymptotic formulas in the problem of diffraction by convex bodies," Vest.Leningrad University **3**(13), 5-21, 1962.

[Da]   G. Darboux, Théorie Générale des surfaces, Chelsea, 1972.

[F]   V. Fock, "Electromagnetic wave propagation and diffraction problem," Pergamon Press, 1965.

[Go]   Gouillon, "Calcul tensoriel," Masson, 1963.

[Iv]   V. I. Ivanov, "Uniform asymptotic behavior of the field produced by a plane wave reflection at a convex cylinder," *USSR Journal of Comput. Math. and Math. Phys.* **2**, 216-232, 1971.

[KL] S. Kaplun and P. A. Lagerstrom, *J. Math Mech.* **6**, 585-593, 1957.

[L] M. A. Lyalinov, "Diffraction of a high frequency electromagnetic field on a smooth convex surface in a nonuniform medium," *Radiofizika*, 704-711, 1990.

[Lo] N. Logan, "General research in diffraction theory," Rapport LMSD 288087, 1959.

[LY] N. Logan and K. Yee, dans "Electromagnetic waves," R.E. Langer, Ed., 1962.

[Pa] P. H. Pathak, "An asymptotic analysis of the scattering of plane waves by a smooth convex cylinder," *Radio Science*, **14**, 419-435, 1979.

[Po] M. M. Popov, "The problem of whispering gallery waves in a neighborhood of a simple zero of the effective curvature of the boundary," *J. Sov. Math.* **11**(5), 791-797, 1979.

[VD] Van Dyke, "Perturbation methods in fluid mechanics," Parabolic Press, 1975.

[Za1] E. Zauderer, "Boundary layer and uniform asymptotic expansions for diffraction problems," *SIAM J. Appl. Math* **19**, 575-600, 1970.

[Za1] E. Zauderer, "Partial Differential Equations of Applied Mathematics," Wiley, 1988.

[Zw] Zworski, thèse, MIT, 1990.

4. Spectral Theory of Diffraction

4.1 Introduction

The concept of Spectral Theory of Diffraction [STD] was introduced by Mittra and his collaborators [MR] in the 1970s to circumvent some of the problems encountered with the GTD. The basic strategy followed in STD is to represent a complex field, which is not a ray field, in terms of a superposition of plane waves. Such fields are often encountered in diffraction problems.

Plane waves are, obviously, solutions of the wave equation, and in view of the linearity of this equation, it follows that a superposition of these waves also is its solution. The plane wave spectrum (PWS) approach to dealing with complex fields was introduced in the 1950s by Clemmow in [Cl]. The above paper deals with the problem of diffraction by two-dimensional objects and constructs the solution to the scattering problem in the form of a spectrum of plane waves which takes the form of an integral.

$$\int_C A(\alpha) \exp\left[ikx\cos[\alpha] + iky\sin[\alpha]\right] d\alpha. \tag{4.1}$$

In (4.1), the contour C is the dotted line that follows the path $-\pi + i\infty$ to π to 0 to $0 - i\infty$ as shown in Fig. 4.3 of Sect. 4.2.3. The portion of the contour C which coincides with the real axis of the complex α plane corresponds to the real angles, i.e., to the homogeneous plane waves that fall in the visible range. On the other hand, the two vertical semi-infinite paths in the same figure are associated with the complex angles, that is with the inhomogeneous plane waves belonging to the invisible range. The principles of the plane wave spectrum approach will be elaborated on in Sect. 4.2.

It is possible to represent the fields diffracted by an object, including those in the boundary-layers where it is not a ray field, in terms of a spectrum of plane waves, as for instance in (4.1). Illustrative PWS representations for various types of fields will be provided in Sect. 4.3.

The plane wave representation is convenient to use when considering the problem of diffraction of a boundary-layer field by an edge or a smooth object. For this reason, the diffraction coefficients for edges, as well as the launching coefficients of the creeping waves along the light-shadow boundaries, are usually calculated for the case of an incident plane wave. The diffraction coefficients $D(\alpha)$ of the constituent plane waves, that comprise the boundary-layer field, depend upon their incident angles α. The diffracted field u_d is therefore also expressed in terms of its plane wave spectrum with its weight coefficient given by the product of the incident field amplitude $A(\alpha)$ and the diffraction coefficient $D(\alpha)$ as follows

$$u_d(x, y) = \int_C A(\alpha)\, D(\alpha)\, exp\big[ikx\, cos[\alpha] + iky\, sin[\alpha]\big]d\alpha. \tag{4.2}$$

The coefficient $D(\alpha)$ is known when α is real. When α is complex, it can be determined via analytical continuation, i.e., simply by inserting the complex·value of the angle α in the expression of the diffraction coefficient $D(\alpha)$ to yield the diffracted field u_d. The appropriate diffraction coefficient is either that obtained from GTD or one of the uniform theories (UAT or UTD), respectively, according to whether the observation point is outside or inside a transition zone of the diffracted field. We can also express the field diffracted by each plane wave in the form of its plane wave spectrum, in which event the solution to the diffraction problem takes the form of a double integral. An example of this technique will be presented in Sect. 4.4. However, the discussion of the relationship between the STD and the *uniform theories* will be deferred until the next chapter, i.e., Chap. 5.

4.2 The Plane Wave Spectrum (PWS)

The plane wave spectrum (PWS) approach has been extensively described in publications by Clemmow [Cl] and by Roubine [Ro]. In this section we will merely attempt to present a summary of the main features of the approach and refer the reader is to the above papers for additional details.

4.2.1 Homogeneous and inhomogeneous plane waves

Let us consider the two-dimensional case which, as we have seen before, can be reduced to a scalar problem. Let us denote the electric or the magnetic field, $\vec{E}$ or $\vec{H}$, by $\vec{u} = u\,\hat{z}$. Next, we write the plane wave as

$$u = exp(i\,\vec{k}\cdot\vec{r}). \tag{4.3}$$

If $\vec{k}$ is a real vector, then the plane wave is said to be homogeneous. However, we can also consider, at least formally, complex $\vec{k}$ vectors, i.e., let

$$\vec{k} = \vec{k}' + i\vec{k}''. \tag{4.4}$$

Hence, u becomes

$$u = \exp (i\vec{k}'\cdot\vec{r}) \, \exp(-\vec{k}''\cdot\vec{r}). \tag{4.5}$$

Since the wave function u must satisfy the wave equation, we require

$$\vec{k}^2 = k^2, \tag{4.6}$$

where k is the wave number, i.e., $k = \omega/c$. From (4.6) we have

$$\vec{k}'^2 - \vec{k}''^2 = k^2, \tag{4.7}$$

and

$$\vec{k}' \cdot \vec{k}'' = 0. \tag{4.8}$$

According to (4.5), the direction of wave propagation is along $\vec{k}'$ and the associated wave number k' satisfies (4.7). It is evident that this wave number is greater than k. On the other hand, the wave exhibits a decaying behavior along $\vec{k}''$, which is orthogonal to $\vec{k}'$ in accordance with (4.8). In other words, the equiphase surfaces are parallel to $\vec{k}''$ and orthogonal to $\vec{k}'$ and vice versa for the equiamplitude surfaces. This is shown in Fig. 4.1.

An alternate representation for inhomogeneous plane waves will now be presented. A homogeneous plane wave can be written in the (x, y) coordinates as

$$u = \exp (ik(x \cos \alpha + y \sin \alpha), \tag{4.9}$$

when α is real. In contrast, by choosing α to be complex, i.e., letting $\alpha = \alpha' + i\alpha''$, we can obtain an inhomogeneous plane wave. For complex α (4.9) becomes

$$u = \exp (ik \cosh\alpha'' (x \cos \alpha' + y \sin \alpha') - k \sinh\alpha'' (-x \sin \alpha' + y \cos \alpha'). \tag{4.10}$$

If $\alpha'' > 0$, the wave propagates along the vector $\hat{k}'(\cos \alpha', \sin \alpha')$ and decays along the direction defined by the orthogonal vector $\hat{k}''(-\sin \alpha', \cos \alpha')$. Along the direction of wave propagation, the wave number is $k \cosh \alpha''$, which is greater than k. Hence, its phase velocity is $c/\cosh \alpha''$ which is less than c.

The above inhomogeneous wave only decays along positive $\hat{k}''$, while it grows exponentially in the opposite direction. Consequently, it represents a physical solution of the wave equation only in an upper half-plane. The Eq. (4.9) may also be written in polar coordinates (r, θ) as follows

$$u = \exp \; (ikr \; cos(\theta - \alpha)). \tag{4.11}$$

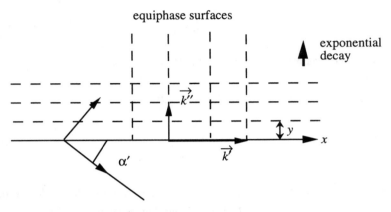

Fig. 4.1. Inhomogeneous plane wave

As r tends to infinity, (4.11) remains physically meaningful only if u does not become unbounded, and this condition is satisfied only if $sin(\theta - \alpha') \sinh \alpha'' \geq 0$. This is consistent with the previous result that (4.11) represents a physical solution only in the half-plane $\alpha' < \theta < \alpha' + \pi$ when $\sinh \alpha'' > 0$.

4.2.2 Superposition of plane waves

Both the homogeneous and inhomogeneous plane waves, that were defined above, are solutions of the wave equation. However, a more general solution can be constructed by superimposing a spectrum of these plane waves and expressing the solution in the form of an integral along a contour in the complex plane as follows

$$u(r, \theta) = \int_\Gamma p(\alpha) \exp \; (ikr \; cos(\theta - \alpha)) \; d\alpha. \tag{4.12}$$

We must realize at this point that the representation in (4.12) is strictly formal in nature and that we must verify the convergence of the integral as well as assure that the result is physically meaningful. If we impose the condition that the positive imaginary part of $cos(\theta - \alpha)$ be greater than zero along the infinite branches of Γ, i.e., if we require that $sin(\theta - \alpha') \sin \alpha'' \geq 0$, we can assure that the integral is convergent for a broad class of functions $p(\alpha)$. If this condition is imposed in the range $\theta \in [0, \pi]$, it leads to the following constraints

$$\alpha' = 0 \quad \text{if} \quad \alpha'' > 0,$$

$$\alpha' = \pi \quad \text{if} \quad \alpha'' < 0.$$

The Sommerfeld contour C (see Fig. 4.3), which was introduced in Sect. 4.1, satisfies the conditions mentioned above. If the growth of $p(\alpha)$ is slow on the infinite branches of this contour, the integral in (4.12) is convergent and we can still apply, formally, the method of stationary phase to evaluate it. Equation (4.12) has a stationary point at $\alpha = \theta$, and the second derivative of the phase term at that point is $-kr$. Upon applying the stationary phase construct to the integral in (4.12), we obtain the result

$$u(q) \approx \sqrt{2\pi} \; e^{-i\pi/4} \, p(\theta) \, e^{ikr} / \sqrt{kr} \, . \tag{4.13}$$

It is evident from (4.13) that u exhibits the anticipated asymptotic behavior of $1/\sqrt{kr}$ for large r, and, hence, under the conditions stipulated above, it represents a solution to the wave equation. Furthermore, we observe that in the range $\alpha \in [0, 2\pi]$ the function p is associated with the scattered far field pattern and that p can then be determined on C via an analytic continuation procedure.

4.2.3 Plane wave spectrum and Fourier transformation

Let us assume that u is solution to the wave equation which is know on the x-axis. We can write a Fourier transform representation of u as follows

$$u(x, 0) = \frac{1}{2\pi} \int_{-\infty}^{+\infty} \tilde{u}(k_x, 0) \, \exp(-ik_x x) \, dk_x. \tag{4.14}$$

The representation in (4.14) can be physically interpreted as a superposition of plane waves, with k_x identified as the projection of the wave vector of a constituent wave along the x-axis and $\tilde{u}(k_x, 0)$ is the weight coefficient of this wave in the representation given in (4.14). For $y \geq 0$, the plane waves have the form

$$\tilde{u}(k_x, 0) \, \exp(-ik_x x) \, \exp\!\left(i\sqrt{k^2 - k_x^2} \, y\right), \tag{4.15}$$

and $u(x, y)$ may be written as

$$u(x, y) = \frac{1}{2\pi} \int_{-\infty}^{+\infty} \tilde{u}(k_x, 0) \, \exp\!\left(-i k_x \, x + i\sqrt{k^2 - k_x^2} \, y\right) dk_x. \tag{4.16}$$

The sign of the square root $\sqrt{k^2 - k_x^2}$ must be determined by imposing certain physical constraints upon the behavior of the solution. For instance, (4.15) must be either: (i) a homogeneous plane wave propagating in the half-space $y > 0$, with $\sqrt{k^2 - k_x^2} > 0$ in the range $-k \leq k_x \leq +k$; or (ii) an inhomogeneous plane wave which is evanescent in the upper half-space $y > 0$. The square root $\sqrt{k^2 - k_x^2}$ is now chosen to be $i\sqrt{k_x^2 - k^2}$ for $k_x^2 > k^2$, assuring that u will be decaying in the upper half-space.

Even with the above restrictions, a considerable amount of flexibility still remains in the choice of the branch cuts for the function $\sqrt{k^2 - k_x^2}$, and it is only necessary to ensure that the branch originating from k and $-k$ be in the upper and lower half-planes, respectively. To evaluate (4.16), it may be necessary to deform the contour of integration from the real axis to the upper half-plane. In this event, we must impose the condition that $Im\sqrt{k^2 - k_x^2} \geq 0$ in the entire upper half-plane in order for the waves to decay for $y > 0$. To satisfy this condition, the branch cut must be aligned with the coordinates axes as shown in Fig. 4.2 (see [Va]).

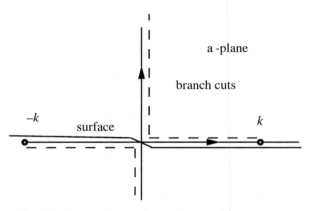

Fig. 4.2. Contour of integration and branch cuts in the k-plane

By letting $k_x = -k \cos \alpha$, $\sqrt{k^2 - k_x^2} = k \sin \alpha$, we can write (4.16) in polar coordinates as

$$u\,(r,\,\theta) = \int_C \sin \alpha \; \frac{\tilde{u}(-k \cos \alpha, 0)}{2\pi} \exp\,(ikr \cos\,(\theta - \alpha))\, d\alpha, \qquad (4.17)$$

where C is the indented Sommerfeld contour as shown in Fig. 4.3.

Hence, we have derived in (4.17), using Fourier transformation, the plane wave spectral representation on the Sommerfeld contour of u.

The solutions given above are valid for $y > 0$, i.e., for $\theta \in [0, \pi]$. By replacing $\sqrt{k^2 - k_x^2}$ with $-\sqrt{k^2 - k_x^2}$ in (4.15), or α with $-\alpha$ in (4.12) and (4.17), we can derive an integral representation of the solution valid for the left half-plane $y < 0$. Finally, we can combine the upper and lower half-plane solutions to get

$$u(r, \theta) = \int_C p(\alpha) \exp(ikr \cos(\theta \pm \alpha))\, d\alpha, \qquad (4.18)$$

where the upper and lower signs, i.e. $\pm$, apply to the upper and lower half-planes respectively. Before closing this section, we remark that while we have chosen to use the Sommerfeld contour in the discussion above, the choice is by no means unique, and we have other options to deform the contour to evaluate the integral in (4.17). This point is further discussed below.

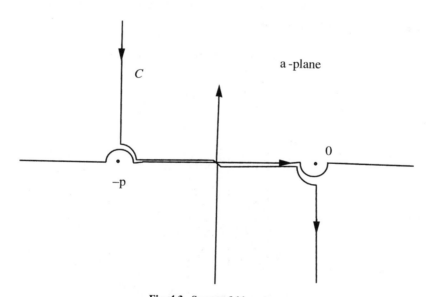

Fig. 4.3. Sommerfeld contour

4.2.4 Choice of the contour of integration

The choice of the Sommerfeld contour is dictated by the condition that the imaginary part of $\cos(\theta - \alpha)$ be positive, a constraint which is imposed in the range $\theta \in [0, \pi]$.

However, if this condition is enforced only for a fixed θ, it would lead to the condition that $\sin(\theta - \alpha')\sinh\alpha'' \geq 0$, i.e.,

$$\theta - \pi + 2n\pi < \alpha' < \theta + 2n\pi \qquad \text{if } \alpha'' > 0,$$

and

$$\theta - 2\pi + 2n\pi < \alpha' < \theta - \pi + 2n\pi \qquad \text{if } \alpha'' < 0,$$

where n is an integer. The above conditions are satisfied so long as the contour is located in one of the cross-hatched regions of Fig. 4.4.

The choice of the contour in the complex-plane provides us with an additional flexibility in the application of the plane wave spectrum approach. A convenient contour of integration is the steepest descent path that passes through the point θ. In the vicinity of the point θ, which is referred to as the saddle point, the real part of the phase is stationary and the imaginary part exhibits a rapid variation. As may be seen from the appendix *Asymptotic Approximation of Integrals*, the integrand decays very rapidly along the SDP on both sides of the saddle point, where it reaches its maximum. The SDP path is defined by the equations

$$\cos(\theta' - \alpha')\cosh(\theta'' - \alpha'') = 1, \qquad (4.19)$$

and

$$\alpha''(\theta - \alpha') > 0. \qquad (4.20)$$

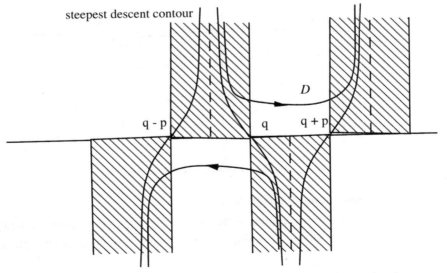

Fig. 4.4. Steepest descent contours and Maliuzhinet's contours in the complex plane

Its asymptotes are the semi-infinite lines given by $\alpha'' = \theta - \pi/2$ for $\alpha'' > 0$ and $\alpha' = \theta + \pi/2$ for $\alpha'' < 0$ (see Fig. 4.4). There exist steepest descent contours that go through $\theta + n\pi$, with the ones for $n = +1$ and -1 depicted in Fig. 4.4. A frequently-chosen path is the Sommerfeld-Maliuzhinets contour D. This contour is symmetrical with respect to the origin and tends towards two steepest descent contours that are shifted by 2π relative to one another, as shown in Fig. 4. On the other hand, the dependence on the location of θ on the contour can be suppressed by carrying out a change of variable $\beta = \alpha - \theta$ in (10), which leads to the following representation for the wave function u

$$u\,(r,\,\theta) = \frac{1}{2\pi i}\int_D p\,(\beta + \theta)\,\exp\,(ikr\cos\beta)\,d\beta. \qquad (4.21)$$

Using this representation, it is possible to represent a wide class of solutions of the wave equation in the form of a plane wave spectrum. We will provide a number of illustrative examples of field representations in terms of the plane waves spectrum.

4.3 Examples of plane wave spectral representation

4.3.1 Surface waves

We seek the solutions $u(x,\,y)$ of the wave equation in the upper half-plane that satisfy an impedance condition on the line $y = 0$. The wave function $u(x,\,y)$ satisfies the wave equation

$$(\Delta + k^2)\,u = 0 \quad \text{for} \quad y > 0, \qquad (4.22)$$

together with the impedance condition

$$\frac{\partial u}{\partial y} + ikZu = 0 \quad \text{for} \quad y = 0, \qquad (4.23)$$

$$lim_{,\,y \to \infty} \quad u = 0. \qquad (4.24)$$

A particular solution to this problem, called surface waves, is given by

$$u(x,\,y) = \exp\,(ik\sqrt{1 - Z^2}\,x)\,\exp\,(-ikZ\,y), \qquad (4.25)$$

which satisfies (4.24) if $Im(Z) < 0$. The square root $\sqrt{1 - Z^2}$ is chosen to have a positive imaginary part in order to assure that the wave decays for $x \to +\infty$. Since the wave function tends to infinity as $x \to -\infty$ if Z is neither real nor pure imaginary, it has a physical meaning only in a certain region of the upper half-plane.

If we assume that Z has the form $-\sin \varphi_s$, then (4.25) becomes

$$u(x, y) = \exp (ikx \cos \varphi_s + iky \sin \varphi_s). \tag{4.26}$$

Thus the surface wave may be interpreted as an inhomogeneous wave propagating in the $\theta = \varphi_s$ direction. Furthermore, we note that if φ_s were real, u would represent a reflected wave. This is not unexpected, however, since, as shown in Chap. 1, the surface wave corresponds to the pole $Z = - \sin \varphi_s$ of the reflection coefficient $R = (\sin \varphi_s - Z)/(\sin \varphi_s + Z)$ of a planar surface with a surface impedance of Z. When the criterion for the existence of the surface wave is fulfilled, the reflected field, which is also equal to the total field, remains finite even as the incident fields tends to zero.

Let us now examine some special cases of the surface impedance. If $Z = -iX$, which may be realized, for instance, by coating a conducting plane with a lossless dielectric, the wave function u given in (4.26) becomes

$$u(x, y) = \exp (ikx \sqrt{1 + X^2} - kyX). \tag{4.27}$$

We see from (4.27) that the wave propagates along the x axis with the velocity $c/\sqrt{1 + X^2}$, which is less than the speed of light, and it decays exponentially along the y axis. Its equiphase and equiamplitude contours are parallel to $0x$ and $0y$.

For the case of a complex impedance given by $Z = -iX + Y$, where Y is positive real, the surface wave has an exponential decay along x. We have discussed this case in previous chapters and have observed that as soon as the real part of the impedance becomes non-zero, the exponentially decaying behavior of the surface waves enables us to neglect them. We will see in Chap. 8 that this occurs when a conducting surface is coated with a lossy material. We note that, if $ImZ \geq 0$, (4.25) yields solutions that no longer have the nature of surface waves. For instance, if Z is real and less than 1, (4.25) represents a wave incident upon the plane at the so-called Brewster's angle for which the surface is reflectionless. Mathematically speaking, a surface wave is a wave propagating along the direction φ_s such that $Z = -\sin \varphi_s$. Consequently, the field diffracted by a surface wave that encounters a discontinuity can be derived by employing the diffraction coefficients of the discontinuity given in Chap. 1, provided that we generalize these diffraction coefficients for the complex angles of the surface waves reaching the discontinuity. Some examples of this procedure will be given in Sect. 4.4.

The surface wave solution is a very special case of plane wave spectral representation, since, for this case, the entire spectrum collapses to a single inhomogeneous wave. We will now present some examples that are more general in nature and require a contour integral type of representation for the field instead of one that has only a single wave component.

4.3.2 Line current

The field radiated by an electric line current is expressed simply in terms of the Hankel function $H_0^{(1)}$. For an electric line current of intensity I and orientation in the z-direction, the plane wave representation of the electric field E_z has been given by Clemmow [Cl]. It reads

$$E_z = -\frac{kZ_0 I}{4\pi} \int_C \exp\left(ikr \cos\left(\theta \pm \alpha\right) d\alpha. \tag{4.28}$$

The – and + signs apply to the upper and lower half-planes, respectively. We note from (4.28) that the spectral weight function is uniform over the entire contour. This result can be readily derived by following the discussion appearing at the end of Sect. 4.2.2, viz., that the far field of a line source is independent of the observation angle θ; hence, the weight function $p(\alpha)$ is a constant.

4.3.3 Arbitrary current source

The plane wave spectrum of an arbitrary current source can be derived from its induced field on a line by using the Fourier transform technique as explained in Sect. 4.2.3. Alternatively, we can first use geometrical optics (GO) to calculate the far field, and then employ (11) in Sect. 7.2.2 which links the far field to the weight function of the plane wave spectrum.

We will now elaborate on this second approach in some detail. Let us assume that the field u, as represented on Fig. 4.5, is known along the x-axis. The field in the θ direction is associated with the geometrical optical rays propagating in the same direction. Let us consider a point M with abscissa x from where the GO ray emanates in the θ direction as the derivative of the phase $kS(x)$ of u is $k \cos \theta$ at this point. The caustic is located at a distance $\rho = |\sin \alpha \, (dx/d\alpha)|$ and θ is a function of x. Depending upon whether $(d\alpha/dx)$ is less than or greater than 0, the caustic is located before or after the point M, respectively. Let N denote a point located on the ray emanating from M and situated at a very large distance t from M. If the value of the field at the point M is $A(x) \exp(ikS(x))$, then the GO field at the point N is given by

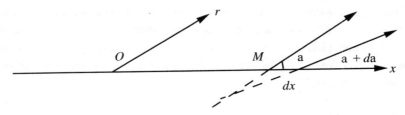

Fig. 4.5. Radiation from a field u which is known along the x-axis

$$u_0(N) = A(x) \exp(ik\,(S(x)+t)) \sqrt{\frac{\rho}{t}}, \tag{4.29}$$

for $(d\alpha/dx) < 0$. For $r \to \infty$, t can be written as

$$t \approx r - x \cos \alpha. \tag{4.30}$$

Using the above expression in (4.29) we can write

$$u_0(N) = A(x) \exp(ik\,(S(x) - x \cos \alpha)) \sqrt{\frac{-\sin\alpha\,dx}{d\alpha}} \frac{\exp(ikr)}{\sqrt{r}}. \tag{4.31}$$

The weight function $p(\alpha)$, can be derived from the formulas (4.11) and (4.13) of Sect. 4.2.2. It reads

$$p(\alpha) = \sqrt{\frac{ik}{2\pi}} \, A(x) \exp(ik\,(S(x) - x \cos \alpha)) \sqrt{\frac{-\sin\alpha\,dx}{d\alpha}}. \tag{4.32}$$

The field u can be written

$$u_0(r,\,\theta) = \sqrt{\frac{ik}{2\pi}} \int A(x) \exp(ik\,(S(x) - x \cos \alpha + r \cos(\theta - \alpha))) \sqrt{\frac{-\sin\alpha\,dx}{d\alpha}}\, d\alpha. \tag{4.33}$$

We have implicitly assumed in the discussion above that all of the rays emanating from arbitrary points on the real axis are real, that is, the absolute value of the first derivative of the phase is less than k over this line. Consequently, the integral in (4.33) is over a real interval $[-\pi/2, \pi/2]$.

Let us now apply (4.33) to a point whose abscissa is x'_0. Introducing a change of variable $p = \cos \alpha$ in the integrand, we obtain

$$u_0(r,\,\theta) = \left(\frac{ik}{2\pi}\right)^{1/2} \int A(x) \sqrt{\frac{dx}{dp}} \exp(ik\,(S(x) - px + px'))\, dp. \tag{4.34}$$

where $(dx/dp) > 0$, and $\sqrt{dx/dp}$ is the usual square root. We will find this result again in Chap. 6 where we will employ the Maslov's method to handle this problem. In fact (4.32) is identical to (6.22) in Sect. 6.1.2, if we introduce a change of notation

$$x \rightarrow x_{1s} \quad p \rightarrow p_1 \quad x' \rightarrow x \quad \text{for the case} \quad \frac{dx_{1s}}{dp_1} > 0.$$

For the case of $(d\alpha/dx) > 0$, the caustic introduces a $-\pi/2$ phase shift and instead of (4.29) we have

$$u_0(N) = - i A(x) \exp (ik (S(x) + t)) \sqrt{\frac{\rho}{t}}. \tag{4.35}$$

After going through the same steps as employed in the derivation of (13), we now get

$$u_0(r, \theta) = \left(\frac{ik}{2\pi} \right)^{1/2} \int (-i)A(x) \sqrt{-\frac{dx}{dp}} \exp (ik (S(x) - px + px')) \, dp. \tag{4.36}$$

The above result, which is derived somewhat intuitively, will be shown to be identical to (6.22) in Sect. 6.1.2 for the case $(dx_{1s}/dp_1) < 0$, where the Maslov's approach will be employed.

Thus far we have only discussed the case of $|(dS/dx)| < 1$, which leads to real rays. It is possible, at least formally, to extend (4.34) and (4.36) for the case where (dS/dx) varies in R. To this end, we notice that (4.34) and (4.36) are identical under the condition that when dx/dp is less than zero, then $\sqrt{dx/dp}$ is replaced by $-i\sqrt{|dx/dp|}$.

We thus see that it is possible to derive the plane wave spectral representation for an arbitrary current source from a knowledge of the induced field on a straight line. Another useful application of this approach is to the problem of diffraction by an object with an edge. We have seen in Chap. 3 that this field was not a ray field, either in the vicinity of the edge or in the proximity of the light-shadow boundary. However, it is possible to represent the field diffracted by a wedge in terms of its plane wave spectrum. In the following section, we will discuss the particular case of diffraction of non-ray fields by a half-plane and show how the PWS approach can be used to derive simple representations of the solutions.

4.3.4 Field diffracted by a perfectly conducting half-plane

The field diffracted by a half-plane can be represented in terms of a plane wave spectrum. Let us consider a TM plane wave with a unit amplitude incident upon a half-plane (see Fig. 4.6).

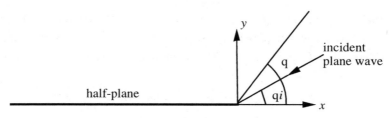

Fig. 4.6. Diffraction by a half-plane

The total field is the sum of the incident field u^i given by

$$u^i = \exp\,(-ikr\,\cos(\theta - \theta^i)),\tag{4.37}$$

and the diffracted field u^d. It can be shown [RM] that u^d also can be written in its plane wave spectral representation with its weight function $p(\alpha, \theta_i)$ given by

$$p(\alpha, \theta_i) = -\frac{i}{\pi}\,\frac{cos(\alpha/2)\,cos(\theta^i/2)}{cos\,\alpha + cos\,\theta^i}.\tag{4.38}$$

Note that $p(\alpha, \theta_i)$ has a pole at $\alpha = \pi - \theta^i$, i.e., at the angle of reflection. The contour of integration is the Sommerfeld contour C, which is indented traverse underneath the pole

$$u^d(r, \theta) = \int_C p(\alpha, \theta_i)\ exp\,(ikr\,cos\,(\alpha - |\theta|)\,d\alpha.\tag{4.39}$$

Equation (4.39) is not only useful for understanding the basic concepts of various uniform theories that have been devised to circumvent the difficulties with GTD (see Chap. 5), but also for investigating the problem of multiple diffraction, e.g., the diffraction by two half-planes. The plane wave spectral representations exist for a wedge, both when it is a perfect conductor and when its surface satisfies an impedance boundary condition. For details on this problem, we refer the reader to two works on this subject [IM], [RM].

4.3.5 Fock field (Fig. 4.7)

We have seen, in Chap. 3, that the field in the vicinity of a point illuminated by a wave at a grazing incidence can be expressed in terms of Fock functions. It is easily verified that this field is not a ray field; however, Michaeli [Mi] has derived a PWS representation by following the procedure described below.

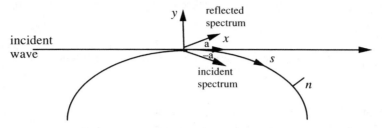

Fig. 4.7. Decomposition of the Fock in terms of a spectrum of plane waves

For the TE polarization, Michaeli [Mi] uses a Taylor expansion for the field in the vicinity of the surface, derived by Pathak. He obtains, for a unit incident field

$$u \approx \int_{-\infty}^{+\infty} exp(iks)\, exp(iks\, \tau/2m^2)\, \frac{1+(kn/m)^2\, \tau/2}{\sqrt{\pi}\, w_1'(\tau)}\, d\tau, \qquad (4.40)$$

where m is the Fock's parameter $(ka/2)^{1/3}$. Next, Michaeli rewrites the term $1+(kn/m)^2$ $\tau/2$ in terms of a hyperbolic cosine as

$$1+(kn/m)^2\, \tau/2 \approx \frac{1}{2}(exp(kn\sqrt{\tau}/m) + exp(-kn\sqrt{\tau}/m)), \qquad (4.41)$$

which, in turn, enables him to evaluate the integral in (4.40) and write it as

$$(1/2\sqrt{\pi}\; w_1'\,(\tau))\, \{(exp\; iks\; (1+\tau/2m^2) - kn\;\sqrt{\tau}/m) +$$

$$+ (exp\; iks\; (1+\tau/2m^2) + kn\;\sqrt{\tau}/m)\}. \qquad (4.42)$$

Michaeli then identifies $i\sqrt{\tau}/m$ with $\sin \alpha$ and $1 + \tau/2m^2$ with $\cos \alpha$, where α is a complex parameter. Furthermore, he identifies the parameters s and n with the Cartesian coordinates x and y in the vicinity of C. After incorporating all of these approximations, (4.42) can be rearranged in the form of a sum of an incident plane wave

$$exp\;(ik\;(x \cos \alpha - y \sin \alpha)),$$

and a reflected wave with the reflection coefficient $R = 1$ (corresponding to a TM wave reflecting from a perfect conductor), i.e.,

$$exp\;(ik\;(x \cos \alpha + y \sin \alpha))$$

Finally, using the above, the field u is written in terms of a plane wave representation as follows

$$u \approx \int_{-\infty}^{+\infty} (1/2\sqrt{\tau} \ w_1'(\tau)) \ \{\exp(ik \ (x \cos \alpha - y \sin \alpha)) + \exp(ik \ (x \cos \alpha + y \sin \alpha)\}d\tau$$

(4.43)

Michaeli also presents the plane wave spectral representation for the field at a point M in the shadow region whose abscissa is s_0 and shows that it is only necessary to multiply the weight function of (4.43) by the factor $\exp(iks_0 \ (1 + \tau/2m^2))$.

For the TM polarization, Michaeli obtains a result analogous to the one given in (4.43) except that the reflection coefficient, which previously was $+1$ for the TE case, is now replaced by -1. While Michaeli's approach to the derivation of the PWS representation of the Fock field is largely heuristic, and thus not strictly rigorous, the representation in (4.43) is nonetheless very useful, as we will see in Sect. 4.4.5, for computing the diffraction of waves at a grazing incidence.

4.3.6 Other examples

We have seen in Chap. 3 that the fields in the vicinity of a shadow boundary can be described, up to the highest order, by a Fresnel function. Clemmow has shown [Cl] that this Fresnel function, in turn, can be written in the form of a plane wave spectrum. As for caustics, we will see in Chap. 6, that Maslov's method is useful in deriving the PWS representation for the fields in their vicinity. We can, in fact, make a general statement that most of the non-ray fields may be expressed, directly or indirectly, in terms of their plane wave spectral representation. As a result, the problem of diffraction of complex fields is often handled by using the plane wave spectrum method. In next section we will provide a number of examples to illustrate this point.

4.4 Diffraction of complex fields diffraction – Case examples

4.4.1 Diffraction of surface waves

Let us consider a surface wave propagating on a planar surface with surface impedance Z, which we assume to be reactive, i.e., $Z = -iX$. Let the incident surface wave be given by

$$u = \exp(ik \sqrt{1 + X^2} \ x - kXy),$$

(4.44)

and let this wave be incident along one of the faces of the wedge as shown in Fig. 4.8. The incident surface wave encounters the edge of the wedge from which it diffracts as shown in Fig. 4.8.

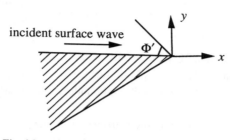

Fig. 4.8. Incident surface wave on a dihedral's edge

According to the discussion in Sect. 4.3.1, the surface wave may be viewed as though it were the sum of two waves, viz., a wave (actually zero) incident upon the edge at an angle $\Phi' = \varphi_s$ with respect to the face of the wedge that represents the lit side, where $\sin \varphi_s = -Z = iX$, and a reflected wave. We can derive the diffracted field u^d generated by this incident surface wave field by dividing it by the factor $1 + R$, where $R(\Phi')$ is the reflection coefficient, and multiply the result with the diffraction coefficient $D(\Phi, \Phi')$ of the edge

$$u^d = \lim_{\Phi \to \varphi_s} \frac{D(\Phi, \Phi')}{1 + R(\Phi')} \frac{exp(ikr)}{\sqrt{r}}. \qquad (4.45)$$

We find, in this case, a result similar to the one obtained by using the method of the hybrid coefficients of diffraction proposed by Albertsen and Christiansen [AC] and described in Chap. 1. For the case of surface wave incidence, both $1 + R(\Phi')$ and $D(\Phi, \Phi')$ simultaneously tend to infinity although their ratio remains bounded. However, as shown below, it is possible to arrive at an alternative interpretation of the diffraction of the surface wave. To this end, we return to (4.44), and interpret it as a representation for a wave incident upon the wedge from the direction $-\varphi_s$. The reflection coefficient is equal to zero so that (4.44) only represents the incident field. The diffracted field is then given by

$$u^d = D(\Phi, -\varphi_s) \frac{exp(ikr)}{\sqrt{r}}. \qquad (4.46)$$

The equivalence of both of these approaches can be easily verified for the case where the surface contains a discontinuity in the curvature. In fact, for this case, we have

$$D(\Phi, -\Phi') = \frac{-\sin(\Phi') + Z}{\sin(\Phi') + Z} \, D(\Phi, \Phi') = \frac{D(\Phi, \Phi')}{R(\Phi')}, \tag{4.47}$$

so that (4.45) and (4.46) yield identical results.

We will next turn to the example of surface wave excitation by a line current located on a plane described by an impedance boundary condition.

4.4.2 Surface wave excitation by a line source over an impedance plane

Let us consider a magnetic current line source of unit strength, located at the origin, and above a plane described by a surface impedance Z (Fig. 4.9a).

Let the line current radiate a magnetic field u, directed along z. This field is considered as the incident field over the plane $y = 0$. The line current is assumed to be just above the impedance plane. Therefore, we must consider the case $y < 0$, or choose the + sign in (4.28) in Sect. 4.3.2. The incident field on the plane is represented by

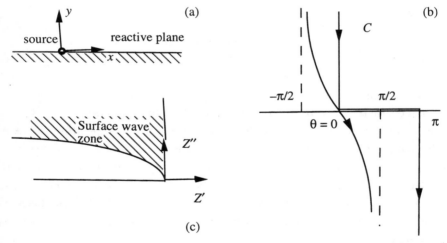

Fig. 4.9. Source on the surface of a reactive plane (a) geometry of the problem, (b) deformation of the surface of integration, (c) region of excitation of the surface wave

$$u^i = -\frac{k}{4\pi Z_0} \int_C \exp\left(ikr \cos\left(\theta + \alpha\right)\right) d\alpha, \tag{4.48}$$

where exp $(ikr \cos(\theta + \alpha))$ is an incident plane wave with angle $\Phi = \alpha$ which generates a reflected wave given by $(\sin \alpha - Z)/(\sin \alpha + Z)$ exp $(ikr \cos(\theta - \alpha))$. Thus, the reflected field may be written as

$$u^r = - \frac{k}{4\pi Z_0} \int_C \frac{\sin \alpha - Z}{\sin \alpha + Z} \exp(ikr \cos(\theta - \alpha)) \, d\alpha. \tag{4.49}$$

Let us consider the fields on the surface of the plane for $x > 0$, that is to say $\theta = 0$. The steepest descent path traversing through the saddle point $\alpha = 0$ is given by $\cos \alpha'$ $\cosh \alpha'' = 1$ and $\sin \alpha' \sin \alpha'' < 0$ (Fig. 4.9b).

Finally, the reflected field may be expressed as a sum of the following two contributions: (a) the saddle point contribution which happens to be the opposite of the incident field; and, (b) the contribution of the pole of the integrand α_p given by $\sin \alpha_p = -Z$, with the latter contribution only existing when the pole is crossed in the process of deforming the contour C from the original path S. The pole $\alpha_p = \alpha'_p + i\alpha''_p$ satisfies the following condition

$$\sin \alpha'_p \cosh \alpha''_p = -Z', \tag{4.50}$$

$$\cos \alpha'_p \sinh \alpha''_p = -Z''. \tag{4.51}$$

For $Z' \geq 0$, (4.50) imposes the constraint $\alpha'_p \geq 0$. If $Z'' > 0$, from (4.51) we have $\alpha''_p < 0$ when $\alpha'_p \in [-\pi/2, \pi/2]$. Under these conditions, the pole is not crossed and there is no surface wave contribution. If $Z'' < 0$, the pole is crossed and the surface wave exists if and only if $\cos \alpha'_p \sinh \alpha''_p > 1$. The transition between these two cases takes place when the curve $(1 + Z'^2)(1 - Z''^2) = 1$ is crossed (see Fig. 4.9c).

For the case where the surface wave is present, i.e., α_p is crossed by the deformed contour, the pole contribution can be obtained by using the residue calculus to yield the following expression for the scattered field

$$u^s = - \frac{ik}{Z_0} \frac{Z}{(1 - Z^2)^{1/2}} \exp ik \, (x \, (1 - Z^2)^{1/2} - y \, Z), \tag{4.52}$$

where the sign of the square root in $(1 - Z^2)^{1/2}$ is chosen such that its imaginary part is positive, if Z is not pure imaginary. In the event that Z is pure imaginary, $(1 - Z^2)^{1/2}$ is chosen to be positive.

For a negative x, we can readily derive the desired results by simply replacing x by $-x$ in (4.52). Thus we see that the plane wave spectrum approach enables us to directly calculate the radiation from a source over a plane surface impedance and, more generally, over a layered dielectric media. For this problem, we can also derive the solution in a more rigorous but less physical way by using the Fourier transform approach (which also forms the basis of the plane wave spectrum method) as shown, for instance, in Vassalo [Va].

Next, we will discuss a simple example of a multiple diffraction problem and derive its solution by using the plane wave spectrum approach.

4.4.3 Diffraction by two half-planes (Fig. 4.10)

Let us consider the particular case where both the half-planes are parallel to the x axis and where the incident field is a unit TM plane wave propagating along the y axis.

The lower half-plane, located at $y = 0$ and $x < 0$, is illuminated by what might be described as a transition field, since its edge is located on the light-shadow boundary where the classical GTD is known to fail. However, u_1, the field diffracted by the

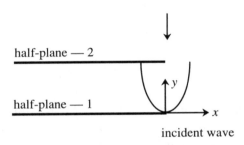

Fig. 4.10. Diffraction by two half planes

upper half-plane located at $y = d$, $x < 0$, may be written in terms of its plane wave spectral representation (see Sect. 4.3.4.) as follows

$$u_1 = u^i (A) \int_C p\, (\alpha, \theta_i) \exp\, (ik\, (x \cos \alpha - (y - d) \sin \alpha))\, d\alpha. \qquad (4.53)$$

The choice of the negative sign for the $(y - d) \sin \alpha$ term in the exponent of (4.53) is dictated by the fact that the field is being calculated for $y < d$. Equation (4.53) is a representation in terms of a superposition of plane waves that propagate in the negative α direction. Thus, these plane waves are incident upon the lower half-plane with an angle $\pi - \alpha$ and each of these waves, in turn, generates a diffracted field which can be expressed in terms of a plane wave spectral representation. For $y < 0$ this representation takes the form

$$\exp\, (ikd \sin \alpha) \int_C p\, (\beta, \pi - \alpha) \exp\, (ik\, (x \cos \beta - y \sin \beta))\, d\beta. \qquad (4.54)$$

Hence the field u_2, which has been diffracted by both the half-planes, can be written as a double integral

$$u_2 = u^i (A) \int_C \int_C p\,(\alpha,\,\theta_i)\, p\,(\beta,\,\pi-\alpha)\, \exp\,(ikd \sin \alpha)\, \exp(ik(x \cos \beta - y \sin \beta))\, d\alpha d\beta$$

$$(4.55)$$

The integral in (4.55) provides a uniform representation of the doubly-diffracted field regardless of the distance between the two half-planes. Mittra and Ramhat-Samii [RM] have dealt with the general case where the angle of incidence is arbitrary, and have asymptotically evaluated the corresponding double integrals.

There is a wealth of literature on many other examples of multiple diffraction that have been treated by the PWS method. The general procedure in all of these cases is to use the plane wave spectral representation of the field diffracted by the first object, and subsequently diffract each of these constituent plane waves from the second object. We refer the reader to [MR, OL, TM, RM] for additional examples that illustrate this approach. In addition, the PWS approach enables us to calculate the diffraction of a wave at grazing incidence by a wedge with curved faces by employing a spectral representation of the Fock field given in Sect. 4.3.5. We will discuss this case in the next section.

4.4.4 Grazing-incidence diffraction by a wedge with curved faces

Let us consider a TM polarized plane wave incident upon a curved wedge at a grazing angle as shown in Fig. 4.11.

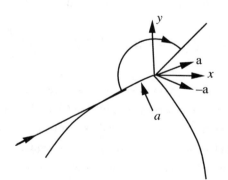

Fig. 4.11. Fock field diffraction by a curved wedge

As a first step, we write the Fock field (see Sect. 4.3.5) on the wedge in terms of a plane wave spectral representation as a sum of an incident wave $u(\alpha)$, at an angle α, and a reflected wave with an angle $-\alpha$ as follows

$$u \approx \int A(\tau) \left(u(\alpha) + u(-\alpha) \right) d\tau, \tag{4.56}$$

with

$$A(\tau) = u^i(Q') \, \frac{exp[i \, s_0(k + \tau m / a)]}{2\sqrt{\pi} \, w_1'(\tau)}. \tag{4.57}$$

Each of the plane wave spectral components diffracts in the direction Φ and is weighted by $D(\pm \alpha, \Phi)$, where D is the diffraction coefficient of the wedge given in Sect. 1.4. This representation is valid outside the transition zone of the diffracted field, i.e., when $\Phi \pm \alpha$ is not close to π. The diffracted field u^d can be written in terms of a superposition integral as

$$u^d \approx \int A(\tau) \left(D(\alpha, \Phi) + D(-\alpha, \Phi) \, d\tau \right) \frac{e^{ikr}}{\sqrt{r}}. \tag{4.58}$$

The above analysis is valid for α small. For this case, we can introduce the approximation $D(\alpha, \Phi) \approx D(-\alpha, \Phi) \approx D(0, \Phi)$ in (4.58), suggested by Michaeli, to yield

$$u^d \approx \frac{u^i(Q') \, exp(iks_0)}{2\sqrt{\pi}} \int \frac{exp(is_0 \, \tau m / a)}{w_1'(\tau)} \, d\tau \, D(0, \Phi) \frac{e^{ikr}}{\sqrt{r}}. \tag{4.59}$$

Recognizing that the second part of (4.59) can be rewritten in terms of the Fock function $g\left((s_0/a) \, m\right)$, we can reorganize (4.59) to read

$$u_d \approx \frac{1}{2} u^i(Q') \, exp(iks_0) \, g\left(\frac{s_0}{a} m\right) D(0, \Phi) \frac{e^{ikr}}{\sqrt{r}}, \tag{4.60}$$

which is the result we were seeking. We remark here that (4.60) can be obtained, heuristically, by following the procedure outlined in Chap. 1 for creeping wave diffraction. In this approach, we regard the Fock field as a grazing incidence field comprising of a combination of incident and reflected fields that are equal in magnitude. Under this assumption, the incident field is half of the total field, and is given by $(1/2) \, u^i(Q') \, exp(iks_0) \, g\left((s_0/a) \, m\right)$. Furthermore, since this field is incident at a grazing angle, its associated diffraction coefficient is $D(0, \Phi)$. Although not as rigorous, this procedure leads to a result that is identical to the one we obtained by using the plane wave spectrum approach.

The method described above can also be applied to the TM polarization case for which the diffracted field may be written [Mi] as

$$u_d \approx \frac{1}{2} u^i (Q') \; \frac{i}{m} f \left(\frac{s_0}{a} m\right) \exp (iks_0) \; \frac{\partial D(0, \Phi)}{\partial \alpha} \frac{e^{ikr}}{\sqrt{r}}. \qquad (4.61)$$

In addition to being useful for the analysis of TE and TM diffraction from wedges, the PWS method can also be applied to a number of other problems, e.g., the diffraction by a discontinuity in the curvature or a discontinuity of the nth derivative of the surface. To obtain the solution to the above problems, it is merely necessary to replace the diffraction coefficient of the wedge in (4.60) or (4.61) by the corresponding coefficient of the appropriate discontinuity. When the surface of the wedge is described by a surface impedance condition, the expression given in (4.60) for the p.e.c. wedge can be generalized to yield

$$u^d \approx u^i (Q') \; \exp (iks_0) \, F_Z \left(\frac{s_0}{a} m\right) \lim_{\alpha \to 0} \frac{D(\alpha, \Phi)}{1 + R(\alpha)} \frac{e^{ikr}}{\sqrt{r}} \qquad (4.62)$$

where F_Z is the Fock function for an impedance Z. We note that for large arguments of the Fock functions, g in (4.60) and f in (4.61), F_Z in (4.62) can be expressed in terms of a set of creeping waves. Thus, for this situation, we can again use the hybrid diffraction coefficients of Albertsen and Christiansen that were discussed in Chap. 1 in connection of the problem of creeping wave diffraction by discontinuities.

We summarize the discussion presented in this section with the statement that while the Geometrical Theory of Diffraction is useful for investigating the problem of diffraction by rays fields, the plane waves spectrum method enables us to handle the more general case of non-ray fields that are often encountered in diffraction analysis. In this section, we have presented some representative examples of the application of the PWS method. Other applications of this method, notably to the problems of diffraction by plane objects, may be found in two review articles by Guiraud [G1, G2].

References

[AC]   N. C. .Albertsen and P. L. Christiansen, "Hybrid diffraction coefficients for first and second order discontinuities of two-dimendionnal scatterers," *SIAM J.Appl. Math*, **34**, 398-414, 1978.

[Cl]   P. C. Clemmow, "The plane wave spectrum representation of electromagnetic fields," Pergamon Press, 1966.

[G1]   J. L. Guiraud, "Introduction à la théorie spectrale de la diffraction," *Revue du CETHEDEC*, **69**, 81-116, 1981.

[G2]   J. L. Guiraud, "Une approche spectrale de la théorie physique de la diffraction," Annales des Télécom, **38**, 145-157, 1983.

[IM]   L. Ivrissimitis and R. Marhefka, "Double diffraction at a coplanar edge configuration," *Radio Science*, **26**, 821-830, 1991.

[Mi]   A. Michaeli, "Transition functions for high-frequency diffraction by a curved perfectly conducting wedge, Part II : A partially uniform solution for a general wedge angle," *IEEE Transactions on Antenna and Propagation*, **37**, 1080-1085, Sept. 1989.

[MR]   R. Mittra, Y. Rahmat-Samii, and W. L. Ko, "Spectral theory of diffraction," *Appl Phys*, **10**, 1-13, 1976.

[Pa]   P. H. Pathak, "An asymptotic analysis of the scattering of plane waves by a smooth convex cylinder," *Radio Science*, **14**, 419-435, 1979.

[OL]   Y. Orlov and V. Legkov, "Diffraction of the half-shadow field by a perfectly conducting smooth convex cylinder," *Radio Eng. Elect. Phys*, **2**, 249-257, 1986.

[RM]   Y. Rahmat-Samii and R. Mittra, "A spectral domain interpretation of high-frequency diffraction phenomena," *IEEE Trans Ant Prop*, **AP-25**, 676-687, Sept. 1977.

[Ro]   E. Roubine and J. C. Bolomey, "Antennes," Masson, 1986.

[TM]   R. Tiberio, G. Manara, G. Pelosi, and R. C. Kouyoumjian, "High-frequency electromagnetic scattering of plane waves from double wedges," *IEEE Trans Ant Prop*, **AP-37**, 1172-1180, 1989.

[Va]   C. Vassalo, "Théorie des guides d'ondes electromagnetiques," *Eyrolles*, 1987.

5. Uniform Solutions

5.1 Definition and properties of a uniform asymptotic expansion

In this chapter we will present the uniform solutions in the context of GTD, and we begin by defining the properties of uniform asymptotic expansions in this section. Let $f(X,\varepsilon)$ be a scalar or vectorial function of the variable $X \in D$, where D is a given domain, depending upon the small parameter $\varepsilon \in R_0$. The expansion

$$F(X,\varepsilon) = \sum_{n=0}^{N} a_n(X) \nu_n(\varepsilon), \qquad (5.1)$$

is said to be *uniform*, if it is uniformly valid in all of the domain D. This implies that for all $X \in D$, the following relationship holds.

$$f(X,\varepsilon) = F(X,\varepsilon) + o(\nu_N(\varepsilon)). \qquad (5.2)$$

This definition implies that the error in f remains on the order of $\nu_N(\varepsilon)$ not only when ε tends to zero for a fixed X, but also when X has possible variations and is given by $X = X(\varepsilon)$, provided that X continues to remain in the domain D. When the condition (5.2) is satisfied by f, the perturbation induced in f by the parameter ε is said to be regular.

Typically, an asymptotic expansion is not uniform in certain regions which are called the regions of non-uniformity. These regions are associated with the boundary layers introduced in Chap. 2 and discussed in Chap. 3.

There may be a number of reasons why an expansion may not be uniform. In diffraction problems, for instance, the non-uniformity is essentially due to the presence of singularities in the coefficients $a_n(X)$ and also to the fact that the domain of X is infinite. The non-uniformity of the expansions in the boundary layers is attributable to the latter cause, where it manifests itself by the presence of the secular terms in X which causes $a_{n+1}/a_n \to \infty$ as $X \to \infty$.

In the vicinity of a point of non-uniformity, which one can choose as the origin to X, the expansion (5.1) is valid as ε tends to zero for a fixed X. However, it ceases to represent the function f to be expanded if both ε and X tend simultaneously to zero. It is thus necessary to choose the order of magnitude of X with respect to ε in order for the uniform expansion to be useful. Let us now define

$$X_1 = \frac{X}{\eta_1(\varepsilon)} \quad , \quad X_2 = \frac{X}{\eta_2(\varepsilon)} . \qquad (5.3)$$

For $X \in D$, one then can define the domains D_1 and D_2 as

$$X_1 \in D_1 \quad , \quad X_2 \in D_2. \qquad (5.4)$$

Suppose the variable X is fixed, and that $F_1(X_1, \varepsilon)$ and $F_2(X_2, \varepsilon)$ are the two approximations of the same function $f(X, \varepsilon)$ that are built up to the orders $v_{N1}(\varepsilon)$ and $v_{N2}(\varepsilon)$, respectively, in the domains D_1 and D_2, i.e.,

$$F_1(X_1,\varepsilon) = \sum_{n=0}^{N_1} v_n(\varepsilon)\, a_n^{(1)}(X_1),$$

$$(5.5)$$

$$F_2(X_2,\varepsilon) = \sum_{n=0}^{N_2} v_n(\varepsilon)\, a_n^{(2)}(X_2).$$

Then the approximation F_2 is said to be contained in F_1 if, after expressing F_1 in the variable X_2, one regains to the order v_{N2} all of the terms of F_2. If an approximation is contained in another, it is referred to as "non-significant"; if not, the approximation is called "significant."

There exist as many asymptotic expansions of a function $f(X, \varepsilon)$ with respect to a given compatible asymptotic sequence as there are possible choices in the order of magnitude of X relative to ε, i.e., an infinite number. However, the approximations, thus obtained, are generally non-significant and an expansion for a given order of magnitude of X is in fact valid in the neighborhood of this order of magnitude. This enables one, in general, to only introduce a finite number of distinct expansions corresponding to domains in which each expansion is uniformly valid.

Generally speaking, when we speak of a *domain* in the context of asymptotic expansions, we are referring to a set of values of the variable that are characterized by their order of magnitude as compared to the small parameter of the problem. More specifically, we denote as *closed domain* $D([\eta_1(\varepsilon), \eta_2(\varepsilon)])$ the set of values of X such that

$$\eta_1 = O(X) \quad , \quad X = O(\eta_2),$$

where $\eta_1(\varepsilon)$ and $\eta_2(\varepsilon)$, which are the boundaries of the domain D, are the two given gage functions. Similarly, we refer to the *open domain* $D(]\eta_1(\varepsilon), \eta_2(\varepsilon)[)$ the set of values of X such that

$$\eta_1 = o(X) \quad , \quad X = o(\eta_2), \tag{5.7}$$

Finally, the domains $D([\eta_1, \eta_2])$ and $D([\eta_1, \eta_2[)$ are defined in an analogous manner.

Extension theorem of KAPLUN

If an asymptotic expansion is uniformly valid in a closed domain $D([\eta_1, \eta_2])$, it continues to remain valid in an open domain $D'(]\eta_1', \eta_2'[)$ containing $D([\eta_1, \eta_2])$. This theorem, which has been proven in Ref. [C], provides a justification for decomposing the outer domain of a diffracting object into a finite number of boundary layers with correspondingly distinct asymptotic expansions. However, this theorem does not specify the limits within which the validity of this expansion can be extended, and no general result of this nature has yet been established. Thus, it is not possible to demonstrate the existence of an intermediate domain, which is sandwiched between an outer and an inner domain, and within which the inner and outer expansions are simultaneously valid. As a consequence of this, it becomes necessary to verify that there exists a common domain of validity which is shared by the two expansions. The expansion in the intermediate domain represents a non-significant approximation of the initial function $f(X, \varepsilon)$, as it is simultaneously included in the outer and inner expansions. This property serves as the foundation of the matched expansion methods presented in Sects. 2.1.4 and 3.10.

5.2 General approach to the derivation of uniform solutions

It is possible to construct a uniform expansion which is valid in the entire domain of the variable X, once we know the outer and inner expansions, as well as the intermediate one which will be defined below. Let $F_e(X, \varepsilon)$ and $F_i(\tilde{X}, \varepsilon)$ be the outer and inner expansions, respectively, given by

$$F_e(X, \varepsilon) = \sum_{n=0}^{Ne} a_n(X) v_n(\varepsilon), \tag{5.8}$$

$$F_i(\tilde{X}, \varepsilon) = \sum_{n=0}^{N_i} b_n(\tilde{X}) \mu_n(\varepsilon), \tag{5.9}$$

where the variable $\tilde{X}$ of the inner domain is the stretched variable defined by

$$\tilde{X} = \frac{X}{\varepsilon^{\alpha}}, \quad \alpha > 0. \tag{5.10}$$

Similarly, one denotes the intermediate variable $\overline{X}$ as

$$\overline{X} = \frac{X}{\eta(\varepsilon)}, \tag{5.11}$$

where $\eta(\varepsilon)$ is a function with an arbitrary order of magnitude ranging between ε^{α} and 1 (intermediate scale) such that

$$\begin{aligned} \eta(\varepsilon) &= o(1), \\ \varepsilon^{\alpha} &= o(\eta(\varepsilon)). \end{aligned} \tag{5.12}$$

Let $F_m(\overline{X}, \varepsilon)$ be the intermediate expansion given below, which is obtained by replacing the variable X by $\overline{X}$ in (5.8) and expanding the terms with respect to the asymptotic sequence $v_n(\varepsilon)$ up to the order $v_{Ne}(\varepsilon)$:

$$F_m(\overline{X}, \varepsilon) = \sum_{n=0}^{Ne} C_n(\overline{X}) v_n(\varepsilon). \tag{5.13}$$

Then, the expression

$$\hat{f}(X, \varepsilon) = F_e(X, \varepsilon) + F_i(\tilde{X}, \varepsilon) - F_m(\overline{X}, \varepsilon), \tag{5.14}$$

is a representation of the initial function $f(X,\varepsilon)$, that is uniformly valid in all D. To demonstrate this result we first write $\hat{f}$, as shown below, as a function of the variable $\overline{X}$, defined in (5.11). The function of order $\eta(\varepsilon)$ is now of order less than or equal to 1, so that $\overline{X}$ describes the entire domain D. The expression for $\hat{f}$ reads

$$\hat{f}(\eta\overline{X}, \varepsilon) = F_e(\eta\overline{X}, \varepsilon) + F_i\left(\eta\frac{\overline{X}}{\varepsilon^{\alpha}}, \varepsilon\right) - F_m(\overline{X}, \varepsilon). \tag{5.15}$$

Now, if $\eta(\varepsilon) = O(\varepsilon^{\alpha})$, then X is in the inner domain, since, by definition $\overline{X} \approx 1$ and $X = \eta(\varepsilon)$. Under these conditions one has

$$F_e(\eta\overline{X}, \varepsilon) = F_m(\overline{X}, \varepsilon) + o(v_{Ne}(\varepsilon)), \tag{5.16}$$

which is a consequence of the manner in which the expansion $F_m(X, \varepsilon)$ was constructed. It follows, then, that $\hat{f}$ reduces to the inner expansion which, according to (5.9), is an approximation of $f(X, \varepsilon)$ in the inner domain with an error that is $o(\mu_{Ni}(\varepsilon))$.

It is important to note that for $\eta(\varepsilon) = O(\varepsilon^\alpha)$, F_e and F_m are no longer approximations of the function $f(X, \varepsilon)$; however, the expansion F_m is still contained in the expansion F_e.

If $\eta(\varepsilon) = O(1)$ and $\varepsilon^\alpha = o(\eta)$, then X is in the outer domain. Furthermore, since F_m is contained in F_i, $F_m(\overline{X}, \varepsilon)$ is also the expansion of $F_i\left(\dfrac{\eta \overline{X}}{\varepsilon^\alpha}, \varepsilon\right)$ with respect to $\nu_n(\varepsilon)$ for a fixed $\overline{X}$, and one can write

$$ F_i\left(\frac{\eta \overline{X}}{\varepsilon^\alpha}, \varepsilon\right) = F_m(\overline{X}, \varepsilon) + o(\nu_{Ne}(\varepsilon)). \tag{5.17} $$

As in the previous case, F_i and F_m no longer provide an approximation of $f(X, \varepsilon)$, although F_m is still contained in F_i. It follows, then, that $\hat{f}$ reduces to the outer expansion which represents the function to be expanded in the outer domain up to the order $\nu_{Ne}(\varepsilon)$.

To summarize, the representation in (5.14) is uniformly valid to within $\nu_{Ne}(\varepsilon)$ in the entire domain D. It is a composite solution, also referred to as *a composite expansion* [N], which can be rewritten as a sum of two terms, as given below, after regrouping the terms F_i and F_m. We then obtain

$$ \hat{f}(X, \varepsilon) = G(\tilde{X}, \varepsilon) + F_e(X, \varepsilon), \tag{5.18} $$

with

$$ G(\tilde{X}, \varepsilon) = F_i(\tilde{X}, \varepsilon) - F_m\left(\frac{\varepsilon^\alpha}{\eta(\varepsilon)} \tilde{X}, \varepsilon\right). \tag{5.19} $$

Having laid the foundation of the uniform theory, we will now apply this method to the problem of computing the fields in the shadow zone due to the diffraction of a scalar wave by a smooth surface. The inner solution, which is valid in the boundary layer and is expressed in terms of the ray coordinates, can be extracted from formula (3.33) of Sect. 3.1 once it has been multiplied by the divergence factor of the creeping rays $[d\eta(o)/d\eta(s)]^{1/2}$. This factor is equal to unity for the case of a cylindrical surface illuminated by a plane wave and, consequently, it is not present in (3.33), referred to above.

Using the same notations as in Sect. 3.1, the inner solution can be written as

$$F_i(\xi, Y) \cong A \left[\frac{\rho(o)}{\rho(s)} \right]^{\frac{1}{6}} \left[\frac{d(o)}{d(s)} \right]^{\frac{1}{2}} \left[\frac{d\eta(o)}{d\eta(s)} \right]^{\frac{1}{2}} W_1(\xi - Y) exp\left(ikl - i\frac{2}{3} Y^{\frac{3}{2}} \right)$$

$$\times exp\left(i\xi\sqrt{Y} + i\frac{k^{\frac{1}{3}}}{2^{\frac{1}{3}}} \int_0^{s^r} \frac{\xi ds}{\rho(s)^{\frac{2}{3}}} \right),$$

(5.20)

where Y is the stretched coordinate given by (3.49) of Sect. 3.1. The outer solution can be derived from (3.201) of Sect. 3.12 by replacing $1/\sqrt{t}$ with $[\rho_d/\{t(\rho_d + t)\}]^{1/2}$, where ρ_d is the distance of the observation point to the caustic of the diffracted rays, and by multiplying the result with the divergence factor of the creeping rays. This leads to the result

$$F_e(l, s^r) \cong U_0(s^r) \sqrt{\frac{\rho_d}{t(\rho_d + t)}} \left[\frac{d\eta(o)}{d\eta(s)} \right]^{\frac{1}{2}} exp\left(ikl + ik^{\frac{1}{3}} \varphi(s^r) \right),$$

(5.21)

where $l = s^r + t$.

In Sect. 3.12, the unknown functions $U_0(s^r)$ and $\varphi(s^r)$ have been determined by means of a matching technique involving an intermediate expansion. The latter is given by an asymptotic expansion, for large Y, of the inner solution, restricted to the dominant term. Proceeding in the same manner as we did in Sect. 3.12, we obtain

$$F_m(\xi, Y) \cong A \left[\frac{\rho(o)}{\rho(s)} \right]^{\frac{1}{6}} \left[\frac{d(o)}{d(s)} \right]^{\frac{1}{2}} \left[\frac{d\eta(o)}{d\eta(s)} \right]^{\frac{1}{2}} Y^{-\frac{1}{4}} e^{i\frac{\pi}{4}} exp\left(ikl + i\frac{k^{\frac{1}{3}}}{2^{\frac{1}{3}}} \int_0^{s^r} \frac{\xi ds}{\rho(s)^{\frac{2}{3}}} \right).$$

(5.22)

This solution can be identified as the inner limit of the outer solution in the above formula, obtained by letting $t \to 0$. If we only retain the dominant order terms we recover the expressions in (3.202) and (3.203) of Sect. 3.12.

We now have at our disposal all of the terms that are needed to construct a composite solution of the type in (5.14), which is uniformly valid in the shadow zone. This solution is given by

$$U \cong F_i(\xi, Y) - F_m(\xi, Y) + F_e(l, s^r).$$

(5.23)

Equation (5.23) can be rewritten in the form

$$U \cong A \left[\frac{\rho(o)}{\rho(s)} \right]^{\frac{1}{6}} \left[\frac{d(o)}{d(s)} \right]^{\frac{1}{2}} \left[\frac{d\eta(o)}{d\eta(s)} \right]^{\frac{1}{2}} \left[f(\xi, Y) - \hat{f}(\xi, Y) \right] + F_e(1, s^r), \quad (5.24)$$

where $f(\xi, Y)$ is a special function and $\hat{f}(\xi, Y)$ the first term of its asymptotic expansion. Using the notations given in Sect. 3.1, we can write

$$f(\xi, Y) = W_1(\xi - Y) exp\left(i\xi\sqrt{Y} - i\frac{2}{3}Y^{\frac{3}{2}} \right) exp\left(i\frac{k^{\frac{1}{3}}}{2^{\frac{1}{3}}} \int_0^{s^r} \frac{\xi ds}{\rho(s)^{\frac{2}{3}}} \right), \quad (5.25)$$

and in view of (3.1) of Sect. 3.1 we have

$$\hat{f}(\xi, Y) = Y^{-\frac{1}{4}} e^{i\frac{\pi}{4}} exp\left(i\frac{k^{\frac{1}{3}}}{2^{\frac{1}{3}}} \int_0^{s^r} \frac{\xi ds}{\rho(s)^{\frac{2}{3}}} \right). \quad (5.26)$$

One can readily verify that (5.24) reduces to $F_e(1, s^r)$ when $Y \to \infty$ and to $F_i(\xi, Y)$ when $t \to 0$.

For the case of a cylindrical surface illuminated by a plane wave, one obtains $F_m(\xi, Y) = F_e(1, s^r)$ and the composite solution (5.24) is identical to the inner solution expressed in terms of the ray coordinates which, consequently, is uniformly valid owing to the coordinate system attached to the rays that we have chosen to express the inner solution. As a matter of fact, the choice of such a coordinate system is optimal in the method for matching the inner and outer solutions. For instance, had we chosen the coordinates (s, n) for matching, the inner solution would not have been equal to the composite solution.

For a three-dimensional surface, the ray coordinate system is again well-suited for the purpose of matching; however, as we have just shown, the matching technique does not directly lead to a uniform solution. We can still arrive at such a solution by slightly modifying the internal solution as follows. If we multiply the expression (5.20) for $F_i(\xi, Y)$ by the factor $D = (\rho_d/\rho_d + t)^{1/2}$, we obtain the identity $F_m(\xi, Y) = F_e(1, s^r)$, which enables us to generalize, for an arbitrary surface, the result which we established previously for the particular case of the cylindrical surface, namely that for a source located far from the surface, the inner solution expressed in terms of the ray coordinates is uniform. Inside the boundary layer, we have $t/\rho_d \ll 1$, and therefore the modification of the inner solution due to the factor D is negligible; however, this factor is indispensable for rendering the solution uniform.

The procedure we just employed to modify the inner solution, viz., the multiplication of this solution by the factor D, is not always applicable, and such a matching technique does not, in general, lead to a uniform solution directly. As a consequence, two expansions are usually needed in practice, and one must make a transition from one to the other when the argument reaches a certain value, which is not always well-defined. This is probably the reason why the composite solution is used more frequently in practice. It should be realized, however, that in order to construct the composite solution of a given diffraction problem, one must carry out in detail, all of the steps of the matching expansion techniques, and these can be quite tedious, especially in the presence of several boundary layers.

Rather than obtaining the inner and outer expansions as a first step, and then matching them to construct a composite expansion, we may begin instead by assuming that the solution is of the form given by (5.18), view it as an Ansatz, and apply all of the boundary conditions to the above equation. The Ansatz generally contains special functions that are characterized by the type of singularities responsible for the non-uniformity of the outer expansion when the variable enters into the inner domain. It is then advantageous to postulate the solution directly in a form where these special functions occur, as for instance in the expression (5.24). This point of view is adopted, for instance, by Kravtsov [Kr] and Ludwig [L] to construct a uniform solution through a caustic (see Sect. 5.7) and by Lewis and Boersma [LB] (see also Ahluwalia [A]) in their uniform asymptotic theory for studying the diffraction by an edge (see Sect. 5.3).

The special functions occurring in the Ansatz can be found by determining the solution to the inner problem, either by using the boundary layer method, or by solving a canonical problem.

With the knowledge of these special functions, one is generally left with some freedom to choose the Ansatz. For example, if F_e in (5.18) is the sum of two asymptotic expansions with respect to different asymptotic sequences, it is possible, under certain conditions, to group together one or the other of these two asymptotic expansions in the function G given in (5.19). This situation arises in the solution of the diffraction problem by a sharp edge, when crossing the shadow boundaries of the incident and reflected fields. In this case, F_e is the sum of the Geometrical Optics field represented by a Luneberg-Kline series, and the diffracted field by the edge whose outer solution is an expansion in terms of the sequence $k^{-(n+1/2)}$. We will see, in Sect. 5.3, that is obtained a uniform solution, called the Uniform Asymptotic Theory (UAT), when the Geometrical Optics solution is incorporated into G. Alternatively, if the solution for the diffracted field by the edge is incorporated into G instead, we are led to a uniform solution referred to as the Uniform Theory of Diffraction (UTD). The method for incorporating these expansions into G is not trivial, however, and depends upon the specific problem under consideration.

One technique, which will be detailed in the following sections, entails the identification of the type of singularity responsible for the non-uniformity of the outer expansion. For the case of a wedge, for instance, we can show that, in the vicinity of the incident and reflected field shadow boundaries, the non-uniformity is associated with the presence of poles near the stationary point in the asymptotic evaluation of the integral defining the exact solution of the canonical problem. Consequently, we seek a uniform asymptotic solution of such an integral which, in turn, provides the Ansatz.

Another example, which will be studied in detail in Sect. 5.7, is concerned with the neighborhood of a regular point of a caustic. We have seen, in Sect. 3.4, that the special function that describes the field at proximity of a caustic is an Airy function, which provides an oscillatory field on one side of the caustic and an evanescent field, which exhibits an exponential decay, on the other side. This behavior is typical of the transition point occurring in a second order differential equation of the type

$$\frac{d^2Y}{dx^2} + \left[\alpha q_1(x) + q_2(x)\right]y = 0,\qquad (5.27)$$

where α is a positive number which is assumed to be large, $q_1(x)$ has a zero at the transition point $X = X_0$, while q_2 is regular at this point. The transition point is also called a turning point because, in classical mechanics, it corresponds to the point where the potential and kinetic energies of an incident particle are equal and, consequently, the particle reverses its path at this point.

We can thus generate an Ansatz for the field in the neighborhood of a regular caustic by following an approach similar to the one employed for the construction of the uniform asymptotic expansion of the solution of an equation of the type (5.27). An alternative approach to generating the Ansatz is to start with the boundary layer solution. In the following sections, we will illustrate the application of these general concepts by considering concrete examples and construct uniform solutions which are found to be very useful in practical applications of the GTD.

5.3   Derivation of solutions that are uniform through the shadow boundaries of the incident field and the field reflected by a wedge

We consider a wedge with curved faces and a curved edge, illuminated by an incident wave which is locally plane and whose direction of propagation is arbitrary (see Fig. 5.1). The principal radii of curvature of the incident wave and of the curved faces of the wedge, as well as the radius of curvature of the edge, are assumed to be large compared to the wavelength of the incident field. We further assume that the faces of the wedge are either perfectly conducting or are characterized by a constant surface impedance on each face.

We seek an asymptotic expansion for the field diffracted by this wedge (see Fig. 5.1), that is uniformly valid as the observation point traverses through the incident and reflected field shadow boundaries.

Keller [K] has derived a non-uniform solution to this problem, which is valid when the observation point is located far from the edge in region 1, is outside the shadow boundaries 2 and 2' of the incident and reflected fields, respectively, and is away from the transition zones 3 and 3' where the diffracted rays in the space are transformed into creeping rays or whispering gallery waves. The solution is given in Sect. 1.2.4.2. in terms of the diffraction coefficients D_s and D_h, which are themselves obtained from the dominant term of the asymptotic expansion of the exact solution of the diffraction problem of a wedge with planar faces that are tangents to the wedge with curved faces at the point of diffraction.

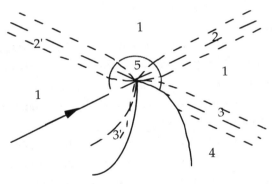

Fig. 5.1. Region of validity of some uniform solutions available in literature for the curved wedge and location of the transistion region.

Keller's solution provides the dominant term of the asymptotic expansion of the field diffracted by the edge with respect to the sequence $k^{-(n+1/2)}$ with $n \geq 0$. The total field is obtained by adding the incident and reflected fields to the field diffracted by the edge. These fields are represented by asymptotic expansions of the Luneberg-Kline type, corresponding to a sequence k^{-n} $(n \geq 0)$. If we limit the expansion of the total field up to $O(k^{-1/2})$, then only the dominant terms of $O(k^0)$ of the incident and the reflected fields appear in this expansion. Outside the transition zones 2 and 2', these terms are given by the standard Geometrical Optics approximation which reads

$$\vec{E}^i_{OG}(P) = \vec{e}^i_0(P)\, U(-\varepsilon_i)\, e^{ikS_i(P)},$$

$$\vec{E}^R_{OG}(P) = \vec{e}^R_0(P)\, U(-\varepsilon_R)\, e^{ikS_R(P)},$$

(5.28)

at an observation point P, where $\vec{e}^i_0(P)$ is the $O(k^0)$ vectorial or complex amplitude of the incident field that would exist at P in the absence of the scatterer, viz., the wedge in this example. Also, $\vec{e}^R_0(P)$ is the $O(k^0)$ vectorial or complex amplitude of the reflected field on one or the other face of the wedge, assuming that each face is continuously extended beyond its edge. In (5.28), U is the Heaviside function whose argument is $\varepsilon_i = +1$, or -1 depending upon whether the observation point is in the shadow or lit zone of the incident field, respectively. Likewise, $\varepsilon_R = +1$ if the point P is in the geometrical shadow of the reflected field, and $\varepsilon_R = -1$ if it is in the region where the reflected field exists. The functions $s_i(P)$ and $s_R(P)$ represent, respectively, the paths traversed by the incident wave and the reflected rays, respectively, to reach the observation point P.

If we add the Geometrical Optics field, given by (5.28), to the diffracted field by the edge to $O(k^{-1/2})$, we obtain the following asymptotic expansion, associated with the outer domain and constructed in accordance with the procedure given in Sect. 5.2. The expansion reads

$$\vec{E}^e(P) = e^{ikS_i(P)}\, \vec{e}^i_0(P)\, U(-\varepsilon_i) + e^{ikS_R(P)}\, \vec{e}^R_0(P)\, U(-\varepsilon_R)$$

$$+ e^{ikS_r(P)}\, \frac{\vec{e}^d_0}{\sqrt{k}}(P) + o\left(k^{-\frac{1}{2}}\right),$$

(5.29)

where $s_r(P)$ is the path traversed by the diffracted ray and $\vec{e}^d_0 / \sqrt{k}$ is the dominant term of the vectorial and complex amplitude of the field diffracted at P. The above solution in (5.29) is not valid in the transition zones 2 and 2', but outside these zones it provides an appropriate asymptotic solution in each sub-domain of the region 1.

Next, we will seek an asymptotic solution which would be valid through the transition regions 2 and 2' and which would yield the solution, defined by (5.29), in each of the sub-domains of the region–1. Later, in Sect. 5.6, this solution would be extended both to region 3 and to the deep shadow zone 4 of the diffracted field.

It was shown in Sect. 5.2, that it is possible to construct a uniform solution once the non-uniform solutions of the outer and inner domains are known. In the present case, the inner domain comprises the regions 2 and 2' that are away from the edge, and the boundary layer 5 close to the edge, the latter being a line caustic for the diffracted rays.

In Sect. 3.5 we have shown that the amplitude of the field in the vicinity of a shadow boundary satisfies a parabolic type of equation, whose particular solution is a

Fresnel function. This solution can be generalized by noticing that, if $F(\tau)$ is a solution of the Fresnel equation

$$F'' - 2i\tau F' = 0,$$ (5.30)

with the definition $\tau = k^{-1/2}u$, then its derivative $F'(\tau)$ is also a solution to $O(k^{-1})$. More generally, all of the combinations of the type

$$AF + k^{-1/2}BF'$$ (5.31)

where A and B are some functions of u, are solutions to within $O(k^{-1})$.

Starting from an Ansatz of type (5.31) for the field amplitude, and imposing this solution to satisfy the matching conditions with the outer solution (5.29) and with the solution valid in the vicinity of the edge caustic, one can, in principle, determine the unknown functions and then construct a composite solution by following the method described in Sect. 5.2. This method has been applied by Borovikov [Bo] to the problem of diffraction by a two-dimensional wedge. The method requires for us to search for an asymptotic expansion which is valid in the boundary layer of the edge caustic. The latter is derived by using the method presented in Sect. 3.6, in which the calculation is carried out up to the $O(k^{-1})$. It should be pointed out that a generalization of this method to the 3D wedge has never been achieved because the procedure is anticipated to be extremely cumbersome.

We will now proceed to present a more direct method leading to solutions which, though less general, are nonetheless extremely useful in their applications. Since our objective is not to construct a uniform solution including region 5, it is sufficient for us to know the solution in the periphery of that region. We note, however, that to the order considered in (5.29), this solution must be identical to that of the tangent wedge at the diffraction point. The solution we seek is to be uniformly valid through the zones 2 and 2', and must consequently match both to the solution of the outer domain given by (5.29) as well as to the solution of the local tangent wedge at the periphery of the boundary layer of the edge caustic. This prompts us to choose as an Ansatz, a solution that has the same structure as the uniform solution of the wedge with planar faces. Furthermore, in order to satisfy all of the conditions stated above, it becomes necessary to modify the argument of the special function as well as the amplitudes of the terms in the representation for the solution as well.

Before implementing this method, we must first solve the canonical problem of the wedge with planar faces, and subsequently extract a uniform asymptotic solution from it. We will see in the next section that, even though the solutions will depend upon the procedure we adopt to derive them, they would nevertheless be equivalent to the order considered in (5.29). This equivalence is however strictly true only in the case of a wedge with planar faces illuminated by a plane wave. In effect, the curvature of the faces and of the edge generate terms of $O(k^{-1/2})$ in the boundary layers 2 and 2'. Terms of the same order also appear when the derivative of the incident field in the direction normal to the shadow boundary does not vanish. These terms, which are still of second order at the periphery of the boundary layer of the caustic edge, are eliminated in the approximation of the tangent plane and, consequently, do not appear in the solution of the wedge with planar faces illuminated by a plane wave; hence, they cannot be controlled in the procedure used to construct a uniform solution. One consequence of this is that various uniform solutions of the canonical problem that are equivalent to each other, would yield, via the generalization to a 3-D wedge, different uniform solutions, that would differ from each other in the transition zones 2 and 2', by certain terms of $O(k^{-1/2})$.

In the Sects. 5.3.2.1 and 5.3.2.2, we will present two of these solutions which have been the subject of numerous publications, viz., the Uniform Theory of Diffraction (UTD) of Kouyoumjian and Pathak [KP1] and the Uniform Asymptotic Theory (UAT) of Lee and Deschamps [LD]. These solutions are verified in Sect. 5.3.2.4 by comparing them with that derived by using the rigorous approach followed by the Spectral Theory of Diffraction (STD) introduced by Rhamat-Samii and Mittra [RM1], which is presented in Sect. 5.3.2.3.

5.3.1. Uniform asymptotic solutions of a perfectly conducting wedge with planar faces

The exact solution of the problem of plane wave scattering by a perfectly conducting wedge with planar faces is available in an integral form (see Appendix 1). If we use the $\exp(-i\omega t)$ time convention and a Cartesian system with the edge of the wedge as the OZ axis, and if the OX axis is contained in one of the faces, we can write the solution for the wedge problem as a plane wave spectral representation

$$\binom{E_z}{\eta H_z} = \frac{e^{ikz\cos\beta_0}}{4\pi in} \int_{L+L'} e^{-ik\rho\sin\beta_0\cos\xi}\left\{cot\,g\left(\frac{\xi+\phi-\phi'}{2n}\right) \pm cot\,g\left(\frac{\xi+\phi+\phi'}{2n}\right)\right\} d\xi,$$

(5.32)

where η is the free-space wave impedance; β the angle of the Keller cone ($0 \le \beta \le \pi/2$); ϕ and ϕ' are the angles between the projections $-\bar{k}^i$ and $-\bar{k}^d$, respectively, of the incident and diffracted wave vectors onto XOY and the OX axis; and, $n\pi$ represent the outer angle of the wedge (see Fig. 5.2).

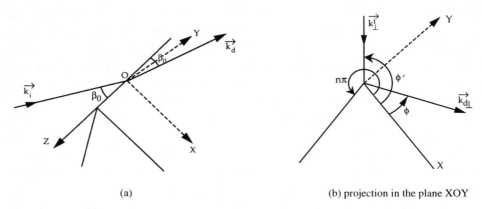

(a)   (b) projection in the plane XOY

Fig. 5.2. Geometry for scattering from a wedge with plane wave incidence (a) coordinate system and angle of oblique incidence with the edge, (b) projection of incident and diffracted ray on a plane perpendicular to the edge, and definition of the angles ϕ and ϕ'

The integration contour is composed of the curves L and L' of the complex plane and the angular spectral variable is ξ (see Fig. 5.3).

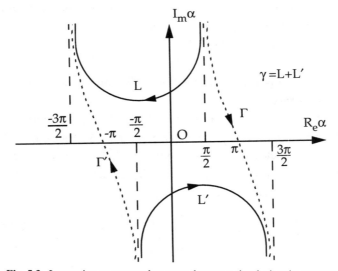

Fig. 5.3. Integration contour and steepest descent paths closing the contour.

The expression in (5.32) defines the solution in its entirety, since the transverse components $\vec{E}_t$, $\vec{H}_t$ are directly derivable from E_z, H_z via the use of the Maxwell's equations.

After closing the integration contour ($\gamma = L + L'$) with the steepest descent paths Γ and Γ' passing through the stationary phase points π and $-\pi$ of the integrand, the integral in (5.32) can be replaced by the integrals along Γ and Γ' to which one must add the residues at the poles of the integrand, which are located at the values of ξ where the cotangents in (5.32) go to zero, viz.,

$$\begin{cases} \xi + \phi - \phi' = 2nN\pi \\ \xi + \phi + \phi' = 2nN\pi \end{cases} \quad N = 0, \pm 1, \pm 2.... \tag{5.33}$$

It is evident that these poles are located on the real axis. In addition, it is easily verified that there exist at most two poles $\xi = r_1$ and $\xi = r_2$ in the interval $[-\pi, \pi]$, and they correspond to the incident and reflected fields. For the more general case of a wedge whose faces are characterized by a surface impedance, other poles can exist in the complex plane inside the closed contour of Fig. 5.3; these correspond to the slow surface waves and leaky waves.

The appearance of the transition zones 2 and 2' directly corresponds to the situation where one of the poles r_1 or r_2 moves close to one of the stationary points $\xi_s = \pm\pi$. As this pole crosses the contour from the interior, the field associated with it (incident or reflected) disappears suddenly and this creates a discontinuity in the representation of the field that is not present in the initial expression of the integral form of the solution given by (5.32). Consequently, there exists a similar jump with an opposite sign of the integral along the contour (Γ or Γ') which has been crossed by this pole. This property, which is a consequence of the residue theorem, must be conserved in the evaluation of the integral along the paths of steepest descent Γ and Γ' in order for the complete solution to be uniformly valid through the transition zones 2 and 2'.

In light of the above remarks, we observe that the search for a uniform asymptotic solution of the diffraction by a wedge with planar faces reduces to that of a search for a uniform asymptotic expansion of an integral whose integrand is oscillatory in nature and which has an isolated pole close to a stationary point of order 1.

When this pole moves close to the steepest descent path, the uniformity of the solution can be assured by staying on the same side of this path. If we do this, we obtain two different uniform expansions depending upon which side of the steepest descent path the pole is located. These two expansions do not match, but have a jump discontinuity which must equal the residue of the pole.

In order to obtain a uniform asymptotic expansion possessing these properties, two different methods have been proposed in the scientific literature. The first of these is the Pauli method [Pau], which has been simplified by Ott [Ot] and generalized by Clemmow [Cl], and the Van der Waerden method [VDW] which has also been investigated by Oberhettinger [Ob]. A comparative study of these two uniform expansions has been carried out by Hutchins and Kouyoumjian [HuK]. For more information on this subject we refer the reader to Bleistein [Bl] and to Chap. 4 of the book authored by Felsen and Marcuvitz [FM].

We are now going to review the foundations of these two methods before applying them to the wedge problem.

5.3.1.1. *The Pauli-Clemmow method*

To introduce the Pauli-Clemmow method we begin by considering a simple integral of the type

$$I(\Omega, \alpha) = \int_{\gamma} g(z, \alpha) e^{\Omega q(z)} \, dz, \tag{5.34}$$

where $\Omega > 0$ is a large real number as compared to unity, α is a complex-valued parameter and $g(z, \alpha)$ and $q(z)$ are holomorphic functions of the integral variable z. We assume that $q(z)$ has a stationary point of order 1 at z_s, which is a solution of $q'(z) = 0$ with $q''(z_s) \neq 0$. Finally, γ is the steepest descent path traversing through z_s. We also assume that the function

$$g(z, \alpha) = \frac{f(z, \alpha)}{z - z_0}, \tag{5.35}$$

has a simple pole at $z_0 = z_0(\alpha)$, and that $z_0 \to z_s$ if $\alpha \to \alpha_0$. Next, we apply the following transformation

$$q(z) = q(z_s) - s^2, \tag{5.36}$$

to (5.34), which, in turn, transforms γ into the real axis of the complex plane; the stationary point s into $s = 0$; and, the pole into $s = s_1 = s_1(\alpha)$. The integral (5.34) then reads

$$I(\Omega, \alpha) = \int_{-\infty}^{+\infty} G(s, \alpha) e^{-\Omega s^2} \, ds, \tag{5.37}$$

where

$$G(s, \alpha) = \overline{G}(s, \alpha) e^{\Omega q(z_s)}, \tag{5.38}$$

and

$$\overline{G}(s, \alpha) = g(z, \alpha) \frac{dz}{ds}. \tag{5.39}$$

In the Pauli-Clemmow method, one sets

$$G(s, \alpha) = \frac{T(s, \alpha)}{s - s_1(\alpha)}. \tag{5.40}$$

The function $T(s)$ is holomorphic and can be expanded as a Taylor series in the neighborhood of $s = 0$ which takes the form

$$T(s) = (s - s_1)\, G(s) = \sum_{n=0}^{\infty} A_n s^n ,$$

where the explicit α-dependence has been suppressed. Equations (5.40) and (5.37) may now be rewritten as

$$G(s) = \sum_{n=0}^{\infty} \frac{A_n s^n}{s - s_1} = \sum_{n=0}^{\infty} \frac{A_n s^n (s + s_1)}{s^2 - s_1^2} , \tag{5.41}$$

and

$$I(\Omega) = \sum_{n=0}^{\infty} A_n \int_{-\infty}^{+\infty} \frac{e^{-\Omega s^2}}{s^2 - s_1^2}\, s^n (s + s_1)\, ds . \tag{5.42}$$

The integrals occurring in (5.42) are of the type

$$\int_{-\infty}^{+\infty} \frac{e^{-\Omega s^2}}{s^2 - s_1^2}\, s^m\, ds , \tag{5.43}$$

and they vanish for m odd. Only the even terms with $m = 2p$ $(p = 0, 1, 2, \ldots)$ survive and we define

$$J_p = \int_{-\infty}^{+\infty} \frac{e^{-\Omega s^2}}{s^2 - s_1^2}\, s^{2p}\, ds .$$

Zero-order term

If we retain only the first term in (5.42), we obtain

$$J_0 = \int_{-\infty}^{+\infty} \frac{e^{-\Omega s^2}}{s^2 - s_1^2}\, ds = -\frac{1}{s_1^2} \sqrt{\frac{\pi}{\Omega}}\, F\!\left(-i\Omega s_1^2\right), \tag{5.44}$$

where $F(X)$ is the transition function defined by Kouyoumjian. It is given by

$$F(X) = -2i\sqrt{X}\, e^{-iX} \int_{\sqrt{X}}^{\infty} e^{it^2}\, dt, \quad -\frac{\pi}{2} < \arg X < 3\frac{\pi}{2}. \tag{5.45}$$

This function is defined on the upper Riemann sheet of the complex X-plane, with a cut along the negative imaginary axis, as shown in Fig. 5.4. The transition function is convergent [FM], [Ro] as $|X| \to \infty$, provided that $I_m X < 0$.

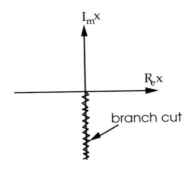

Fig. 5.4. Branch cut in the complex plane defining uniquely Kouyoumjian's transition function for a complex arguement.

Inserting (5.44) into (5.42) and retaining terms up to $O(\Omega^{-1/2})$, we obtain the uniform asymptotic solution *a la* Pauli-Clemmow, which reads

$$I(\Omega) = A_0 s_1 \left[-\frac{1}{s_1^2} \sqrt{\frac{\pi}{\Omega}} F\left(-i\Omega s_1^2\right) \right].$$

Since $A_0 = T(0) = -s_1 \, G(0)$, it follows that

$$I(\Omega) = \sqrt{\frac{\pi}{\Omega}} \, G(0) \, F(\Omega a) \tag{5.46}$$

where

$$a = -is_1^2 = i\,[q(z_0) - q(z_s)] \tag{5.47}$$

It is possible to verify that (5.46) has indeed the desired properties we were seeking in the uniform solution. In effect, when the pole s_1 crosses the path of steepest descent, we have, by definition, $I_m q(z_0) = I_m q(z_s)$. Hence, by using (5.36), we can obtain the result

$$I_m \left(-s_1^2\right) = 0, \ s_1 \in \gamma,$$

and, consequently,

$$R e \left(-i\Omega s_1^2\right) = 0, \ s_1 \in \gamma. \tag{5.48}$$

On the other hand, denoting the residue of the integrand of (5.37) by R_1, we get

$$R_1 = T(s_1) e^{-\Omega s_1^2}$$
(5.49)

In order for R_1 to converge as $\Omega \to \infty$, it is necessary that $R_e s_1^2 > 0$,, which leads to the condition that $I_m (-i\Omega s_1^2) < 0$. Under these conditions, and in view of the definition of the branch cut along the negative imaginary axis of the complex plane X (see Fig. 5.4), $\sqrt{X}$ and, consequently, $F(X)$ exhibit a jump discontinuity as the pole crosses the contour γ.

Since $R_e X = I_m s_1^2 = 0$ on γ, and $R_e s_1^2 > 0$, it follows that $I_m s_1 = 0$ on γ. If we choose the following definition for $\sqrt{X}$ where $R_e X > 0$.

$$\sqrt{X} = \sqrt{-i\Omega s_1^2} = e^{-i\frac{\pi}{4}} \sqrt{\Omega}\, s_1 .$$
(5.50)

Then, for $R_e X < 0$ we have

$$\sqrt{X} = e^{3i\frac{\pi}{4}} \sqrt{\Omega}\, s_1 .$$
(5.51)

Using (5.45), we have the result that, when $R_e (-i\Omega s_1^2)$ passes through zero from the negative side, the jump $\delta F(-i\Omega s_1^2)$ in $F(X)$ has the value

$$\delta F(-i\Omega s_1^2) = -2i s_1 \sqrt{\pi\Omega}\, e^{-\Omega s_1^2} .$$

As a result, the jump $\delta I(\Omega)$ of the solution in (5.46) is given by

$$\delta I(\Omega) = 2\pi i(-s_1)\, G(0) = 2\pi i\, T(0) e^{-\Omega s_1^2} .$$
(5.52)

When the pole crosses the steepest descent path at $s_1 = 0$, we obtain $2\pi i R_1 = 2\pi i T(0)$. Thus, the jump in the expression given in (5.46) exactly compensates for the loss of the residue R_1 given by (5.49).

Furthermore, when $|Re(\Omega a)| \gg 1$, the function $F(\Omega a)$ can be replaced by its asymptotic expansion

$$F(\Omega a) = 1 - i\frac{1}{2\Omega a} - \frac{3}{4}\frac{1}{(\Omega a)^2} + O(\Omega^{-3}) .$$
(5.53)

Using (5.53) we can show that far from the steepest descent path we recover I_{nu}, the first term of the non-uniform asymptotic expansion of the integral in (5.37). It is given by

$$I_{nu}(\Omega) = \sqrt{\frac{\pi}{\Omega}}\, G(0) + O(\Omega^{-\frac{3}{2}}) .$$
(5.54)

In view of the above, we have the result that the solution given in (5.46) is uniform with respect to the position of the pole on each side of the steepest descent path γ, and on γ, it exhibits a jump discontinuity that equals the residue at the pole of the integrand. An immediate consequence of this is that the expression

$$S_u^0(\Omega) = 2\pi i R_1 U\left[-R_e(-i\Omega s_1^2)\right] + \sqrt{\frac{\pi}{\Omega}}\, G(0)\, F(\Omega a) + O(\Omega^{-\frac{3}{2}}), \qquad (5.55)$$

where U denotes the Heaviside function, is an asymptotic solution, which is uniformly valid regardless of the position of the pole with respect to the steepest descent contour.

Higher-order terms

To derive the higher-order terms, we begin by uniting the integral in (5 43) as follows

$$J_p = (-1)^p\, \frac{d^p}{d\Omega^p} \int_{-\infty}^{+\infty} \frac{e^{-\Omega s^2}}{s^2 - s_1^2}\, ds \qquad (5.56)$$

Taking (5.44) and (5.45) into account, we obtain for $p = 1$

$$J_1 = \sqrt{\frac{\pi}{\Omega}}\left[1 - F(-i\Omega s_1^2)\right], \qquad (5.57)$$

and

$$\begin{aligned}
I(\Omega) &= -\frac{A_0}{s_1}\sqrt{\frac{\pi}{\Omega}}\, F(-i\Omega s_1^2) + (A_1 + s_1 A_2)\sqrt{\frac{\pi}{\Omega}}\left[1 - F(-i\Omega s_1^2)\right] \\
&= \sqrt{\frac{\pi}{\Omega}}\,\frac{1}{s_1}\left\{(A_0 + s_1 A_1 + s_1^2 A_2)\left[1 - F(-i\Omega s_1^2)\right] - A_0\right\}
\end{aligned} \qquad (5.58)$$

We note that to derive the results for the order $p = 1$, we need to calculate the expansion coefficients, given in (5.41), up to the order $n = 2$. Due to the complexity of the integrand in (5.32), an explicit derivation of the coefficients A_1 and A_2 is quite cumbersome, and for orders $p > 1$, the Pauli-Clemmow approach ceases to be practicable; however, we may still calculate, formally, the sum of the terms of the series in (5.42) corresponding to each order $\Omega^{-(n+1/2)}(n \geq 0)$. Explicit expressions for these terms may be found in [GP].

We see from the expressions (5.46) and (5.58), that for $p = 0$ and $p = 1$, we obtain only the terms of $O(\Omega^{-1/2})$. For $p > 1$, the authors just cited above have shown that, in addition to some new terms of $O(\Omega^{-1/2})$, additional higher-order terms also appear in the right-hand side of (5.58), viz.,

$$A_0 + s_1 A_1 + s_1^2 A_2,$$

which are the first terms of the Taylor expansion of $T(s)$ for $s = s_1$. This property holds true to any order in p and enables us to replace the sum of the terms of $O(\Omega^{-1/2})$ by the expression

$$I(\Omega) = \sqrt{\frac{\pi}{\Omega}} \frac{1}{s_1} \left\{ T(s_1)\left[1 - F(-i\Omega s_1^2)\right] - T(0) \right\}. \tag{5.59}$$

To prove (5.59), we start with the expression in (5.42) which can be written in the form

$$I(\Omega) = \sum_{p=0}^{\infty} A_{2p} s_1 J_p + A_{2p+1} J_{p+1}. \tag{5.60}$$

We then use the identity

$$J_{p+1} = -\frac{d}{d\Omega} J_p. \tag{5.61}$$

Starting from the expression of J_1, which reads

$$J_1 = \int_{-\infty}^{+\infty} e^{-\Omega s^2} ds + s_1^2 J_0,$$

and applying (5.60) to it, we obtain the following recursion relations

$$J_2 = \int_{-\infty}^{+\infty} s^2 e^{-\Omega s^2} ds + s_1^2 J_1,$$

$$J_3 = \int_{-\infty}^{+\infty} s^4 e^{-\Omega s^2} ds + s_1^2 J_2,$$

$$\text{-----} \quad \text{------------------------}$$

$$J_p = \int_{-\infty}^{+\infty} s^{2p} e^{-\Omega s^2} ds + s_1^2 J_{p-1}.$$

The integral representations above provide contributions of order higher than $\Omega^{-1/2}$. If we restrict ourselves to the terms of $O(\Omega^{-1/2})$ and denote as $\overline{J}_p$ the truncated version of J_p to the terms of this order, we get

$$\overline{J}_p = s_1^{2(p-1)} \overline{J}_1.$$

Now since $\bar{J}_1 = J_1$, (5.59) becomes

$$I(\Omega) = A_0 s_1 J_0 + A_1 J_1 + J_1 \sum_{p=1}^{\infty} A_{2p} s_1^{2p-1} + A_{2p+1} s_1^{2p}$$

$$= A_0 s_1 J_0 + \frac{J_1}{s_1} \left[T(s_1) - T(0) \right]$$

Finally, by inserting the expressions of J_0 and J_1, as given by (5.44) and (5.57), we are led to the desired result in (5.59), which can be written in the form

$$I(\Omega) = \sqrt{\frac{\pi}{\Omega}} \left[\frac{T(s_1) - T(0)}{s_1} - \frac{T(s_1)}{s_1} F(-i\Omega s_1^2) \right], \qquad (5.62)$$

and by taking into account the behavior of the transition function $F(x)$ as the pole crosses the steepest descent path, we obtain

$$I(\Omega) = \begin{cases} \sqrt{\dfrac{\pi}{\Omega}} \left[\dfrac{T(s_1) - T(0)}{s_1} - \dfrac{T(s_1)}{s_1} F(X^+) \right], & R_e X < 0 \\[4mm] \sqrt{\dfrac{\pi}{\Omega}} \left[\dfrac{T(s_1) - T(0)}{s_1} - \dfrac{T(s_1)}{s_1} F(X^+) \right] + 2\pi i R_1, & R_e X > 0 \end{cases}, \qquad (5.63)$$

where $Re\, X^+ = 0^+$. From the results given above it can be shown that the expression

$$Su(\Omega) = 2\pi i R_1 U \left[-R_e(-i\Omega s_1^2) \right] + \sqrt{\frac{\pi}{\Omega}} \left[\frac{T(s_1) - T(0)}{s_1} - \frac{T(s_1)}{s_1} F(X) \right], \quad (5.64)$$

is uniform as a function of the location of the pole regardless of where the pole crosses the path of steepest decent. In contrast, the uniformity in (5.55) is achieved only if the pole crosses the path of steepest descent at the stationary point.

By writing (5.62) in the form

$$I(\Omega) = \sqrt{\frac{\pi}{\Omega}} \left\{ \frac{T(s_1) - T(0)}{s_1} \left[1 - F(-i\Omega s_1^2) \right] - \frac{T(0)}{s_1} F(-i\Omega s_1^2) \right\}, \qquad (5.65)$$

and comparing it with (5.46), we find that the latter is an incomplete asymptotic expansion to the $O(\pi^{-1/2})$, the missing term being

$$\sqrt{\frac{\pi}{\Omega}} \frac{T(s_1) - T(0)}{s_1} \left[1 - F(-i\Omega s_1^2) \right]. \qquad (5.66)$$

This term vanishes when the pole is far from the steepest descent path. However, it is not negligible in the vicinity of this path, and, in particular, it is proportional to the derivative of the function $T(s)$ when s_1 tends to zero. By using these expressions in (5.32), we show in Sect. 5.3.1.3 that this term is negligible for the case of the perfectly conducting wedge with planar faces, illuminated by a plane wave. However, in many practical situations, some of which are listed in Sect. 5.3.5, this term must be included as its contribution is by no means negligible.

5.3.1.2. Van der Waerden's method

We will now turn to an alternate approach, viz., the Van der Waerden method, deriving the uniform formula for the field diffracted by a wedge. In this method we begin with the following expression for $G(s)$:

$$G(s) = \frac{A}{s - s_1} + H(s),\qquad(5.67)$$

where $H(s)$ is a holomorphic function. Inserting (5.67) in (5.37), one gets

$$I(\Omega) = \int_{-\infty}^{+\infty} \frac{A}{s - s_1} e^{-\Omega s^2}\, ds + \int_{-\infty}^{+\infty} H(s)\, e^{-\Omega s^2}\, ds.\qquad(5.68)$$

The first integral can be evaluated to yield

$$A \int_{-\infty}^{+\infty} \frac{s + s_1}{s^2 - s_1^2} e^{-\Omega s^2}\, ds = A s_1 \int_{-\infty}^{+\infty} \frac{e^{-\Omega s^2}}{s^2 - s_1^2}\, ds = -\frac{A}{s_1}\sqrt{\frac{\pi}{\Omega}}\, F(-i\Omega s_1^2)$$

An asymptotic evaluation of the second integral, to $O(\pi^{-1/2})$, gives

$$\sqrt{\frac{\pi}{\Omega}}\, H(0) = \left[G(0) + \frac{A}{s_1} \right] \sqrt{\frac{\pi}{\Omega}} + O\!\left(\Omega^{-\frac{3}{2}} \right).$$

Next, we note that $A = T(s_1)$ and $G(0) = -\dfrac{T(0)}{s_1}$ Hence, after grouping together the two terms, we obtain

$$I(\Omega) = \sqrt{\frac{\pi}{\Omega}}\, \frac{1}{s_1}\left\{ T(s_1)\left[1 - F(-i\Omega s_1^2) \right] - T(0) \right\} + O\!\left(\Omega^{-\frac{3}{2}} \right).\qquad(5.69)$$

The above expression is seen to be identical to (5.59). Thus the Van der Waerden method provides the complete asymptotic expansion of (5.37), to $O(\Omega^{-1/2})$, in a direct manner. Thus, at this point one might raise a question regarding the utility, if any, of the Pauli-Clemmow method. Experience shows, however, that the numerical calculation of the expression in (5.69) is very sensitive since the terms inside the braces vanish to first order in s_1. As $s_1 \to 0$, one is forced to extract, accurately, the second-order terms in order to use this expression for numerical computation. This problem does not occur in the Pauli-Clemmow solution for the orders $p = 0$ and 1, as may be seen by the expression in (5.46). Moreover, there are other more physical reasons, discussed in Sect. 5.3.5, which show the relative advantage of the Pauli-Clemmow method.

The solution given in (5.69) can be written in an alternate form which is more convenient for the construction of an Ansatz for the solution of the diffraction problem of a wedge with curved faces. To derive this new form we define

$$V(s_1) = -\frac{1}{s_1} \frac{1}{2\pi i} \sqrt{\frac{\pi}{\Omega}} \, F(-i\Omega s_1^2) \, e^{\Omega s_1^2} \, , \tag{5.70}$$

$$\hat{V}(s_1) = -\frac{1}{s_1} \frac{1}{2\pi i} \sqrt{\frac{\pi}{\Omega}} \, e^{\Omega \frac{?}{i}} \, , \tag{5.71}$$

where $\hat{V}(s_1)$ is the first term of the asymptotic expansion of $V(s_1)$. Using (5.54), we can write (5.69) in the form

$$I(\Omega) = 2\pi i R_1 \, [V(s_1) - \hat{V}(s_1)] + I_{nu}(\Omega). \tag{5.72}$$

We have seen in Sect. 5.3.1.1 that the function $F(-i \, \Omega \, s_1^2)$ exhibits a jump discontinuity as the pole crosses the path of steepest descent γ. It follows, therefore, that the same is true for the function $V_1(s_1)$ as well. Inserting (5.50) and (5.51) into (5.70), and using (5.45), we can show that the jump of $V(s_1)$ is equal to -1. As a consequence, $V(s_1)$ can be written as

$$V(s_1) = \overline{F}\left(e^{-i\frac{\pi}{4}} \sqrt{\Omega} \, s_1\right) - U\left[-R_e(-i\Omega s_1^2)\right], \tag{5.73}$$

where $\overline{F}(u)$ is the Fresnel function

$$\overline{F}(u) = \frac{e^{-i\frac{\pi}{4}}}{\sqrt{\pi}} \int\limits_{u}^{\infty} e^{it^2} \, dt \tag{5.74}$$

By letting

$$\tau = e^{-i\frac{\pi}{4}} \sqrt{\Omega} \, s_1 \tag{5.75}$$

the solution in (4.75) can be rewritten as

$$I(\Omega) = 2\pi i R_1 \left[\overline{F}(\tau) - \hat{\overline{F}}(\tau) \right] + I_{nu}(\Omega) - 2\pi i R_1 U(-I_m s_1), \qquad (5.76)$$

with

$$\hat{\overline{F}}(\tau) = \frac{1}{2\tau\sqrt{\pi}} e^{i\left(\tau^2 + \frac{\pi}{4}\right)}. \qquad (5.77)$$

Finally, by adding the contribution of the pole given by $2\pi i R_1 U(-R_e(-i\Omega s_1^2)]$ to the solution in (5.76) the uniform solution can be is written as

$$S_u(\Omega) = 2\pi i R_1 \left[\overline{F}(\tau) - \hat{\overline{F}}(\tau) \right] + I_{nu}(\Omega) \qquad (5.78)$$

For real τ and $|\tau| \to \infty$, we get

$$\overline{F}(\tau) = U(-\tau) + \hat{\overline{F}}(\tau) + O(\Omega^{-\frac{3}{2}}). \qquad (5.79)$$

Hence, when the pole is far from the steepest descent path, the uniform solution in (5.78) reduces to

$$S_u(\Omega) = 2\pi R_1 \, U(-\tau) + I_{nu}(\Omega) + O(\Omega^{-\frac{1}{2}}). \qquad (5.80)$$

This completes the derivation of the various uniform solutions, and we turn next, in the following section, to the application of these two methods to a wedge with planar faces.

5.3.1.3. *Application of uniform theories to the problem of diffraction by a wedge with planar faces*

The integrand of (5.32) has two stationary points, viz., $\xi_s = \pm\pi$. Consequently, the diffracted field by the wedge is given by

$$\begin{pmatrix} E_Z^D \\ \eta H_Z^D \end{pmatrix} = \frac{e^{-ikz\cos\beta_0}}{4\pi in} \int_{\Gamma+\Gamma'} e^{-ik\rho\sin\beta_0\cos\xi} \left[\cot g\left(\frac{\xi+\phi-\phi'}{2n}\right) \mp \cot g\left(\frac{\xi+\phi+\phi'}{2n}\right) \right] d\xi$$

$$(5.81)$$

where Γ and Γ' are the steepest descent paths passing through the points $\xi_s = \pm\pi$ (see Fig. 5.3).

THE METHOD OF PAULI-CLEMMOW

Zero-order term

The contribution of the diffracted field to the uniform asymptotic solution of order zero, given by (5.55), is represented by the term given in (5.46). Thus we need to calculate $G(0)$ and the argument "a" of the transition function F to construct the uniform solution in the Pauli-Clemmow's method.

Using (5.38) and (5.39), and replacing z by ξ, we get

$$G(0) = g(\xi_s, \alpha)\, e^{\Omega q(\xi_s)} \left(\frac{d\xi}{ds}\right)_{s=0}. \tag{5.82}$$

The zeros of the cotangent function in the integrand in (5.81) are given by

$$q(\xi) = -i\cos\xi, \quad q''(\xi) = i\cos\xi, \tag{5.83}$$

From (5.83) we have

$$q(\pm\pi) = i, \quad q''(\pm\pi) = -i, \tag{5.84}$$

and

$$\left(\frac{d\xi}{ds}\right)_{s=0} = \left|\sqrt{\frac{2}{q''(\xi_s)}}\right| e^{i\alpha} = \sqrt{2}\, e^{i\alpha}, \tag{5.85}$$

where α is the phase of (dz/ds) which is identical to the phase of (dz) since ds is real. From the values of the angles of the tangent to the contours Γ and Γ' at $\xi_s = \pi$ and $\xi_s = -\pi$, respectively, we can obtain

$$e^{i\alpha} = \begin{cases} e^{3i\frac{\pi}{4}} = ie^{i\frac{\pi}{4}} & when\ \xi_s = \pi \\ e^{-i\frac{\pi}{4}} = -ie^{i\frac{\pi}{4}} & when\ \xi_s = -\pi \end{cases}. \tag{5.86}$$

Using the expression for "a" given in (5.47), we have, from (5.84) and (5.33)

$$a = 1 + iq(z_0) = 1 + iq(-\beta + 2nN^{\pm}\pi) = \overset{\pm}{a}(b), \tag{5.87}$$

where

$$\beta = \phi \pm \phi' \equiv \overset{\pm}{\beta}, \tag{5.88}$$

and $N^{\pm}$ are the nearest integers that satisfy the criterion

$$2nN^{\pm}\pi - \beta = \pm \pi.$$

(5.89)

Utilizing (5.82) through (5.89) and (5.46) in conjunction with (5.81), the contribution of the diffracted field to the uniform asymptotic solution of order zero can be written as

$$\begin{pmatrix} E_Z^D \\ \eta H_Z^D \end{pmatrix} \equiv \frac{e^{ik(z\cos\beta_0 + \rho\sin\beta_0)}}{2n\sqrt{2\pi k\rho\sin\beta_0}} \left\{ \cot g\left(\frac{\pi+\overline{\beta}}{2n}\right) F\left[K \overset{+}{a}(\overline{\beta})\right] + \cot g\left(\frac{\pi-\overline{\beta}}{2n}\right) F\left[K \overline{a}(\overline{\beta})\right] \right.$$
$$\left. \mp \left(\cot g\left(\frac{\pi+\overset{+}{\beta}}{2n}\right) F\left[K \overset{++}{a}(\beta)\right] + \cot g\left(\frac{\pi-\overset{+}{\beta}}{2n}\right) F\left[K \overset{-+}{a}(\beta)\right] \right) \right\}$$

,

where we have used the notation $K = k\rho\sin\beta_0$.

Higher-order terms

We turn next to the higher-order terms whose sum is given by (5.56). For the case of the wedge with planar faces illuminated by a plane wave, the higher-order contributions turn out to be negligible. To show this let us consider one of the terms in (5.81), say

$$I = \int_{\Gamma} e^{-ik\rho\sin\beta_0\cos\xi} \cot g\left(\frac{\xi+\phi-\phi'}{2n}\right) d\xi.$$

(5.91)

Using (5.36), (5.83), (5.84) and (5.86), we obtain for the case where the contour Γ passes through $\xi_s = \pi$,

$$\frac{d\xi}{ds} = \frac{\sqrt{2}}{\sin\frac{\xi}{2}} e^{3i\frac{\pi}{4}}$$
$$s = \sqrt{2}\cos\frac{\xi}{2} e^{i\frac{\pi}{4}}$$

(5.92)

From the above, it follows that

$$T(s) = G(s)(s-s_1) = 2\frac{\left(\cos\frac{\beta}{2} - \cos\frac{\xi}{2}\right)}{\sin\frac{\xi}{2}} \cot g\left(\frac{\xi+\beta}{2n}\right) = Q(\xi),$$

(5.93)

where $\beta = \phi - \phi'$.

If we now take the limit of (5.93) as $s \to s_1$, which is tantamount to making ξ tend to the pole $\xi_1 = -\beta$, we find the result

$$T(s_1) = Q(-\beta) = 2n. \tag{5.94}$$

On the other hand, for $s = 0$, or $\xi = \xi_s = \pi$, (5.93) leads to

$$T(0) = Q(\pi) = 2 \cos\frac{\beta}{2} \cot g\left(\frac{\pi + \beta}{2n}\right). \tag{5.95}$$

As $s_1 \to 0$, the pole tends to the stationary point $\xi_s = \pi$ and, consequently, β tends to $-\pi$. The expansion of (5.95) in the vicinity of $\beta = -\pi$ reads

$$T(0) \approx 2n - \frac{n\varepsilon^2}{12}\left(1 - \frac{2}{n^2}\right) + \dots . \tag{5.96}$$

Since, from (5.92) we have

$$s_1 \approx \frac{\sqrt{2}}{2} e^{i\frac{\pi}{4}} \varepsilon, \tag{5.97}$$

we can derive the result

$$\frac{T(s_1) - T(0)}{s_1} \approx n\varepsilon \frac{\sqrt{2}}{12} e^{-i\frac{\pi}{4}}\left(1 - \frac{2}{n^2}\right). \tag{5.98}$$

Thus, the sum of the higher-order terms given in (5.66) vanishes when $s_1 \to 0$ (or $\varepsilon \to 0$). It follows, therefore, that the sum of their contribution is zero on the transition line, i.e., at the shadow boundary, and remains negligible in the transition zone.

The same type of analysis can also be carried out for the three other terms of (5.81) leading to the same conclusion.

Vectorial form of the solution: Dyadic Diffraction Coefficients

The transverse components of the diffracted field can be deduced from the longitudinal components given by (5.32) by invoking the Maxwell's equations.

If we apply the operator $\nabla_t = \hat{\rho}\dfrac{\partial}{\partial\rho} + \dfrac{\hat{\phi}}{\rho}\dfrac{\partial}{\partial\phi}$ to the integrand in (5.32), and retain only the dominant term as k becomes large, we obtain the final form.

$$\begin{pmatrix} \overrightarrow{\nabla}_t E_z \\ \eta\overrightarrow{\nabla}_t H_z \end{pmatrix} = \hat{\rho}\frac{e^{ikz\cos\beta_0}}{4\pi in} \int\limits_{L+L'} (-i)k\sin\beta_0\cos\xi \left[cot\,g\left(\frac{\xi+\phi-\phi'}{2n}\right) \right.$$
$$\left. \mp cot\,g\left(\frac{\xi+\phi+\phi'}{2n}\right) \right] e^{-ik\rho\sin\beta_0\cos\xi}d\xi \qquad (5.99)$$

After separating the diffracted field, as in (5.81), and performing the uniform asymptotic expansion limited to the zero order, we obtain the relationship

$$\begin{pmatrix} \overrightarrow{\nabla}_t E_z^D \\ \eta\overrightarrow{\nabla}_t H_Z^D \end{pmatrix} = \hat{\rho}k\sin\beta_0 \begin{pmatrix} E_7^D \\ \eta H_Z^D \end{pmatrix}. \qquad (5.100)$$

In (5.32) as well as in (5.90), we have assumed that $E_z^i = \eta H_z^i = 1$. If $\vec{E}^i, \vec{H}^i$ denotes the electromagnetic field of an incident plane wave, then (5.90) becomes

$$\begin{pmatrix} E_Z^D \\ \eta H_Z^D \end{pmatrix}_0 = \begin{pmatrix} E_z^i \\ \eta H_z^i \end{pmatrix}\sqrt{\sin\beta_0}\ \underset{h}{D_s}\ \frac{e^{ik(z\cos\beta_0+\rho\sin\beta_0)}}{\sqrt{\rho}}, \qquad (5.101)$$

where we have used the following notations introduced by Kouyoumjian and Pathak [KP1].

$$\underset{h}{D_s} = \frac{1}{\sin\beta_0}\left(\overset{+}{d}(\overset{-}{\beta}, n)\ F\left[K\overset{+}{a}(\overset{-}{\beta})\right] + \overset{-}{d}(\overset{+}{\beta}, n)\ F\left[K\overset{-}{a}(\overset{+}{\beta})\right]\right.$$
$$\left. \mp\left\{\overset{+}{d}(\overset{+}{\beta}, n)\ F\left[K\overset{+}{a}(\overset{+}{\beta})\right] + \overset{-}{d}(\overset{-}{\beta}, n)\ F\left[K\overset{-}{a}(\overset{+}{\beta})\right]\right\}\right) \qquad (5.102)$$

and

$$\overset{\pm}{d}(\beta, n) = -\frac{e^{i\frac{\pi}{4}}}{n\sqrt{2\pi k}}\frac{1}{2}cot\,g\left(\frac{\pi\pm\beta}{2n}\right). \qquad (5.103)$$

It is convenient to express the field solutions in a dyadic form by using the system of unit vectors attached to the incident and diffracted rays. Towards this end we introduce the ray-fixed co-ordinate systems for the incident and diffracted rays, viz., $(\hat{i}, \hat{\phi}', \hat{\beta}')$ and $(\hat{d}, \hat{\phi}, \hat{\beta})$, respectively, as defined in Sect. 1.3.4.2. If we now express the incident and diffracted fields $(\vec{E}^i, \vec{H}^i)$ and $(\vec{E}^D, \vec{H}^D)$, respectively, in the form

$$\vec{E}^i = \hat{\beta}' E_{//}^i + \hat{\phi}' E_{\perp}^i \quad , \quad \vec{H}^i = \hat{\beta} H_{\perp}^i + \hat{\phi}' H_{//}^i,$$

$$\vec{E}^D = \hat{\beta} E_{//}^D + \hat{\phi} E_{\perp}^D \quad , \quad \vec{H}^D = \hat{\beta} H_{\perp}^D + \hat{\phi} H_{//}^D, \tag{5.104}$$

where $\left(H_{//}^i, H_{\perp}^i \right)$ and $\left(H_{//}^D, H_{\perp}^D \right)$ are the components, respectively, of the incident and diffracted magnetic fields associated with the corresponding electric field components $\left(E_{//}^i, E_{\perp}^i \right)$ and $\left(E_{//}^D, E_{\perp}^D \right)$. Then, from (5.101) and (5.100), we can derive the following expression for the zero order term of the asymptotic expansion

$$\begin{pmatrix} E_z^D \\ H_z^D \end{pmatrix}_0 \cong \begin{pmatrix} E_{//}^i \\ H_{\perp}^i \end{pmatrix} \sin \beta_0 \; D_{\substack{s \\ h}} \; \frac{e^{iks}}{\sqrt{s}}, \tag{5.105}$$

$$\begin{pmatrix} \vec{\nabla} E_z^D \\ \vec{\nabla} H_z^D \end{pmatrix}_0 \cong \hat{\rho} \begin{pmatrix} E_{//}^i \\ H_{\perp}^i \end{pmatrix} ik \sin^2 \beta_0 \; D_{\substack{s \\ h}} \; \frac{e^{iks}}{\sqrt{s}}, \tag{5.106}$$

where ρ and z have been replaced, respectively, by $s \sin\beta$ and $s \cos\beta$, s being the distance of the observation point to the diffraction point on the edge. For the translationally invariant problem, Maxwell's equations enable one to express the transverse components as functions of the longitudinal components as follows

$$\vec{E}_t = ik_z \frac{\vec{\nabla}_t E_z}{k^2 - k_z^2} - ik\eta \frac{\hat{z} \times \vec{\nabla}_t H_z}{k^2 - k_z^2}$$

$$\vec{H}_t = ik_z \frac{\vec{\nabla}_t H_z}{k^2 - k_z^2} - ik\eta \frac{\hat{z} \times \vec{\nabla}_t E_z}{k^2 - k_z^2} \tag{5.107}$$

where $k_z = k\cos\beta_0$. Using (5.104) through (5.107), and noting that $\hat{\rho} = \hat{s} \; \sin\beta_0 + \hat{\beta} \; \cos\beta_0$, $\hat{z} = \hat{s} \; \cos\beta_0 - \hat{\beta} \sin\beta_0$, and $\hat{z} \times \hat{\rho} = \hat{\phi}$, we obtain

$$\vec{E}_0^D = \vec{E}_t^D + \hat{z} E_z^D \cong \left(-\hat{\beta} E_{//}^i D_s - \hat{\phi} E_{\perp}^i D_h \right) \frac{e^{iks}}{\sqrt{s}},$$

and since $E_{//}^i = \vec{E}_i \cdot \hat{\beta}'$ and $E_{\perp}^i = \vec{E}_i \cdot \hat{\phi}'$, we can write

$$\vec{E}_0^D = \left(-\hat{\beta}\hat{\beta}'D_s - \hat{\phi}\hat{\phi}'D_h\right) \cdot \vec{E}_i \, \frac{e^{iks}}{\sqrt{s}} \, . \tag{5.108}$$

Similarly, we can show that

$$\vec{H}_0^D = \left(-\hat{\beta}\hat{\beta}'D_h - \hat{\phi}\hat{\phi}'D_s\right) \cdot \vec{H}_i \, \frac{e^{iks}}{\sqrt{s}} \, . \tag{5.109}$$

Finally, we can define the dyads $\overline{\overline{D}}_E$ and $\overline{\overline{D}}_H$ as follows

$$\begin{aligned} \overline{\overline{D}}_E &= -\hat{\beta}\hat{\beta}'D_s - \hat{\phi}\hat{\phi}'D_h \\ \overline{\overline{D}}_H &= -\hat{\beta}\hat{\beta}'D_h - \hat{\phi}\hat{\phi}'D_s \end{aligned}, \tag{5.110}$$

as the dyadic diffraction coefficients for the electric and magnetic diffracted fields.

VAN DER WAERDEN'S METHOD

The general expression for the uniform solution derived by using the Van der Waerden's method was given in (5.78). This expression directly yields the total field and to apply it to the integral in (5.32), we need to calculate the residue R_1 and the parameter τ. Let us assume that the incident field illuminates the OX face (see Fig. 5.2). For this case, we have $\phi' < \pi$ and the integrand in (5.32) has two poles at $\xi_1 = \phi'-\phi$ and $\xi_2 = -(\phi + \phi')$. The corresponding residues R_1 and R_2 at these poles are given by

$$R_1 = \frac{1}{2\pi i} \, e^{ik[z\cos\beta_0 - \rho\sin\beta_0\cos(\phi-\phi')]}, \tag{5.111}$$

$$R_2 = \mp \frac{1}{2\pi i} \, e^{ik[z\cos\beta_0 - \rho\sin\beta_0\cos(\phi+\phi')]}, \tag{5.112}$$

We observe that $2\pi i R_1$ and $2\pi i R_2$ correspond to the direct and reflected fields, respectively. Since the residues R_1 and R_2 are defined by (5.111) and (5.112), regardless of whether the poles ξ_1 and ξ_2 are located on the real axis of the complex plane ξ, we must extend the notion of incident and reflected fields to all space, whether they are inside or outside of the closed contour 3. This poses no difficulty for the incident field, because it is identical to the field that exists at the observation point in absence of the wedge. However, to generalize the reflected field, we must enlarge the face from which the reflection occurs and extend the standard definition by means of the continuity of the phase as shown in Fig. 5.5. The use of (5.78) does not require us to extend the definition of the incident and reflected fields in the region interior to the wedge. Consequently, we restrict the domain of definition of these fields only to the exterior region of the wedge.

The determination of the parameter τ is based upon its definition given in (5.75) with the quantity s_1 replaced by its expression as given in (5.47). Using (5.32) we obtain the following expressions for the two values of the parameter τ

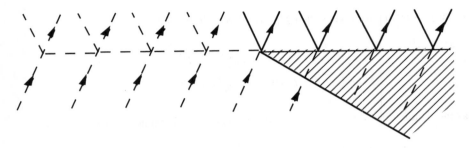

Fig. 5.5. Extension of the field reflected on a straight wedge outside its physical domain.

$$\tau_1^2 = 2k\rho\sin\beta_0\cos^2\left(\frac{\phi-\phi'}{2}\right), \tag{5.113}$$

$$\tau_2^2 = 2k\rho\sin\beta_0\cos^2\left(\frac{\phi+\phi'}{2}\right), \tag{5.114}$$

where τ_1 and τ_2 correspond to the poles ξ_1 and ξ_2. It is possible to provide the following geometrical interpretation of the parameters τ_1 and τ_2. If we assume that the phase reference is at the diffraction point, and denote the phases of the incident and diffracted fields at the observation point $M(\rho, \phi, z)$ by kS_i and kS_r, respectively, we can write

$$kS_i(M) = -k\rho\sin\beta_0\cos(\phi-\phi') + kz\cos\beta_0$$

$$kS_r(M) = +k\rho\sin\beta_0 + kz\cos\beta_0 \tag{5.115}$$

Hence, we have

$$k[S_r(M) - S_i(M)] = 2k\rho\sin\beta_0\cos^2\left(\frac{\phi-\phi'}{2}\right) = \tau_1^{\,2} \tag{5.116}$$

We now observe that (5.116) corresponds to the *detour* that the phase makes along the path followed by the diffracted ray. As for the reflected field, we have

$$k[S_r(M) - S_R(M)] = 2k\rho \sin\beta_0 \cos^2\left(\frac{\phi+\phi'}{2}\right) = \tau_2^2 \qquad (5.117)$$

In view of the above, τ_1 and τ_2 have been termed the *detour parameters* by Lee and Deschamps [LD]. The sign of these parameters can be defined by imposing the constraint on $U(-\tau)$, occurring in (5.80), that it be set equal to 1 when the observation point is reached either by the incident field or the reflected field of Geometrical Optics, and to zero elsewhere. Incorporating this condition in the expressions for τ_1 and τ_2, we get

$$\begin{aligned}
\tau_1 &= \varepsilon_1(\bar{\rho})\sqrt{2k\rho \sin\beta_0}\left|\cos\left(\frac{\phi-\phi'}{2}\right)\right| \\
\tau_2 &= \varepsilon_2(\bar{\rho})\sqrt{2k\rho \sin\beta_0}\left|\cos\left(\frac{\phi+\phi'}{2}\right)\right|
\end{aligned} \qquad (5.118)$$

where $\varepsilon_1(\bar{\rho})$ is set equal to 1 if the observation point is in the shadow region of the GO incident field while it equals -1 if the point is in the lit zone. Likewise, $\varepsilon_2(\bar{\rho})$ is equal to 1 if the observation point is in the geometrical shadow of the reflected field and -1 if it is in the region where the reflected field exists.

Inserting (5.111) and (5.112) into (5.80), we can extract the following uniform asymptotic solution

$$\begin{aligned}
\begin{pmatrix} E_z \\ \eta H_z \end{pmatrix} &= e^{ik(z\cos\beta_0 - \rho\sin\beta_0 \cos(\phi-\phi'))}\left[F(\tau_1) - \hat{F}(\tau_1)\right] \\
&\mp e^{ik(z\cos\beta_0 - \rho\sin\beta_0 \cos(\phi+\phi'))}\left[F(\tau_2) - \hat{F}(\tau_2)\right] \\
&+ \sqrt{\frac{\sin\beta_0}{\rho}}\, D_{\substack{s\\h}}\, e^{ik(z\cos\beta_0 + \rho\sin\beta_0)}
\end{aligned} \qquad (5.119)$$

where $E z^i = \eta H_z^i = 1$, and $D_{s,h}$ is the Keller coefficient given in Sect. 1.5. The latter is calculated from (5.102) by replacing the Fresnel functions by unity.

The vectorial expression of the uniform solution is obtained by calculating the transverse components by means of (5.107). In order to determine $\vec{\nabla}_t E_z$ and $\vec{\nabla}_t H_z$, we first apply the operator $\vec{\nabla}_t$ inside the integral in (5.32), and then proceed next, as previously, to calculate the residue at the poles ξ_1 and ξ_2 and insert them in (5.80).

This leads to the following vectorial expression for the uniform asymptotic solution of the total field at the point $M(\rho, \phi)$:

$$\vec{E}(M) = \vec{E}^i(M)\left[F(\tau_1) - \hat{F}(\tau_1)\right] + \vec{E}^R(M)\left[F(\tau_2) - \hat{F}(\tau_2)\right]$$

$$\left(-\hat{\beta}\hat{\beta}' D_s - \hat{\phi}\hat{\phi}' D_h\right) \cdot \vec{E}^i(Q) \frac{e^{ik|\overrightarrow{QM}|}}{\sqrt{QM}}$$

(5.120)

The quantities D_s and D_h in (5.120) above have the same expressions as in (5.119), and $\vec{E}^i(M)$ and $\vec{E}^R(M)$ are, respectively, the incident and reflected fields extended over the entire space outside the wedge, as shown in Fig. 5.5.

5.3.2. Uniform asymptotic solutions for a perfectly conducting wedge with curved faces

5.3.2.1. *UTD solution*

At the observation point P, the total field is written in the form

$$\vec{E}(P) = \vec{E}^i_{OG}(P) + \vec{E}^R_{OG}(P) + \vec{E}^D(P)$$

(5.121)

where $\vec{E}^i_{OG}$ and $\vec{E}^R_{OG}$ are the incident and the reflected Geometrical Optics fields given by (5.28), and where $\vec{E}^D$ is the field diffracted by the edges. To construct the uniform solution, Kouyoumjian and Pathak [KP2] began with an Ansatz for the field $\vec{E}^D$, which they derived from the uniform solution of the wedge with planar faces via an application of the Pauli-Clemmow method. By considering a wedge which is locally tangent to the wedge with curved faces at the diffraction point (Q), and by adapting the divergence factor of the diffracted wave to the case of a curved edge illuminated by a locally planar wave, we can write (5.108) in the form

$$\vec{E}^D(P) = \left(-\hat{\beta}\hat{\beta}' D_s - \hat{\phi}\hat{\phi}' D_h\right) \cdot \vec{E}^i(Q) \sqrt{\frac{\rho}{s(\rho + s)}} e^{iks}$$

(5.122)

where ρ is the radius of curvature of the diffracted wave in the plane of diffraction, and $\vec{E}^i(Q)$ is the incident field at the diffraction point Q. In order for the diffracted field $\vec{E}^D$ to compensate for the jump discontinuity of the incident and reflected fields at the crossing of the respective shadow boundaries, it is necessary to adapt the arguments of the Fresnel functions in the expressions of the diffraction coefficients $D_{s,h}$ given by (5.102). Towards this end, we state our Ansatz in the form given by (5.122), where

$$D_{\substack{s \\ h}} = \frac{1}{\sin\beta_0}\left(\overset{+}{d}(\overline{\beta},n)\, F\left[kL^i\, \overset{+}{a}(\overline{\beta})\right] + \overline{d}(\overline{\beta},n)\, F\left[kL^i\, \overline{a}(\overline{\beta})\right]\right.$$

$$\left.\mp\left\{\overset{++}{d}(\overline{\beta},n)\, F\left[kL^{ro}\, \overset{++}{a}(\overline{\beta})\right] + \overline{d}(\overline{\beta},n)\, F\left[kL^{rn}\, \overline{a}(\overline{\beta})\right]\right\}\right) \tag{5.123}$$

L^i, L^{ro} and L^{rn} are three unknown parameters which, for the case of a wedge with planar faces illuminated by a plane wave, are equal to the distance $L = \rho\sin\beta_0$. The parameter L^i is associated with the incident wave while the parameters L^{ro} and L^{rn} are associated with the reflected waves on the faces $\phi = 0$ and $\phi = n\pi$. To determine these parameters, we can choose the jump discontinuity in the Fresnel function to compensate for, precisely, the discontinuity in the incident or the reflected terms.

Using the general expression for the discontinuity of the function $F(X)$, which was given just below (5.50) and (5.51), we obtain

$$\delta F[kLa] = 2e^{-i\frac{\pi}{4}}\sqrt{\pi kLa} \tag{5.124}$$

Hence, from (5.123), (5.87) and (5.103), the discontinuity in $D_{s,h}$ can be written as

$$\delta D_{s,h} = \begin{cases} \dfrac{1}{\sin\beta_0}\sqrt{L^i}, & \text{for } L = L^i \\[2mm] \mp\dfrac{1}{\sin\beta_0}\sqrt{L^r}, & \text{for } L = L^r \end{cases} \tag{5.125}$$

where L^r is used to denote either L^{ro}, or L^{rn}. The discontinuities of the incident and reflected fields are given by

$$\delta\vec{E}_{OG}^i(P) = \vec{E}^i(Q)\sqrt{\frac{\rho_1^i\rho_2^i}{(\rho_1^i+s)(\rho_2^i+s)}}\, e^{iks} \tag{5.126}$$

$$\delta\vec{E}_{OG}^R(P) = \vec{E}^i(Q)\left(\hat{u}_{//}^i\hat{u}_{//}^r R_h + \hat{u}_\perp\hat{u}_\perp R_s\right)\sqrt{\frac{\rho_1^r\rho_2^r}{(\rho_1^r+s)(\rho_2^r+s)}} \tag{5.127}$$

where $R_h = 1$ and $R_s = -1$ for a perfect conductor.

On the other hand, on the shadow boundaries of the incident and reflected fields, we have

$$\hat{u}_{//}^i = -\hat{\phi}' \cdot \hat{u}_{//}^r = \hat{\phi} \cdot \hat{u}_\perp = \hat{\beta} = -\hat{\beta}' \tag{5.128}$$

From (5.122), (5.125) and (5.126), it follows that the continuity of the incident field at the crossing of the shadow boundary is satisfied if we choose L^i as

$$L^i = \frac{s(\rho+s)\rho_1^i \rho_2^i \sin^2 \beta_0}{\rho(\rho_1^i + s)(\rho_2^i + s)} \tag{5.129}$$

Also, by identifying the expression in (5.127) with the jump of the field given in (5.122), and taking (5.128) into account, we find that the continuity of the reflected field at the crossing of the shadow boundaries, associated with each face of the wedge, is satisfied if we choose

$$L^r = \frac{s(\rho+s)\rho_1^r \rho_2^r \sin^2 \beta_0}{\rho(\rho_1^r + s)(\rho_2^r + s)} \tag{5.130}$$

The parameters L^i and L^r, given by (5.129) and (5.130), are constants if the distances of the diffraction point to the observation point and the radius of curvature of the diffracted wave in the plane of diffraction are calculated on the incident and reflected shadow boundaries, respectively. However, we can also calculate s and r as functions of the location of the observation point P. In this case, the parameters L^i and L^r vary as functions of the position of the observation point P. In any event, we must enforce the continuity of the total field at the crossing of the shadow boundaries of the incident and reflected fields.

To summarize, the uniform UTD solution has been provided in (5.122), (5.123), (5.129) and (5.130) above, and the total field, obtained by adding the GO incident and reflected fields to this solution, are continuous as their respective shadow boundaries are crossed, although their derivatives are not continuous. This deficiency, which must be corrected by introducing a slope diffraction coefficient (see [KP2]), does not exist in the UAT solution.

5.3.2.2. UAT solution

The UAT uniform solution, derived by Lee and Deschamps [LD], relies upon an Ansatz for the total field $\vec{E}$, which is constructed from the uniform solution of the wedge with planar faces by using the Van der Waerden method. By adapting the divergence factor of the diffracted wave to the case of a curved edge illuminated by a *locally* plane wave, and generalizing the detour parameters τ_1 and τ_2 to the 3D geometries of the wedge and the incident wave, we can write (5.120) in the form

$$\vec{E}(P) = \vec{E}^i(P)\left[F(\tau_i) - \hat{F}(\tau_i)\right] + \vec{E}^R(P)\left[F(\tau_R) - \hat{F}(\tau_R)\right]$$

$$\left(-\hat{\beta}\hat{\beta}' D_s - \hat{\phi}\hat{\phi}' D_h\right) \cdot \vec{E}^i(Q)\sqrt{\frac{\rho}{s(\rho + s)}}\, e^{iks} \qquad (5.131)$$

The quantities $D_{s,h}$ are the non-uniform Keller diffraction coefficients and ρ and s are defined in Sect. 5.3.2.1.

According to (5.116) and (5.117), the detour parameters extended to the 3D case read

$$\tau_i = \varepsilon_i(\vec{r})\sqrt{k\left|S_r(P) - S_i(P)\right|} \qquad (5.132)$$

$$\tau_R = \varepsilon_R(\vec{r})\sqrt{k\left|S_r(P) - S_R(P)\right|} \qquad (5.133)$$

where $\varepsilon_i = +1$ if the observation point P is in the shadow of the incident field and it is -1 if it is in the lit zone. Likewise, $\varepsilon_R = +1$ if the point P is in the geometrical shadow of the reflected field; it is set equal to -1 if it is in the region where the reflected field exists.

The field $\vec{E}^i(P)$ is the incident field existing at the point P in the absence of the wedge. In the region where the Geometrical Optics field exists, $\vec{E}^i(P)$ is identical to $\vec{E}_{OG}^i(P)$. In the shadow of the incident field of Geometrical Optics $\vec{E}_{OG}^i = 0$, while $\vec{E}^i \neq 0$.

The field $\vec{E}^R(P)$ is an extension of the Geometrical Optics reflected field in the region where the latter does not exist, through the continuation of the phase and the amplitude. For the case of wedges with curved faces, we first extend each of the faces by enforcing the condition of continuity of the function and its derivatives, and then replace the reflected field on the real surface by the one reflected by the fictitious surfaces we have just defined. Finally, this field must be extended to the domain located on the other side of these surfaces by the continuation of the phase and the amplitude.

Figure 5.6 provides an illustration of this procedure for the field reflected from the face 0. A similar picture can be provided for the reflected field from the face *n*.

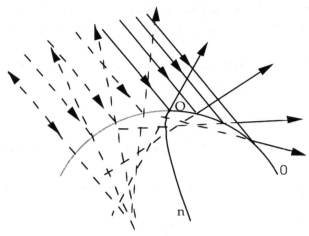

Fig. 5.6. Extension of the field reflected on a curved wedge outside its physical domain.

The extended reflected field satisfies the laws of Geometrical Optics along the fictitious rays. Consequently, it is not defined on their envelope, that is to say on the caustic of these rays.

Outside the transition zones and the caustics of the fictitious rays, the Ansatz (5.131) yields the solution (5.29), which is valid in the outer domain. This result follows directly from the property of the function F described in (5.79).

When $s/\rho << 1$ and $ks >> 1$, the Ansatz (5.131) yields the uniform asymptotic solution of the wedge with planar faces tangent to the wedge with curved faces at the diffraction point given by (5.120).

It remains to show that the Ansatz in (5.131) leads to a continuous field through the shadow boundaries of the direct and reflected fields, s/ρ being arbitrary with $ks >> 1$, $k\rho >> 1$.

Let us first consider an observation point P located in the neighborhood of the shadow boundary of the direct field. This point is defined by its coordinates (s, β_0, ϕ) where β_0 is the half-angle of the Keller cone $(\beta_0 \le \pi/2)$, s the distance of the observation point to the diffraction point Q located on the edge, and ϕ is the angle between the plane of diffraction and a reference plane. Let P_1 be the point on the Keller cone located on the incident ray passing through Q such that $QP_1 = s$; hence, the coordinates of P_1 are (s, β_0, ϕ'). Since P_1 and P are on the same cone, and are located at a distance from Q which is the same, they are to be associated with the same diffracted wave front. We then have $S_r(P) = S_r(P_1)$, and, since P_1 is located on the incident ray passing through Q, we also have $S_r(P_1) = S_i(P_1)$. From (5.132), it follows that

$$\tau_i = \varepsilon_i(\vec{r})\sqrt{k|S_i(P_1) - S_i(P)|} \qquad (5.134)$$

and we note that the calculation of τ_i entails the calculation of the phase difference between two points adjacent to the incident beam. Let us choose the origin of the coordinates O at the point Q; denote $\hat{x}_1, \hat{x}_2$ as the two unit vectors of the principal directions of the incident wave front passing through Q; and, $\hat{x}_3$ as the unit vector carried by the axial ray. We can then write

$$S_i(P) = S_i(O) + x_3 + \frac{1}{2}\left[\frac{x_1^2}{\rho_1^i + x_3} + \frac{x_2^2}{\rho_2^i + x_3}\right] \qquad (5.135)$$

where the point P has the curvilinear coordinates (x_1, x_2, x_3) associated with the orthogonal system $(\hat{x}_1, \hat{x}_2, \hat{x}_3)$ (see Eq. (1.76) of Sect. 1.4). We also have

$$S_i(P_1) = S_i(O) + s \qquad (5.136)$$

As $P \to P_1$, $\tau_i \to 0$, and $\hat{F}(\tau_i) \to \infty$. In order for the solution (5.131) to be continuous, it is then necessary that the singularity of $F(\tau_i)$ be compensated for by the corresponding singularities of the diffraction coefficients D_s and D_h that are located at $\phi - \phi' = \pm\pi$. We should point out that there is no sign ambiguity once we choose to orient the angles ϕ' and ϕ toward the shadow zone. For instance, we have $\phi - \phi' = \pi$ on the shadow boundary; $\phi - \phi' = \pi - |\psi^i|$ in the lit zone; and, $\phi - \phi' = \pi + |\psi^i|$ in the shadow zone, where $|\psi^i|$ is the inner angle between the planes of incidence and diffraction (see Fig. 5.7).

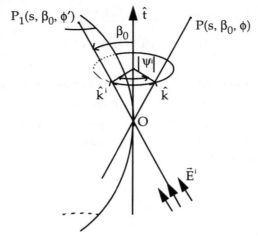

Fig. 5.7. Geometrical representation of the angle $|\psi|$ between the plane of incidence and the plane of diffraction.

Now, if we set

$$\psi^i = \phi - \phi' - \pi \qquad (5.137)$$

we see that sgn $\psi^i = \varepsilon_i(\vec{r})$. In the vicinity of the incident shadow boundary, τ_i and $D_{s,h}$ can be replaced by their limited expansion with respect to ψ^i and this is done by expressing (x_1, x_2, x_3) as functions of ψ^i. Denoting Ω^i as the angle between the plane of incidence $(\hat{t}, \hat{k}^i)$ and the plane $(\hat{x}_1, \hat{x}_3)$ (see Fig. 5.8), and by using elementary geometrical relations, we can show that

$$x_1 = s\ sin\beta_0\ [sin\ \Omega^i\ sin\ \psi^i + cos\beta_0\ cos\ \Omega^i\ (1 - cos\ \psi^i)]$$

$$x_2 = -s\ sin\beta_0\ [cos\ \Omega^i\ sin\ \psi^i + cos\beta_0\ cos\ \Omega^i\ (1 - cos\ \psi^i)] \qquad (5.138)$$

$$x_3 = s + s\ sin^2\ \beta_0\ (cos\ \psi^i - 1)$$

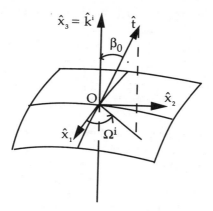

Fig. 5.8. Principal directions of the wavefront and plane of reflection.

Inserting (5.138) into the expression $S_i(P_1) - S_i(P)$, obtained by subtracting (5.135) from (5.136), we can show that the term of $O(\psi^i)$ is zero and that the term of order $(\psi^i)^2$ takes the form

$$S_i(P_1) - S_i(P) = s\left[\sin^2\beta_0\right]\frac{(\psi^i)^2}{2}\left[1 - s\left(\frac{\sin^2\Omega^i}{\rho_1^i + s} + \frac{\cos^2\Omega^i}{\rho_2^i + s}\right)\right]$$

Next, by using the relationship

$$\frac{1}{\rho^i} = \frac{\cos^2\Omega^i}{\rho_1^i} + \frac{\sin^2\Omega^i}{\rho_2^i} \tag{5.139}$$

where ρ^i is the curvature radius of the incident wave in the plane of incidence, we get

$$S_i(P_1) - S_i(P) = s\left[\sin^2\beta_0\right]\frac{(\psi^i)^2}{2}\frac{\left(1 + \frac{s}{\rho^i}\right)}{\left(1 + \frac{s}{\rho_1^i}\right)\left(1 + \frac{s}{\rho_2^i}\right)} \tag{5.140}$$

and

$$\tau_i = \frac{1}{2}\sqrt{2ks}\,\sin\beta_0\,\psi^i\sqrt{\frac{1 + \frac{s}{\rho^i}}{\left(1 + \frac{s}{\rho_1^i}\right)\left(1 + \frac{s}{\rho_2^i}\right)}} \tag{5.141}$$

Finally, by using

$$\vec{E}^i(P_1) = \vec{E}^i(Q)\left[\left(1 + \frac{s}{\rho_1^i}\right)\left(1 + \frac{s}{\rho_2^i}\right)\right]^{-\frac{1}{2}}e^{ikS_i(P_1)} \tag{5.142}$$

we find the expression for the singular term of the first line in the right-hand side of (5.131). It reads

$$-\vec{E}^i(P_1)\,\hat{F}(\tau_1) \cong -\frac{e^{i\frac{\pi}{4}}}{\sqrt{2k\pi}\,\psi^i\,\sin\beta_0}\left(1+\frac{s}{\rho^i}\right)^{-\frac{1}{2}}\vec{E}^i(Q)\frac{e^{ikS_i(P_1)}}{\sqrt{s}} \qquad (5.143)$$

Next, by expanding the coefficients $D_{s,h}$ as functions of the ψ^i given in (5.137), we find that their singular terms are given by

$$D_{\substack{s\\h}} \cong \frac{e^{i\frac{\pi}{4}}}{\sqrt{2k\pi}\,\psi^i\,\sin\beta_0} \qquad (5.144)$$

However, since on the incident shadow boundary we have

$$\hat{\beta}' = -\hat{\beta}\ ,\quad \hat{\phi}' = -\hat{\phi}\ ,\quad \rho = \rho^i$$

it follows that

$$\left(-\hat{\beta}\hat{\beta}' - \hat{\phi}\hat{\phi}'\right)\cdot\vec{E}^i(Q) = \vec{E}^i(Q)$$

and

$$\left(-\hat{\beta}\hat{\beta}'D_s - \hat{\phi}\hat{\phi}'D_h\right)\cdot\vec{E}^i(Q)\sqrt{\frac{\rho}{s(\rho+s)}}\,e^{iks} \cong \frac{e^{i\frac{\pi}{4}}}{\sqrt{2k\pi}\,\psi^i\,\sin\beta_0}\left(1+\frac{s}{\rho^i}\right)^{-\frac{1}{2}}\vec{E}^i(Q)\frac{e^{ikS_i(P_1)}}{\sqrt{s}}$$
$$(5.145)$$

Upon comparing (5.144) with (5.143), we find that the terms of order $(\psi^i)^{-1}$ compensate each other. Since the higher order terms of the limited expansion of the solution (5.131) around $\psi^i = 0$ tend to zero as $\psi^i \to 0$, the continuity of the solution is demonstrated.

A similar procedure can be followed to show that a field at a point P located in the neighborhood of the reflective shadow boundary it is free of singularities. For this case, we choose a point P_1 on the reflected ray passing through Q such that $|\vec{QP_1}| = |\vec{QP}| = s$. The quantity τ_R is now calculated from the phase difference between two points adjacent to the reflected ray, with the ray QP playing the role of an axial ray. All of the other steps involved in the calculation of the field are identical, the limited expansions being carried out with respect to $\psi^R = \phi + \phi' - \pi$.

We have thus shown that the solution in (5.131) is continuous across both the incident and reflected shadow boundaries, and yet it becomes identical to the non-uniform solution (5.29) far from the shadow boundaries. Therefore, it fulfills the desired characteristics of the uniform asymptotic solution we were seeking.

Since $F(0) = 1/2$, on the shadow boundaries, where $\tau_i = 0$ and $\tau_R = 0$, the dominant term of the asymptotic expansion is given by $\frac{1}{2}\vec{E}^i$ or $\frac{1}{2}\vec{E}^R$, which are of $O(k^0)$. In order to extract the $O(k^{-1/2})$ terms, we need to find the $O(1)$ terms (independent of $\psi^{i,r}$) in the limited expansion of the right-hand side of (5.131). This is achieved by carrying out the expansions of $\tau_{i,r}$ up to the order $(\psi^{i,r})^2$, since

$$\tau = \psi\,[A + B\psi + O(\psi^2)]$$

and $1/\tau$ can be expressed as

$$\frac{1}{\tau} \cong \frac{1}{A\psi}\left[1 - \frac{B\psi}{A} + O(\psi^2)\right]$$

The results of these calculations are given in the reference [LD].

The terms of order higher than $k^{-1/2}$ cannot be calculated since the solution in (5.131) is an asymptotic expansion that includes only the terms up to $O(k^{-1/2})$.

In the vicinity of the shadow boundaries, the solution in (5.131) can be replaced by its limited expansion. The latter can be derived with respect to ψ and its derivatives are continuous at all orders. It follows, then, that the total field given by (5.131) is continuous across the shadow boundaries and that its derivatives are continuous.

We might recall at this point that the continuity of the derivatives of the field are not generally satisfied by the UTD solution.

5.3.3. Solution based on Spectral Theory of Diffraction (STD)

The spectral theory of diffraction, or STD, was described in Chap. 4 where it was applied to a number of representative scattering problems. In this section, we consider its application to the problem of diffraction by an arbitrary incident field, which is not necessarily a plane wave, by a half-plane or a wedge with planar faces. As we mentioned in the introduction of Sect. 5.3, the two uniform solutions, viz., UTD and UAT, are not equivalent in the transition zones close to the incident and reflection shadow boundaries, for the case of a wedge with planar faces illuminated by a plane wave. If either the wedge geometry is curved, or the incident field is not a plane wave, neither the UTD nor the UAT Ansatz correctly accounts for the $O(k^{-1/2})$ terms in the transition zones.

In contrast, STD provides an exact integral representation for the problem of diffraction of an arbitrary incident wave by a half-plane or a wedge, as long as the incident field of a plane wave spectrum is known. It is thus possible to obtain an exact asymptotic expansion of this solution and to compare the same with UTD and UAT.

Such an approach has been followed by Rhamat-Samii and Mittra [RM1,2] (see also Boersma and Rhamat-Samii [BR] who have applied the STD to the problem of diffraction by a half-plane illuminated by an incident wave whose spectral representation is of the type given in (4.1) of Sect. 4.1. The authors have derived complete asymptotic expansions both in and out of the transition zones, and, have shown that these expansions consist of terms that depend upon the first and second derivatives of the field in the direction normal to the shadow boundary. For the details of these derivations the reader is referred to the works cited above. However, some observations regarding the UTD and UAT solutions are summarized in the next section.

5.3.4. Comparison between UTD, UAT and STD solutions

The analysis presented in [RM1] and [BR] has revealed that the UTD solution does not correctly yield the terms of $O(k^{-1/2})$ in the transition zones near the incident and reflected shadow boundaries. Specifically, as mentioned above, the terms that depend upon the derivative of the incident field in the direction normal to the shadow boundary are not present in the UTD solution. To remedy this defect, Hwang and Kouyoumjian [HwK] have added a corrective term, the so-called "slope diffraction coefficient," to the UTD solution. In contrast, the UAT does not exhibit a similar behavior and it correctly yields the terms of $O(k^{-1/2})$ found in the STD.

For the case of a 3-D wedge, the curvature of the faces generates other terms of $O(k^{-1/2})$. In the UAT solution, one finds a term that depends upon the derivative of the reflected field in the direction normal to the shadow boundary (see [LD]). This term plays a role that is similar to that of the derivative of the incident field, and, consequently, it can be conveniently introduced in the UTD formulation via the slope diffraction coefficient of the reflected field.

If we augment the UTD with slope diffraction coefficients of the incident and reflected fields, we obtain a solution equivalent to that given by the UAT. For the general case, however, it is not possible to prove that the terms of $O(k^{-1/2})$ in UAT are complete. Before closing we would also like to mention that, with appropriate modifications, UTD is often the preferred choice for practical implementation in computer codes as its format is better suited for numerical computation.

We have seen that UAT fails to reproduce the field on a caustic of fictitious reflected rays. This situation occurs, for instance, at the image point of a point location source illuminating the edge of a half-plane. It becomes critical when the image of the source is in the vicinity of the shadow boundary of the direct field. In addition, as far as practical applications are concerned, UTD is more convenient to use than the UAT, since it is neither necessary to extend the surface nor to look for fictitious rays when the UTD is used.

Although, the use of STD enables one to treat in great detail the diffraction of an arbitrary wave by a half-plane or a wedge with planar faces, it can not be applied to a wedge with curved faces, and the search of terms of $O(k^{-1/2})$ in the transition zones.

For the normal incidence case, it is possible to extend UTD and UAT to a 2-D wedge with curved faces, characterized by a constant surface impedance, by following the same procedure as the one used for the case of a perfectly conducting wedge.

5.4 UAT solution for a line discontinuity in the curvature

5.4.1 Statement of the problem and details of resolution

The diffraction of a plane wave by a line discontinuity in the curvature has been studied by Weston [W] who, in the two-dimensional case, has derived an asymptotic solution of the integral equation satisfied by the currents on the surface of a diffracting body with the above type of discontinuity. He has also derived the expression for the creeping rays generated by such a singularity. Senior [Se] has extended this method to the calculation of the diffracted field and has provided the expression for the diffraction coefficients of a line discontinuity in the curvature. The same result has been derived independently by Kaminetsky and Keller [KK] who have applied the boundary layer theory to this problem. They have also extracted the diffraction coefficients, not only for a discontinuity in the curvature of a surface with Neuman or Dirichlet boundary conditions, but more generally with a mixed boundary condition of the Leontovitch type [Le].

Furthermore, these authors have derived the diffraction coefficients for the case of linear singularities of higher order (discontinuity in the derivatives of order greater than two).

The range of validity of the coefficients has been given in [Se] and [KK]; it has been shown that this range is limited to a certain region of space surrounding the diffracting body. In the two-dimensional case, it is limited to region-1 in Fig. 5.9, outside of the transition zone 2 of the reflected field of Geometrical Optics. The latter is discontinuous when the reflection point crosses the discontinuity line in the curvature located at O. Consequently, the asymptotic solution obtained by the authors cited previously is not uniform as a function of the direction of observation, and the resulting diffraction coefficients are not useful in the neighborhood (region-2) of the transition surface whose trace is denoted by OR in Fig. 5.9. The reflected field is discontinuous when crossing that trace and the coefficients in question tend to infinity when the observation point approaches this surface. There exist additional regions where the asymptotic solution given in references [Se] and [KK] is not valid, e.g., the neighborhood of the tangent plane to the surface at the diffraction point (region-3) and the neighborhood of the line discontinuity in the curvature (region-4). Region-3 is a transition zone for the diffracted wave whose character transitions from a spatial wave, not attached to the surface, to a wave guided by this surface, and the latter can be represented in the form of a series of creeping rays. Region-4 is the neighborhood of a caustic. In effect, all of the diffracted rays pass through the line discontinuity in the curvature which, consequently, forms a line caustic for these rays.

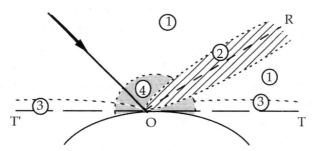

Fig. 5.9. Boundary layers in the problem of diffraction by a discontinuity in curvature and region of validity of known asymptotic solutions.

The objective of this section is to construct an asymptotic solution of the UAT-type which would be uniform through the regions 1 and 2 of Fig. 5.9. This solution has been derived by [Mo1] by using a method similar to that described in Sect. 5.3. On one hand, the uniform solution we are looking for must match the solution in the outer domain (region-1); on the other hand, it must be valid at the periphery of region-4.

Just as we did for the wedge problem, we start with a solution in the form of an Ansatz having the same structure as the uniform solution valid at the periphery of boundary layer-4, then adapt the argument of the special function and the amplitudes of the terms in such a way that the conditions stated above are satisfied. The canonical problem of the wedge is replaced here by the solution obtained by [KK] in the boundary layer and serves as a starting point of the developments that follow.

5.4.1.1. Solution in the boundary layer

We now derive the solution in the boundary layer, and, in order to simplify the formalism, we consider a two-dimensional surface and assume that the direction of the incident plane wave is orthogonal to the generatrices. The knowledge of the solution of this problem enables one to derive the expressions of the diffraction coefficients and to construct, via the use of standard techniques of Geometrical Theory of Diffraction, the asymptotic solution for a line discontinuity on an arbitrary three-dimensional surface.

The implementation of the boundary layer theory for the solution of this problem requires three boundary layers (regions 2, 3 and 4 of Fig. 5.9). The deepest layer, layer-4, is a boundary layer which decomposes itself into three sub-layers as indicated in Fig. 5.10 below. The region-4a is a simple boundary layer arising from the line caustic of the diffracted rays associated with the discontinuity in the curvature. The layers 4b and 4c are, respectively, the intersections of the boundary layer associated with the previous line caustic with the boundary layer in the vicinity of the surface of discontinuity (with trace OR) of the reflected field of Geometrical Optics and with the boundary layer in the vicinity of the shadow boundary of the diffracted rays (with trace OT' and OT).

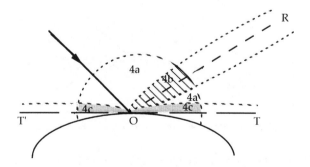

Fig. 5.10. Sub-domains in the caustic boundary layer of a discontinuity in curvature.

A fundamental step in the application of the boundary layer theory entails the choosing of the exponent α in the transformation which enables us to stretch the coordinates (see Sect. 2.1.4.). The limiting surface towards which the boundary layer

4a tends when k → ∞ merges into the line discontinuity in the curvature, denoted here as D. Using the cylindrical coordinates with an axis D, we obtain

$$\vec{r}' = k^{\alpha}\vec{r}\,,\tag{5.146}$$

where $\vec{r}$ is the position vector in the plane orthogonal to D.

   If we apply the stretching condition, defined in Sect. 2.1.4, to Eq. (2.9) of Sect. 2.2, expressed with the stretched coordinates we obtain $\alpha = 1$. If we use this value of α and apply the transformation (5.146) just defined above, we find that the three terms in Eq. (2.9) of Sect. 2.2 are of zero order with respect to $1/k$, so that the eikonal equation of Geometrical Optics is no longer useful. Furthermore, since this transformation places all the terms of (2.9) of Sect. 2.2 at the same level, it can be extended to regions 4b and 4c and, consequently, to the entire region 4.

   Kaminetsky and Keller have solved the internal problem of the boundary layer by using the stretched coordinates. A detailed description of their method can be found in reference [KK] and we only discuss, in what follows, the main steps of the approach followed by these authors. Let Oxy be a Cartesian co-ordinate system in the plane of a cross-section of the cylindrical structure, whose origin O is taken at the discontinuity in the curvature of the contour C of this cross-section and the ox axis is tangent to the curve C at the point O (Fig. 5.11).

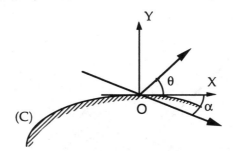

Fig. 5.11. Geometry for the derivation of the solution.

The total field at a point $M(x,y)$ outside the surface satisfies the Helmholtz equation

$$(\nabla^2 + k^2)\,U(x,y) = 0,\tag{5.147}$$

where ∇^2 is the two-dimensional Laplacian $\nabla^2 = \dfrac{\partial^2}{\partial x^2} + \dfrac{\partial^2}{\partial y^2}$

We assume that the boundary condition on the surface can be described by

$$\frac{\partial U}{\partial v} + ik\eta U = 0, \qquad y = f(x), \tag{5.148}$$

where $\frac{\partial U}{\partial v}$ is the derivative of U along the normal to the surface directed outward from the diffracting body (propagation medium) and η is the impedance along the contour C of equation $y = f(x)$, this impedance being measured with respect to the propagation medium.

When the transformation defined by (5.146), and $\alpha = 1$, is applied to (5.147) and (5.148), it leads to the equations

$$(\nabla'^2 + 1)\, U(k^{-1}x',\, k^{-1}y') = 0, \tag{5.149}$$

$$(\hat{v} \cdot \vec{\nabla}' + i\eta)\, U(k^{-1}x',\, k^{-1}y') = 0, \qquad y' = kf(k^{-1}x'), \tag{5.150}$$

with $x' = kx$, $y' = ky$, $\hat{v}$ = unit vector in the direction of the normal to the contour (c) and $\hat{i}$ and $\hat{j}$ are the unit vectors on the axes ox and oy, respectively.

In order to solve (5.149) with the requirement that condition (5.150) be satisfied, we apply the perturbation method with respect to the small parameter $1/k$. This is achieved by setting

$$u(x',\, y') = U(k^{-1}x',\, k^{-1}y') = \sum_{m=0}^{\infty} k^{-m}\, U_m(x',\, y'), \tag{5.151}$$

and inserting (5.151) into (5.149) and (5.150). By identifying the terms with the same order with respect to $1/k$, we obtain a hierarchy of equations for the new unknowns, U_1, U_2, ...U_m, these equations being coupled through the boundary conditions. For the orders 0 and 1, we get

Order 0

$$\begin{cases} (\nabla'^2 + 1)U_0(x',y') = 0 \\ \dfrac{\partial}{\partial y'}U_0(x',y') + i\eta U_0(x',y') \Big|_{y'=0} = 0 \end{cases}, \tag{5.152}$$

Order 1

$$\begin{cases} (\nabla'^2 + 1)U_1(x',y') = 0 \\ \left.\dfrac{\partial}{\partial y'}U_1(x',y') + i\eta U_1(x',y')\right|_{y'=0} = q(x') \end{cases} \tag{5.153}$$

with

$$q(x') = \frac{x'^2}{2} f^{(2)}(0^+_-)\left[-\frac{\partial^2 U_0}{\partial y'^2}(x',y') - i\eta\frac{\partial}{\partial y'}U_0(x',y')\right]_{y'=0} + x'f^{(2)}(0^+_-)\frac{\partial U_0}{\partial x'}(x',y')\bigg|_{y'=0}$$

$$\tag{5.154}$$

where $f^{(2)}(0\pm)$ is the second derivative of $f(x)$ at $x = 0^+$ and 0^-. We note that the boundary conditions in (5.152) are the same to the left and to the right of the curvature discontinuity so that U_0 is independent of this discontinuity. In fact, according to (5.152), the zero-order term of the solution satisfies the Helmholtz equation and boundary conditions similar to those of an incident wave on a plane with a relative impedance of η. To zero order, the solution is consequently given by the sum of the incident and reflected fields. The incident field entering inside the boundary layer-4 is obtained by expressing the incident field of the outer domain (region-1) with the stretched coordinates $\vec{r}' = k\vec{r}$, and by keeping only the zero-order term with respect to $1/k$. For the case of an incident plane wave with the direction of propagation normal to the generatrices $U^i(x, y) = U_0 exp(i\vec{k}.\vec{r})$, we simply obtain

$$U_0^i(k^{-1}\vec{r}') = U_0 e^{i(x'\cos\alpha - y'\sin\alpha)}, \tag{5.155}$$

where α is the angle between the direction of propagation of the incident wave and the ox-axis. The zero-order term in the boundary layer is given by

$$U_0(x', y') = U_0 e^{i(x'\cos\alpha - y'\sin\alpha)} + RU_0 e^{i(x'\cos\alpha + y'\sin\alpha)}, \tag{5.156}$$

where R is the reflection coefficient

$$R = \frac{\sin\alpha - \eta}{\sin\alpha + \eta}. \tag{5.157}$$

It can been seen from (5.153) and (5.154) that, to order 1, the boundary conditions depend on the discontinuity in the curvature through the function $q(x')$ which is discontinuous at the point $(x' = 0, y' = 0)$.

The solution to (5.153) has been given by Sommerfeld [So]

$$U_1(x',y') = -\int_{-\infty}^{+\infty} q(\xi)G(\xi,0|x',y')d\xi, \qquad (5.158)$$

where G is the Green function of the impedance plane

$$G(\xi,\tau|x',y') = v(\xi,\tau|x',y') + v(\xi,-\tau|x',y') - 2i\eta \int_{-\infty}^{-y'} e^{-i\eta(\zeta+y')} v(\xi,\tau|x',\zeta)d\zeta, \qquad (5.159)$$

and

$$v(\xi,\tau|x',y') = \frac{i}{4}H_0^{(1)}\left(\sqrt{(x'-\xi)^2 + (y'-\tau)^2}\right), \qquad (5.160)$$

where $H_0^{(1)}(\cdot)$ is the cylindrical Hankel function of order zero.

The external boundary of the boundary layer-4 corresponds to the values $r' = (x'^2 + y'^2)^{1/2}$, which are large compared to unity. Consequently, in the matching zone with the outer solution, we obtain the expression for the inner solution by performing the asymptotic expansion of this solution for large r'. This is accomplished by replacing, in (5.158), the Hankel functions by their Debye expansion, which is given by

$$H_0^{(1)}(u) = \left(\frac{2}{\pi u}\right)^{\frac{1}{2}} e^{-iu+i\frac{\pi}{4}}. \qquad (5.161)$$

From (5.158) and (5.160) it follows that

$$U_1(x',y') \approx I_1(x',y') + I_2(x',y'), \qquad (5.162)$$

where

$$I_1(x',y') = -\frac{e^{i\frac{\pi}{4}}}{\sqrt{2\pi}} J(x',y'), \qquad (5.163)$$

$$I_2(x',y') = \eta \frac{e^{-i\frac{\pi}{4}}}{\sqrt{2\pi}} \int_{-\infty}^{-y'} e^{-i\eta(\zeta+y')} J(x',\zeta)d\zeta, \qquad (5.164)$$

and

$$J(x',y') = \int_{-\infty}^{+\infty} q(\xi) \frac{e^{i\left[(x'-\xi)^2+y'^2\right]^{\frac{1}{2}}}}{\left[(x'-\xi)^2+y'^2\right]^{\frac{1}{4}}} \, d\xi . \tag{5.165}$$

The expression for $q'(\xi)$ is derived by inserting (5.155) into (5.154), and this leads to the result

$$q(\xi) = \frac{2iu_0 \sin\alpha}{\sin\alpha+\eta} Q(\xi) f^{(2)}(0_-^+) e^{i\xi \cos\alpha}, \tag{5.166}$$

where

$$Q(\xi) = \xi \cos\alpha - i(\sin^2\alpha - \eta^2)\frac{\xi^2}{2}. \tag{5.167}$$

To carry out the asymptotic expansion of the integrals I_1 and I_2 it is convenient to extract, explicitly, the discontinuity of $q(\xi)$. Denoting by $\bar{f}^{(2)}(0)$ and $[f^{(2)}(0)]$, the mean value and the jump of the second derivative of f at the origin, respectively, we obtain the relationship

$$f^{(2)}(0_-^+) = \bar{f}^{(2)}(0)_-^+ \frac{1}{2}\left[f^{(2)}(0)\right] \tag{5.168}$$

Inserting (5.168) into (5.163) and (5.164) and using the polar coordinates $x' = r'\cos\theta$, $y' = r'\sin\theta$, we obtain, after introducing the change of variables, $u = \frac{\xi}{r'}$, and $v = \frac{\zeta}{r'}$

$$\begin{aligned}I_1(r',\theta) = C_1 \int_{-\infty}^{+\infty} \bar{f}^{(2)}(0)F(u,\sin\theta;r') e^{ir'q(u,\sin\theta)} \, du \\ + \frac{C_1}{2}\left(\int_0^\infty - \int_{-\infty}^0\right)\left[f^{(2)}(0)\right] F(u,\sin\theta;r') e^{ir'q(u,\sin\theta)} \, du\end{aligned} \tag{5.169}$$

$$\begin{aligned}I_2(r',\theta) = C_2 \int_{-\infty}^{-\sin\theta} e^{-i\eta r'(v+\sin\theta)} \, dv \Bigg\{ \int_{-\infty}^{+\infty} \bar{f}^{(2)}(0)F(u,v;r') e^{ir'q(u,v)} \, du \\ + \frac{1}{2}\left(\int_0^\infty - \int_{-\infty}^0\right)\left[f^{(2)}(0)\right] F(u,v;r') e^{ir'q(u,v)} du \Bigg\}\end{aligned} \tag{5.170}$$

with

$$C_I = -\frac{2i}{\sqrt{2\pi}} \frac{u_0 \sin\alpha}{\sin\alpha+\eta} e^{i\frac{\pi}{4}} r'^{\frac{3}{2}},$$ (5.171)

$$C_2 = \frac{2i}{\sqrt{2\pi}} \eta \frac{u_0 \sin\alpha}{\sin\alpha+\eta} e^{i\frac{\pi}{4}} r'^{\frac{5}{2}},$$ (5.172)

$$F(u,v;r') = \frac{Q(u;r')}{\left[(\cos\theta-u)^2+v^2\right]^{\frac{1}{4}}} \quad , \quad Q(u;r') = u\cos\alpha - i(\sin^2\alpha-\eta^2)r'\frac{u^2}{2},$$ (5.173)

$$q(u,v) = u\cos\alpha + \left[(\cos\theta-u)^2+v^2\right]^{\frac{1}{2}} \quad , \quad q(u,\sin\theta) = q(u,v=\sin\theta).$$ (5.174)

5.4.1.2. Non-uniform asymptotic expansion

For large r', the asymptotic evaluation of the simple integrals occurring in (5.169) leads to the equation

$$I_1 \approx I_1^e + I_1^s$$

where I_1^e is the contribution of the neighborhood of the point $u = 0$, and I_1^s that of the neighborhood of the stationary phase point, which is assumed to be located far from the end point $u = 0$. Similarly, by replacing the inner integral in the right hand side of (5.170) by its asymptotic non-uniform expansion, we are led to the result

$$I_2 \approx I_2^e + I_2^s,$$

where I_2^e is the contribution of the end point $u = 0$, and I_2^s that of the stationary phase point, with I_1^s and I_2^s corresponding to the reflected field while $I_1^e + I_2^e$ representing the field diffracted by the discontinuity in the curvature. The inner expansion associated with the diffracted field is then given by

$$u^d(r',\theta) = \frac{1}{k}\left[I_1^e(r',\theta) + I_2^e(r',\theta)\right],$$ (5.175)

where the factor $1/k$ arises from the application of perturbation procedure limited to $O(1)$. If we restrict ourselves to the first term of the asymptotic expansion we find

$$u^d(r',\theta) = \left(\frac{2}{\pi}\right)^{\frac{1}{2}} \frac{1}{ik} e^{i\frac{\pi}{4}} u_0 [f^{(2)}0] \frac{\sin\alpha\sin\theta}{(\sin\alpha+\eta)(\sin\theta+\eta)} \frac{1-\cos\alpha\cos\theta-\eta^2}{(\cos\theta-\cos\alpha)^3} \frac{e^{ir'}}{\sqrt{r'}}.$$ (5.176)

The outer solution, valid in region-1, is provided by the Geometrical Theory of Diffraction. It reads

$$u_e^d(r,\theta) = \frac{1}{\sqrt{k}}\frac{e^{ikr}}{\sqrt{r}}\frac{D(\theta,\alpha)}{k}u_0, \tag{5.177}$$

where $D(\theta, \alpha)$ is the diffraction coefficient of the discontinuity in the curvature (See Sect. 5.2.2).

By using the stretched coordinates, $r' = kr$, the expression in (5.177) can be written as

$$u_e^d(k^{-1}r',\theta) = \frac{e^{ir'}}{\sqrt{r'}}\frac{D(\theta,\alpha)}{k}u_0, \tag{5.178}$$

from which we can derive the expression of $D(\theta, \alpha)$, after identifying the right-hand sides of (5.176) and (5.178). It reads

$$D(\theta,\alpha) = \left(\frac{2}{\pi}\right)^{\frac{1}{2}}\frac{1}{i}e^{i\frac{\pi}{4}}[f^{(2)}0]\frac{\sin\alpha\sin\theta}{(\sin\alpha+\eta)(\sin\theta+\eta)}\frac{1-\cos\alpha\cos\theta-\eta^2}{(\cos\theta-\cos\alpha)^3}. \tag{5.179}$$

We observe that $D(\theta, \alpha)$ becomes infinite when the observation point tends to the surface $\theta = \alpha$, its trace being denoted by OR in Fig. 5.10, through which the reflected field is discontinuous. The origin of the infinity can be traced directly to the method used in the asymptotic evaluation of the integrals I_1 and I_2. In effect, it has been assumed that the stationary phase point is not close to the end point $u = 0$, or, in other words, that the reflection point is not close to the diffraction point 0. Because of this, the inner expansion so derived is not valid in the region-4b. Similarly, the outer expansion given by (5.177) is not valid in region-2, and, consequently, the validity of the diffraction coefficient (5.179) is limited to region-1 of Fig. 5.9.

In order for the solution to be valid in region-2 outside the boundary layer-4, we need to derive the expansions valid in regions 4b and 2 and follow this up by matching them in the domain where they overlap. The latter is located at the periphery of the boundary layer-4 corresponding to a large r', and we obtain the inner expansion via an uniform asymptotic expansion of the integrals I_1 and I_2.

We will now present a general method for providing the uniform expansion of this type of integral and follow this up by applying it to the calculation of the integrals I_1 and I_2.

5.4.1.3   Search for a uniform solution at the periphery of the most internal boundary layer

GENERAL METHOD

We will now describe a method which is attractive because it clearly shows how a uniform solution can be constructed from the knowledge of the non-uniform solution.

Let us consider the following integral for $I(\Omega)$, where Ω is a parameter which is large compared to unity

$$I(\Omega) = \int_0^\infty F(u)\, e^{i\Omega q(u)}\, du, \tag{5.180}$$

and introduce the following change of variables

$$q(u) = q(u_s) + s^2, \tag{5.181}$$

where u_s is the stationary phase point, with $q'(u_s) = 0$ and $q''(u_s) \neq 0$. Thus, we have

$$I(\Omega) = e^{i\Omega q(u_s)} \int_{s_a}^\infty G(s)\, e^{i\Omega s^2}\, ds, \tag{5.182}$$

with

$$G(s) = F(u)\,\frac{du}{ds}, \quad s_a = \mp\sqrt{q(0) - q(u_s)}\,.$$

The (-) sign is used when the stationary phase point is in the domain of integration; otherwise the (+) sign is chosen. As u_s approaches the lower end point $u = 0$ of the integral (5.180), s_a tends to zero; then integrating (5.182) by parts leads to terms of the type $G(s_a)/s_a$ which are singular when $s_a \to 0$.

To circumvent this difficulty, we set

$$G(s) = G(0) + G(s) - G(0), \tag{5.183}$$

and integrate the term $G(s)-G(0)$ by parts. This leads to the result

$$I(\Omega) = e^{i\Omega q(u_s)} \left\{ G(0)\int_{s_a}^\infty e^{i\Omega s^2}\, ds - \frac{1}{2i\Omega}\left[\frac{G(s_a) - G(0)}{s_a}\right] e^{i\Omega s_a^2} \right.$$
$$\left. -\frac{1}{2i\Omega}\int_{s_a}^\infty \frac{d}{ds}\left[\frac{G(s) - G(0)}{s}\right] e^{i\Omega s^2}\, ds \right\} \tag{5.184}$$

Next, integrating the last integral in the right-hand side of (5.184) by parts, we obtain
the uniform asymptotic expansion of $I(\Omega)$, whose first terms can be written in the form

$$
\begin{aligned}
I(\Omega) = e^{i\Omega q(u_s)} \Bigg\{ & G(0) \int_{s_a}^{\infty} e^{i\Omega s^2} ds - \frac{1}{2i\Omega} \left[\frac{G(s_a) - G(0)}{s_a} \right] e^{i\Omega s_a^2} \\
& - \frac{1}{2i\Omega} H(0) \int_{s_a}^{\infty} e^{i\Omega s^2} ds + \frac{1}{(2i\Omega)^2} \left[\frac{H(s_a) - H(0)}{s_a} \right] e^{i\Omega s_a^2} \\
& + \frac{1}{(2i\Omega)^2} K(0) \int_{s_a}^{\infty} e^{i\Omega s^2} ds - \frac{1}{(2i\Omega)^3} \left[\frac{K(s_a) - K(0)}{s_a} \right] e^{i\Omega s_a^2} + \ldots \Bigg\}
\end{aligned}
$$
, (5.185)

with

$$
\begin{aligned}
H(s) &= \frac{d}{ds} \left[\frac{G(s) - G(0)}{s} \right] \\
K(s) &= \frac{d}{ds} \left[\frac{H(s) - H(0)}{s} \right]
\end{aligned}
$$

Equation (5.185) explicitly shows the contribution of the end point s_a as it appears in
the non-uniform asymptotic expansion of the integral. To show this, we rewrite
(5.185), limited to three integrations by parts, as follows :

$$
\begin{aligned}
I(\Omega) = e^{i\Omega q(u_s)} \Bigg\{ & G(0) \left[F(\Omega) - \hat{F}_3(\Omega) \right] - \frac{H(0)}{2i\Omega} \left[F(\Omega) - \hat{F}_2(\Omega) \right] \\
& + \frac{K(0)}{(2i\Omega)^2} \left[F(\Omega) - \hat{F}_1(\Omega) \right] \Bigg\} + I_d(\Omega)
\end{aligned}
$$
,   (5.186)

where

$$
\begin{aligned}
I_d(\Omega) = & \left\{ -\frac{1}{2i\Omega} \left[\frac{G(s_a) - G(0)}{s_a} \right] + \frac{1}{(2i\Omega)^2} \left[\frac{H(s_a) - H(0)}{s_a} \right] - \frac{1}{(2i\Omega)^3} \left[\frac{K(s_a) - K(0)}{s_a} \right] \right\} e^{i\Omega s_a^2} \\
& + G(0) \hat{F}_3(\Omega) - \frac{H(0)}{2i\Omega} \hat{F}_2(\Omega) + \frac{K(0)}{(2i\Omega)^2} \hat{F}_1(\Omega)
\end{aligned}
$$

(5.187)

and $F(\Omega)$ is the Fresnel function defined by

$$
F(\Omega) \cong \int_{s_a}^{\infty} e^{i\Omega s^2} ds.
$$
    (5.188)

Here $\hat{F}_n(\Omega)$ is the asymptotic expansion limited to the order n for s_a positive and $\sqrt{\Omega}\ s_a \gg 1$. It is given by

$$\hat{F}_n(\Omega) = -e^{i\Omega s_a^2}\frac{1}{2i\Omega}\frac{1}{s_a}\pi^{-\frac{1}{2}}\sum_{p=0}^{n-1}\frac{\Gamma\left(p+\frac{1}{2}\right)}{\left(i\Omega s_a^2\right)^p}, \tag{5.189}$$

where

$$\Gamma\left(p+\frac{1}{2}\right)=\sqrt{\pi}\times\frac{1}{2}\times\frac{3}{2}\times...\times\frac{2p-1}{2}.$$

When $s_a < 0$, the asymptotic expansion of (5.188), limited to the order n for $\sqrt{\Omega}\ s_a \gg 1$, is written as

$$F(\Omega) =\sqrt{\frac{\pi}{\Omega}}\,e^{-i\frac{\pi}{4}} + \hat{F}_n(\Omega). \tag{5.190}$$

Inserting (5.190) into (5.186) we find

$$I(\Omega) = e^{i\Omega q(u_s)}\sqrt{\frac{\pi}{\Omega}}\,e^{-i\frac{\pi}{4}}\left[G(0)-\frac{H(0)}{2i\Omega}+\frac{K(0)}{(2i\Omega)^2}\right]+I_d(\Omega). \tag{5.191}$$

The first term in the right-hand side of (5.191) is nothing but the contribution of the stationary phase point. In effect, if we perform a Taylor expansion of $G(s)$ at $s = 0$, and integrate (37) term by term, we obtain the contribution of the stationary point which is assumed to be far from the end point ($\sqrt{\Omega}\ s_a \gg 1$).

$$I(\Omega) = e^{i\Omega q(u_s)}\sqrt{\frac{\pi}{\Omega}}\,e^{-i\frac{\pi}{4}}\left[G(0)-\frac{1}{2i\Omega}\frac{G''(0)}{2}+\frac{1}{(2i\Omega)^2}\frac{G^{(4)}(0)}{8}\right].$$

On the other hand, from the definition of H(s) and K(s) we have

$$H(0) = \frac{G''(0)}{2}, \quad K(0) = \frac{G^{(4)}(0)}{8},$$

which justifies the statement we made above. In addition, since the asymptotic expansion (5.191) has been obtained for $\sqrt{\Omega}\ s_a \gg 1$, it is identical to the non-uniform asymptotic expansion of the integral. It follows then, that $I_d(\Omega)$ in (5.191) represents the contribution of the end point such as it appears in the non-uniform asymptotic expansion of (5.180). This result shows that it is possible to represent, directly, the uniform asymptotic expansion of the integral given in (5.186) if one knows its non-uniform asymptotic expansion given in (5.191). We can also show, via a direct calculation, that $I_d(\Omega)$ depends only on the end point s_a of the integral. In fact, if we replace the terms $\hat{F}_1(\Omega)$, $\hat{F}_2(\Omega)$, and $\hat{F}_3(\Omega)$ in (5.187) by their explicit expressions given by (5.189), we find that the terms

$$\frac{G(0)}{2i\Omega s_a}e^{i\Omega s_a^2} \quad , \quad -\frac{H(0)}{(2i\Omega)^2 s_a}e^{i\Omega s_a^2} \quad , \quad \frac{K(0)}{(2i\Omega)^3 s_a}e^{i\Omega s_a^2} \quad ,$$

disappear in (5.187). The same is also true for the terms depending on $G(0)$ and $H(0)$ in $H(s_a)$ and $K(s_a)$, and the latter takes the form

$$H(s_a) = \frac{1}{s_a}\left(\frac{dG}{ds}\right)_{s=s_a} - \frac{1}{s_a^2}G(s_a) + \frac{1}{s_a^2}G(0)$$

$$K(s_a) = \frac{1}{s_a}\left(\frac{dH}{ds}\right)_{s=s_a} - \frac{1}{s_a^2}H(s_a) + \frac{1}{s_a^2}H(0)$$

$$-\frac{3}{s_a^3}\left(\frac{dG}{ds}\right)_{s=s_a} + \frac{1}{s_a^2}\left(\frac{d^2G}{ds^2}\right)_{s=s_a} + \frac{3}{s_a^3}G(s_a)$$

$$+\frac{1}{s_a^2}H(0) - \frac{3}{s_a^2}G(0)$$

Finally, for the $O(\Omega^{-3})$ we are left with the expression

$$I_d(\Omega) = e^{i\Omega q(u_s)}\, e^{i\Omega s_a^2} \times \left\{-\frac{1}{2i\Omega}\frac{G(s_a)}{s_a} + \frac{1}{(2i\Omega)^2}\left[\frac{1}{s_a^2}\left(\frac{dG}{ds}\right)_{s=s_a} - \frac{1}{s_a^3}G(s_a)\right]\right.$$

$$\left. -\frac{1}{(2i\Omega)^3}\left[\frac{1}{s_a^3}\left(\frac{d^2G}{ds^2}\right)_{s=s_a} - \frac{3}{s_a^4}\left(\frac{dG}{ds}\right)_{s=s_a} + \frac{3}{s_a^4}G(s_a)\right]\right\}$$

$$(5.192)$$

We observe that $I_d(\Omega)$ depends only on the end point s_a and we can show that the expression in (5.192) corresponds to the non-uniform asymptotic expansion obtained by means of successive integration by parts of (5.182), with respect to the end point s_a.

In practice, it is better to express $I_d(\Omega)$ directly as a function of $F(u)$ and $q(u)$. This is accomplished by integrating (5.180) by parts, and this yields, to the $O(\Omega^{-3})$, the result

$$I_d(\Omega) = e^{i\Omega q(0)} \left\{ -\frac{1}{i\Omega} \frac{F(0)}{q'(0)} + \frac{1}{(i\Omega)^2} \frac{1}{q'(0)} \left[\frac{d}{du} \left(\frac{F(u)}{q'(u)} \right) \right]_{u=0} \right.$$
$$\left. -\frac{1}{(i\Omega)^3} \frac{1}{q'(0)} \left(\frac{d}{du} \left[\frac{1}{q'(u)} \frac{d}{du} \left(\frac{F(u)}{q'(u)} \right) \right] \right)_{u=0} \right\} .$$

(5.193)

It is also necessary to express the quantities $G(0)$, $H(0)$ and $K(0)$ as functions of $F(u_s)$ and $q(u_s)$, and the general formulas for $G(0)$ and $H(0)$ are given below
1

$$G(0) = \left(\frac{2}{q''(u_s)} \right)^{\frac{1}{2}} F(u_s),$$

(5.194)

$$H(0) = \left(\frac{2}{q''(u_s)} \right)^{\frac{3}{2}} \left\{ \frac{1}{2} \left[F''(u_s) - \frac{q^{(3)}(u_s)}{q''(u_s)} F'(u_s) + \frac{1}{8} F(u_s) \left(\frac{5}{3} \frac{q^{(3)}(u_s)}{q''(u_s)} - \frac{q^{(4)}(u_s)}{q''(u_s)} \right) \right] \right\} .$$

(5.195)

ASYMPTOTIC EVALUATION OF $I_1(r',\theta)$

According to (5.169), $I_1(r',\theta)$ can be written in the form

$$I_1(r',\theta) = C_1 \left\{ i_0(r',\theta) \bar{f}^{(2)}(0) + \frac{1}{2} \left[f^{(2)}(0) \right] \left(i_1(r',\theta) - i_2(r',\theta) \right) \right\},$$

(5.196)

where i_0, i_1, and i_2 are three integrals given by

$$\begin{cases} i_0(r',\theta) = \int_{-\infty}^{+\infty} F(u,\sin\theta;r') \, e^{ir'q(u,\sin\theta)} du \\ i_1(r',\theta) = \int_0^{\infty} F(u,\sin\theta;r') \, e^{ir'q(u,\sin\theta)} du \\ i_2(r',\theta) = \int_0^{\infty} F(-u,\sin\theta;r') \, e^{ir'q(-u,\sin\theta)} du \end{cases},$$

(5.197)

and where the coefficient C_1 and the functions $F(u,v;r')$ and $q(u,v)$ are defined by the relations (5.171) to (5.174). The integral i_0 has no end points at finite distances and, consequently, its asymptotic expansion for large r' reduces to the contribution of the stationary phase point. Hence, if we restrict ourselves to $O(1/r'^2)$, we find, according to (5.191), that

$$i_0(r',\theta) \approx e^{ir'q(u_s)} \sqrt{\frac{\pi}{r'}} e^{-i\frac{\pi}{4}} \left[G(0) - \frac{1}{2ir'} H(0) + \frac{1}{(2ir')^2} K(0) \right].$$

(5.198)

The asymptotic expansions of the integrals i_1 and i_2 are given by expressions similar to (5.186). The effect of changing u into $-u$ is to change the sign of s_a in the corresponding expansions. Hence, we can write

$$
\begin{aligned}
I_1(r',\theta) = C_1 e^{ir'\cos(\theta-\alpha)} & \Bigg(\sqrt{\frac{\pi}{r'}}\, e^{-i\frac{\pi}{4}} \Big[G(0) - \frac{1}{2ir'} H(0) + \frac{1}{(2ir')^2} K(0) \Big] \bar{f}^{(2)}(0) \\
& + \frac{1}{2}[f^{(2)}(0)]\Big\{ G(0)\big([F(r',s_a)-\hat{F}_3(r',s_a)] - [F(r',-s_a)-\hat{F}_3(r',-s_a)] \big) \\
& - \frac{1}{2ir'} H(0)\big([F(r',s_a)-\hat{F}_2(r',s_a)] - [F(r',-s_a)-\hat{F}_2(r',-s_a)] \big) \\
& - \frac{1}{(2ir')^2} K(0)\big([F(r',s_a)-\hat{F}_1(r',s_a)] - [F(r',-s_a)-\hat{F}_1(r',-s_a)] \big) \Big\} \Bigg) + C_1 i_d(r',\theta),
\end{aligned}
$$

(5.199)

where $s_a = \sqrt{2}\sin\left(\frac{\theta-\alpha}{2}\right)$ and $i_d = C_1^{-1} I_{1d}$, and where I_{1d} is the non-uniform asymptotic expansion of I_1 that is terminated at the $O(r^{-3})$. Applying (5.193) to the integrals i_1 and i_2, we find

$$
i_d = i_d = \left(-\frac{1}{r'^2}\frac{1-\cos\alpha\cos\theta-\eta^2}{(\cos\alpha-\cos\theta)^3} + \text{term } O(\frac{1}{r'^3}) \right) e^{ir'}.
$$

(5.200)

The quantities $G(0)$, $H(0)$ and $K(0)$ can be expressed as functions of the angles θ, α, and of the function $Q(u;r')$, which is defined in (5.173). Using (5.194) and (5.195), as well as the general expression of $K(0)$ as a function of $F(u_s)$ and $q(u_s)$, which may be found in [Mo2] or [Mo3], we obtain

$$
\begin{cases}
G(0,v) = \dfrac{i\sqrt{2}}{\sin\alpha} Q(u_s) \\[2mm]
H(0,v) = \dfrac{i\sqrt{2}|v|}{\sin^4\alpha}\left[Q''(u_s) - \dfrac{2\sin\alpha\cos\alpha Q'(u_s)}{|v|} + \dfrac{1}{4}\dfrac{\sin^4\alpha}{v^2} Q(u_s) \right] \\[2mm]
K(0,v) = \dfrac{i4\sqrt{2}}{\sin^5\alpha}\Bigg\{ \dfrac{3}{2}\left(\cos^2\alpha+\dfrac{3}{8}\sin^2\alpha\right)Q''(u_s) \\[2mm]
\qquad \dfrac{3}{4}\dfrac{\sin\alpha\cos\alpha Q'(u_s)}{|v|}\left(11\cos^2\alpha-\dfrac{3}{2}\sin^2\alpha-1\right) \\[2mm]
\qquad + \dfrac{\sin^2\alpha}{8v^2}\left[3\left(\sin^4\alpha-\dfrac{1}{4}\right)-\dfrac{7}{2}\left(\sin^2\alpha-\dfrac{\cos^4\alpha}{8}+\dfrac{3}{2}\cos^2\alpha\right)\right]Q(u_s)\Bigg\}
\end{cases}
$$

(5.201)

where $Q(u_s) = Q(u_s; r')$, $u_s = \dfrac{\sin(\alpha-\theta)}{\sin\alpha}$, $|v| = \sin\theta$.

ASYMPTOTIC EVALUATION OF $I_2(r',\theta)$

According to (5.170), $I_2(r',\theta)$ can be written in the form

$$I_2(r',\theta) = C_2 \left\{ j_0(r',\theta) \bar{f}^{(2)}(0) + \frac{1}{2} [f^{(2)}(0)] (j_1(r',\theta) - j_2(r',\theta)) \right\}, \quad (5.202)$$

where j_0, j_1 and j_2 are three double integrals given by

$$j_0(r',\theta) = \int_{-\infty}^{-\sin\theta} e^{-i\eta r'(v+\sin\theta)} dv \int_{-\infty}^{+\infty} F(u,v;r') e^{ir'q(u,v)} du$$

$$j_1(r',\theta) = \int_{-\infty}^{-\sin\theta} e^{-i\eta r'(v+\sin\theta)} dv \int_{0}^{+\infty} F(u,v;r') e^{ir'q(u,v)} du \qquad (5.203)$$

$$j_2(r',\theta) = \int_{-\infty}^{-\sin\theta} e^{-i\eta r'(v+\sin\theta)} dv \int_{0}^{+\infty} F(-u,v;r') e^{ir'q(-u,v)} du,$$

and where the coefficient C_2 and the functions $F(u,v;r')$ are defined by the relations given in (5.172) to (5.174).

To evaluate these integrals asymptotically, we replace the integrals whose integration variable is u by their uniform asymptotic expansions. The latter are identical to the uniform asymptotic expansions of the integrals i_0, i_1 and i_2, provided in the preceding section. However, the quantities $G(0)$, $H(0)$ and $K(0)$, as well as u_s and $q(u_s)$, are now functions of the integration variable v of the external integral. According to (5.174), the stationary phase point u_s of the inner integrals of j_0 and j_1, given by $dq/du(u, v) = 0$, can be written as

$$u_s = \cos\theta - \frac{\cos\alpha}{\sin\alpha} |v| \quad (v < \infty). \qquad (5.204)$$

Hence,

$$q(u_s(v), v) = \cos\alpha \cos\theta - v \sin\theta \quad (v < \infty). \qquad (5.205)$$

As for the integral j_2, $q(u_s(v), v_s)$ appearing in this integral remains unchanged despite a change in the sign of u_s. However, according to (5.174), we have

$$q(0,v) = (\cos^2\theta + v^2)^{1/2}. \qquad (5.206)$$

From the preceding results, we see that the uniform asymptotic expansion of each of the inner integrals of j_1 and j_2 is composed of two terms. The first of these has the phase given by (5.205), while the phase of the second, which corresponds to the non-uniform contribution of the end point of integration, is expressed by (5.206).

Let turn us now to the integral j_0. We see that the second term is absent in its asymptotic expansion and the first reduces to the contribution of the stationary phase point given by (5.205). The external integrals corresponding to each of the terms whose phase is given by (5.205), have no stationary phase points, since the total phase of the integrand is a linear function of v. In contrast, the external integrals corresponding to the terms whose phase is given by (5.206) do have stationary phase points, although they are now complex since the relative impedance η is, in general, a complex number. When $\eta = 0$, $v_s = 0$. Under these conditions, the end point of the variable of integration $v = -\sin \theta$ tends to the stationary phase point v_s when θ tends to 0 or π, and this corresponds to an observation point in the zone 4c of Fig. 5.10, a situation that cannot be handled with the present method. When the observation point is outside this zone, and when $\eta = 0$, v_s is always far from the end point $v = -\sin \theta$, and the same is true when $\eta \neq 0$, owing to the condition $R_e\eta \geq 0$. For the reasons given above, the asymptotic expansions of the outer integrals in (5.203) are always derived via successive integrations by parts. (The details of the calculation are not reproduced here.)

Finally, in view of (5.172), to extract all of the terms of I_2 up to $O(1/\sqrt{r'})$ one has to retain, in the expansion of the integrals in (5.197), terms up to $O(r'^{-3})$ of the asymptotic expansions of the inner integrals. These integrals depend on the functions $G(0,v)$, $H(0,v)$ and $K(0,v)$, and on the products $G(0,v)\Delta F_3$, $H(0,v)\Delta F_2$ and $K(0,v)\Delta F_1$, where we have defined

$$\Delta F_n = F(r',s_a) - \hat{F}_n(r',s_a) - \left[F(r',-s_a) - \hat{F}_n(r',-s_a) \right]. \qquad (5.207)$$

According to (5.201) and (5.173), $G(0,v)$, $H(0,v)$ and $K(0,v)$ only contain terms of $O(0)$ and $O(1)$ with respect to r', and, consequently, one can show that, to $O(r'^{-3})$, the asymptotic expansions of the integrals in (5.203) depend only upon the derivatives $\partial G \partial v$ $(0,v) = L(0,v)$, $\partial^2 F/\partial v^2$ $(0,v) = M(0,v)$ and $\partial H/\partial v$ $(0,v) = N(0,v)$, whose expressions as functions of u_s are

$$L(0, v) = \frac{i\sqrt{2}\,\cos\alpha\,Q'(u_s)}{\sin^2\alpha}$$

$$M(0, v) = -\frac{i\sqrt{2}}{\sin^4\alpha}\left[(1 + 2\cos^2\alpha)\,Q''(u_s) - \frac{1}{4}\frac{\sin^3\alpha\cos\alpha}{\sin\theta}Q'(u_s) - \frac{1}{4}\frac{\sin^4\alpha}{\sin^2\theta}Q(u_s)\right]$$

$$N(0, v) = i\sqrt{2}\,\frac{\cos^2\alpha}{\sin^3\alpha}\,Q''(u_s)$$

$$(5.208)$$

where u_s is given by (5.204).

Because of these remarks, the asymptotic expansion of $I_2(r',\theta)$, terminated with the $O(1/\sqrt{r'})$ term, can finally be written in the form

$$I_2(r',\theta) = -C_2\,\frac{e^{ir'\cos(\theta-\alpha)}}{ir'(\eta+\sin\alpha)}\left(\sqrt{\frac{\pi}{r'}}\,e^{-i\frac{\pi}{4}}\,\bar{f}^{(2)}(0)\left\{G(0) + \frac{1}{ir'}\left(\frac{L(0)}{\eta+\sin\alpha} + \frac{H(0)}{2}\right)\right.\right.$$

$$+\frac{1}{(ir')^2}\left(\frac{N(0)}{(\eta+\sin\alpha)^2} + \frac{M(0)}{2(\eta+\sin\alpha)}\right) + \frac{1}{(2ir')^3}K(0)\right\}$$

$$+\left\{\left[G(0) + \frac{1}{ir'}\frac{L(0)}{\eta+\sin\alpha} + \frac{1}{(ir')^2}\frac{N(0)}{(\eta+\sin\alpha)^2}\right]\Delta F_3\right.$$

$$+\left[\frac{1}{2ir'}H(0) + \frac{1}{(ir')^2}\frac{M(0)}{2(\eta+\sin\alpha)}\right]\Delta F_2 + \frac{K(0)}{(2ir')^2}\Delta F_1\right\}\frac{1}{2}\left[f^{(2)}(0)\right]\right)$$

$$+ C_2\,j_d(r',q)$$

$$(5.209)$$

with $j_d = C_2^{-1}I_{2d}$ where I_{2d} is the non-uniform asymptotic expansion of I_2 whose expression is given in [KK]. The expressions of $L(0)$, $M(0)$, and $N(0)$ are deduced from $L(0,v)$, $M(0,v)$ and $N(0,v)$, respectively, by setting v equal to the value of the end point $v = -\sin\theta$, which is equivalent to replacing u_s in (63) by $u_s = \sin(\alpha-\theta)/\sin\alpha$. We observe that the asymptotic expansion of I_2 has the same structure as that of I_1.

UNIFORM ASYMPTOTIC SOLUTION AT THE PERIPHERY
OF THE BOUNDARY LAYER

According to (5.151), the solution of the diffraction problem inside the boundary layer as extracted from the perturbation method with respect to the small parameter $1/k$, takes the form

$$u(x',y') = u_0(x',y') + \frac{1}{k}u_1(x',y')$$

$$(5.210)$$

In the above expression, we have used the stretched coordinates and have restricted ourselves to terms up to $O(1)$. The function $u_0(x',y')$ is given by (5.156).

At the periphery of the boundary layer, the uniform asymptotic expansion of $u_1(x',y')$ is given as the sum of the asymptotic expansions of the integrals $I_1(x',y')$ and $I_2(x',y')$ given by (5.150) and (5.209). As a result, we find the following expression for the uniform asymptotic expansion of $u(x',y')$ at the periphery of the boundary layer :

$$u(r',\theta) = u_0\, e^{ir'\cos(\theta+\alpha)} + u_0\, \frac{\sin\alpha - \eta}{\sin\alpha + \eta}\, e^{ir'\cos(\theta-\alpha)} + \frac{1}{k} u_1(r',\theta), \qquad (5.211)$$

where

$$U_1(r',\theta) = -\left(\frac{2}{\pi}\right)^{\frac{1}{2}} ie^{i\frac{\pi}{4}} U_0 \frac{\sin\alpha}{\sin\alpha + \eta}\, r'^{\frac{3}{2}}\, e^{ir'\cos(\theta-\alpha)}$$

$$\left\{ \sqrt{\frac{\pi}{r'}}\, e^{-i\frac{\pi}{4}}\, \bar{f}^{(2)}(0)\left[A_1(r',\theta) + \frac{1}{2ir'} A_2(r',\theta) + \frac{1}{(2ir')^2} A_3(r',\theta) \right] \right.$$

$$\left. + \frac{1}{2}\left[\bar{f}^{(2)}(0)\right]\left[A_1(r',\theta)\Delta F_3 + \frac{1}{2ir'} A_2(r',\theta)\Delta F_2 + \frac{1}{(2ir')^2} A_3(r',\theta)\Delta F_3 \right] \right\}$$

$$+ I_d(r',\theta)$$

$$(5.212)$$

$$I_d(r',\theta) = -\left(\frac{2}{\pi}\right)^{\frac{1}{2}} ie^{i\frac{\pi}{4}} u_0\, \frac{\sin\alpha}{\sin\theta + \eta}\, \frac{\sin\theta}{\sin\theta + \eta}\, \frac{1 - \cos\alpha\cos\theta - \eta^2}{(\cos\theta - \cos\alpha)^3}\left[f^{(2)}(0)\right]\frac{e^{ir'}}{\sqrt{r'}},$$

$$+ \text{term } O(1/r'^{3/2})$$

$$(5.213)$$

$$\begin{cases} A_1(r',\theta) = \dfrac{\sin\alpha}{\sin\alpha + \eta} G(0) - \dfrac{1}{ir'}\dfrac{\eta}{(\sin\alpha + \eta)^2} L(0) - \dfrac{1}{(ir')^2}\dfrac{\eta}{(\sin\alpha + \eta)^3} N(0) \\[2mm] A_2(r',\theta) = \dfrac{\sin\alpha}{\sin\alpha + \eta} H(0) - \dfrac{1}{ir'}\dfrac{\eta}{(\sin\alpha + \eta)^2} M(0) \\[2mm] A_3(r',\theta) = \dfrac{\sin\alpha}{\sin\alpha + \eta} K(0) \end{cases} ,$$

$$(5.214)$$

and ΔF_n, $n = 1, 2, 3$ was defined by (5.207). Finally, the coefficients $G(0)$, $H(0)$, $K(0)$, $L(0)$, $M(0)$, and $N(0)$ are given by (5.201) and (5.208).

5.4.1.4 Physical meaning of the terms occurring in the asymptotic solution

For $\sqrt{r}$ $|s_a|$ large, we have

$$\nabla F_n \approx \pm \sqrt{\frac{\pi}{r'}} e^{-i\frac{\pi}{4}}, \tag{5.215}$$

with the (+) sign corresponding to the case where s_a is negative. For this case, $u_1(r',\theta)$, given by (5.212), reduces to the non-uniform solution

$$u_1(r',\theta) = -\sqrt{2}\, iu_0 r' \frac{\sin\alpha}{\sin\alpha+\eta} e^{ir'\cos(\theta-\alpha)} A(r',\theta) f^{(2)}(0\pm) + I_d(r',\theta), \tag{5.216}$$

where

$$A(r',\theta) = A_1(r',\theta) + \frac{1}{2ir'} A_2(r',\theta) + \frac{1}{(2ir')^2} A_3(r',\theta), \tag{5.217}$$

and where $f^{(2)}$ $(0\pm)$ are defined by (5.168), and denote the limit of the second derivative on each side of the discontinuity, for $x' > 0$ and $x' < 0$, respectively.

From (5.214), (5.201), (5.208) and (5.173), we see that $A(r',\theta)$ is composed of terms proportional to u_s^2 and u_s, that vanish with u_s on the transition $\theta = \alpha$, and of terms that are independent of u_s and non-vanishing for $\theta = \alpha$.

Putting all of the non-vanishing terms of $A(r',\ \theta)$ together when $\theta = \alpha$, we see that they are of $O(0)$ and $O(1/r')$ with respect to r'. It follows, then, that for $\theta = \alpha$ the first term in the right-hand side of (5.216), denoted here by $u_1{}^s(r',\theta)$, has the form

$$u_1{}^s(r',\theta) = ar' + b. \tag{5.218}$$

Let us first consider the term containing r'. A straightforward calculation yields

$$a = -u_0\, e^{ir'\cos(\theta-\alpha)} \frac{1}{\sin\alpha} \frac{\sin\alpha-\eta}{\sin\alpha+\eta} f^{(2)}(0\pm). \tag{5.219}$$

Next, inserting this term into (5.211), we obtain

$$u(r',\theta) = u_0\, e^{ir'\cos(\theta+\alpha)} + u_0 R \left[1 - \frac{r'}{k\sin\alpha} f^{(2)}(0\pm) \right] e^{ir'\cos(\theta-\alpha)} + I_d(r',\theta),$$

$$\tag{5.220}$$

where R is the reflection coefficient of the plane with an impedance η.

In the second term of the right-hand side of (5.220) we recognize the small r expansion of the reflected field of Geometrical Optics (first term of the Luneberg-Kline series), on each side of the discontinuity in the curvature. In effect, for θ close to α, we can express the first term of the non-uniform asymptotic expansion of the reflected field in the region 2 of Fig. 5.9 in the form

$$u_1^R(r',\theta) = u_0 \sqrt{\frac{\rho}{\rho+r}} \, e^{ikr\cos(\theta-\alpha)} \, ,$$

where ρ is the curvature radius of the reflected wave at the reflection point. For small r, we have

$$\sqrt{\frac{\rho}{\rho+r}} \approx 1 - \frac{r}{2\rho} = 1 - \frac{r'}{2k\rho} \, .$$

On the other hand, at the discontinuity point (see Fig. 5.12) we have

$$\frac{1}{2\rho} = \frac{f^{(2)}(0-)}{\sin\alpha} \quad , \quad \theta = \alpha + \varepsilon \quad , \quad \varepsilon > 0$$

$$\frac{1}{2\rho} = \frac{f^{(2)}(0+)}{\sin\alpha} \quad , \quad \theta = \alpha - \varepsilon \quad , \quad \varepsilon > 0$$

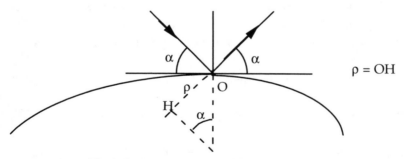

$\rho = OH$

Fig. 5.12. Geometry for the reflection at the curvature discontinuity.

Hence, it follows that

$$u_1^R(k^{-1}r',\theta) = u_0 R \left[1 - \frac{r'}{k\sin\alpha} f^{(2)}(0\pm) \right] e^{ir'\cos(\theta-\alpha)}$$

Similarly, we can show that the constant term (with respect to r') in (5.218) is directly related to the second term in the Luneberg-Kline series of the reflected field. In the vicinity of the transition line in region 2, this term reads (see ref. [KLS] and the corrections mentioned in ref. [GU])

$$u_2^R(r',\theta) = \frac{1}{ik} u_0 \sqrt{\frac{\rho}{\rho+r}} \; f^{(2)}(0\pm) \frac{1}{\sin\alpha+\eta} e^{ikr\cos(\theta-\alpha)}$$

$$\left[a_0 + \frac{\rho}{\rho+r} a_1 + \left(\frac{\rho}{\rho+r}\right)^2 a_2 + \left(\frac{\rho}{\rho+r}\right)^3 a_3\right] \, , \qquad (5.221)$$

with

$$a_0 = \frac{1}{2^4(\sin\alpha+\eta)^2}\left[3\sin^2\alpha + 11\eta\sin\alpha + 8 - 11\eta^2 - \left(\frac{8\eta+3\eta^2}{\sin\alpha}\right) + \frac{8\eta^2}{\sin^2\alpha} + \frac{8\eta^3}{\sin^3\alpha}\right]$$

$$a_1 = \frac{2}{\sin\alpha} \frac{1}{2^5(\sin\alpha+\eta)^2}\left[3\sin^2\alpha + 3\eta\sin^2\alpha - (1+11\eta^2)\sin\alpha - 9\eta - 3\eta^3 + \frac{9\eta^2}{\sin\alpha} + \frac{\eta^3}{\sin^2\alpha}\right]$$

$$a_2 = \left(\frac{2}{\sin\alpha}\right)^2 \frac{1}{2^6(\sin\alpha+\eta)}\left[3\sin^3\alpha - (3\eta^2+2)\sin\alpha - 8\eta + \frac{2\eta^2}{\sin\alpha} + \frac{8\eta}{\sin^3\alpha}\right]$$

$$a_3 = \left(\frac{2}{\sin\alpha}\right)^3 \frac{15}{2^7}\left[-\sin^2\alpha + \eta\sin^2\alpha + \sin\alpha - \eta\right]. \qquad (5.222)$$

For small r with respect to ρ, we obtain

$$\left[a_0 + \left(\frac{\rho}{\rho+r}\right)a_1 + \left(\frac{\rho}{\rho+r}\right)^2 a_2 + \left(\frac{\rho}{\rho+r}\right)^3 a_3\right] \times \sqrt{\frac{\rho}{\rho+r}}$$

$$\approx a_0 + a_1 + a_2 + a_3 - \frac{r}{2\rho}(a_0 + 3a_1 + 5a_2 + 7a_3) + O\left(\frac{r}{\rho}\right)^2 .$$

We can show by means of straightforward calculations, which are not reproduced here, that the constant term b in (5.218), given below,

$$b = \frac{1}{ik} u_0 e^{ir'\cos(\theta-\alpha)} f^{(2)}(0\pm) \frac{1}{\sin\alpha+\eta} (a_0 + a_1 + a_2 + a_3),$$

is exactly equal to the term of order zero with respect to (r/ρ) in (5.221). If we use the stretched coordinate $r' = kr$, the term of $O(1)$ with respect to (r/ρ) in (5.221) gives a term of $O(1/k) \cdot (r'/k\rho)$. Since the solution inside the boundary layer has been terminated at $O(1/k)$, this term does not appear in the non-uniform solution (5.216). To extract its contribution we need to carry out the perturbation calculation (see (5.151)) up to the $O(1/k^2)$.

The parameter $\sqrt{r'}\, s_a$, which is present in (5.212) inside the terms ΔF_n, also has a simple physical interpretation. According to the definition of s_a just given after (5.182), we get

$$s_a = \mp\sqrt{q(0)-q(u_s)} = \sqrt{2}\,\sin\!\left(\frac{\theta-\alpha}{2}\right) \qquad (5.223)$$

Using the stretched coordinates and denoting the difference between the paths traveled by the diffracted ray $S_\infty OM$ and the reflected ray $S_\infty QM$ (see Fig. 5.13 below) by Δ^2, we get

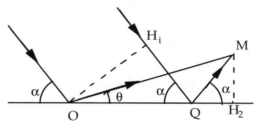

Fig. 5.13. Geometrical representation of the path difference between the reflected ray and the ray diffracted by the curvature discontinuity.

$$\Delta = OM - (H_IQ + QM) = 2r'\sin^2\!\left(\frac{\theta-\alpha}{2}\right) = r's_a{}^2\,.$$

However, using the definitions of $F(r',\,s_a)$ and $\hat{F}_n(r',\,s_a)$, given by (5.188) and (5.189), we can write

$$F(r',s_a) = \int_{s_a}^{\infty} e^{ir's^2}\,ds = \frac{1}{\sqrt{r'}}\int_{\sqrt{r'}\,s_a}^{\infty} e^{iu^2}\,du$$

$$\hat{F}_n(r',s_a) = \frac{1}{\sqrt{r'}}\,e^{ir's_a^2}\,\frac{1}{2i\sqrt{r'}\,s_a}\sum_{p=0}^{n-1}\frac{\Gamma\!\left(p+\dfrac{1}{2}\right)\cdot}{i\left(r's_a^2\right)^p} \qquad (5.224)$$

Consequently, we see that within the factor $1/\sqrt{r'}$, the argument of both the functions $F(r',\,s_a)$ and $\hat{F}_n(r',\,s_a)$ is $\sqrt{r'}\,s_a = \sqrt{\Delta}$, which can be written in unstretched coordinates as

$$\sqrt{r'}s_a = \sqrt{kr}\,\sin\!\left(\frac{\theta-\alpha}{2}\right) = \varepsilon(\vec{r})\sqrt{k\,|\,s(\vec{r})-s_R(\vec{r})|} = \xi(\vec{r}), \qquad (5.225)$$

where $s(r)$ and $s_R(r)$ are, respectively, the phases at the observation point $M(\vec{r})$ of the diffracted wave and of the reflected wave, and where $e(\vec{r}) = +1$ if $q > \alpha$ and -1 otherwise. The parameter $\xi(\vec{r})$ is called the *detour parameter* because it corresponds to the detour of the optical path followed by the diffracted wave from that of the reflected wave. This parameter is identical to the one introduced earlier for the case of the wedge.

5.4.2 Expression of the Uniform Solution

In Sect. 5.4.1.3, we have derived a uniform asymptotic solution valid at the periphery of the boundary layer, in the domains 4a and 4b of Fig. 5.10. This expansion represents the so-called *inner expansion* in the theory of the boundary layer. An outer expansion in region-1 of Fig. 1 has also been developed in Sect. 5.4.1.2, and is given by

$$u_e(r,\theta) = u_i(r,\theta) + u_R(r,\theta) + u_e{}^d(r,\theta) , \qquad (5.226)$$

where u_i and u_r are the incident and reflected fields of Geometrical Optics, respectively, and where $u_e{}^d$ is the field diffracted by the discontinuity in the curvature and is given by (5.177). An initial matching in the overlapping domain of the regions 4a and 1, between the term $I_d(r',\theta)$ of (5.216) corresponding to the diffraction by the line discontinuity in the curvature and the term $u_e{}^d(r,\theta)$ of (5.226), enabled us to determine, in Sect. 5.4.1.2, the diffraction coefficient appearing in (5.177), and to define, completely, the term $u_e{}^d(r,\theta)$ in region-1 of Fig. 5.9. A similar matching can be performed between the inner expansion associated with the reflected field and the outer expansion associated with this field in the domain where the regions 4a and 1 overlap.

In accordance with the physical interpretation given in Sect. 5.4.1.4, the inner expansion associated with the reflected field includes the first two terms of the Luneberg-Kline expansion. To match this expansion to the outer one, it is essential to retain in the latter the first two terms of the Luneberg-Kline series.

This leads us to write the reflected field in region-1 in the form

$$u_R(r,\theta) = e^{iksR(r,\theta)}\left[A_0(r,\theta) + \frac{1}{ik}A_1(r,\theta)\right], \qquad (5.227)$$

where A_0 and A_1 are the first two terms of the Luneberg-Kline series. For a two-dimensional surface A_0 and A_1 are given by

$$\begin{cases} A_0(r,\theta) = R\sqrt{\dfrac{\rho}{\rho+s}}u_0 \\ A_1(r,\theta) = \sqrt{\dfrac{\rho}{\rho+s}}\dfrac{1}{a(Q)}\dfrac{1}{\sin\beta+\eta}\left[a_0 + \left(\dfrac{\rho}{\rho+s}\right)a_1 + \left(\dfrac{\rho}{\rho+s}\right)^2 a_2 + \left(\dfrac{\rho}{\rho+s}\right)^3 a_3\right]U_0 \end{cases}.$$

$$(5.228)$$

The coefficients a_0, a_1, etc., in the above equation are given by (5.222), in which α is replaced by the angle β between the incident direction and the plane tangent to the surface at the reflection point Q, where $a(Q)$ and ρ are the radii of curvature of the surface and of the reflected wave, respectively, at the point of reflection. The length $s = |\vec{QM}|$ is the distance between the observation point M and the reflection point Q.

In Sect. 5.4.1.4, the non-uniform asymptotic solution (5.227) was found to match exactly the non-uniform inner asymptotic solution in the neighborhood of the transition straight line $\theta = \alpha$. In the matching zone, both the inner and outer non-uniform solutions are identical. Because the inner uniform solution is derived from the non-uniform solution by using the technique presented in Sect. 5.4.1.3 (see relations (5.186) and (5.191)), and because the inner and outer solutions in the matching zone of the non-uniform solutions are identical, the outer uniform solution shows a structure similar to the inner solution and is constructed in the same manner. Consequently, setting $U^D = U_R + U_e^d$, we can write in region 2, outside the boundary layer 4,

$$U^D(r,\theta) = e^{iks R_1\,(r,\theta)} \left\{ A_0^{(1)}(r,\theta)\left[F(\xi_1) - \hat{F}_1(\xi_1)\right] + \frac{1}{ik} A_1^{(1)}(r,\theta)\left[F(\xi_1) - \hat{F}_2(\xi_1)\right] \right\}$$
$$+ e^{iks R_2\,(r,\theta)} \left\{ A_0^{(2)}(r,\theta)\left[F(\xi_2) - \hat{F}_1(\xi_2)\right] + \frac{1}{ik} A_1^{(2)}(r,\theta)\left[F(\xi_2) - \hat{F}_2(\xi_2)\right] \right\} + U_e^d(r,\theta)$$

$$(5.229)$$

where the functions $F(\Omega)$ and $\hat{F}_n(\Omega)$ are given by (5.188) and (5.189), respectively, and have been replaced by $F(\tau)$ and $\hat{F}_n(\tau)$. The latter are given by

$$F(\tau) = \sqrt{\frac{\Omega}{\pi}}\; e^{i\frac{\pi}{4}}\; F(\Omega)$$
$$\hat{F}_n(\tau) = \sqrt{\frac{\Omega}{\pi}}\; e^{-i\frac{\pi}{4}}\; \hat{F}_n(\Omega) \;,\quad \tau = \sqrt{\Omega}\; s_a$$

where Ω is equal to k.

The indices 1 and 2 in (5.229) refer to the regions to the right and left of the discontinuity in the curvature, respectively, and the functions $F(\xi_1) - \hat{F}_n(\xi_1)$, n = 1, 2, correspond to $F(kr, s_a) - \hat{F}_n(kr, s_a)$ within a scale factor, while the functions $F(\xi_2) - \hat{F}_n(\xi_2)$ correspond to $F(kr, -s_a) - \hat{F}_n(kr, -s_a)$. The arguments ξ_i of these functions are defined as in (5.225) and are given by

$$\xi_i = \varepsilon_i(\vec{r})\sqrt{k\,|s(\vec{r}) - s_{R_i}(\vec{r})|}\,, \qquad (5.230)$$

where $s(\vec{r})$ and $s_{R_i}(\vec{r})$ are the phases at the observation point of the wave diffracted by the discontinuity in the curvature, and of the reflected wave, respectively. The sign function $\varepsilon_i(\vec{r})$ of ξ_i is defined as follows. It is chosen equal to -1 if the reflection point is located on the surface with index i and $+1$, otherwise.

For $|\xi_i| >> 1$, $U^D(r,\theta)$ tends to $U_a{}^D$, which is given by

$$U_a^D = e^{iks R_1 (r,\theta)} \left[A_0^{(1)} + \frac{1}{ik} A_1^{(1)} \right] \Theta(-\varepsilon_1)$$
$$+ e^{iks R_2 (r,\theta)} \left[A_0^{(2)} + \frac{1}{ik} A_1^{(2)} \right] \Theta(-\varepsilon_2) + U_e^d \quad , \qquad (5.231)$$

where $\Theta(X)$ is the Heaviside function. Thus, one regains the non-uniform solution (5.147) when $|\xi_i|$ becomes large compared to unity, or equivalently, when the observation point leaves region-2 to enter the region-1. The asymptotic solution (5.229) is then uniform in regions-1 and 2 of Fig. 5.9. Since it is constructed in such a manner that it matches also the uniform solution in the boundary layer, it is an asymptotic solution of the diffraction problem of a plane wave by a line discontinuity in the curvature, uniformly valid in regions-1 and 2 of Fig 5.9. In contrast to the case of the wedge with curved faces treated in Sect. 5.3, the asymptotic expansion at the periphery of the inner layer is complete to the order considered. The uniform solution (5.229) can thus be extended to the case of an incident wave which is more general than a plane or locally-plane wave, which is represented, for instance, by an asymptotic expansion of the Luneberg-Kline type.

5.4.3 Numerical application

The plates 1 and 2 contain the results of a numerical application of (5.229) to the problem of a junction with a continuous tangent plane between two circular cylinders of different radii, matched along a generatrix and illuminated by an incident plane wave whose direction of propagation is normal to the generatrices.

Plate 1 corresponds to a perfectly conducting surface illuminated by a TE wave, while plate 2 to that of an imperfectly conducting surface with relative impedance $Z = 0.15 - i\, 0.15$. The curves drawn on each plate represent the variation of the total field given by the uniform solution, the reflected field of GO, and the diffracted field as provided by the non-uniform solution of [KK]. One observes that the transition zone is relatively spread out and that the angular domain, where the non-uniform solution provides a correct description, is rather limited.

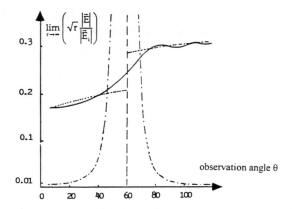

Plate 1 Diffraction by a line discontinuity of the curvature at the junction of two perfectly conducting cylinders of different radii $a_1 = 2$ m, $a_2 = 1$ m. Incident plane wave, angle of propagation $\alpha = 120°$, observation at large distance r. Frequency 1 GHz, polarization TE. Solid line = uniform solution, dotted line = reflected field, mixed line = non uniform solution of Keller and Kaminetsky.

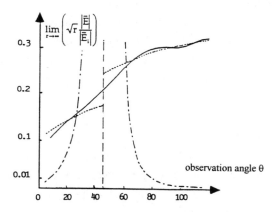

Plate 2 Diffraction by a line discontinuity of the curvature at the junction of two imperfectly conducting cylinders of different radii $a_1 = 1$ m, $a_2 = 0.5$ m. Incident plane wave, angle of propagation $\alpha = 135°$, observation at large distance r. Frequency 1 GHz, polarization TE, relative impedance $Z = 0.15$-i 0.15. Solid line = uniform solution, dotted line = reflected field, mixed line = non uniform solution of Keller and Kaminetsky.

5.5 Uniform solution through the shadow boundary and the boundary layer of a regular convex surface

High frequency diffraction by a regular convex surface has been the subject of numerous works in the past, e.g., Fock 1946 [Fo1], 1965 [Fo2], Levy and Keller 1959 [LK], Franz and Klante 1959 [FK], Logan and Yee 1962 [LY], Cullen 1958 [C], Wait and Conda 1958 [WC], Ivanov 1960 [I], Babich and Kirpicnikova 1979 [BK], Hong 1967 [H], Pathak 1979 [P], Mittra and Safavi-Naini 1979 [MSN], and Pathak et al. 1981 [PWBK]. This topic continues to be an important area of investigation in recent years, with special attention being given to determining the terms that depend upon the torsion of the geodesics and on the radius of curvature in the direction orthogonal to the geodesics (see Michaeli [Mi], Bouche [Bo], and Lafitte [L]).

In Chap. 3, we have derived a variety of different solutions that are appropriate for different regions of space. However, the challenge at hand is to find a uniform analytical solution that reduces to the known solutions in the different sub-domains of space, and yet provides a continuous transition between them. A classification of the sub-domains is given in [P] and a graphic representation is shown in Fig. 5.14. Geometrical Optics is valid in region I, whereas region III corresponds to the deep shadow beyond the surface boundary layer (region V). Region II is the transition zone between I and III, and it is divided into two parts, viz., a lit and an illuminated shadow zone, also called a semi-lit region. Regions IV and VI are the intersections of the transition zone II with the boundary layer in the vicinity of the surface.

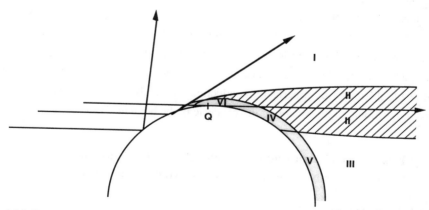

Fig. 5.14. Boundary layers in the diffraction by a smooth convex body and regions of validity of known asymptotic functions.

For the three-dimensional case, uniform solutions have been reported in [PWBK] for the two following cases:

(1) Both the source and the observation points are far from the surface

(2) The source is far from the surface and the observation point is located on it, and vice-versa.

The solution of the first problem (case-1) is based upon canonical problems associated with the cylinder illuminated by a plane wave at normal incidence. A uniform asymptotic solution for this problem has been derived by Pathak [P] for a perfectly conducting cylinder. In Sect. 5.5.1, we apply the method followed by [P] to the case of an imperfectly conducting cylinder. In Sect. 5.5.2, we present the three-dimensional form of this solution, and in Sect. 5.5.3 we derive a completely uniform solution, including one for case-2.

The method followed to treat case-1 is to seek a uniform solution of the circular cylinder problem illuminated by an incident plane wave. This canonical problem has an exact solution, which is written in the form of an infinite series of Bessel and Hankel functions. This series is transformed into an integral via the Watson transformation. The resulting integral is further transformed in a manner such that the special functions appear in the description of the field in the transition zone. To transition from this solution to the case of an arbitrary cylinder, one modifies the arguments of these functions in a way such that one reproduces the known asymptotic solutions in the GO zone and in the deep shadow zone, while simultaneously imposing the continuity of the solution on the shadow boundary. The same constraints also lead to a uniform solution for the case of a general convex surface.

Section 5.5.3 is devoted to the direct implementation of the method of the composite solution described in Sect. 5.2 which we applied to the problem of diffraction of a scalar wave by a smooth cylinder. This method is extended to an electromagnetic wave diffracted by a perfectly conducting surface by taking into account of the terms proportional to the torsion of the geodesic in the boundary layer.

5.5.1   Uniform asymptotic solution through the shadow boundary of an imperfectly conducting surface – two-dimensional case

5.5.1.1 Details of the solution

We consider a normally-incident plane wave illuminating an infinite circular cylinder with a surface impedance Z, or with a coating of one or more homogeneous layers of materials. The radius of the outermost surface is denoted as b and we assume that $kb \gg 1$.

The incident field can be decomposed into a TE (electric field parallel to the axis) and a TM wave (magnetic field parallel to the axis). The diffraction of each wave can be treated separately since, at normal incidence, they satisfy the boundary conditions on the surface of the cylinder independently. Consequently, the problem reduces to a scalar one described by the equation

$$(\nabla^2 + k^2)U_t(\rho,\varphi) = 0, \tag{5.232}$$

$$\frac{\partial U_t}{\partial \rho} + ik\zeta U_t = 0 \quad , \quad \rho = b, \tag{5.233}$$

where $\zeta = Z$, $U_t = H_z$ for a TM wave and $\zeta = 1/Z$, $U_t = E_z$ for a TE wave, the axis OZ being chosen along the axis of the cylinder.

We need to add the Sommerfeld radiation condition at infinity to the conditions (1) and (2), given below.

$$\lim_{\rho \to \infty} \rho^{\frac{1}{2}}\left(\frac{\partial U^d}{\partial \rho} - ikU^d\right) = 0, \tag{5.234}$$

where the index d refers to the diffracted field. The general solution to this problem can be written in the form

$$U_t = U + U_{OR}, \tag{5.235}$$

where U_{OR} is the contribution of the creeping rays reaching the observation point after having left, tangentially, the surface in the deep shadow and where U is given by

$$U = U_0 \int_{-\infty+i\varepsilon}^{\infty+i\varepsilon} d\nu \left[J_\nu(k\rho) - \frac{\Omega J_\nu(kb)}{\Omega H_\nu^{(1)}(kb)} H_\nu^{(1)}(k\rho) \right] e^{i\nu\psi}. \tag{5.236}$$

U_0 denotes the complex amplitude of the incident plane wave and $\psi = |\phi| - \frac{\pi}{2}$ with $|\phi| < \pi$ or $|\psi| < \frac{\pi}{2}$, ψ being negative in the lit region.

The operator Ω is defined as follows

$$\Omega = \frac{\partial}{\partial \rho} + ik\zeta. \tag{5.237}$$

For the case of a cylinder coated with one or more layers of materials, $\zeta = \zeta(v)$ is a function of the index v whose expression for each type of wave (TM or TE) is known. The above expression is derived from the corresponding modal impedance or admittance via the Watson transformation.

Replacing $J_v(x)$ by

$$J_v(x) = \frac{1}{2}\left[H_v^{(1)}(x) + H_v^{(2)}(x)\right].$$ (5.238)

(5.236) can be written as

$$U = \frac{U_0}{2}\int_{-\infty+i\varepsilon}^{\infty+i\varepsilon} dv \left[H_v^{(2)}(k\rho) - \frac{\Omega H_v^{(2)}(kb)}{\Omega H_v^{(1)}(kb)} H_v^{(1)}(k\rho)\right] e^{iv\psi}.$$ (5.239)

In the transition zone and for large kb, the contribution of the integrals (5.236) or (5.239) is more important when v is of the same order as kb. Far from the transition zone, all of the Hankel functions occurring in (5.239) can be replaced by the first term of their Debey's asymptotic expansion. When this is done, a stationary phase point in the integrand of (5.239) can be derived from

$$\psi + 2\,Arc\,cos\left(\frac{v}{kb}\right) - Arc\,cos\left(\frac{v}{k\rho}\right) = 0\,.$$

Assuming $\rho \gg b$, we have Arc cos $(v/k\rho) \approx \pi/2$. It follows then, that when the observation point tends to the shadow boundary ($\psi \to \pi/2$), $v \to kb$. For $kb \gg 1$, the Debye expansions of $H_v^{1,2}(kb)$ remain valid very close to $v = kb$ and, consequently, the most important contribution to the integral comes from the neighborhood of $v = kb$. The same argument applies to the integral in (5.236). Inside the transition zone, this leads to the following change of variables

$$v = kb + m\tau \qquad m = \left(\frac{kb}{2}\right)^{\frac{1}{3}}.$$ (5.240)

Next, we replace the Bessel and Hankel functions by their expansion of the Olver type and insert the same change of variables. Retaining only the first terms in the above expansion we have

$$J_{v(\tau)}(kb) = (m\sqrt{\pi}\,)^{-1}\,V(\tau),$$ (5.241)

$$H_{v(\tau)}^{(1,2)}kb) = \mp i(m\sqrt{\pi}\,)^{-1}\,W_{1,2}(\tau),$$ (5.242)

where $V(\tau)$ and $W_{1,2}(\tau)$ are Airy's functions defined by

$$2iV(\tau) = W_1(\tau) - W_2(\tau), \qquad (5.243)$$

$$W_{1,2}(\tau) = \frac{1}{\sqrt{\pi}} \int_{\Gamma_{1,2}} e^{\tau t - \frac{t^3}{3}} \, dt .$$

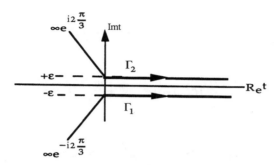

Fig. 5.15. Integration paths for the Miller-type Airy functions.

Inserting (5.241) and (5.242) in (5.236) and (5.239), we get

$$U = mU_0 \int_{-\infty+i\varepsilon}^{\infty+i\varepsilon} d\tau \left[J_{\nu(\tau)}(k\rho) - i \frac{\tilde{\Omega}V(\tau)}{\tilde{\Omega}W_1(\tau)} H^{(1)}_{\nu(\tau)}(k\rho) \right] e^{i(kb+m\tau)\psi} , \qquad (5.244)$$

$$U = \frac{mU_0}{2} \int_{-\infty+i\varepsilon}^{\infty+i\varepsilon} d\tau \left[H^{(2)}_{\nu(\tau)}(k\rho) - \frac{\tilde{\Omega}W_2(\tau)}{\tilde{\Omega}W_1(\tau)} H^{(1)}_{\nu(\tau)}(k\rho) \right] e^{i(kb+m\tau)\psi} , \qquad (5.245)$$

where

$$\tilde{\Omega} = -\frac{1}{m}\frac{\partial}{\partial\tau} + i\zeta . \qquad (5.246)$$

In addition, since, according to (9) $k\rho \approx \nu - m\tau$ in the vicinity of $\rho = b$, we have the relation $\partial/\partial\rho \approx -k/m \, \partial/\partial\tau$, and we have utilized this in (5.244) and (5.245).

It is now possible to express u by combining (5.244) and (5.245). In fact, the transformation that enables us to go from (5.244) to (5.245) is only applied to the integrand and not on the contour of integration. It follows, then, that the integrals in (5.244) and (5.245) are identical along an arbitrary path of integration. In particular, the integral in (5.244) along the path C_1 starting at $-\infty + i\varepsilon$ and ending at $0 + i\varepsilon$ is identical to the integral in (5.245) along the same path. Likewise, the two preceding integrals are identical along the path C_2 starting at $0 + i\varepsilon$ and ending at $\infty + i\varepsilon$. Thus one can express U as the sum of (5.244) and (5.245), integrated along the paths C_2 and C_1, respectively.

After reordering the terms and using (5.238), we obtain

$$U = I_1 + I_2, \tag{5.247}$$

where

$$I_1 = mU_0 \int_{C_1+C_2} d\tau \, J_{\nu(\tau)}(k\rho) \, e^{i\nu(\tau)\psi} - \frac{mU_0}{2} \int_{C_1} d\tau \, H^{(1)}_{\nu(\tau)}(k\rho) \, e^{i\nu(\tau)\psi}, \tag{5.248}$$

$$I_2 = -imU_0 \int_{C_2} d\tau \, \frac{\tilde{\Omega}V(\tau)}{\tilde{\Omega}W_1(\tau)} H^{(1)}_{\nu(\tau)}(k\rho) \, e^{i\nu(\tau)\psi} + \frac{mU_0}{2} \int_{C_1} d\tau \, \frac{\tilde{\Omega}W_2(\tau)}{\tilde{\Omega}W_1(\tau)} H^{(1)}_{\nu(\tau)}(k\rho) \, e^{i\nu(\tau)\psi}, \tag{5.249}$$

We observe that I_1 is totally independent of the boundary conditions on the cylinder, and that the only difference between the expression in (5.249), and the corresponding one for the perfectly conducting cylinder derived by Pathak [P], is the form of the operator $\tilde{\Omega}$. Since the integral I_1 is independent of the boundary conditions, its asymptotic expansion is identical to the one derived by Pathak. We also note that the asymptotic expansion of I_2 differs from that of I_1 only because the expression of $\tilde{\Omega}$ is different, and we can write it directly from the formulas given in Pathak. The discussion given above leads us to the results given in the following for several different situations as explained below.

(i) <u>Part of the transition zone located in the shadow region</u>
For this case I_1 is given by

$$I_1 = U_0 \frac{e^{i\frac{\pi}{4}}}{\sqrt{2\pi k}} \frac{e^{ikb\theta}}{\theta} F[kL\tilde{a}] \frac{e^{iks}}{\sqrt{s}}, \tag{5.250}$$

where

$$F[kL\tilde{a}] = -2i\sqrt{kL\tilde{a}} \, e^{-ikL\tilde{a}} \int_{\sqrt{kL\tilde{a}}}^{\infty} e^{i\tau^2} d\tau, \qquad L = s \quad, \quad \tilde{a} = \frac{\theta^2}{2},$$

and

$$s = \rho^2 - b^2 = PQ_2 \quad, \qquad \theta = \psi - \beta_s \quad, \qquad \beta_s = Arc\,cos\frac{b}{\rho}.$$

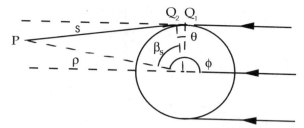

Fig. 5.16. Geometry for scattering from a circular cylinder with planewave incidence, observation in the shadow region.

Also, I_2 takes the form

$$I_2 = -U_0 \frac{e^{i\frac{\pi}{4}}}{\sqrt{2\pi k}} \frac{e^{ikb\theta}}{\theta} \frac{e^{iks}}{\sqrt{s}} - U_0 m \sqrt{\frac{2}{k}} \, e^{ikb\theta} \, \tilde{P}(\xi) \frac{e^{iks}}{\sqrt{s}} , \qquad (5.251)$$

and $\tilde{P}(\xi)$ is expressed as

$$\tilde{P}(\xi) = \frac{e^{i\frac{\pi}{4}}}{\sqrt{\pi}} \int_{-\infty}^{\infty} \frac{\tilde{\Omega}V(\tau)}{\tilde{\Omega}W_1(\tau)} \, e^{i\xi\tau} d\tau , \qquad (5.252)$$

where $\xi = m\theta \geq 0$. We point out that the function $\tilde{P}(\xi)$ is a modified Pekeris function which is not reducible to the tabulated Pekeris functions that arise in the solution of the perfectly conducting case.

(ii) <u>Illuminated part of the transition zone</u>

In the illuminated part of the transition zone the solution is written as

$$I_1 = U_i(P) + U_0 m \frac{e^{i\frac{\pi}{4}}}{\sqrt{2\pi k}\xi'} \, e^{-2ikb\cos\theta^i + i\frac{(\xi')^3}{12}} \, F[kL'\tilde{a}'] \frac{e^{ik(l+b\cos\theta^i)}}{\sqrt{l}} , \quad (5.253)$$

where

$$L' = 1 \quad , \quad \tilde{a}' = \frac{(\xi')^2}{2m^2} = 2\cos^2\theta^i \quad , \quad \xi' = -2m\cos\theta^i.$$

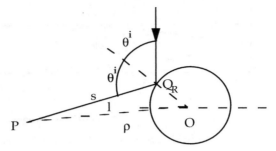

Fig. 5.17. Geometry for scattering from a circular cylinder with plane wave incidence : observation in the lit region.

$$I_2 = -U_0 m \sqrt{\frac{2}{kL}}\, e^{ik(1+b\cos\theta^i)}\, e^{-i2kb\cos\theta^i + i\frac{(\xi')^3}{12}} \left[\frac{e^{i\frac{\pi}{4}}}{2\sqrt{\pi}\,\xi'} + \tilde{P}(\xi') \right], \quad (5.254)$$

and the function $\tilde{P}$ is given by (5.252).

5.5.1.2 Expression of the uniform solution for the circular cylinder

Using I_1 and I_2, given by (5.250) and (5.251), we can write the following representations for the solution in the shadow and lit regions.

Shadow region

The solution in the shadow region is written as

$$U(P) = -U_i(Q_1) m \sqrt{\frac{2}{k}}\, e^{ikt} \left[\frac{e^{i\frac{\pi}{4}}}{2\xi\sqrt{\pi}}(1 - F[kL\tilde{a}]) + P(\xi) \right] \frac{e^{iks}}{\sqrt{s}}, \quad (5.255)$$

where $U^i(Q_1)$ is the incident field at the point Q_1 on the cylinder located at the light-shadow boundary associated with the transition zone and where $t = b\theta$ and $\xi = m\theta \geq 0$.

Lit region

In the lit region the field expression reads

$$U(P) = U^i(P) + U^i(Q_R) \left[\sqrt{\frac{-4}{\xi'}}\, e^{i\frac{(\xi')^3}{12}} \left\{ \frac{e^{i\frac{\pi}{4}}}{2\xi'\sqrt{\pi}}(1 - F[kL'\tilde{a}']) + \tilde{P}(\xi') \right\} \right] \sqrt{\frac{\rho_a}{\rho_a + l}}\, e^{ikl},$$

$$(5.256)$$

where $U^i(Q_R) = U_0\, e^{-ikb\cos\theta^i}$ is the incident field at the reflection point Q_R, $\rho_c = (b\cos\theta^i)/2$ is the distance between the reflection point and the caustic of the reflected rays, and the functions $F[X]$ and $\widetilde{P}(X)$ were defined by (5.250) and (5.252). However, the expression for $\widetilde{P}(X)$ is new and is explicitly written in the form

$$\widetilde{P}(\xi) = \frac{e^{i\frac{\pi}{4}}}{\sqrt{\pi}} \int_{-\infty}^{+\infty} \frac{V'(\tau) - im\zeta V(\tau)}{W_1'(\tau) - im\zeta W_1(\tau)} e^{-\xi\tau} d\tau . \tag{5.257}$$

An efficient procedure to numerically evaluate this function has been established by Pearson [Pe].

5.5.1.3 Uniform solution for any cylinder

The field expressions (5.255) and (5.256) can be transposed directly to the case of the cylinder with an arbitrary cross-section. Following Pathak, the parameters ξ, t, m and $\tilde{a}$, occurring in the solution for an observation point located in the shadow region, are generalized as follows

$$\xi = \int_{Q_1}^{Q_2} \frac{m(t')}{\rho_g(t')} dt' \quad , \quad m(t') = \left[\frac{k\rho_g(t')}{2} \right]^{\frac{1}{3}} ,$$

$$t = \int_{Q_1}^{Q_2} dt' \quad , \quad \tilde{a} = \frac{\xi^2}{2m(Q_1)m(Q_2)} \quad , \quad L = s , \tag{5.258}$$

where $\rho_g(t')$ is the local curvature radius of the geodesic followed by the creeping ray on the cylinder. If we insert into (5.255) the expressions for the parameters ξ, t, m, and $\tilde{a}$ given in (5.258), we can write the field at the point P in the shadow region as

$$U(P) = U^i(Q_1) \left[-\sqrt{m(Q_1)m(Q_2)} \, e^{ikt} \sqrt{\frac{2}{k}} \left\{ \frac{e^{i\frac{\pi}{4}}}{2\xi\sqrt{\pi}} \left(1 - F[kL\tilde{a}] \right) + \widetilde{P}(\xi) \right\} \right] \frac{e^{iks}}{\sqrt{s}} . \tag{5.259}$$

The above expression for the diffracted field differs from that obtained by Pathak for a perfectly conducting cylinder only through the form of the function $\widetilde{P}(\xi)$ given by (5.257).

In the lit zone, the parameters ξ', t, m, and $\tilde{a}'$ are generalized as follows

$$\xi' = -2m(Q_R)\cos\theta^i \quad , \quad m(Q_R) = \left[\frac{k\rho_g(Q_R)}{2} \right]^{\frac{1}{3}} ,$$

$$\rho_c = \frac{\rho_g(Q_R)\cos\theta^i}{2} \quad , \quad \tilde{a}' = 2\cos^2\theta^i \quad , \quad L' = s . \tag{5.260}$$

Thus we can write

$$U(P) = U^i(P) + U^i(Q_R) \left[\sqrt{\frac{-4}{\xi'}} \, e^{i\frac{(\xi')^3}{12}} \right] \left\{ \frac{e^{i\frac{\pi}{4}}}{2\xi'\sqrt{\pi}} \left(1 - F[kL'\tilde{a}']\right) + \tilde{P}(\xi') \right\} \sqrt{\frac{\rho_c}{\rho_c + l}} \, e^{ikl} \, .$$

$$(5.261)$$

for the field in the lit region.

5.5.2   Uniform asymptotic expansion through the shadow boundary of an imperfectly conducting surface – three-dimensional case

On a three-dimensional surface, which is convex and regular, we need to take into account of the divergence of the creeping rays on the surface, and of the divergence of the spatial waves that leave the surface tangentially. In addition, it is necessary to account for the vectorial nature of the field when the problem is three-dimensional.

For the case of a perfect conductor ($Z = 0$), or if its relative impedance is not close to 1, we have shown in Sect. 3.2 that, to the zero-order of the asymptotic expansion, the electric field components along the normal to the surface at the excitation point Q' of the creeping ray, and along the binormal to the geodesic Q', stay uncoupled along the entire trajectory of the creeping ray. We can thus apply (28) to each of the components of the electric field along $\hat{n}_{Q'}$ and $\hat{b}_{Q'}$ and obtain the following expression for the field at an observation point located in the shadow region.

$$\vec{E}(P) = \vec{E}^i(Q') \cdot \left[\hat{b}_{Q'}\hat{b}_Q D_S + \hat{n}_{Q'}\hat{n}_Q D_h \right] e^{ikQ'Q} A_d e^{ik|\vec{QP}|} \sqrt{\frac{d\eta(Q')}{d\eta(Q)}} \, , \qquad (5.262)$$

where $\hat{n}_{Q'}$, $\hat{n}_Q$ are the unit vectors along the outward normal to the surface S at the points Q' and Q; $\hat{b}_{Q'}$, $\hat{b}_Q$ are the unit vectors tangent to the wavefront of the surface wave at Q' and Q, whose orientations are defined by

$$\hat{b}_{Q'} = \hat{t}_{Q'} \times \hat{n}_{Q'} \, , \quad \hat{b}_Q = \hat{t}_Q \times \hat{n}_Q \, . \qquad (5.263)$$

In (5.263) $\hat{t}_{Q'}$ and $\hat{t}_Q$ are the unit vectors tangent to the geodesic followed by the wave, at Q' and Q, respectively, and are oriented in the direction of propagation.

The diffraction coefficients D_s and D_h are defined by

$$D_{\substack{s\\h}} = -\sqrt{m(Q)m(Q')} \sqrt{\frac{2}{k}} \left\{ \frac{e^{i\frac{\pi}{4}}}{2\sqrt{\pi}\xi} \left(1 - F\left[kL^d\tilde{a}\right]\right) + \tilde{P}_{\substack{s\\h}}(\xi) \right\} \, . \qquad (5.264)$$

The functions $F(x)$ and $\tilde{P}(\xi)$ are identical to those used in the two-dimensional case and their arguments are defined by (5.258). However, the parameter L^d in the argument of F is modified and will be determined by imposing the continuity condition on the total field across the shadow boundary.

The expression (5.261) is generalized by using the fact that it tends to the GO field when the point P is sufficiently far from the shadow boundary. Consequently, the solution in the lit zone can be written in the form

$$\vec{E}(P) = \vec{E}^i(P) + \vec{E}^R(P), \tag{5.265}$$

where

$$\vec{E}^R(P) = \vec{E}(Q_R) \cdot \left[\hat{e}_\perp \hat{e}_\perp R_s + \hat{e}^i_{//} \hat{e}^r_{//} R_h \right] A \, e^{ik|\overrightarrow{Q_R P}|}, \tag{5.266}$$

and

$$R_{s_h} = -\sqrt{\frac{-4}{\xi'}} \, e^{i\frac{(\xi')^3}{12}} \left\{ \frac{e^{i\frac{\pi}{4}}}{2\sqrt{\pi}\xi'} \left[1 - F(kL^r \tilde{a}') \right] + \tilde{P}_{s_h}(\xi') \right\}. \tag{5.267}$$

The arguments of the functions $F(X)$ and $\tilde{P}(\xi')$ are defined by (5.260), although the parameter L^r is different from L'; the vectors $\hat{e}_\perp, \hat{e}^i_{//}, and \, \hat{e}^r_{//}$ are the unit vectors perpendicular and parallel, respectively, to the incidence plane, and orthogonal to the direction of propagation of the incident and reflected waves. Finally, the divergence factor A is given by (see Sect. 1.3.4)

$$A = \sqrt{\frac{\rho_1^r \rho_2^r}{\left(\rho_1^r + s_1\right)\left(\rho_2^r + s_1\right)}}, \tag{5.268}$$

where $s_1 = |\overrightarrow{Q_R P}|$ and ρ_1^r, ρ_2^r are the principal radii of curvature of the reflected field at Q_R.

The components $R_{s,h}$ tend to the GO reflection coefficients when the observation point leaves the transition zone in the vicinity of the shadow boundary, and consequently, $\vec{E}^R(P)$ tends to the GO reflected field. It only remains to impose the continuity of the total field across the shadow boundary.

For small X we can employ the small argument approximation of F to write

$$F(X) \cong \sqrt{\pi X} \, e^{-i\left(\frac{\pi}{4} + X\right)}. \tag{5.269}$$

Since the term

$$\tilde{P}_{s_h}(Y) + \frac{e^{i\frac{\pi}{4}}}{2\sqrt{\pi Y}}, \tag{5.270}$$

is well defined at Y = 0, owing to the fact that the singular part of $\tilde{P}_{s_h}(Y)$ compensates, exactly, the singularity of the second term (5.270) (see Pearson [Pe]), we can show that the jump discontinuity in $D_{s,h}$ is equal to $-\frac{1}{2}\sqrt{L^d}$. On the other hand, on the shadow boundary, we have

$$\rho_1^r \approx \frac{\rho_g(Q')\cos\theta^i}{2} \tag{5.271}$$

$$\rho_2^r \approx \rho_2^i,$$

and, using the expression for ξ' given in (5.260) we obtain

$$\rho_1^r \approx \sqrt{\frac{\rho_g}{m(Q')}} \sqrt{\frac{-\xi'}{4}} = m(Q')\sqrt{\frac{2}{k}} \sqrt{\frac{-\xi'}{4}}. \tag{5.272}$$

It follows, then, that the jump of $R_{s,h}\sqrt{\rho_1^r}$ is equal to $-\frac{1}{2}\sqrt{L^r}$.

The divergence factor A_d appearing in (5.262) is given by

$$A_d = \sqrt{\frac{\rho_2^d}{(\rho_2^d + s)s}}, \tag{5.273}$$

where s = $|\vec{QP}|$ and ρ_2^d is the radius of curvature of the wavefront of the surface wave at Q. On the shadow boundary we have the relation $\rho_2^d = \rho_2^r$, and if we set

$$A_r = \sqrt{\frac{\rho_2^r}{(\rho_2^r + s)s}}, \tag{5.274}$$

we see that the jumps of the diffracted and reflected fields are given, respectively, by

$$-\frac{1}{2}\vec{E}^i(Q')\sqrt{L^d}\,A_d e^{iks} \quad , \quad -\frac{1}{2}\vec{E}^i(Q')\sqrt{L^r}\,A_r e^{iks} \tag{5.275}$$

Since the jump of the direct field is equal to $\vec{E}^i(Q')A_i e^{iks}$ with

$$A_i = \sqrt{\frac{\rho_1^i \rho_2^i}{\left(\rho_1^i + s\right)\left(\rho_2^i + s\right)}} \, , \qquad (5.276)$$

where ρ_1^i and ρ_2^i are the principal radii of curvature of the incident wave at Q', the continuity of the total field is achieved if we let

$$L^d = L^r = \left(\frac{A_i}{A_d}\right)^2 = \left(\frac{A_i}{A_r}\right)^2 , \qquad (5.277)$$

on the shadow boundary. Thus, by setting

$$L^r = \frac{\rho_1^i \rho_2^i}{\left(\rho_1^i + s_1\right)\left(\rho_2^i + s_1\right)} \frac{s_1\left(\rho_2^r + s_1\right)}{\rho_2^r} , \qquad (5.278)$$

$$L^d = \frac{\rho_1^i \rho_2^i}{\left(\rho_1^i + s\right)\left(\rho_2^i + s\right)} \frac{s\left(\rho_2^d + s\right)}{\rho_2^d} , \qquad (5.279)$$

we see that the condition (5.277) is indeed satisfied.

Inserting (5.278) into (5.266) and (5.279) into (5.262), we obtain two asymptotic solutions that match on the shadow boundary. The solution so defined is continuous on the shadow boundary, although its derivative with respect to the observation angle is generally discontinuous.

5.5.3 Totally uniform asymptotic solution

5.5.3.1 Definition of the problem

In Sect. 3.2 we have presented an asymptotic solution which is valid in the boundary layer at the proximity of the surface, through the zones V, IV and VI of Fig. 5.14 of a convex surface. In Sect. 5.5.3, an asymptotic solution uniformly valid through the regions I, II and III has been presented for an observation point located far from this surface. We will now combine these two solutions and construct one that is uniformly valid through the entire space, encompassing the regions I through VI. The treatment of the problem will be restricted to the case of a perfectly conducting convex surface illuminated by an incident plane wave.

5.5.3.2 Method of solution

We will now apply the method of the composite solution, described in Sect. 5.2, to construct the desired uniform solution. The inner expansion is given by the solution in the boundary layer and it is written with the semi-geodesic coordinates (n,t,b) at a point S of the boundary layer of a perfectly conducting surface [Mi]. The solution is written as

$$\vec{E}^i(s) = \left[\frac{d\eta(Q')}{d\eta(Q_P)}\right]^{\frac{1}{2}} \left[\frac{a(Q')}{a(Q_P)}\right]^{\frac{1}{6}} e^{ikt}$$

$$\vec{E}^{inc}(Q') \cdot \left[\hat{b}'\hat{b}_P V_2(\xi,\zeta) - \hat{n}'\hat{b}_P \frac{iT_0}{m(Q_P)} \frac{\partial V_1}{\partial \zeta}(\xi,\zeta) + \hat{n}'\hat{n}_P V_1(\xi,\zeta) \right. , \quad (5.280)$$

$$\left. + \hat{b}'\hat{n}_P \frac{iT_0}{m(Q_P)} \frac{\partial V_2}{\partial \zeta}(\xi,\zeta) + \hat{n}'\hat{t}_P \frac{i}{m(Q_P)} \frac{\partial V_1}{\partial \zeta}(\xi,\zeta) \right]$$

where the unit vectors $\hat{b}$, $\hat{n}$ and $\hat{t}_p$ form a rectangular frame associated with the projection Sp of S on the surface, $\hat{n}_p$ being normal to the surface and $\hat{t}_p$ being tangent to the geodesic passing through Sp (see Fig. 5.18).

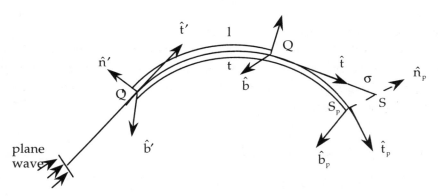

Fig. 5.18. Ray coordinates for creeping waves.

The functions $V_1(\xi, \zeta)$ are the Fock functions of two variables ξ and ζ, which are the reduced coordinates of the point S and are given by

$$\zeta = \frac{kn}{m(S_P)} \quad , \quad \xi = \int_0^t \frac{m(t')}{\rho_g(t')} dt' \quad (5.281)$$

where $n = S_p S$, $m = \left(\frac{k\rho_g}{2}\right)^{\frac{1}{3}}$ and ζ corresponds to the variable ν of Chap. 3.

Finally, the Fock functions are expressed as

$$V_1(\xi,\zeta) = \frac{i}{2\sqrt{\pi}} \int\limits_{\infty e^{2i\frac{\pi}{3}}}^{\infty} e^{i\xi\tau} \left[W_2(\tau-\zeta) - \frac{W_2'(\tau)}{W_1'(\tau)} W_1(\tau-\zeta) \right] d\tau ,$$

$$V_2(\xi,\zeta) = \frac{i}{2\sqrt{\pi}} \int\limits_{\infty e^{2i\frac{\pi}{3}}}^{\infty} e^{i\xi\tau} \left[W_2(\tau-\zeta) - \frac{W_2(\tau)}{W_1(\tau)} W_1(\tau-\zeta) \right] d\tau ,$$

(5.282)

where W_1, W_2 are the Airy functions defined in Sect. 3.1.4.

In order to match the solution (5.280) to the outer expansion, it is important to express it in terms of the ray coordinates (l, σ) where l is the geodesic arc $Q'Q$, and σ is the distance measured along the ray from Q' to S (Fig. 5.18).

In the boundary layer, $\zeta < 1$, and, consequently, $\sigma/a = O(m^{-1})$ where $a = a(Q)$ is the radius of curvature of the geodesic at Q. We also note that the projection S_p of S on the surface can be considered as being located on the extension of the geodesic along the surface. In effect then, the departure from the geodesic is due to a torsion of the geodesic and is $O(\sigma/a)^2 \approx m^{-2}$. Under these conditions we can derive the relationships

$$t \approx 1 + \sigma - \frac{\sigma^3}{3a^2}, \qquad n \approx \frac{\sigma^2}{2a},$$

(5.283)

from which it follows that

$$\zeta = \left(\frac{m\sigma}{a} \right)^2 = X^2 \quad , \quad \xi \approx \xi_0 + \frac{m\sigma}{a} = \xi_0 + X,$$

(5.284)

where

$$\xi_0 = \int\limits_0^1 \frac{m(t')}{\rho_g(t')} dt'$$

(5.285)

Furthermore, to $O(m^{-2})$, we get

$$\begin{cases} \hat{b}_p = \hat{b} + T_0 \dfrac{\sigma}{a} \hat{n} \\[2mm] \hat{n}_p = \hat{n} - T_0 \dfrac{\sigma}{a} \hat{b} + \dfrac{\sigma}{a} \hat{t} . \\[2mm] \hat{t}_p = \hat{t} - \dfrac{\sigma}{a} \hat{n} \end{cases}$$

(5.286)

Inserting (5.284) and (5.286) into (5.280) and neglecting the terms of $O(m^{-2})$, we obtain

$$\vec{E}^{(i)}(s) = \left[\frac{d\eta(Q')}{d\eta(Q)}\right]^{\frac{1}{2}}\left[\frac{a(Q')}{a(Q)}\right]^{\frac{1}{6}} e^{ik\left(l+\sigma-\frac{\sigma^3}{3a^2}\right)}$$

$$\vec{E}^{inc}(Q') \cdot \left(\hat{b}'\hat{b}V_2(\xi,\zeta) - \hat{n}'\hat{b}\left[\frac{iT_0}{m(Q)}\frac{\partial V_1}{\partial\zeta}(\xi,\zeta) + T_0\frac{\sigma}{a}V_1(\xi,\zeta)\right]\right.$$

$$+\hat{n}'\hat{n}V_1(\xi,\zeta) + \hat{b}'\hat{n}\left[\frac{iT_0}{m(Q)}\frac{\partial V_2}{\partial\zeta}(\xi,\zeta) + T_0\frac{\sigma}{a}V_2(\xi,\zeta)\right]$$

$$+\hat{n}'\hat{t}\left[\frac{i}{m(Q)}\frac{\partial V_1}{\partial\zeta}(\xi,\zeta) + \frac{\sigma}{a}V_1(\xi,\zeta)\right]\right) \qquad (5.287)$$

The intermediate expansion is derived from the inner expansion in (5.287) by replacing the Fock functions with their asymptotic expansions in the limit $\zeta \to \infty$. Next, if we use the asymptotic expansion of W_1 and W_2 given by (3.198) of Sect. 3.12, we obtain the following efxpression for the dominant term.

$$V_1(\xi,\zeta) \approx exp\left(\frac{2iX^3}{3}\right)\left[F(\sqrt{X}\,\xi_0) - \frac{e^{i\frac{\pi}{4}}}{\sqrt{X}}q(\xi_0)\right]$$

$$\frac{\partial V_1}{\partial\zeta}(\xi,\zeta) \approx iXV_1(\xi,\zeta) = im\frac{\sigma}{a}V_1(\xi,\zeta)$$

$$\frac{\partial V_1}{\partial\zeta}(\xi,\zeta) \approx iXV_1(\xi,\zeta) = im\frac{\sigma}{a}V_1(\xi,\zeta) \qquad (5.288)$$

$$\frac{\partial V_2}{\partial\zeta}(\xi,\zeta) \approx iXV_2(\xi,\zeta) = im\frac{\sigma}{a}V_2(\xi,\zeta).$$

Hence, the limit of the inner expansion is given by

$$\vec{E}^{(m)}(s) = \left[\frac{d\eta(Q')}{d\eta(Q)}\right]^{\frac{1}{2}}\left[\frac{a(Q')}{a(Q)}\right]^{\frac{1}{6}} e^{ik(l+\sigma)}\,\vec{E}^{inc}(Q') \cdot \left(\hat{b}'\hat{b}\left[F(\sqrt{X}\,\xi_0) - \frac{e^{i\frac{\pi}{4}}}{\sqrt{X}}P(\xi_0)\right]\right.$$

$$\left.+\hat{n}'\hat{n}\left[F(\sqrt{X}\,\xi_0) - \frac{e^{i\frac{\pi}{4}}}{\sqrt{X}}q(\xi_0)\right]\right)$$

$$(5.289)$$

where we have set

$$F(u) = exp\left(-iu^2 - i\frac{\pi}{4}\right)\frac{1}{\sqrt{\pi}}\int_u^\infty e^{it^2}\,dt\,, \tag{5.290}$$

$$P(\xi_0) = \frac{i}{2\sqrt{\pi}}\int_{\infty e^{2i\frac{\pi}{3}}}^\infty e^{i\xi_0\tau}\frac{W_2(\tau)}{W_1(\tau)}\,d\tau \tag{5.291}$$

$$q(\xi_0) = \frac{i}{2\sqrt{\pi}}\int_{\infty e^{2i\frac{\pi}{3}}}^\infty e^{i\xi_0\tau}\frac{W_2'(\tau)}{W_1'(\tau)}\,d\tau\,. \tag{5.292}$$

We can show that the Fresnel function, given in (5.290), is related to the Pathak and Kouyoumjian transition function F_{PK} through the relation

$$F_{PK}(u^2) = -2i\sqrt{\pi}\ u\ e^{i\frac{\pi}{4}}\ F(u). \tag{5.293}$$

The Fock-Pékéris functions $P(\xi_0)$ and $q(\xi_0)$ correspond, respectively, to the limits of the function defined in (5.257) for $\zeta \to 0$ and $\zeta \to \infty$, since, in the present case, the surface is assumed to be perfectly conducting.

Using (5.280), (5.282) and (5.290) through (5.293), the outer expansion at a point S located in the shadow region may be written

$$\vec{E}^{(e)}(s) = \left[\frac{d\eta(Q')}{d\eta(Q)}\right]^{\frac{1}{2}}\left[\frac{a(Q')}{a(Q)}\right]^{\frac{1}{6}}\left(\frac{\rho^d}{\rho^d+\sigma}\right)^{\frac{1}{2}}e^{ik(l+\sigma)}\ \vec{E}^{inc}(Q')\cdot\left(\hat{b}'\hat{b}\left[F(\sqrt{X}\ \xi_0)-\frac{e^{i\frac{\pi}{4}}}{\sqrt{X}}P(\xi_0)\right]\right.$$

$$\left.+\hat{n}'\hat{n}\left[F(\sqrt{X}\ \xi_0)-\frac{e^{i\frac{\pi}{4}}}{\sqrt{X}}q(\xi_0)\right]\right) \tag{5.294}$$

It is important to note that for a plane wave we have $\rho_1^i = \rho_2^i = \infty$, and $\rho_2^d = \infty$ on the shadow boundary, which, according to (5.279), leads to the results that $L^d = s = \sigma$ and $kL^d\tilde{a} = (m\sigma/a)\ \xi_0^2 = X^2$. We see that the inner limit of the outer expansion (5.294), which results from the limit σ tending to zero, is equal to the outer limit of the inner expansion given by (5.289).

We now have at our disposal all the elements necessary to construct the composite solution according to the definition given in Sect. 5.2. This uniform solution is given by

$$\vec{E}^u(s) = \vec{E}^{(i)}(s) - \vec{E}^{(m)}(s) + \vec{E}^{(e)}(s).$$ (5.295)

By following the suggestion we provided toward the end of Sect. 5.2, we can simplify the uniform solution (5.295) by introducing the following factor in the inner solution given in (5.287), viz.,

$$A_d = \sqrt{\frac{\rho_2^d}{\rho_2^d + \sigma}},$$

which does not affect this solution since $(\sigma/\rho_2^d) \ll 1$ in the boundary layer. Under these conditions, we have $\vec{E}^{(e)}(s) = \vec{E}^{(m)}(s)$, and, consequently, $\vec{E}^u(s) = \vec{E}^{(i)}(s)$, which is the desired result we were seeking.

5.5.3.3 Expressions for the totally uniform solution
The asymptotic solution which is completely uniform has the form

$$\vec{E}^u(s) = \left[\frac{d\eta(Q')}{d\eta(Q)}\right]^{\frac{1}{2}} \left[\frac{a(Q')}{a(Q)}\right]^{\frac{1}{6}} \left(\frac{\rho_2^d}{\rho_2^d + \sigma}\right)^{\frac{1}{2}} e^{ik\left(l+\sigma-\frac{\sigma^3}{3a^2}\right)}$$

$$\vec{E}^{inc}(Q') \cdot \left(\hat{b}'\hat{b}V_2(\xi,\zeta) - \hat{n}'\hat{b}\left[\frac{iT_0}{m(Q)}\frac{\partial V_1}{\partial \zeta}(\xi,\zeta) + T_0\frac{\sigma}{a}V_1(\xi,\zeta)\right]\right.$$

$$+\hat{n}'\hat{n}V_1(\xi,\zeta) + \hat{b}'\hat{n}\left[\frac{iT_0}{m(Q)}\frac{\partial V_2}{\partial \zeta}(\xi,\zeta) + T_0\frac{\sigma}{a}V_2(\xi,\zeta)\right]$$ (5.296)

$$\left.+\hat{n}'\hat{t}\left[\frac{i}{m(Q)}\frac{\partial V_1}{\partial \zeta}(\xi,\zeta) + \frac{\sigma}{a}V_1(\xi,\zeta)\right]\right)$$

where $a = a(Q)$. Michaeli [Mi] was the first to obtain this solution by using a method which is similar to the one presented here. If the point S is on the surface ($\zeta = 0$), then $V_2(\xi, 0) = 0$, $\partial V_1/\partial \zeta (\xi, 0) = 0$. Using the fact that the Wronskian $[W_1, W_2] = 2i$, we get

$$[V_1(\xi,\zeta)]_{\zeta=0} = g(\xi_0),$$

$$\left[\frac{\partial V_2}{\partial \zeta}(\xi,\zeta)\right]_{\zeta=0} = \tilde{g}(\xi_0),$$ (5.297)

where $g(\xi_0)$ and $\widetilde{g}(\xi_0)$ are the usual Fock functions defined by

$$g(\xi_0) = \frac{1}{\sqrt{\pi}} \int_{\infty e^{2i\frac{\pi}{3}}}^{\infty} \frac{e^{i\xi_0\tau}}{W_1'(\tau)} d\tau \,,$$

$$(5.298)$$

$$\widetilde{g}(\xi_0) = \frac{1}{\sqrt{\pi}} \int_{\infty e^{2i\frac{\pi}{3}}}^{\infty} \frac{e^{i\xi_0\tau}}{W_1(\tau)} d\tau \,,$$

We point out to the reader that in the appendix 5, these functions are denoted by f and g instead of g and $\widetilde{g}$. Next, using (5.297) and (5.298), and the fact that if $\zeta = 0$ then $\sigma = 0$, we can rewrite (5.296) in the form

$$\vec{E}(s) = \left[\frac{d\eta(Q')}{d\eta(Q)}\right]^{\frac{1}{2}} \left[\frac{a(Q')}{a(Q)}\right]^{\frac{1}{6}} e^{ikl}$$

$$\vec{E}^{inc}(Q') \cdot \left(\hat{n}'\hat{n}g(\xi_0) + \hat{b}'\hat{n}T_0\left[\frac{i}{m(Q)}\widetilde{g}(\xi_0)\right]\right)$$

$$(5.299)$$

The expression in (5.299) is a uniform solution through the shadow boundary for a source located on the surface and an observation point far away from it. It is identical to the results derived by Pathak et al. [PWBK] by using the GTD.

5.5.3.4 Numerical applications

The Plates 1 and 2 plot the real and imaginary parts of the Fock functions of two variables $V_1(\xi, \zeta)$ and $V_2(\xi, \zeta)$, defined in (5.282), as functions of s expressed in terms of the wavelength, for a constant path length $1 = Q'Q$ traversed by the creeping ray.

The Plates 3 and 4 depict the diffracted fields in the shadow for a circular cylinder illuminated by an incident plane wave with TM polarization, as functions of the distance expressed in terms of the wavelength for different angles of observation θ (Fig. 5.19).

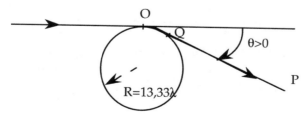

Fig. 5.19. Geometry for the numerical application of the asymptotic solution of the circular cylinder.

The dashed curves correspond to the uniform solution given in (5.296) while the solid curves have been obtained with the solution in (5.294), which is valid far from the surface. We observe that the difference between the two solutions is large when the observation point is close to the surface.

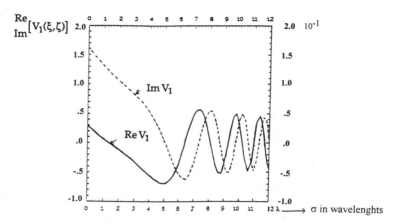

Plate 1 Real part and imaginary part of the Fock function $V_1(\xi, \zeta)$ as a function of the distance σ expressed in terms of normalized wavelength, for $\theta = 45°$, $\sigma = a\sqrt{\zeta}/m$, $\xi = m\theta + \sqrt{\zeta}$, $a = 1$ meter, $F = 4$ GHz.

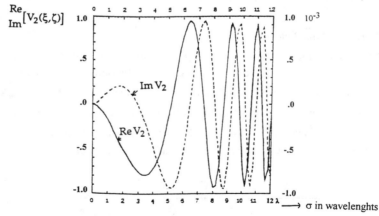

Plate 2 Real part and imaginary part of the Fock function $V_2(\xi, \zeta)$ as a function of the distance σ expressed in terms of normalized wavelength, for $\theta = 45°$, $\sigma = a\sqrt{\zeta}/m$, $\xi = m\theta + \sqrt{\zeta}$, $a = 1$ meter, $F = 4$ GHz.

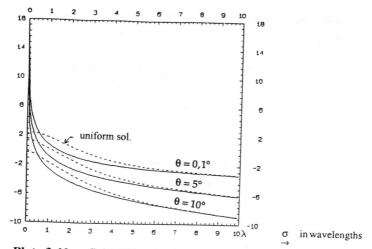

Plate 3 Near-field diffraction by a cylinder for TM incidence

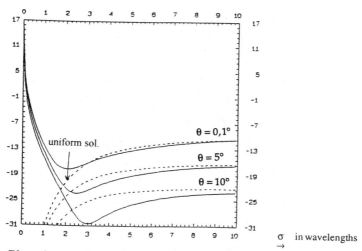

Plate 4 Near field diffraction by a cylinder for TE incidence

5.6 Partially- and totally-uniform solutions for a wedge with curved faces, including creeping rays

5.6.1 Classification of asymptotic solutions for a wedge with curved faces

The analytical form of the asymptotic solutions for the field diffracted by the edge of a wedge with curved faces depends upon the region where the field is observed. A complete representation of the field in the space outside the wedge, beyond the surface boundary layers and the neighborhood of the edge, requires us to juxtapose several asymptotic solutions, each of them having a limited domain of validity in different regions of space. We distinguish five different zones, numbered 1 through 5, as shown in Fig. 5.20.

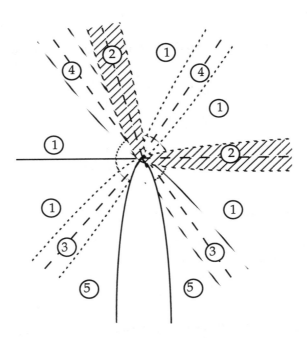

Fig. 5.20. Regions of validity of known asymptotic solutions and location of transition regions.

Region-1 corresponds to the domain where the standard GTD is valid. The field is given by the non-uniform solution published by [K] in 1957.

Region-2 corresponds to transition zones located in the vicinity of the shadow boundaries of the direct and reflected fields of Geometrical Optics. A uniform solution has been derived by Kouyoumjian and Pathak [KP] in 1974 for the case where the incident ray is in region-1 and the diffracted is in region-2. This solution is currently called the UTD solution where UTD refers to the Uniform Theory of Diffraction. It reduces to the non-uniform GTD solution when the observation point traverses into region-1 from region-2. Another solution, known as the Uniform Asymptotic Theory (UAT) has been published by Lee and Deschamps [LD] in 1976. Both of these solutions were presented and analyzed in Sect. 5.3.

Region-5 corresponds to the deep shadow region. A ray incident from region-1 generates creeping rays which are diffracted into region-5. Likewise, a creeping ray, excited by a spatial wave originating from region-5, is diffracted into region-1 in the form of a spatial wave. Asymptotic solutions that enable one to describe these diffraction phenomena were first proposed by Albertsen [A] in 1974 and by Albertsen and Christiansen [AC] in 1978. These results have been presented in Chap. 1.

Region-3 is the transition zone in the vicinity of the shadow boundaries of spatial rays diffracted by the edge. A solution for a ray incident from region-1 and diffracted into region-3 has been developed by Molinet [Mo] in 1977 who employed hybrid diffraction coefficients including a Fock function. A similar solution, based upon the edge equivalent currents which play the role of secondary sources radiating in the presence of each of the faces of the wedge, was presented by Pathak and Kouyoumjian [KP] in 1977.

More recently, Hill and Pathak [HP] have provided a solution for a ray incident from region-3 and diffracted into region-1. This solution is an extension to the semi-lit region of the Albertsen-Christiansen method.

Region-4 is associated with the grazing incidence on one of the faces of the wedge. If the incident ray is in region-4, the regions 2 and 3 are superimposed and the field in the transition region where the two zones superpose can no longer be described in terms of the usual Fock function. A new universal function was introduced by Michaeli [Mi] in 1989 to represent the field in that region. It was obtained by replacing the Airy function in the integrand of the Fock function by an incomplete Airy function. This new transition function can be justified by the work of Idemen and Felsen [IF]. In 1981 these authors solved a new canonical problem, viz., the diffraction of a plane wave by a truncated cylindrical shell. Michaeli [Mi] obtained an asymptotic expansion of their solution in which the new transition function appeared explicitly.

The two-dimensional configurations which have been treated by Michaeli are:

(a) incident ray in region-3, observation in region-4,
(b) incident ray in region-5, observation in region-4,
(c) incident ray in region-3 associated with one of the faces and observation in the same region or region-3 associated with the other face.

The solutions of the problems (a) and (b) are new, whereas the solution of problem (c) confirms the previous heuristic solutions.

    For the problems (a) and (b), alternate solutions have been independently developed by Chuang and Liang [CL] and Liang [L] in 1988 for the 2-D case (See also Liang, Pathak and Chuang [LPC]). The solutions make use of more complex transition functions which reduce to those given by Michaeli for incidence and observation angles that are close to that of the tangent on one of the faces of the wedge. In particular, the integrand is not expressed in terms of a standard incomplete Airy function. The solution [LPC] has the advantage of being uniform while the solution [Mi] is not. However, by means of a modification of the arguments of the transition function, and the introduction of a multiplication coefficient which tends to one on the transition surface, Michaeli was later able to construct a uniform solution. The method employed to construct this solution is described in Sect. 5.6.4. The uniform solution [LPC] is presented in Sect. 5.6.5. It is important to point out that, in practice, none of these solutions is able to replace all of the partially-uniform solutions described previously because their calculation is very tedious, especially in region-1. Thus, they are used only in region-4 and are replaced by non-uniform solutions elsewhere. Going from one region to the other with a change of the asymptotic solution does not lead to a visible discontinuity in the numerical results when the solution is uniform. Notice that these uniform expansions are valid in region-1 to first order with respect to the curvature. The reader is referred to the works by Filippov [Fi], Borovikov [Bo] and Bernard [Be] for additional details concerning the second order term.

5.6.2 Solution which is valid close to the grazing incidence: Michaeli's approach (2-D case)

We now consider the case of a source located in region-3 at P′ and an observation point at P in region-4 (Fig. 5.21).

    For this case, the spectral diffraction coefficients in the integrals in (4.58) of Sect. 4.4.5 are no longer slowly-varying as functions of τ. In addition, the approximation of the tangent plane which has been used in the spectral method is no longer

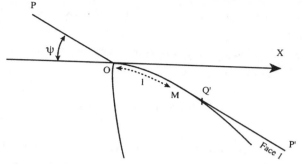

Fig. 5.21. Geometry of scattering by a curved wedge near grazing incidence with the edge in the shadow region.

valid. Since the direction of diffraction was assumed to be close to the grazing angle on face-1, the dominant terms of the asymptotic expansion can be evaluated by integrating the radiation due to the currents on the surface as predicted by Physical Optics, and a fringe or Ufimtsev type of correction can subsequently be added to this current to refine the solution. The field radiated by the PO currents is given by

$$U_h^0(P) = \pm \frac{i}{4} \int_0^\infty \Omega_1 \frac{H_0^{(1)}}{2}\left(k|\vec{r} - \vec{\rho}|\right) \Omega_2 \, U_h^\infty(l,0) \, dl , \qquad (5.300)$$

where $\Omega_1 = \dfrac{\partial}{\partial n}$, $\Omega_2 = 1$, U_h^0 is the field which would exist at a point M of face-1 if the edge were absent (PO approximation), and where we have set (Fig. 5.21):

$$\overrightarrow{OP} = \vec{r} = r\hat{r} \quad , \quad \hat{r} = -x \cos \psi + y \sin \psi \quad , \quad l = \overset{\frown}{OM}$$

The Debye approximation of the Hankel function $H_0^{(1)}\left(k|\vec{r} - \vec{\rho}|\right)$ for large $k|\vec{r} - \vec{\rho}|$

$$\frac{1}{4} H_0^{(1)}\left(k|\vec{r} - \vec{\rho}|\right) \approx \Pi(r) \, e^{ik\hat{r}\cdot\vec{p}} \qquad (5.301)$$

where $\Pi(r)$ is given by

$$\Pi(r) = exp\left(ikr + i\frac{\pi}{4}\right)(8\pi kr)^{-\frac{1}{2}} \qquad (5.302)$$

cannot be used to calculate the integral-1 via the stationary phase method. Indeed, there is a caustic effect along the direction of observation due to the fact that, at grazing incidence on face-1, the surface currents near the edge radiate in phase toward the point

P. Consequently, there is no stationary point in this case as the phase behavior is essentially constant.

There exists a procedure that allows us to circumvent this difficulty, however. It entails the rewriting of the integral in (5.253) in the form

$$\int_0^\infty (\cdot)\, dl = \int_0^{-\infty} (\cdot)\, dl + \int_{-\infty}^{+\infty} (\cdot)\, dl,\qquad (5.303)$$

where the integration is carried out on face-1, extended beyond the edge through continuity.

The second integral in the right-hand side of (5.303) corresponds to the diffraction by a regular surface. The asymptotic evaluation of this integral provides the uniform asymptotic solution published by Pathak [P] for the diffraction by a smooth convex cylinder from which one must subtract the incident field. Lit and shadow region representations of this solution for a perfectly conducting surface are reproduced below. They have been derived from the expressions given in Sect. 5.5.1.3 by substituting the values $\zeta = 0$ (case h) and $\zeta = \infty$ (case s) and by using the notation $U^s = U^\infty - U^i$ where U^∞ is the total field.

Lit zone

In the lit zone we have

$$U_s^s(P_L) = U_s^i(Q_R)\left[-\sqrt{\frac{-4}{\xi^L}}\, exp\left(\frac{i(\xi^L)^3}{3}\right)\left\{\frac{e^{i\frac{\pi}{4}}}{2\sqrt{\pi}\, \xi^L}\left[1 - F(X^L)\right] + \hat{P}_s(\xi^L)\right\}\right]\sqrt{\frac{\rho}{s+\rho}}\, e^{iks},$$
$$\underset{h}{}\qquad\underset{h}{}\qquad\qquad\qquad\qquad\qquad\qquad\qquad\underset{h}{}$$

$$\qquad\qquad\qquad\qquad\qquad\qquad\qquad\qquad\qquad\qquad\qquad (5.304)$$

$$\xi^L = -2m(Q_R)\cos\theta^i,\qquad X^L = 2ks\cos^2\theta^i,\qquad \rho = \frac{\rho_g(Q_R)\cos\theta_i}{2},$$

where ρ denotes the principal radius of curvature of the reflected wave at Q_R, and θ^i is the reflection angle.

Shadow zone

In the shadow zone the field expression reads

$$U_s^s(P_S) = -U_s^i(P_S) - U_s^i(Q')\sqrt{m(Q)m(Q')}\sqrt{\frac{2}{k}}\left\{\frac{e^{i\frac{\pi}{4}}}{2\sqrt{\pi}\, \xi^d}\left[1 - F(X^d)\right] + \hat{P}_s(\xi^d)\right\}e^{ikt}\frac{e^{iks}}{\sqrt{s}},$$
$$\underset{h}{}\qquad\underset{h}{}\qquad\underset{h}{}\qquad\qquad\qquad\qquad\qquad\qquad\qquad\qquad\qquad\underset{h}{}$$

$$\qquad\qquad\qquad\qquad\qquad\qquad\qquad\qquad\qquad\qquad\qquad (5.305)$$

$$\xi^d = \int_{Q'}^{Q} \frac{m(t')}{\rho(t')} dt' \quad , \quad t = \int_{Q'}^{Q} dt'$$

$$X^d = \frac{ks(\xi^d)^2}{2m(Q)m(Q')} \qquad , \qquad (5.306)$$

and the functions F(.) and P(.) are defined by

$$F(X) = -2i \sqrt{X} e^{iX} \int_{\sqrt{X}}^{\infty} e^{it^2} dt ,$$

$$(5.307)$$

$$\hat{P}_h(X) = \frac{e^{i\frac{\pi}{4}}}{\sqrt{\pi}} \int_{-\infty}^{+\infty} \frac{\tilde{Q}V(\tau)}{\tilde{Q}W_1(\tau)} e^{iX\tau} d\tau ,$$

where $\tilde{Q} = \dfrac{\partial}{\partial \tau}$ for the *hard* and $Q = 1$ for the *soft* case.

Let us now return to Eq. (5.303). As $1 \to -\infty$, the field on the regular surface obtained by extending the face-1 beyond the edge 0 decreases exponentially. As a consequence of this, the dominant contribution of the first integral on the right-hand side of (5.303) arises from the neighborhood of $1 = 0$. Te effect of the caustic disappears in this case and we can use the Debye approximation (5.301) for the Hankel function $H_0^{(1)}$ in this integral to get

$$U_h^{eo}(P) = \pm \frac{i}{4} \int_0^{-\infty} \Omega_1 H_0^{(1)}\left(k|\vec{r} - \vec{\rho}|\right) \Omega_2 U_h^{\infty}(l,0) dl$$

$$= \Pi(r) \int_0^{-\infty} \begin{Bmatrix} -ik(\hat{r} \cdot \hat{n}) \\ 1 \end{Bmatrix} e^{ik\hat{r}\cdot\vec{\rho}} \; \Omega_2 U_h^{\infty}(l,0) dl \qquad (5.308)$$

This expression corresponds to the diffraction effect associated with the truncated surface and thus describes the diffraction by the edge.

The variation of the phase of the integrand in (5.308) is described more accurately by using the cylindrical rather than the tangent plane approximation (see Fig. 5.22).

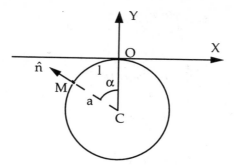

Fig. 5.22. Local cylindrical approximation of the surface.

For this case we have

$$\left\{\begin{array}{l} \alpha = \dfrac{l}{a} \quad , \quad l < 0 \\[2mm] \vec{\rho} = \overrightarrow{OM} = \hat{x} a \sin \dfrac{l}{a} - \hat{y} a\left(1 - \cos \dfrac{l}{a}\right) \\[2mm] \hat{n} = \hat{x}\sin \dfrac{l}{a} - \hat{y}\cos \dfrac{l}{a} \end{array}\right\},$$

It follows then

$$\hat{r}\cdot\hat{n} = \sin\left(\psi - \dfrac{l}{a}\right) \quad , \quad \hat{r}\cdot\vec{\rho} = a\left[\sin\left(\psi - \dfrac{l}{a}\right) - \sin\psi\right],$$

Next, we set $\eta = \psi - l/a$ and replace $U_{h,s}^\infty$ in (5.308) by its expression in (4.40), given in Sect. 4.3.5, and its counterpart for the TE polarization (case s), evaluated for $n = 0$ after having performed the differentiation with respect to the normal coordinate n. Interchanging the order of the integrations we get for U_h^{eo}

$$U_h^{eo}(P) = \Pi(r)U^i(Q')e^{ikl'_0}e^{ika(\sin\eta - \psi)}\frac{1}{\sqrt{\pi}}\int_\Gamma \frac{e^{ikl'_0\tau\frac{m}{a} - ika\psi\frac{\tau}{2m^2}}}{W'_1(\tau)}\,d\tau$$

$$\int_\psi^\infty ika\sin\eta\, e^{ika(\sin\eta - \eta) + ika\tau\frac{\eta}{2m^2}}\,d\eta$$

and a similar expression for $U_s^{eo}(P)$.

The surface field decreases rapidly as $(-l)$ increases from zero. If ψ is assumed to be small, we can write

$$\sin \eta \cong \eta \text{ in the amplitude terms,}$$

$$\left.\begin{array}{l} \sin \eta \cong \eta - \dfrac{\eta^3}{6} \\[2mm] \sin \psi \cong \psi - \dfrac{\psi^3}{6} \end{array}\right\} \text{ in the phase terms.}$$

By letting

$$\frac{l_0' m}{a} = m\beta' = \sigma_0' \quad , \quad m\psi = \sigma_0 ,$$

and noting that

$$\frac{ka}{2} m^{-2}\psi = m\psi = \sigma_0 \quad , \quad ka\frac{\psi^3}{6} = \frac{(m\psi)^3}{3} = \frac{\sigma_0^3}{3} ,$$

we obtain

$$U_h^{eo}(P) = \Pi(r)U^i(Q')\, e^{ikl_0'}\, e^{-i\frac{\sigma_0^3}{3}} \frac{1}{\sqrt{\pi}} \int_\Gamma \frac{e^{ik(\sigma_0'-\sigma_0)\tau}}{W_1'(\tau)} \, d\tau \int_\psi^\infty ika\,\eta\, e^{i\frac{(m\eta)^3}{3}+im\eta\tau}\, d\eta .$$

Introducing the change of variable $m\eta = s$, the internal integral may be written

$$2im\int_{\sigma_0}^\infty s\, e^{i\frac{s^3}{3}+is\tau}\, ds = 2m\frac{\partial}{\partial\tau} I_1(\tau,\sigma_0) ,$$

where I_1 is given by

$$I_1(\tau,\sigma_0) = \int_{\sigma_0}^\infty e^{i\frac{s^3}{3}+is\tau}\, ds , \tag{5.309}$$

and is an incomplete Airy function.

Finally, following the notations of Michaeli and the PO approximation, the diffracted field by the edge can be written as

$$U_h^{eo}(P) = \Pi(r)U^i(Q')e^{ikl_0'}\, S_{h\atop s}(\sigma_0',\sigma_0) , \tag{5.310}$$

where

$$S_{h\atop s}(\sigma_0',\sigma_0) = 4m\sqrt{\pi}\, e^{-i\frac{\sigma_0^3}{3}} M_{h\atop s}(\sigma_0'-\sigma_0,\sigma_0) , \tag{5.311}$$

and the new transition functions M_s^h are defined by

$$M_h(X,Y) = \frac{1}{2\pi} \int_\Gamma \frac{e^{iX\tau}}{W_1'(\tau)} \frac{\partial}{\partial \tau} I_1(\tau,Y) \, d\tau , \tag{5.312}$$

$$M_s(X,Y) = \frac{1}{2\pi} \int_\Gamma \frac{e^{iX\tau}}{W_1'(\tau)} I_1(\tau,Y) \, d\tau . \tag{5.313}$$

We can also derive slightly simpler expressions by using the modified incomplete Airy function and its derivative, in place of the usual incomplete Airy function defined by (5.309) and its derivative. The expressions are

$$I_1^{(m)}(\tau,Y) = e^{-i\frac{Y^3}{3}-iY\tau} \, I_1(\tau,Y), \tag{5.314}$$

$$\left[\frac{\partial}{\partial \tau} I_1(\tau,Y)\right]^{(m)} = e^{-i\frac{Y^3}{3}-iY\tau} \left[\frac{\partial}{\partial \tau} I_1(\tau,Y)\right]^{(m)} . \tag{5.315}$$

These new functions generate modified transition functions $M_{h,s}^{(m)}$ (X,Y) which are constructed by replacing $I_1(\tau,Y)$ and $\partial/\partial\tau \, I_1(\tau,Y)$, in (5.312) and (5.313), by their modified expressions. Then $\underset{s}{S_h}(\sigma_0',\sigma_0)$ takes the form

$$\underset{s}{S_h}(\sigma_0',\sigma_0) = 4m\sqrt{\pi} \, \underset{s}{M_h^{(m)}}(\sigma_0',\sigma_0). \tag{5.316}$$

We will have occasion to use the transition functions $\underset{s}{\tilde{M}_h}(X,Y)$ defined by

$$\underset{s}{\tilde{M}_h}(\sigma_0',\sigma_0) = e^{-i\frac{\sigma_0^3}{3}} \, \underset{s}{M_h}(\sigma_0'-\sigma_0,\sigma_0), \tag{5.317}$$

and if we use the relation

$$U^i(Q') \, e^{ikl_0'} = U^i(O) \, e^{+i\frac{\sigma_0^3}{3}}, \tag{5.318}$$

we can finally write (5.310) in the form

$$\underset{s}{U_h^{eo}}(P) = U^i(O) \, \underset{s}{S_h}(\sigma_0',\sigma_0)\Pi(r), \tag{5.319}$$

where

$$S_h(\sigma_0',\sigma_0) = 4m\sqrt{\pi}\, e^{+i\frac{\sigma_0^3}{3}}\, \tilde{M}_h(\sigma_0',\sigma_0).$$

$$\underset{s}{} \qquad\qquad\qquad\qquad\qquad \underset{s}{} \qquad\qquad\qquad (5.320)$$

The asymptotic solution (5.310) or (5.319) has been derived by assuming that P' is located in that area of region-3 which is below the tangent to face-1 and that P is in region-4 above the tangent to face-1 ($\psi > 0$). Thus, if P moves below the tangent to face-1 ($\psi < 0$), or if P' enters the area of region-3 above that tangent ($\phi' > 0$), we must account for the existence of a field diffracted by the fictitious surface which differs from the one due to the extremity, and which is associated with the extension of the face-1.

Let us assume that $\psi < 0$ and $\phi' < 0$. In this case, the diffracted ray QP becomes a pseudo-ray (see Fig. 5.23) whose contribution must be included in the solution since it is part of the asymptotic expansion of the second integral in the right-hand side of (5.303), to which we must add the incident field to express the total field.

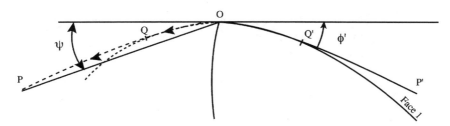

Fig. 5.23. Pseudo-creeping ray P'Q'QP with edge in the shadow region.

The contribution of the pseudo-ray $P'Q'QP$ is equal to $U_h^\infty(P) = U_h^s(P) + U_h^i(P)$ where $U_h^s(P)$ is given by (5.305).

If $\phi' > 0$ and if $\psi < -\phi'$, the excitation point Q' of the surface wave is also on the fictitious surface (Fig. 5.24.), and the contribution of the pseudo-ray $P'Q'QP$ still equals $U_h^\infty(P)$. If $\phi' > 0$ and $|\psi| < \phi'$, there is a virtual reflection point on the fictitious surface. In this case, the reflected field by the fictitious surface must be included in the solution since it appears in the asymptotic evaluation of the integral that yields the field radiated by the currents on the fictitious surface. Since the total field U^∞ also includes the incident field, the reflected field is given by $U^\infty - U_i$. It is important to note that the incident field is associated with the diffraction effect on the edge only if it is nonexistent in the Geometrical Optics solution. By collecting all of the solutions corresponding to the different positions of P and P', we obtain

$$U^{teo} = U^{eo} + H(\phi' + |\phi'| - 2\psi)\, U - - H(\phi' - |\psi|)\, U^i, \tag{5.321}$$

where U^{eo} is the solution given by (5.310) or (5.319). Setting $U^s = U^\infty - U_i$, we can also write (5.321) in the form

$$U^{teo} = U^{eo} + H(\phi' + |\phi'| - 2\psi)\, U^s + [H(\phi' + |\phi'| - 2\psi) - H(\phi' - |\psi|)]\, U^i,$$

which reduces to

$$U^{teo} = U^{eo} + H(\phi' + |\phi'| - 2\psi)\, U^s + H(-\phi' - |\phi'| - 2\psi)\, U^i, \tag{5.322}$$

where U^s is given by (5.304) and (5.305).

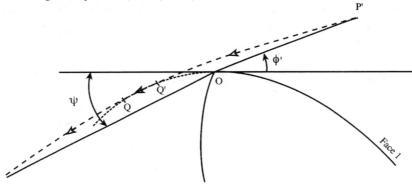

Fig. 5.24. Pseudo-creeping ray P′Q′QP with edge in the lit region.

The asymptotic solution in (5.322) relies upon the Physical Optics approximation which entails the approximation of the electromagnetic field on face-1 and its fictitious extension by the field that would exist on this surface if the edge were absent. In order to derive the complete solution, one must add the Ufimtsev correction, that is to say the contribution of the fringe current, which can be obtained from the spectral theory. It is sufficient to replace, in the integral in (4.58) of Sect. 4.4.5, the diffraction coefficient of a plane wave by an edge as given by Keller's GTD with the fringe coefficient $D^u_{s,h}$ of Ufimtsev, defined by

$$D^u_{s,h}\,(\phi,\xi) = D_{s,h}(\phi,\xi) - D^{OP}_{s,h}\,(\phi,\xi). \tag{5.323}$$

We can then write

$$U_s^u(P) = \Pi(r) \int_\Gamma A_s(\tau) D_s^u(\phi,\xi),$$
(5.324)

where the h subscripts appear under U, A, D.

$$U_h^u(P) = \frac{1}{2} U_h^i(Q') D_h^u(\phi,0) g(\sigma_0') e^{ikl_0} \Pi(r),$$
(5.325)

$$U_s^u(P) = \frac{1}{2} U_s^i(Q') \frac{i}{m} f(\sigma_0') e^{ikl_0} \frac{\partial}{\partial \xi} D_s^u(\phi,0) \Pi(r),$$
(5.326)

$$D_s(\phi,\phi') = \frac{2 \sin \frac{\pi}{N}}{N} \left[\frac{1}{\cos \frac{\pi}{N} - \cos\left(\frac{\pi - \phi' - \psi}{N}\right)} \mp \frac{1}{\cos \frac{\pi}{N} - \cos\left(\frac{\pi + \phi' - \psi}{N}\right)} \right],$$
(5.327)

$$D_s^{OP}(\phi,\phi') = \pm \frac{2}{\cos \phi' - \cos \psi} \left\{ \begin{array}{c} \sin \phi' \\ \sin \psi \end{array} \right\},$$

where ϕ has been replaced by $\pi - \psi$.

Finally, the complete asymptotic solution for the diffracted field by the edge, for the case of a source in region-3 and an observation point in region-4, is given by

$$U^{te}(P) = U^e(P) + H(\phi' + |\phi'| - 2\psi) U^s + H(-\phi' - |\phi'| - 2\psi) U^i(P),$$
(5.328)

where

$$U^e(P) = U^{eo}(P) + U^u(P),$$
(5.329)

and U^s is the uniform solution for the complete surface; U^{eo} is the cut-off effect; and, U^u is the contribution of the fringe current.

5.6.3 Uniform asymptotic solution – Michaeli's approach (2-D case)

The asymptotic solution as provided by (5.328) and (5.329) is not uniform. Actually, if $|\psi| >> m_0^{-1}$ and $|\phi'| >> m_0^{-1}$, we do not recover the partially uniform solution derived in Sect. 4.4.5 and, when $|\psi|$ and $|\phi'|$ increase simultaneously in a manner such that the inequalities $|\psi| >> m_0^{-1}$ and $|\phi'| >> m_0^{-1}$ are satisfied simultaneously, we neither regain the UTD solution of Sect. 5.3.1, nor the Keller solution far from the shadow boundaries of the direct and reflected fields. This result is not surprising since the solution in question is valid only inside the boundary layer associated with the region where the transition zones 2 and 3 of Fig. 5.20 overlap. The asymptotic expansion of this solution, for $m_0|\psi|$ and $m_0|\phi'|$ large as compared to unity, is valid at the periphery of this boundary layer. Since we know the outer solution, we could construct a composite uniform solution by augmenting the outer solution with an inner one derived by using the method described in Sect. 5.2. An alternative would be to start with an Ansatz whose structure is the same as that of the inner solution and which contains a sufficient number of degrees of freedom so that it is possible to match it continuously to the outer solution, a procedure that has been adopted by Michaeli. The Ansatz is extracted from the inner solution which is valid for an edge illuminated by the incident field ($\phi' > 0$) and it is then extended to the case of $\phi' < 0$.

5.6.3.1 Construction of the Ansatz

For $0 < \phi' \leq 1/m_0$ and $|\psi| \leq \phi'$, the solution in (5.328) reads

$$U_s^{te}(P) = U_s^{eo}(P) + U_s^u(P) + U_s^s(P), \qquad (5.330)$$
$$\quad {}_h \qquad\qquad {}_h \qquad\qquad {}_h \qquad\qquad {}_h$$

where U^{eo} is given by (5.319) and (5.320), the corresponding ray being the pseudo-ray $P'Q'OP$ of Fig. 5.25.

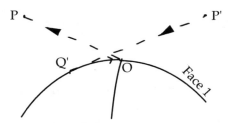

Fig. 5.25. Pseudo-ray P'Q'OP.

The same pseudo-ray also describes the evolution of the phase of U^u. Consequently, we can replace the quantity $U^i(Q')e^{ikl_0}$ in (5.325) and (5.326) with its expression in (5.318). Likewise, the field U^s can be described by the pseudo-ray $P'Q'QP$ of Fig. 5.26.

Fig. 5.26. Pseudo-ray P'Q'QP.

The field U^s, given by (5.304), can also be written in an alternate form as

$$U_{s,h}^s(P) \approx U^i(Q')e^{ik(l_0' - l_0)}\Pi(Q,P)\left[\frac{2F\left(kL\dfrac{\theta^2}{2}\right)}{\theta} - 4m\sqrt{\pi}\ P_{h,s}(m\theta)\right],\ (5.331)$$

where $\theta = -\phi' - \psi$ and where $F(\cdot)$ is defined by (5.307). The function $P_{h,s}(\cdot)$ is related to the Pékéris function $\hat{P}_{h,s}(\cdot)$ through the relation

$$P_{h,s}(X) = e^{-i\frac{\pi}{4}}\ \hat{P}_{h,s}(X) + \frac{1}{2\sqrt{\pi}\ X}.\qquad (5.332)$$

By using the notations $\sigma' = -m\phi'$, $\sigma = m\psi$, the terms inside the bracket in (5.331) can also be written slightly differently as follows

$$[\ ] = -4m\sqrt{\pi}\left[P_{h,s}(\sigma' - \sigma) - \frac{F\left(\Omega^{+2}(\sigma' - \sigma)^2\right)}{2\sqrt{\pi}\ (\sigma' - \sigma)}\right],$$

with

$$\Omega^+ = \left(\frac{kr}{2}\right)^{\frac{1}{2}}\frac{1}{m_0}.$$

After setting

$$\tilde{F}(X) = e^{-iX^2 - i\frac{\pi}{4}}\int_X^\infty e^{it^2}dt,\qquad (5.333)$$

we get

$$F(X) = -2iX e^{i\frac{\pi}{4}}\ \tilde{F}(X),$$

which leads to the result

$$\left[\ \right] = -4m\sqrt{\pi}\left[P_{h,s}(\sigma'-\sigma) + sgn(\sigma-\sigma')\frac{\Omega^+}{\sqrt{\pi}}\,e^{-i\frac{\pi}{4}}\,\tilde{F}\big(\Omega^+|\sigma-\sigma'|\big)\right],$$

where we have used the notation

$$\frac{\sigma-\sigma'}{|\sigma-\sigma'|} = sgn(\sigma-\sigma').$$

Next we insert the expression inside the brackets into (5.331) and use the expression in (5.318) and the approximation

$$\Pi(Q,P) \approx \Pi(O,P)\,e^{-i\frac{\sigma^3}{3}}$$

which is valid when $|\psi| \le 1/m_0$. This yields

$$
\begin{aligned}
U_{s,h}^s(P) &\approx U^i(O)\,e^{i\frac{\sigma'^3}{3}}\,e^{-i\frac{\sigma^3}{3}}\,\Pi(O,P)(-4m\sqrt{\pi}) \\
&\quad \left[P_{h,s}(\sigma'-\sigma) + sgn(\sigma-\sigma')\frac{\Omega^+}{\sqrt{\pi}}\,e^{-i\frac{\pi}{4}}\,\tilde{F}\big(\Omega^+|\sigma-\sigma'|\big)\right]
\end{aligned}
\tag{5.334}
$$

The explicit expressions of the different terms in the right-hand side of (5.330) can be combined to yield

$$U_s^{te}(P) = U^i(O)\left[S_s^+ + \frac{1}{2}\,e^{i\frac{\sigma'^3}{3}}\left\{\begin{matrix}f(\sigma')/i\sigma'\\g(\sigma')\end{matrix}\right\} \cdot D_s^u(\phi',\psi)\right]\Pi(r),
\tag{5.335}$$

where

$$
\begin{aligned}
S_{s}^{+} &= 4m_0\sqrt{\pi}\,e^{i\frac{\sigma'^3}{3}}\left\{\tilde{M}_s(\sigma',\sigma) - H(\phi'-\psi)\,e^{-i\frac{\sigma^3}{3}}\right.\\
&\quad \left.\left[P_s(\sigma'-\sigma) + sgn(\sigma-\sigma')\frac{\Omega^+}{\sqrt{\pi}}\,e^{-i\frac{\pi}{4}}\,\tilde{F}\big(\Omega^+|\sigma-\sigma'|\big)\right]\right\}
\end{aligned}
\tag{5.336}
$$

and $\Pi(r) = \Pi(O,P)$, the polar coordinate of P' is $(r, \pi - \psi)$, and $U(\cdot)$ is the Heaviside function.

The total field at P is given by

$$U^t(P) = U^{te}(P) + H(\phi' + \psi)\, U^i(P) + H(\psi - \phi')\, U^r(P), \qquad (5.337)$$

where U^r is the field reflected by face-1. We recall that the field reflected by the fictitious surface extending the face-1 is already accounted for in $U^{te}(P)$. When ϕ' and $|\psi|$ increase such that they approach unity, three conditions must be satisfied:

(i)   Outside the transition zones around the shadow boundaries of the direct and reflected field, the expression inside the bracket in (5.335) must reduce to the Keller diffraction coefficient $D_{s,h}$, given by (1.92) in Sect. 1.5, so that we recover the conventional GTD result.

(ii)   The total field given by (5.337) must be continuous through the incident field shadow boundary.

(iii)   The same continuity condition must be satisfied through the shadow boundary of the reflected field.

When $\phi' \approx 1$ and $-\sigma' \gg 1$, the second term inside the bracket in (5.335) asymptotically reduces to $D_{s,h}^u$ because of the relationships

$$\left. \begin{aligned} \frac{f(\sigma')}{i\sigma'}\, e^{i\frac{\sigma'^3}{3}} &\cong 2 \\[2mm] g(\sigma')\, e^{i\frac{\sigma'^3}{3}} &\cong 2 \end{aligned} \right\} \ |\sigma'| \gg 1 .$$

However $S_{s,h}^+$, given by (5.336), does not reduce to $D_{s,h}^{OP}$, and it follows that the condition (i) is not satisfied.

This situation could be remedied by multiplying $S_{s,h}^+$ with the factors $C_{s,h}^+(\phi',\psi)$ which are determined by using condition (i). However, this procedure alone would not enable us to satisfy the conditions (ii) and (iii), and for this reason we replace ϕ' and ψ in σ' and σ by $v(\phi')$ and $v(\phi)$, respectively, where $v(\cdot)$ is an unknown function. $C_{s,h}^+$ and v must satisfy the conditions

$$\lim_{\psi \to 0} \frac{v(\psi)}{\psi} = 1 \quad , \quad \lim_{\phi' \to 0} C_{s,h}^+(\phi',\psi) = 1, \qquad (5.338)$$

so that the results derived previously are not invalidated. We also have the flexibility in replacing the distance r in Ω^+ by a characteristic length L^+, provided that L^+ remains large for the range of values of ϕ' and ψ of interest. It is important to note that it is not sufficient to multiply the Michaeli special function $\tilde{M}_{s,h}$ by $M_{s,h}^+$ because the Pékéris-Clemmow function $P_{s,h}(\sigma' - \sigma)$ contributes also to the asymptotic expansion of $S_{s,h}^+$

when $|\sigma'| \gg 1$ (see [Mi] part III Chap. IV B). Consequently, for $\phi > 0$, we start with the following Ansatz

$$
S_{s,h}^+ = 4m\sqrt{\pi}\ C_{s,h}^+\ e^{i\frac{\sigma'^3}{3}}\left\{\tilde{M}_{s,h}(\sigma',\sigma) - H(\phi' - \psi)\,e^{-i\frac{\sigma^3}{3}}\right.
$$
$$
\left.\left[P_{s,h}(\sigma' - \sigma) + sgn(\sigma - \sigma')\frac{\Omega^+}{\sqrt{\pi}}\,e^{-i\frac{\pi}{4}}\,\tilde{F}\big(\Omega^+|\sigma - \sigma'|\big)\right]\right\},
$$

(5.339)

where

$$
\sigma' = -m_0 v(\phi')\quad,\quad \phi' > 0
$$
$$
\sigma = m_0 v(\psi).
$$

Next, we consider the case of a wedge with curved faces whose edge is in the shadow of the direct field ($\phi' < 0$). For this case, by extending the solution of Sec. 5.6.2 to the case of varying curvature along face-1, we obtain

$$
U_s^{te}(P) = U^i(Q')\,e^{ikt'}\left[S_s^- + \frac{1}{2m_0}f(\sigma')\frac{\partial}{\partial\alpha}D_s^u(\alpha,\psi)\big|_{\alpha=0}\right]\left[\frac{a(Q')}{a_0}\right]^{\frac{1}{6}}\Pi(r),
$$

(5.340)

$$
U_h^{te}(P) = U^i(Q')\,e^{ikt'}\left[S_h^- + \frac{1}{2}g(\sigma')D_h^u(0,\psi)\right]\left[\frac{a(Q')}{a_0}\right]^{\frac{1}{6}}\Pi(r),
$$

(5.341)

with

$$
S_{s\atop h}^- = 4m_0\sqrt{\pi}\left\{\tilde{M}_{s,h}(\sigma',\sigma) - H(-\sigma)\,e^{-i\frac{\sigma^3}{3}}\left[P_{s,h}(\sigma' - \sigma) - \frac{\Omega^-}{\sqrt{\pi}}\,e^{-i\frac{\pi}{4}}\,\tilde{F}\big(\Omega^-|\sigma - \sigma'|\big)\right]\right\},
$$

(5.342)

$$
t' = Q'O\quad,\quad \phi' < 0
$$

$$
\sigma' = \int_{Q'}^{O}\frac{m(\tau)}{a(\tau)}\,d\tau\quad,\quad m(\tau) = \left[\frac{ka(\tau)}{2}\right]^{\frac{1}{3}},
$$

(5.343)

where $a(\tau)$ is the radius of curvature of face-1 as a function of the arc length τ measured from point Q'. The expression of Ω^- in (5.342) is given by

$$
\Omega^- = \left(\frac{kL^-}{2}\right)^{\frac{1}{2}}\frac{1}{m_0}.
$$

(5.344)

Ω^- is chosen in a way such that (5.340) continuously matches the expression (5.335) for $\phi' = 0$, $|\psi| \leq m^{-1}$, and the total field is continuous at $\psi = 0$, when $\phi' < 0$ and $|\phi'| \leq m^{-1}$. The first condition implies that

$$\Omega^-\big|_{\phi'=0} = \Omega^+. \tag{5.345}$$

The second involves the creeping ray $P'Q'QP$ whose field at P takes the form

$$U_{\substack{s \\ h}}^{sd}(P) \approx -U^i(Q')e^{ik(t'-t)}4\sqrt{m(Q)m(Q')}\left[P_{s,h}(\zeta^d) - \frac{\Omega^d}{\sqrt{\pi}}e^{-i\frac{\pi}{4}}\tilde{F}\left(\Omega^d\zeta^d\right)\right]\Pi(r), \tag{5.346}$$

where Q is the point from which the spatial ray originates and leaves the surface tangentially. The length of the arc OQ is denoted by t. We have also introduced the notations

$$\zeta^d = \int_{Q'}^{Q}\frac{m(\tau)}{a(\tau)}d\tau$$

$$\Omega^d = \left(\frac{kL^d}{2}\right)^{\frac{1}{2}}[m(Q')m(Q)]^{-\frac{1}{2}}, \tag{5.347}$$

where

$$L^d = \frac{ss'}{s+s'}. \tag{5.348}$$

In the latter expression, s is the length of the ray QP and s' the radius of curvature of the incident wave at Q'. When $\psi = 0$, Q coincides with O, $t = 0$, and $r = s$. As a consequence of this, the continuity of $U^t(P) = U^e(P) + H(\psi)\,U^{sd}(P)$ leads to the relation (see (5.342) and (5.346))

$$\Omega^-\big|_{\psi=0} = \Omega^d\big|_{\psi=0}. \tag{5.349}$$

A simple way to assure that (5.349) is consistent with (5.344) is to set

$$\Omega^- = \left(\frac{kL^-}{2}\right)^{\frac{1}{2}}[m(Q)m(Q')]^{-\frac{1}{2}}, \tag{5.350}$$

with

$$L^- = \frac{rs}{s+s'}, \tag{5.351}$$

where L^- will be specified below. The above manipulations are carried out to assure the continuity of the Michaeli solution given in Sect. 5.6.2, which is valid in the boundary layer.

When $|\psi| \gg 1/m_0$, and $\phi' < 0$, the solution given in (5.340) must revert to the hybrid solution in Sect. 4.4.5 extended to the case of a variable curvature. Consequently, the functions must satisfy

$$
\begin{aligned}
S_s^- &\approx \frac{i}{2m_0} f(\delta') \frac{\partial}{\partial \alpha} D_s^{OP}(\alpha, \psi)_{|\alpha=0} \\
S_h^- &\approx \frac{1}{2} g(\delta') D_h^{OP}(0, \psi)
\end{aligned}
\tag{5.352}
$$

In order to impose these conditions we multiply $S_{s,h}^-$ by the factor $C_{s,h}^-(\phi',\psi)$. Next, to establish the continuity at $\phi' = 0$ of the solution thus modified to that constructed for $\phi' > 0$, we replace ψ by $v(\psi)$, and impose the condition

$$
C_{s,h}^-(0, \psi) = C_{s,h}^+(0, \psi).
\tag{5.353}
$$

An Ansatz, which is sufficiently general for the case of an edge located in the shadow of the incident field ($\phi' < 0$), is given by (5.340) and (5.341), where the expression for $S_{s,h}^-$ in (5.342) is multiplied by the unknown function $C_{s,h}^-(\phi',\psi)$. Furthermore $\sigma = m_0 v(\psi)$, σ', and s' are defined by (5.343), and $v(\psi)$ is an unknown function identical to the one introduced in the Ansatz given in (5.339).

5.6.3.2 Determination of the unknown functions

The unknown functions $C_{s,h}^\pm(\phi',\psi)$ and $v(x)$ are determined by imposing the conditions (i) to (iii) to the solutions modified by these functions. Let us first consider the solution associated with the case of an edge illuminated by the incident field. An inspection of the asymptotic behavior of the incomplete Airy function $I_1(\tau,Y)$ and of the Michaeli function $M_{s,h}(X,Y)$ enables us to derive the relationship

$$
S_{s,h}^+ \approx \pm 4 C_{s,h}^+ \left[v^2(\phi') - v^2(\psi)\right]^{-1} \begin{Bmatrix} v(\phi') \\ v(\psi) \end{Bmatrix}.
\tag{5.354}
$$

The condition (i), i.e., $S_{s,h}^+ \approx D_{s,h}^{OP}$, then leads to

$$
C_{s,h}^+(\phi',\psi) = \frac{v^2(\phi') - v^2(\psi)}{2(\cos\psi - \cos\phi')} \begin{Bmatrix} \sin\phi'/v(\phi') \\ \sin\psi/v(\psi) \end{Bmatrix}.
\tag{5.355}
$$

We see from (5.355), that the condition $v^2(-\phi') = v^2(\phi')$ is necessary in order for the function $C_{s,h}^+$ to tend to a finite limit on the boundary of the incident field ($\psi = -\phi'$). Hence, according to (5.338) we have

$$v(-\phi') = -v(\phi').$$
(5.356)

If we assume that this relation is satisfied and we tend to the limit in the right-hand side of (5.355) when ψ tends to ϕ', we obtain

$$C_{s,h}^+(\phi',-\phi') = \frac{dv(\phi')}{d\phi'}.$$
(5.357)

We still need to satisfy the conditions (ii) and (iii) and we begin with the former. We observe that, on the shadow boundary of the incident field ($\psi = -\phi'$ or $\sigma' = \sigma$), the field $U^{te}(P)$ given by (5.335) exhibits a jump discontinuity because of the presence of the term

$$sign(\sigma - \sigma')\frac{\Omega^+}{\sqrt{\pi}}\, e^{-i\frac{\pi}{4}}\, \tilde{F}\big(\Omega^+ \,|\,\sigma - \sigma'|\big),$$
(5.358)

in S^+ and this jump in the expression above is equal to $\Omega^+\, e^{-i\frac{\pi}{4}}$. We also note here that $H(\phi' - \psi) = H(2\phi') = 1$.

If we insert (5.357) into (5.339), and use the resulting expression of S^+ in (5.335), we see that when $\sigma' = \sigma$, the jump of $U^{te}(P)$ is given by

$$-U^i(O)\, 4m_0\, \sqrt{\pi}\, \Omega^+\, e^{-i\frac{\pi}{4}}\Pi(r)\frac{dv(\phi')}{d\phi'}.$$

The above jump discontinuity in U^{te} must compensate the one in $U^i(P)$ at the shadow boundary. For the case of a linear source located at a distance r' from O, we can write

$$U^i(P) = A\frac{e^{ik(r'+r)}}{\sqrt{r'+r}} = A\frac{e^{ikr'}}{\sqrt{r'}}\, e^{ikr}\left(\frac{r'}{r+r'}\right)^{\frac{1}{2}} = U^i(O)\left(\frac{r'}{r+r'}\right)^{\frac{1}{2}}e^{ikr}.$$

Thus the compensation of the jump discontinuity is achieved if we impose the condition

$$4m_0\, \sqrt{\pi}\, \Omega^+\, e^{-i\frac{\pi}{4}}\Pi(r)\frac{dv(\phi')}{d\phi'}\left(\frac{r'}{r+r'}\right)^{\frac{1}{2}}.$$
(5.359)

Next, we replace Ω^+ by

$$\Omega^+ = \left(\frac{kL^+}{2}\right)^{\frac{1}{2}} \frac{1}{m_0}, \tag{5.360}$$

and use the expression (3) of $\Pi(r)$. Then (60) reduces to

$$L^+ = L_0^+ \left[\frac{dv(\phi')}{d\phi'}\right]^{-2}, \tag{5.361}$$

where

$$L_0^+ = \frac{rr'}{r+r'}. \tag{5.362}$$

The above result is imposed on the shadow boundary of the direct field. It can be extended to the entire range of $\psi \neq \phi'$.

On the shadow boundary of the reflected field, we have the relationship $\phi' = \psi$ or $\sigma' = -\sigma$. In order to apply the condition (iii), we need to identify the discontinuous term in the expression for $U^{te}(P)$ at $\phi' = \psi$, and to equate it to the reflected field. When $\phi' = \psi$ (or $\sigma' = -\sigma$), the discontinuous term in (5.335) is the one multiplying the Heaviside function $H(\phi' - \psi)$. In view of (5.339), this term can be written as

$$-U^i(0)\,\Pi(r)\,4m_0\,\sqrt{\pi}\,e^{2i\frac{\sigma'^3}{3}}\left[\underset{h}{P_s}(2\sigma') + \frac{\Omega^+}{\sqrt{\pi}}\,e^{-i\frac{\pi}{4}}\,\tilde{F}\!\left(-2\sigma'\Omega^+\right)\right]. \tag{5.363}$$

An expression for the reflected field on a regular convex surface, that is uniformly valid from the semi-lit to the Geometrical Optics zone, has been established by Pathak [P]. By employing the same notations as in the expression for U^{te}, this solution, as given by (5.304), can be written as

$$\underset{s}{U_h^r}(P) = -U^i(Q_R)\left(\frac{\rho^r}{\rho^r+1}\right)^{\frac{1}{2}} e^{ikl}\,e^{i\frac{(\zeta^L)^3}{12}}\,e^{i\frac{\pi}{4}}\left(-\frac{4}{\zeta^L}\right)^{\frac{1}{2}}\left[\underset{h}{P_s}(\zeta^L) + \frac{\Omega^L}{\sqrt{\pi}}\,e^{-i\frac{\pi}{4}}\,\tilde{F}\!\left(-\Omega^L\zeta^L\right)\right], \tag{5.364}$$

where Q_R is the reflection point, l the length of the ray Q_RP (Fig. 5.27) and ρ^r the curvature radius of the reflected wavefront at Q_R.

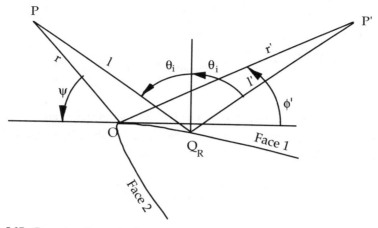

Fig. 5.27. Geometry of scattering by a curved wedge with reflected and edge-diffracted rays.

We can also show that

$$\rho^r = \frac{l'a(Q_R)\cos\theta^i}{2l' + a(Q_R)\cos\theta^i}, \qquad (5.365)$$

where l' is the curvature radius of the incident wave at Q_R, $a(Q_R)$ is the curvature radius of the face-1 of the wedge at Q_R, and θ^i is the incidence angle. Next, for convenience we introduce the notation

$$\Omega^L = \left(\frac{kL^L}{2}\right)^{\frac{1}{2}}\frac{1}{m(Q_R)}, \qquad (5.366)$$

where L^L, given in (5.368) below, is the characteristic length defined as the width of the semi-lit zone on the side of the region directly illuminated by the incident field and

$$m(Q) = \left(\frac{ka(Q)}{2}\right)^{\frac{1}{3}}, \qquad (5.367)$$

$$\begin{aligned}\zeta^L &= \zeta_0^L = -2m(Q_R)\cos\theta^i\\ L^L &= L_0^L = \frac{ll'}{l+l'}\end{aligned}. \qquad (5.368)$$

On the reflection shadow boundary, the point Q_R coincides with the point O, and we have

$$l' = r' \ , \quad l = r \ , \quad \psi = \phi' = \frac{\pi}{2} - \theta^i \ , \quad U^i(Q_R) = U^i(O).$$

Hence,
$$L_0^L = L_0^+, \quad \Omega^L = \Omega^+$$

$$\frac{\rho^r}{\rho^r + l} = \frac{r' a_0 \sin\phi'}{(r + r') a_0 \sin\phi' + 2rr'} = \frac{1}{2r}\left(\frac{a_0 \sin\phi'}{1 + \frac{a_0}{2L_0^+}\sin\phi'}\right).$$

Furthermore, the modifications introduced in U^{te} generate modifications in the parameters ζ^L and L^L. These modifications must, however, satisfy the following constraints at grazing incidence ($\theta^i = \pi/2$).

$$\lim_{\theta^i \to \frac{\pi}{2}}\left(\frac{\zeta^L}{\zeta_0^L}\right) = \lim_{\theta^i \to \frac{\pi}{2}}\left(\frac{L^L}{L_0^L}\right) = 1. \tag{5.369}$$

When Q_R coincides with O, we have from (5.368)

$$\zeta^L = -2m_0\sin\phi' \cong -2m_0\phi' = 2\sigma'.$$

Consequently, on the reflection shadow boundary we choose

$$\zeta^L = -2m_0 v(\phi') = 2\sigma'. \tag{5.370}$$

Inserting (5.370) into (5.364), we obtain

$$U_h^r(P) = -U^i(O)\frac{1}{\sqrt{2r}} a_0 \frac{1}{2}\left(\frac{\sin\phi'}{1 + \frac{a_0}{2L_0^+}\sin\phi'}\right)^{\frac{1}{2}} e^{2i\frac{\sigma'^3}{3}} e^{i\frac{\pi}{4}}$$

$$\times \left(\frac{2}{m_0 v(\phi')}\right)^{\frac{1}{2}}\left[P_h(2\sigma') + \frac{\Omega^+}{\sqrt{\pi}}e^{-i\frac{\pi}{4}}\tilde{F}(-2\sigma'\Omega^+)\right]$$

Hence, by identifying this expression with (5.363) we get

$$\frac{dv(\phi')}{d\phi'} = \frac{1}{\sqrt{v(\phi')}}\left(\frac{\sin\phi'}{1 + \frac{a_0}{2L_0^+}\sin\phi'}\right)^{\frac{1}{2}}. \tag{5.371}$$

The solution of (5.371) is given by

$$v(\phi') = \left(\int_0^{\phi'} \left[\frac{\sin \phi'}{1 + \frac{a_0}{2L_0^+} \sin \phi'} \right]^{\frac{1}{2}} d\alpha \right)^{\frac{2}{3}} . \tag{5.372}$$

The above solution can be extended to the negative values of the angles ϕ' or ψ by taking the condition $v(-\phi') = -v(\phi')$ into account, which results from the application of condition (ii). The solution which is valid for both the positive and negative angles is given by

$$v(\psi) = sign(\psi) \left(\int_0^{|\psi|} \left[\frac{\sin \alpha}{1 + \frac{a_0}{2L_0^+} \sin \alpha} \right]^{\frac{1}{2}} \right)^{\frac{2}{3}} . \tag{5.373}$$

Inserting (5.373) into (5.355), we obtain the explicit expressions for the coefficients $C_{s,h}^+$. Similarly, by inserting (5.372) into (5.361) and (5.360), we obtain L^+ and Ω^+. If Q_R is distinct from O, we can extend the expression in (5.370) of ζ^L by setting

$$\zeta^L = -2m(Q_R) \, v^L\left(\frac{\pi}{2} - \theta^i\right), \tag{5.374}$$

where $v^L(\psi)$ can be deduced from the expression of $v(\psi)$ given in (5.373) by replacing a_0 by $a(Q_R)$ and L_0^+ by L_0^L, which was defined in (5.368).

Similarly, one can extend the expression of L_0^L, by setting

$$L^L = L_0^L \left(1 + \frac{a(Q_R)\cos\theta^i}{2L_0^L} \right) \frac{v^L\left(\frac{\pi}{2} - \theta^i\right)}{\cos\theta^i}$$

$$= L_0^L \left[\frac{dv^L(\alpha)}{d\alpha} \right]_{\alpha = \frac{\pi}{2} - \theta^i}^{-2} . \tag{5.375}$$

We point out that the parameters occurring in the calculation of $v^L(\psi)$ are to be taken at the point Q_R, and that it is not necessary to determine, a priori, the parameters on the shadow boundary of the reflected field when calculating the reflected field given by

(5.364) at a point Q_R distinct from the diffraction point O. This is convenient when a computer code is developed on the basis of the uniform solution. When Q_R tends to 0, $v^L(\psi)$, ζ^L and L_0^L tend to $v(\psi)$, ζ and L_0^+, respectively, and consequently, the total field is continuous.

It is important to note that the condition $\Omega^L = \Omega^+$ on the shadow boundary, where Ω^+ is defined via the relations (5.360) and (5.362), leads to a reflected field (given by (5.364) in the transition zone close to this shadow boundary) which differs from the one described by the Pathak solution [P] in which Ω^L is defined by (5.366) and (5.367). However, since the two solutions tend to the GO field outside the transition region and since the modification introduced in the Pathak solution vanishes according to (5.338) at the grazing incidence $\phi' = \psi = 0$, this modification does not affect the numerical results.

For the case of an edge located in the shadow region of the incident field we obtain, after applying (5.353) through (5.355),

$$C_s^-\big|_{\phi'=0} = \frac{v^2(\psi)}{2(1-\cos\psi)} \quad , \quad C_h^-\big|_{\phi'=0} = \frac{v(\psi)\sin\psi}{2(1-\cos\psi)}. \qquad (5.376)$$

From the asymptotic expansion of $\tilde{M}(\sigma',\sigma)$ for $|\sigma| \gg 1$ and $\sigma' > 0$, given in [Mi], we obtain

$$S_{\substack{s\\h}}^- = 2iC_{\substack{s\\h}}^-[v(\phi)]^{-1}\begin{Bmatrix} f(\sigma')/\sigma \\ ig(\sigma') \end{Bmatrix}. \qquad (5.377)$$

Comparing (5.377) to (5.352), and using (5.327) we have

$$C_s^- = \frac{v^2(\psi)}{2(1-\cos\psi)} \quad , \quad C_h^- = \frac{v(\psi)\sin\psi}{2(1-\cos\psi)}. \qquad (5.378)$$

These expressions are in agreement with (5.376) which, consequently, continues to apply when $\phi' < 0$. They define the uniform solution for $\phi' < 0$, and we realize that the function $v(\psi)$ is identical to that determined for the case $\phi' > 0$, and is given by (5.373).

5.6.3.3 Numerical application

The plates 1 and 2 show the numerical results for the uniform asymptotic solution described in Sects. 5.6.3.1 to 5.6.3.3, for the case of an ogival cylinder whose dimensions, expressed in normalized wavelength, are given in Fig. 5.28.

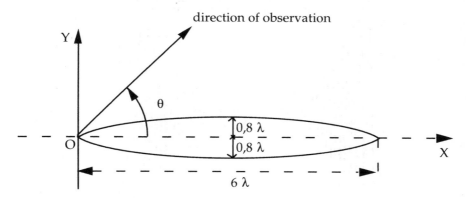

Fig. 5.28. Cross section of an ogival cylinder with length of the ogive 6λ, and width 1.6λ.

Let the incident wave be generated by a line source parallel to the generatrix of the cylinder of electric (TM case) or magnetic (TE case) type, and let its position be defined by the coordinates of its trace S' in a plane of a cross-section of the cylinder. Plates 1 and 2 plot the curves of the variation of the intensity of the reflected field for both TM and TE polarizations. Plate 1 corresponds to $S_x = -10\lambda$, $S_y = 0$ and plate 2 corresponds to $S_x = 0$, $S_y = -3.5\lambda$. This variation is expressed in dB and plotted as a function of the observation angle θ defined on Fig. 5.28. The solid curve was obtained by using the uniform asymptotic solution that only includes the singly-diffracted rays from the sharp edges. The uniform solution is compared to a reference curve (dashed curve) which represents the result obtained by numerically solving the diffraction problem via the Method of Moments. We observe that the curves corresponding to the uniform asymptotic solution are in excellent agreement with the reference curves. Even for the case presented in plate 2, where the edge B is illuminated at grazing incidence, the disagreement is negligible.

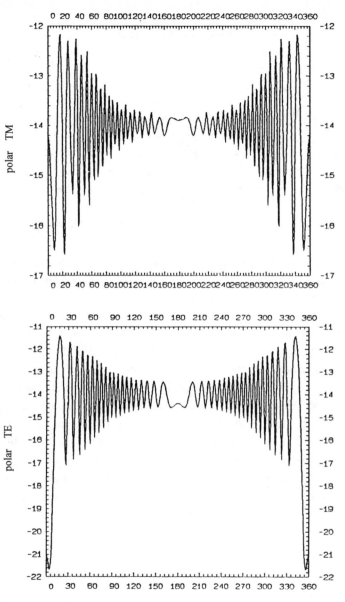

Plate 1   Two-dimensional structure with symmetrical ogival-shape cross-section
Source defined by $S_x = -10\lambda$, $S_y = 0\lambda$ Observation direction ranging from 0° to 360°.
_____ GTD
----------- Moment method

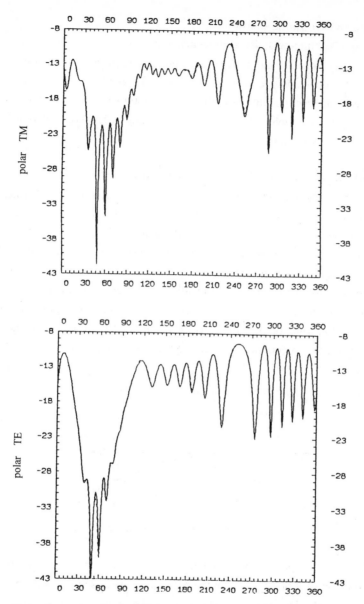

Plate 2 Two-dimensional scatterer with symetrical ogival-shaped cross-section Source defined by $S_x = 0\lambda$, $S_y = -3.5\lambda$ Observation direction ranging from 0° to 360°.

——————— GTD

----------- Moment method

5.6.4   Uniform asymptotic solution:
Liang, Chuang and Pathak approach (2-D case)

In this section we consider the uniform asymptotic solution of a perfectly conducting 2-D wedge with convex curved faces, that has been derived by Liang [L]. It relies upon an Ansatz derived directly from the uniform asymptotic solution of the curved screen, which was originally developed by Chuang and Liang [CL]. We will briefly review their approach in this section.

5.6.4.1 Uniform asymptotic solution of the curved screen

The space outside the curved screen is decomposed into five regions defined in Fig. 5.29. The incident wave is assumed to be either a cylindrical or plane wave such that its direction of propagation is orthogonal to the generatrix of the 2-D screen. We pay special attention to the case where the source is either in region-III or -IV, while the observation point is either in region-I or -II, because there is an overlap of several transition regions for the combinations mentioned above. In all other cases, except in the reciprocal situation that we have just described, the transition regions are separated and the standard asymptotic solutions, viz., UTD and UAT presented in Chap. 5.3, or the hybrid solutions mentioned in Sect. 5.6.1 and discussed in Sect. 4.4.5, are applicable.

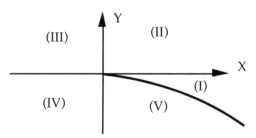

Fig. 5.29. Curved screen and decomposition of the space in sub-domains.

The method employed by [CL] and [L] to derive a uniform asymptotic solution for the diffraction of a plane wave by the edge of a curved screen relies upon the fact that, for an incident plane wave originating from the region $x < 0$ (see Fig. 5.30), the total field close to the edge in the above region can be approximated by the Sommerfeld solution [S] corresponding to the diffraction by a half-plane. Furthermore, we need to consider at least the second derivative of the fictitious surface extending the screen beyond the edge, and we assume that the field on this extended surface is known. By employing the equivalence principle, we can then replace the fictitious surface by a perfectly conducting surface which supports an equivalent magnetic current $\vec{M} = \vec{E} \times \hat{n}$, where $\hat{n}$ is the normal directed outward from the surface (Fig. 5.30), and $\vec{E}$ is the total electric field $\left(\vec{E} = \vec{E}^i + \vec{E}^d \right)$, which is given by the Sommerfeld solution and is identical to the UTD solution for the half-plane. It is important to note that the expression of the field remains exact even when the observation point comes close to the edge of the half plane.

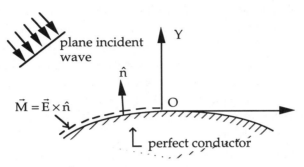

Fig. 5.30. Equivalence principle applied to a curved screen.

The problem can be divided into two parts (Figs. 5.31a and 5.31b), viz., (i) the diffraction of a plane wave by a regular, perfectly conducting, surface; and (ii) the radiation by a sheet of magnetic current, also on a regular, perfectly conducting, surface.

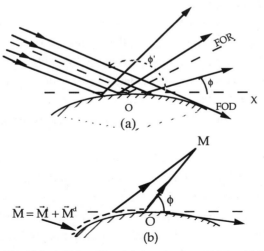

Fig. 5.31. Decomposition of the initial problem into two simpler problems (a) Scattering of the incident wave by the regularized surface, (b) Radiation of equivalent magnetic currents.

The solution to both of these problems can be derived by using the uniform asymptotic solutions presented in Sec. 5.5. In particular, the solution to problem (i) can be directly obtained by starting with the formulas presented in Sect. 5.5.1, and setting $Z = 0$, where Z designates the surface impedance. For problem (ii), the field radiated by an electric or magnetic dipole is given by (5.296) of Sec. 5.5, where the roles of the observation point and the source must be interchanged by invoking reciprocity, and where we have to take the limit $\zeta = 0$ for a source located on the surface. Furthermore, since the direction of the radiation is orthogonal to the generatrix of the cylinder, the terms arising from the torsion of the geodesics are zero. The Fock functions $g(\xi_0)$ and $\tilde{g}(\xi_0)$, as well as the phase term, depend upon the position of the dipole and, consequently, they appear in the integrals involving the sheet of magnetic current carried by the fictitious surface. For both of the polarizations of the incident wave, the authors [CL] have decomposed these integrals into different terms that involve new transition functions. To derive the final result, it is necessary to extract the contribution of the direct incident field from the magnetic current $\vec{M}$ by writing $\vec{M} = \vec{M}^i + \vec{M}^d$ where $\vec{M}^i = \vec{E}^i \times \hat{n}$, $\vec{M}^d = \vec{E}^d \times \hat{n}$.

In effect, the current $\vec{M}^d$ decreases rapidly when the observation point recedes from the origin O. The asymptotic expansion of the corresponding integral then reduces to a contribution arising from the end-point of the path of integration. On the other hand, the current $\vec{M}^i$ does not decrease with an increase of the observation distance and, in general, the integral does not reduce to a contribution from the end point. In order to treat such an integral, it is important to group it with certain terms of the diffraction problem depicted in Fig. 5.31a.

We note that the solution to the problem (i) consists not only of the contributions of the reflected and creeping rays on the fictitious surface which are attributed to the diffraction by the edge, but also of the contribution of the reflected and creeping rays on the real surface. However, only the latter exists if the direction of propagation of the incident wave is above the tangent to the curved screen at $O(\pi/2 < \phi' < \pi)$, and if the direction of observation is below the shadow boundary of the reflected field $(\phi < \pi - \phi')$.

The ray field reflected by the fictitious surface is complemented by the radiation from the current sheet $\vec{M}^i$ in the direction of the reflection. The same is also true for the creeping rays on the fictitious surface. For this reason, the problem sketched in Fig. 5.31a will be solved only when the point of reflection or the launching point of the creeping ray is on the real surface. In all other cases, we apply the induction theorem which enables us to replace the diffraction problem of Fig. 5.31a by the radiation problem shown in Fig. 5.32.

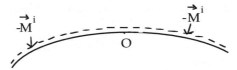

Fig. 5.32. Application of the induction theorem to Fig. 5.31a (where $\vec{M}^i = \vec{E}^i \times \hat{n}$).

By superposing the current sheet in Fig. 5.32 on that in Fig. 5.31b, we observe that the field diffracted by the edge is given by the radiation of a current sheet $\vec{M}^d$ residing on a fictitious surface radiating in conjunction with the current sheet $-\vec{M}^i$ on the real surface.

In summary, when $\pi/2 < \phi' < \pi$ and $\phi < \pi - \phi'$, the original problem is decomposed into two, viz., a diffraction problem and a radiation problem, as shown in Figs. 5.31a and 5.31b. The radiation from the current sheet $\vec{M}^i$ of Fig. 5.31b has been derived by [CL], for both the TE and TM polarizations, in a form that involves the Fresnel functions and a new transition function.

For all other cases, the original problem is replaced by that of the radiation of two sheets of magnetic current, viz., $\vec{M}^d$ on the fictitious surface and $-\vec{M}^i$ on the real surface (see Fig. 5.33).

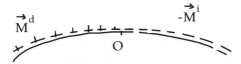

Fig. 5.33. Replacement of the scattering problem by a curved screen by a radiation problem.

As in the preceding case, the field radiated by the sheet of current $-\vec{M}^i$ can be expressed in terms of new transition functions. Although the latter have a similar nature they are not identical to those corresponding to the preceding case, viz., problem (i).

5.6.4.2 Expression for the new transition functions

We will refrain here from providing the complete expressions for the field which can be found in Liang's thesis [L]. However, as an example, we present the new transition functions for the case $\pi/2 < \phi' < \pi, \phi < 0$ and show that they reduce to the Michaeli transition functions when $m|\phi| \leq 1$ and $m|\pi-\phi'| \leq 1$.

When $\pi/2 < \phi' < \pi$ and $\phi < 0$, the Liang transition functions read

$$
L_{s,h}^s(\phi', \xi_1, m) = m^2 \frac{e^{-i\frac{\pi}{4}}}{\sqrt{2\pi k}} \int_0^\infty e^{i\left[\frac{\xi^3}{3} + 2m^2\xi(1+\cos\phi') + 2m\xi^2\cos\frac{\phi'}{2}\right]}
$$

$$
\left\{ \begin{Bmatrix} S(\xi + \xi_1, m) \\ \left(-\frac{\xi}{m} - 2\cos\frac{\phi'}{2}\right)H(\xi + \xi_1) \end{Bmatrix} d\xi + \frac{e^{i\frac{\pi}{4}}}{\sqrt{2\pi k}} \frac{\cos\frac{\phi'}{2}}{2\left(1 + \sin\frac{\phi'}{2}\right)} H(\xi_1) \begin{Bmatrix} 0 \\ 1 \end{Bmatrix} \right\}
$$

$$
\tag{5.379}
$$

with

$$
\left. \begin{aligned} S(x,m) &= \frac{i}{m}\tilde{g}(x) \\ H(x) &= g(x) \end{aligned} \right\} \quad x > 0, \tag{5.380}
$$

where $\tilde{g}$ and g are Fock functions defined in (5.298) of Sect. 5.5, and where the arguments ξ and ξ_1 have the following form

$$
\xi = \int_{Q_C}^0 \frac{m(Q)}{\rho(Q)} dl \quad , \quad \xi_1 = \int_0^{Q_A} \frac{m(Q)}{\rho(Q)} dl . \tag{5.381}
$$

In (5.381), $\rho(Q)$ designates the radius of curvature of the geodesic at the point Q and the points Q_C, O and Q_A are the positions of the element of magnetic current, the edge and the point where the creeping ray exits the surface (Fig. 5.34), respectively.

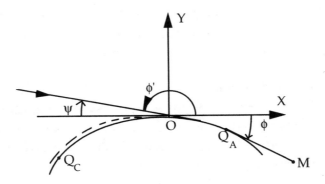

Fig. 5.34. Geometry of the radiation by equivalent magnetic currents.

We set $\psi = \pi - \phi'$, and assume that $m\psi \le 1$, $m\phi \le 1$, which is tantamount to saying that $\psi << 1$, $\phi << 1$. With these assumptions, we arrive at the result

$$1 + \cos \phi' \cong \frac{\psi^2}{2} \quad , \quad \cos \frac{\phi'}{2} \cong \frac{\psi}{2} \quad , \quad \xi_1 = m\phi,$$

and the integral in the right-hand side of (5.379) takes the form

$$I_{s,h} = \int_0^\infty e^{\left[i\left[\frac{\xi^3}{3} + m^2\xi\psi\left(\psi+\frac{\xi}{m}\right)\right]\right]} \left\{ \begin{matrix} S(\xi + m\phi, m) \\ -\left(\psi + \dfrac{\xi}{m}\right) H(\xi + m\phi) \end{matrix} \right\} d\xi. \tag{5.382}$$

If we introduce the change of variable, $\eta = \psi + \dfrac{\xi}{m}$, the integral becomes

$$I_{s,h} = e^{-i\frac{(m\psi)^3}{3}} \int_\psi^\infty e^{i\frac{(m\eta)^3}{3}} \left\{ \begin{matrix} S[m\eta + m(\phi - \psi), m] \\ -\eta H[m\eta + m(\phi - \psi)] \end{matrix} \right\} m\,d\eta. \tag{5.383}$$

Hence, using (5.380) and (5.298) of Sect. 5.5, and interchanging the order of integration after introducing the change of variable $s = m\eta$ we get

$$I_s = i\frac{e^{-i\frac{(m\psi)^3}{3}}}{m\sqrt{\pi}} \int_\Gamma \frac{e^{im(\phi-\psi)\tau}}{W_1(\tau)} d\tau \int_{m\psi}^\infty e^{i\frac{s^3}{3} + is\tau} ds, \tag{5.384}$$

$$I_h = i\frac{e^{-i\frac{(m\psi)^3}{3}}}{m\sqrt{\pi}} \int_\Gamma \frac{e^{im(\phi-\psi)\tau}}{W_1'(\tau)} d\tau \frac{\delta}{\delta\tau} \int_{m\psi}^\infty e^{i\frac{s^3}{3} + is\tau} ds. \tag{5.385}$$

In the right-hand side of (5.384) and (5.385) we recognize the presence of the incomplete Airy functions defined in (5.309), and the Michaeli transition functions defined in (5.312) and (5.313). Setting $m\psi = \sigma_0$, $m\phi = \sigma'_0$, we obtain

$$I_s = I_s(\sigma'_0 - \sigma_0, \sigma_0) = \frac{2i\sqrt{\pi}}{m} e^{i\frac{\sigma_0^3}{3}} M_s(\sigma'_0 - \sigma_0, \sigma_0), \tag{5.386}$$

$$I_h = I_h(\sigma'_0 - \sigma_0, \sigma_0) = \frac{2i\sqrt{\pi}}{m} e^{i\frac{\sigma_0^3}{3}} M_h(\sigma'_0 - \sigma_0, \sigma_0), \tag{5.387}$$

where $M_{s,h}(X, Y)$ are the Michaeli transition functions.

5.6.4.3 Uniform asymptotic solution of the wedge with curved faces

The UTD solution of the wedge with curved faces is not valid in the vicinity of the edge. A consequence of this is that it is no longer possible to use the same method as in the case of the curved screen. This difficulty has been circumvented by Liang [L] who has written the uniform asymptotic solution of the curved screen in a form where the Keller diffraction coefficients D^K appear explicitly. These coefficients are the only quantities which depend on the wedge angle. It is sufficient to perform this operation for the case where the source is in the regions III and IV, and the observation point in I and II, since the solution is of the UTD or hybrid type, and consequently it has the desired structure for all other cases. In the region where several zones overlap, the terms that depend on the new transition function reduce to the diffraction coefficient D^{OP} outside this region. The diffraction coefficient corresponds to the contribution associated with the end-point of the Geometrical Optics integral. These terms do not depend upon the wedge angle. In contrast, the complementary term does depend upon this angle through the difference $D^K - D^{OP}$. Let us recall here that these terms in the Liang solution play the role of the Ufimtsev correction in the Michaeli solution.

The asymptotic solution of the curved screen, in which the Keller diffraction coefficients are adapted to a wedge with an exterior angle equal to $n\pi$, is used as an Ansatz for the curved wedge, which recovers to the solution for the curved screen for $n = 2$. Liang has verified that the solution constructed in this way is continuous and tends to the UTD solution when the incident and reflected shadow boundaries are far away from the faces.

5.6.4.4 Numerical application

The curves of plate 3 depict the variation of the intensity of the diffracted field as a function of the observation angle for the case of a linear source located in the symmetry plane of a semi-circular cylinder closed by a plane on the side of the curved face. The dimensions, expressed in normalized wavelength, are given in Fig. 5.35.

The curves presented here are extracted from the reference [LPC]. The dashed curves correspond to the uniform asymptotic solution while the solid curves were generated by using the Method of Moments and are treated here as a reference solution. Perfect agreement between the two sets of results is evident.

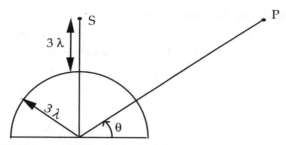

Fig. 5.35. Application of the asymptotic solution to a half cylinder.

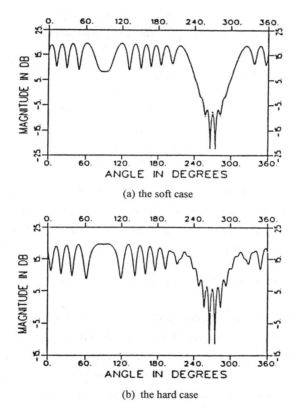

(a) the soft case

(b) the hard case

Plate 3 Comparison between the total field calculated with the moment method (solid curve) and with UTD (dashed curve) of a line source located at $\rho' = 6\lambda$ and $\phi' = 90°$ in the presence of a semi-circular cylinder whose curvature radius is $a = 3\lambda$.

5.7 Uniform solutions for the caustics

In Sects. 3.4 and 3.14 we have presented a solution, derived using the boundary layer method, and valid in the neighborhood of a two-dimensional caustic. This solution has been expressed in the (s,n) coordinate system, where s is the curvilinear abscissa along the caustic and n is the distance from it. The solution thus obtained is not uniform, and, in fact, we do not recover the Geometrical Optics solution as we recede from the caustic. To find a uniform solution, we will proceed, as in [Kr1] and [Lu], where the authors have employed an Ansatz that reduces to the solution given by (3.225) and (3.233) in Sect. 3.14 for points very close to the caustic and which tends to the Geometrical Optics solution for points far removed from the caustic.

To the dominant order in k, the solution can be written in the form (see (3.225) in Sect. 3.14) in the vicinity of the caustic as

$$u \approx exp(iks) \, A(s) \, Ai(-v), \qquad (5.388)$$

where v is the stretched coordinate, and r is the curvature radius of the caustic.

$$v = k^{2/3} \left(\frac{2}{\rho}\right)^{1/3} n. \qquad (5.389)$$

The higher-order terms arranged as decreasing powers of $k^{1/3}$, are expressed in terms of the Airy function and its derivative represented by polynomials with respect to v. If we collect all of the terms associated with Ai on one group, and those with Ai' in another, we can write the solution in the form

$$u \approx exp(iks) \, (P(s, v) \, Ai(-v) + k^{-1/3} Q(s, v) \, Ai' \, (-v)), \qquad (5.390)$$

where $P(s, v)$ and $Q(s, v)$ are polynomials of v, whose coefficients are functions of s and k. Somehow these polynomials are limited expansions of more general functions in the vicinity of the caustic. As explained in Sect. 5.1, the principle difference between the boundary layer and the uniform methods is the way they introduce these functions in the Ansatz. The boundary layer method can then be considered as a first step, enabling one to postulate the Ansatz for the uniform solution. For the case of the caustic, (5.390) suggests an Ansatz of the type [Kr, Lu]

$$u = exp(ikS) \, (g_0 \, Ai \, (-k^{2/3}q) + ik^{-1/3} \, g_1 \, Ai' \, (-k^{2/3} \, q)). \qquad (5.391)$$

In (5.391), S, q, g_0 and g_1 are functions which depend on the position of the point. Just as in the boundary layer method, the above functions are determined by requiring that

(5.391) satisfies the Helmholtz equation and matches the Geometrical Optics solution far from the caustic. Following the same procedure as in Chap. 3, we introduce the Ansatz in the Helmholtz equation and arrange the resulting expression in terms of decreasing powers of $k^{1/3}$. For the six first terms of the expansion we obtain the following conditions that must be satisfied [Lu]

$$(\vec{\nabla}S)^2 + q(\vec{\nabla}q)^2 = 1, \tag{5.392}$$

$$2\vec{\nabla}S \cdot \vec{\nabla}q = 0 \tag{5.393}$$

we have also

$$2\vec{\nabla}S \cdot \vec{\nabla}q + \Delta q \, g_0 + 2q\vec{\nabla}q \cdot \vec{\nabla}g_1 + q\Delta q \, g_1 + \left(\vec{\nabla}q\right)^2 g_1 = 0 \tag{5.394}$$

$$2\vec{\nabla}q \cdot \vec{\nabla}g_0 + \Delta q \, g_0 + 2\vec{\nabla}S \cdot \vec{\nabla}g_1 + \Delta S \, g_1 = 0 \tag{5.395}$$

After manipulation, (5.392) and (5.393) may be interpreted as the eikonal equations, and (5.394) and (5.395) as the transport equations.

Let us first consider (5.392) and (5.393). By forming the sum and the difference of these equations and setting $S_+ = S + 2/3 \, q^{3/2}$, $S_- = S - 2/3 \, q^{3/2}$, we can recover the eikonal equations of Geometrical Optics (see Chap. 2), viz.,

$$(\vec{\nabla}S \pm \sqrt{q} \, \vec{\nabla}q)^2 = 1, \tag{5.396}$$

or

$$(\vec{\nabla}S_+)^2 = (\vec{\nabla}S_-)^2 = 1. \tag{5.397}$$

S_+ and S_- are then the eikonals for the Geometrical Optics rays which will be further discussed later.

Let us use the same procedure for the transport Eqs. (5.394) and (5.395). This time we define $g_+ = g_0 + \sqrt{q} \, g_1$ and $g_- = g_0 - \sqrt{q} \, g_1$. Next, we form the sum and difference of (5.394) and the product of (5.395) times $\sqrt{q}$ to get

$$2\vec{\nabla}S_+ \cdot \vec{\nabla}g_+ + \left(\Delta S_+ - \frac{1}{2}q^{-\frac{1}{2}}\left(\vec{\nabla}q\right)^2\right)g_+ = 0 \tag{5.398}$$

$$2\vec{\nabla}S_- \cdot \vec{\nabla}g_- + \left(\Delta S_- + \frac{1}{2}q^{-\frac{1}{2}}\left(\vec{\nabla}q\right)^2\right)g_- = 0 \tag{5.399}$$

Without the third term, these equations would represent the transport equations of g_+ and g_-. We regain the transport equations by introducing the functions $h_+ = q^{-1/4}g_+$ and $h_- = q^{-1/4}g_-$. The result is :

$$2\vec{\nabla}S_+ \cdot \vec{\nabla}h_+ + (\Delta S_+)h_+ = 0, \qquad (5.400)$$

$$2\vec{\nabla}S_- \cdot \vec{\nabla}h_- + (\Delta S_-)h_- = 0. \qquad (5.401)$$

These are the transport equations of Geometrical Optics for h_+ and h_-. According to (5.396) and (5.397), the functions S_+ and S_- represent then eikonals of Geometrical Optics and, according to (5.400) and (5.401), the functions h_+ and h_- are amplitudes of the Geometrical Optical ray. We note that, due to the presence of the factor $q^{-1/4}$, the functions h_+ and h_- are singular on the caustic where $q = 0$. It remains now to identify these Geometrical Optics rays.

It is possible to identify S_+ with the eikonal and, to within a constant, h^+ with the amplitude of the ray leaving the caustic. S_- and h_- are then the eikonal and the amplitude (to within a constant) of the ray originating toward the caustic (Fig. 5.36).

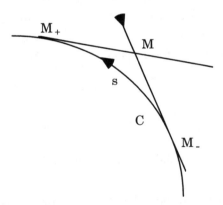

Fig. 5.36. Geometry of Caustic Surface

This can be verified by calculating the asymptotic expansion of (5.391) for large q. After replacing Ai and Ai' by their asymptotic expansions for large negative arguments, we get

$$u(M) \approx \frac{e^{i\pi/4}}{2\pi^{1/2}k^{1/6}}\left[h_- \, exp(iks_-) + e^{-i\pi/2} \, h_+ \, exp(iks_+)\right]. \qquad (5.402)$$

The first term in (5.402) is the ray MM_- (see Fig. 5.36) that approaches the caustic with the phase kS_- and the amplitude ch_-, while the second term is associated with the ray M_+M receding from the caustic. The phase of the latter ray is kS_+ and its amplitude is ch_+ where $c = e^{i\pi/4}/2\pi^{1/2} k^{1/6}$. At the point of contact with the caustic, this ray undergoes a $-\pi/2$ phase-shift (see Sect. 3.14). Since the Geometrical Optics solution is known, the coefficients S_+, S_-, h_+, h_- are completely defined. We might recall from the definitions given earlier, that

$$S_+ = S + 2/3\ q^{3/2}, \tag{5.403a}$$

$$h_+ = q^{-1/4}(g_0 + \sqrt{q}\ g_1), \tag{5.403b}$$

$$S_- = S - 2/3\ q^{3/2}, \tag{5.403c}$$

$$h_- = q^{-1/4}(g_0 - \sqrt{q}\ g_1). \tag{5.403d}$$

From (5.403) we can determine the functions s, q, g_0, g_1 appearing in the Ansatz. In fact, from (5.403a) and (5.403c) we can obtain for S and q

$$S = 1/2\ (S_+ + S_-), \tag{5.404}$$

$$q = \left[3/4(S_+ + S_-)\right]^{2/3}. \tag{5.405}$$

Likewise from (5.403b) and (5.403d) we can obtain, for g_0 and g_1, the results

$$g_0 = 1/2\ q^{1/4}(h_+ + h_-), \tag{5.406}$$

$$g_1 = 1/2\ q^{-1/4}(h_+ - h_-). \tag{5.407}$$

The phase on the caustic C is associated with the curvilinear abscissa s on C. Accordingly the difference $S^+ - S^-$ is equal to the difference between the sum of the two ray lengths, viz., $(M_+ M + MM_-)$, and the length $M_+ M_-$ of the arc on C. Thus $S^+ - S^-$ is always positive, and (5.405) always defines a positive function q. This function q vanishes on C and C can be considered as the curve defined by the equation $q = 0$.

One can show [Lu], that if C is regular, then S, q, g_0 and g_1 are also regular. We refer to [Lu] for these demonstrations and we will restrict ourselves to providing here only the expressions of the relevant functions, detailed previously, in the vicinity of the caustic, using the coordinates (s, n) introduced in Chap. 3.

We obtain $S \approx s$, $q \approx \left(\frac{2}{\rho}\right)^{1/3} n$, $g_0 \approx 2\sqrt{\pi} \; e^{-i\pi/4} \left(\frac{k}{2}\right)^{1/6} r^{-1/3} B(s)$ where $B(s)$ is defined by (3.230) of Sect. 3.14.

Inserting these results in Ansatz (5.391) and retaining only the terms of highest degree, one obtains :

$$u \approx exp(iks) \; 2\sqrt{\pi} \; e^{-i\pi/4} \left(\frac{k}{2}\right)^{1/6} r^{-1/3} B(s) \; Ai(-v), \qquad (5.408)$$

where v is the stretched coordinate defined by (5.389).

We see that we recover the boundary layer solution (see (3.225) and (3.233) of Sect. 3.14), derived in Chap. 3, as the limit of the uniform solution in the vicinity of the caustic.

The uniform solution then contains the boundary layer solution and, consequently, it is more general than the latter. However, the boundary layer solution is a necessary first step for constructing the uniform solution. In particular, it serves to determine the functions that occur in the Ansatz of the uniform solution. In passing, we also notice that, according to (5.405), q, i.e., n, is proportional to the two-thirds power of the difference of the eikonal. When the difference of phase is of $O(1)$, the difference of the eikonal is of $O(k^{-1})$, and consequently, n is of order $k^{-2/3}$. The boundary layer, with a thickness proportional to $k^{-2/3}$, is defined as the set of points such that the phase difference between the rays passing through these points is of $O(1)$. Thus, we recover the heuristic criterion used in Chap. 3 to determine the thickness and the stretching of the boundary layer coordinates.

In this section, we have treated the two-dimensional case which reduces to the Helmholtz equations. The three-dimensional case, involving the Maxwell's equations can be solved in a similar way [Kr2], and identical results can be obtained. The Ansatz for the fields $\vec{E}$ and $\vec{H}$ are similar to (5.391).

$$\vec{E} = exp(ikS) \, (Ai \, (-k^{2/3}q) \, \vec{E}_0 + ik^{-1/3} \, A_i' \, (-k^{2/3}q) \, \vec{E}_1), \qquad (5.409a)$$

$$\vec{H} = exp(ikS) \, (Ai \, (-k^{2/3}q) \, \vec{H}_0 + ik^{-1/3} \, A_i' \, (-k^{2/3}q) \, \vec{H}_1). \qquad (5.409b)$$

Inserting (5.409) in the Maxwell equations, and performing the same linear combination as in the Helmholtz case, we can verify that, to within $O(1/k)$

$$exp(ikS_+) \, \vec{E}_+ \; and \; exp(ikS_+) \, \vec{H}_+ \; where \; S_+ = S + 2/3 \; q^{3/2},$$

$$\vec{E}_+ = q^{-1/4} \, (\vec{E}_0 + \sqrt{q} \, \vec{E}_1), \; \vec{H}_+ = q^{-1/4} \, (\vec{H}_0 + \sqrt{q} \, \vec{H}_1), \qquad (5.410)$$

are solutions of the Geometrical Optics type of the Maxwell equations. The same is true for $\exp(ikS_-)\,\vec{E}_-$ and $\exp(ikS_-)\,\vec{H}_-$ where

$$S_- = S - 2/3\, q^{3/2}, \quad \vec{E}_- = q^{-1/4}(\vec{E}_0 - \sqrt{q}\,\vec{E}_1), \quad \vec{H}_- = q^{-1/4}(\vec{H}_0 - \sqrt{q}\,\vec{H}_1). \quad (5.411)$$

We can then identify the asymptotic expansion of (5.407) for large values of q with the Geometrical Optics solution, and this helps us to associate ($\vec{E}_+$, $\vec{H}_+$) with the fields propagated by the ray going away from the caustic, as well as ($\vec{E}_-$, $\vec{H}_-$) with the fields by the ray propagating toward the caustic. The expressions in (5.410) and (5.411) define the functions that play a role in the Ansatz given in (5.409) in precisely the same way as they do in the scalar case.

Using the above approach, the problem is solved on the lit side of the caustic. In contrast, it is evident that there are no real rays on the shadow side. Although in principle, one can always proceed by using analytical continuation, this approach is delicate to handle. A simple way to obtain an approximate result for the field at a point M in the shadow zone is to perform the calculation at a point in the lit zone, by reflecting the point M with respect to the caustic and then changing the argument in the Airy function. However, this is only an approximation, and to our knowledge, there is no convenient way to find a uniform solution in the shaded zone.

Finally, we point out that the preceding results are valid close to the regular point on C. By using a similar method, it is also possible to derive solutions in the vicinity of a cusp line and more generally close to the singular points of C [Ko]. These solutions are expressed in terms of the functions occurring in the theory of catastrophe and of their partial derivatives with respect to their arguments. For instance, for the case of the beak of a caustic this Ansatz takes the form

$$u \approx exp(ikS)\,(g_0\,P + ik^{-1/4}\,g_1\,\frac{\partial p}{\partial q} + ik^{-1/2}\,g_2\frac{\partial p}{\partial q}), \quad (5.412)$$

where $P = P(q, r)$ is the Pearcey function [Ko]; S, q, r are the phase functions; and g_0, g_1, g_2 are the amplitude functions. Similar to the case of the conventional caustic, the above constants are determined by identifying the asymptotic expansion of (5.412) for large values of q and r to the Geometrical Optics solution. The phase of the Pearcey function is a polynomial of the fourth degree which leads to three stationary phase points that are real in the lit zone and the Geometrical Optics solution is the sum of the three rays. The above leads then to three equations each for the determination of the phases and the amplitudes which completely define the unknown functions of the

Ansatz. However, in contrast to the case of a regular point, these functions cannot be written explicitly as functions of the amplitudes and the eikonal of rays, and we must resort to a numerical solution. Moreover, P is not as well-tabulated as are the Airy functions. Both of these difficulties, which also appear in the case of more complex caustics, make these uniform solutions relatively sensitive to handle, and it is often more convenient to use the integral representations of the solution instead. In the following two chapters we will present some of the most widely-used methods which provide these integral representations in Chaps. 6 and 7.

In Chap. 6 we will present the Maslov method and the method of integration on a wave front, the later being used in Theoretical Physics. Following this we will go on to Chap. 7 and present the Physical Theory of Diffraction (PTD) which is widely-used in the industry.

References

Sects. 5.1 and 5.2

[A]   D. S. Ahluwalia, "Uniform asymptotic theory of diffraction by the edge of a three-dimensional body," *Siam J. Appl. Math.*, **18**(2), 287-301, 1970.

[C]   J. Cole, *Perturbation methods in applied mathematics*, Blaisdell, 1968.

[Kr]   Y. A. Kravtsov, "Asymptotic solutions of Maxwell's equations near a caustic," *Radiofizika*, **7**, 1049-1056 (in Russian), 1964.

[L]   D. Ludwig, "Uniform asymptotic expansions at a caustic," *Comm. Pure Appl. Math.*, **19**, 215-250, 1966.

[LB]   R. H. Lewis and J. Boersma, "Uniform asymptotic theory of edge diffraction," *J. Math. Phys.*, **10**(12), 2291-2306, 1969.

[N]   A. H. Nayfeh, *Perturbation Methods*, Chap. 4, Wiley-Interscience Publication, 1973.

Sect. 5.3

[Bl]   N. Bleistein, "Uniform asymptotic expansions of integrals with many nearby stationary points and algebraic singularities," *J. Math. Mech.*, **17**, 533-559, 1967.

[Bo]   V. A. Borovikov, "Diffraction by a wedge with curved faces," *Sov. Phys. Acoust.* **25**(6), 465-471, 1979.

[BR]   J. Boersma and Y. Rahmat-Samii, "Comparison of two leading uniform theories of edge diffraction with the exact uniform asymptotic solution," *Radio Sci.*, **15**(6), 1179-1194, 1980.

[Cl]   P. C. Clemmow, "Some extensions of the method of integration by steepest descents," *Quart. J. Mech. Appl. Math.* **3**, 241-256, 1950.

[FM]   L. B. Felsen and N. Marcuvitz, *Radiation and Scattering of Waves*, Englewood Cliffs, NJ, Prentice Hall, 1973.

[GP]   C. Gennarelli and L. Palumbo, "A uniform asymptotic expression of a typical diffraction integral with many coalescing simple pole singularities and a first-order saddle point," *IEEE Trans. Ant. Prop.*, **AP-32**, 1122-1124, 1984.

[HuK] D. L. Hutchins and R. G. Kouyoumjian, "Asymptotic series describing the diffraction of a plane wave by a wedge," Report 2183-3, ElectroScience Laboratory, Department of Electrical Engineering, The Ohio State University; prepared under Contract AF 19 (638)-5929 for Air Force Cambridge Research Laboratories.

[HwK] Y. M. Hwang and R. G. Kouyoumjian, "A dyadic diffraction coefficient for an electromagnetic wave which is rapidly varying at an edge," paper presented at USNC/URSI Annual Meeting, Boulder, Colorado, 1974.

[K]   J. B. Keller, "Geometrical theory of diffraction," *J. Opt. Soc. AM.*, **52**, 116-130, 1962.

[KP1] R. G. Kouyoumjian and P. H. Pathak, "A uniform geometrical theory of diffraction for an edge in a perfectly conducting surface," *Proc. IEEE*, **62**, 1448-1461, 1974.

[KP2] R. G. Kouyoumjian and P. H. Pathak, *A uniform GTD approach to EM scattering and radiation*, in *Acoustic Electromagnetic and Elastic Wave Scattering — High and Low Frequency Asymptotics*, vol. II, edited by Varadan and Varadan, North Holland Publishers, 1986.

[LD]   S. W. Lee and G. A. Deschamps, "A uniform asymptotic theory of electromagnetic diffraction by a curved wedge," *IEEE Trans. Ant. Prop.*, **AP-24**, 25-34, 1976.

[Ob]   F. Oberhettinger, "An asymptotic series for functions occurring in the theory of diffraction of waves by wedges," *J. Math. Phys.*, **34**, 245-255, 1955.

[Ot]   H. Ott, "Die Sattelpunktsmethode in der Umgebung eines Poles mit Anwendung an die Wellenoptik und Akustik," *Annalen der Physik*, **43**, 393-403, 1943.

[Pau] W. Pauli, "On the asymptotic series for functions in the theory of the diffraction of light," *Phys. Rev.*, **54**, 924-931, 1938.

[RM1] Y. Rahmat-Samii and R. Mittra, "A spectral domain interpretation of high frequency phenomena," *IEEE Trans. Ant. Prop.*, **AP-25**, 676-687, 1977.

[RM2] Y. Rahmat-Samii and R. Mittra, "Spectral analysis of high frequency diffraction of an arbitrary incident field by a half plane-comparison with four asymptotic techniques," *Radio Science*, **13**(1), 31-48, 1978.

[Ro]   R. G. Rojas, "Comparison between two asymptotic methods," *IEEE Trans. Ant. Prop.*, **AP-35**(12), 1489-1492, 1987.

[VDW]   Van der Waerden, "On the method of saddle points," *Appl. Sci. Research*, **B2**, 33-45, 1951.

Sect. 5.4

[GU]   J. George and H. Uberall, "Approximate methods to describe the reflections from cylinders and spheres with complex impedance," *J. Acoust., Soc. Am.* **65**(1), 15-24, 1979.

[KK]   L. Kaminetzky and J. B. Keller, "Diffraction coefficients for higher order edges and vertices," *SIAM J. Appl. Math.*, **22**(1), 109-134, 1972.

[KLS]   J. B. Keller, R. M. Lewis and B. D. Seckler, "Asymptotic solution of some diffraction problems," *Comm. Pure Appl. Math.*, **9**, 207-265, 1956.

[Le]   M. A. Leontovitch, "Investigations of propagation of radio waves," *Soviet Radio*, Moscow, 1948.

[Mo1]   F. Molinet, *Geometrical Theory of Diffraction*, IEE APS Newsletter, part II, pp. 5-16, 1987.

[Mo2]   F. Molinet, "Etude de la diffraction par une ligne de discontinuité de la courbure," *Rapport MOTHESIM M* **51**, 1982.

[Mo3]   F. Molinet, "Uniform asymtotic solution for the diffraction by a discontinuity in curvature,"*Annales des Télécom.*, 50 (5, 6), 523-535, 1995.

[Se]   T. B. A. Senior, "The diffraction matrix for a discontinuity in curvature," *IEEE Trans. Ant. Prop.*, **AP-20**, 326-333, 1972.

[So]   A. Sommerfeld, *Partial Differential Equation in Physics*, Academic Press, New York, 1964.

[W]   V. H. Weston, "The effect of a discontinuity in curvature in high frequency scattering," *IRE Trans. Ant. Prop.*, **AP-10**, 775-780, 1962.

Sect. 5.5

[BK]   V. M. Babich and N. Y. Kirpicnikova, *The boundary-layer method in diffraction problems*, Springer-Verlag, Berlin, Heidelberg, New York, 1979.

[Bo]   D. Bouche, *La méthode des courants asymptotiques*, Thése de Docteur, Université de Bordeaux I, 1992.

[C]   J. A. Cullen, "Surface currents induced by short-wave length radiation," *Phys. Rev.*, **109**, 1863-1867, 1958.

[FK]   W. Franz and K. Klante, "Diffraction by surfaces of variables curvature," *IRE Trans. Ant. Prop.*, **AP-7**, 568-570, 1959.

[Fo1]   V. A. Fock, "The field of a plane wave near the surface of a conducting body," *J. Phys. USSR*, **10**, 399-409, 1946.

[Fo2]   V. A. Fock, *Electromagnetic Diffraction and Propagation*, Pergamon, New York, 1965.

[H]   S. Hong, "Asymptotic theory of electromagnetic and a caustic diffraction by smooth convex surfaces of variable curvature," *J. Math. Phys.*, **8**, 1223-1232, 1967.

[I]   V. I. Ivanov, "Diffraction d'ondes électromagnétiques planes courtes en incidence oblique sur un cylindre convexe lisse," *Radiotecknica i Electronica*, **5**, 524-528, 1960.

[L]   O. Lafitte, *Thése de Docteur*, Université de Paris Sud, 1993.

[LK]   B. R. Levy and J. B. Keller, "Diffraction by a smooth object," *Comm. Pure Appl. Math.*, **12**, 159-209, 1959.

[LY]   N. A. Logan and K. S. Yee, *A mathematical model for diffraction by convex surfaces*, in *Electromagnetic Waves*, edited by R. E. Langer, pp. 139-180, The University of Wisconsin Press, Madison, Wisconsin, 1962.

[Mi]   A. Michaeli, "High-frequency electromagnetic fields near a smooth convex surface in the shadow region," private communication, 1991.

[MSN]   R. Mittra and Safavi-Naini, "Source radiation in the presence of smooth convex bodies," *Radio Science*, **14**, 217-237, 1979.

[P]   P. H. Pathak, "An asymptotic analysis of the scattering of plane waves by a smooth convex cylinder," *Radio Science*, **14**, 419-435, 1979.

[Pe]   L. W. Pearson, "A schema for automatic computation of Fock-type integrals," *IEEE Trans. Ant. Prop.*, **AP-35**(10), 1111-1118, 1987.

[PWBK]   P. H. Pathak, N. Wang, W. Burnside and R. Kouyoumjian, "A uniform GTD solution for the radiation from sources on a convex surface," *IEEE Trans Ant. Prop.*, **AP-29**, 609-622, 1981.

[WC]   J. R. Wait and A. M. Conda, "Pattern of an antenna on a curved loosy surface," *IRE Trans. Ant. Prop.*, **AP-6**, 348-359, 1958.

Sect. 5.6

[A]   N. C. Albertsen, "Diffraction of creeping waves," Report LD 24, Electromagnetic Institute, Technical University Denmark, 1974.

[AC]   N. C. Albertsen and P. L. Christiansen, "Hybrid diffraction coefficients for first and second order discontinuities of two-dimensional scatterers," *SIAM J. Appl. Math.*, **34**, 398-414, 1978.

[Be]   J. M. L. Bernard, "Diffraction par un dièdre à faces courbes non parfaitement conducteur," *Rev. Tech. THOMSON-CSF*, **23**(2), 321-330, 1991.

[Bo]   V. A. Borovikov, "Diffraction by a wedge with curved faces," *Akust. Zh.*, **25**(6), 825-835.

[CL]   C. W. Chuang and M. C. Liang, "A uniform asymptotic analysis of the diffraction by an edge in a curved screen," *Radio Science*, **23**(5), 781-790, 1988.

[Fi]   V. B. Filippov, "Diffraction by a curved half-plane," *Zap. Nauchn. Sem. LOMI* (Leningrad), **42**, 244, 1974.

[HP]   K. C. Hill and P. H. Pathak, "A UTD analysis of the excitation of surface rays by an edge in an otherwise smooth perfectly-conducting convex surface," URSI Radio Science meeting, Blacksburg, Virginia, 1987.

[IF]   M. Idemen and L. B. Felsen, "Diffraction of a whispering gallery mode by the edge of a thin concave cylindrically curved surface," *IEEE Trans. Ant. Prop.*, **AP-29**, 571-579, 1981.

[K]   J. B. Keller, "Diffraction by an aperture," *J. Appl. Phys.*, **28**, 426-444, 1957.

[KP]   R. G. Kouyoumjian and P. H. Pathak, "A geometrical theory of diffraction for an edge in a perfectly conducting surface," *Proc. IEEE*, **62**, 1448-1474, 1974.

[L]   M. C. Liang, *A generalized uniform GTD ray solution for the diffraction by a perfectly conducting wedge with convex faces*, Ph. D Thesis, Ohio State University, 1988.

[LD]   S. N. Lee and G. A. Deschamps, "A uniform asymptotic theory of electromagnetic diffraction by a curved wedge," *IEEE Trans. Ant. Prop.*, **AP-24**, 25-34, 1976.

[LPC]   M. C. Liang, P. H. Pathak and C. W. Chuang, "A generalized uniform GTD ray solution for the diffraction by a wedge with convex faces," Congre,` s URSI, Prague, Août 1990.

[Mi]   A. Michaeli, "Transition functions for high-frequency diffraction by a curved perfectly conducting wedge; Part I: Canonical solution for a curved sheet; Part II: A partially uniform solution for a general wedge angle; Part III: Extension to overlapping transition regions," *IEEE Trans. Ant. Prop.*, **27**, 1073-1092, 1989.

[Mo]   F. Molinet, "Diffraction d'une onde rampante par une ligne de discontinuité du plan tangent," *Annales des Télécom.*, **32**(5-6), 197, 1977.

[P]   P. H. Pathak, "An asymptotic analysis of the scattering of plane waves by a smooth convex cylinder," *Radio Science*, **14**(3), 419-435, 1979.

[PK]   P. H. Pathak and R. G. Kouyoumjian, "On the diffraction of edge excited surface rays," Paper presented at the 1977 USNC/URSI Meeting, Stanford University, Stanford, CA, 22-24, June 1977.

[S]   A. Sommerfeld, "Mathematische Theorie der Diffraktion," *Math. Ann.*, **47**, 317-374, 1986.

Sect. 5.7

[KO]   Y. A. Kravtsov and Y. I. Orlov, "Caustics, catastrophes and wavefields," *Sov. Phys. Usp.* **26**, 1983.

[K1]   Y. A. Kravtsov, *Radiofizika*, **7**, 664, 1964.

[K2]   Y. A. Kravtsov, *Radiofizika*, **7**, 1049, 1964.

[Lu]   D. Ludwig, *Comm. Pure Appl. Math.* **19**, 215, 1966.

6. Integral Methods

The method of matched asymptotic expansions, presented in Chap. 3, and especially its uniform version developed in Chap. 5, enable us to calculate the field at almost all locations in space. However, its implementation becomes extremely involved in the zones where there is an overlap of several boundary layers. In particular, difficulties arise on the cusp lines of the caustic surface where the formulas of Sect. 5.7 no longer apply. A particular boundary layer exists in the neighborhood of these lines. Fortunately, there are integral representations of the solution that are valid simultaneously at the ordinary point on the caustic as well as near the cusp lines, which helps circumvent the problem of cumbersome matching between the boundary layers. The integral representation of the solution is not unique, and several methods have been proposed for constructing different representations for the solution. The only requirement on the solution is that we must recover the results derived by the ray method when applying the stationary phase approximation to the integral representation; the user remains free to choose the integral representation that is most convenient to him. Mathematicians, who are specialists in the area of differential partial equations, have made significant advances in this area, and their efforts have led to the Fourier-Maslov-Hörmander's theory of integral operators.

In Sect. 6.1, we present the Maslov version of this method which, though less rigorous, is better suited for practical applications than is the theory of the Fourier operators. The Maslov method provides integral representations of the solution in the form of a plane wave spectrum, that are uniformly valid, and a uniform asymptotic evaluation of these integrals provides the field in the vicinity of an arbitrary point on the caustic. The solution can be written in terms of an Airy function near a regular point on the caustic and expressed as a Pearcey function close to an ordinary point on the cusp line of this surface. In the vicinity of the singular points of the cusp line, e.g., a swallowtail, the description of the solution requires us to use the functions occurring in the *catastrophe theory*. The Maslov method provides a powerful tool for calculating the field in the vicinity of the caustic. However, it is based on notions that are not very intuitive, which may be the reason why it has not been extensively employed in electromagnetic applications.

In Sect. 6.2, we propose a more intuitive approach, which represents a generalization of the method employed by Airy to evaluate the field in the neighborhood of a regular caustic. The field at a point located on the caustic is extracted from the radiation integral associated with the field on the wavefront, with the latter calculated using the geometrical optics. This method also provides a simpler and alternate interpretation of the procedure for calculating the field near a caustic, and it highlights the relationship between the geometry of the wavefront and the surface defined by the centers of curvature, i.e., the caustic. Although we can present a number of other integral methods for handling the caustic problem, for instance, the method of Gaussian rays used in geophysics, we will refrain from doing so because our principal concern here is not to provide an exhaustive discussion of these techniques but to present, instead, the essential ideas upon which they are based. Finally, we will present, in Chap. 7, the so-called Physical Theory of Diffraction, which is another integral method designed to compute the diffracted field by associating it with the radiated surface currents on the diffracting object. We will first describe the Maslov method, starting with a presentation of the mathematical notions indispensable for its implementation.

6.1   The Maslov method

During the sixties, V. P. Maslov [Ma] devised a method for deriving uniform asymptotic solutions of partial differential equations, and since then the method has both been improved and justified from a mathematical point of view. Our principal objective here is to provide a presentation of this method in simple and physical terms and also to discuss some practical aspects associated with its implementation in practice. In Sect. 6.1.1 we will introduce the geometrical concepts, such as the phase space and the Lagrangian manifold, that are useful for the understanding of the method. In Sects. 6.1.2 and 6.1.3. we will show how the method enables us to calculate the field in the vicinity of the caustics and to provide explicit integral representations of the solution that are valid close to the caustics. Furthermore, using the example of the circular caustic, we will explain how, from these formulas, it is possible to extract, directly, the field solution derived in Chap. 3 using the boundary layer method. In Sect. 6.1.4, we present a useful variant of the Arnold method and apply it to a case which is somewhat difficult to treat using other methods, namely the field computation at the cusp of the caustic. In Sect. 6.1.5 the method is introduced from a different point of view which shows how it can be viewed as a type of geometrical optics in mixed space.

In Sect. 6.1.6 we demonstrate how the method can be directly generalized such that it applies to Maxwell's equations and provide the limits of its applicability in Sect. 6.1.7.

6.1.1 Preliminary concepts

6.1.1.1 Geometrical optics in phase space
We have seen in Chap. 2 that the solution of the eikonal equation requires the determination of the characteristics of differential equations. For a homogeneous medium, these equations take a very simple form, and are given by

$$\frac{dx_i}{dt} = p_i, \tag{6.1a}$$

$$\frac{dp_i}{dt} = 0, \tag{6.1b}$$

where t is the abscissa on a ray; $\vec{x} = (x_1, x_2, x_3)$ indicates the position of the point on the ray; and $\vec{p} = (p_1, p_2, p_3)$ is the gradient of the eikonal S along the ray. In Chap. 2 we have solved (2.1) in the x-space, which is also referred to as the configuration space. The solution is readily found in the form

$$\vec{x}\,(u, v, t) = \vec{x}\,(u, v, 0) + t\vec{p}\,(u, v, 0), \tag{6.2a}$$

$$\vec{p}\,(u, v, t) = \vec{p}\,(u, v, 0), \tag{6.2b}$$

where $\vec{x}\,(u, v, 0)$ and $\vec{p}\,(u, v, 0)$ denote the initial values of $\vec{x}$ and $\vec{p}$ at $t = 0$, and $\vec{x}\,(u, v, 0)$ is a parametric representation of the initial surface Γ upon which the eikonal is known. The vector $\vec{p}\,(u, v, 0)$ provides the direction of the outgoing ray at the point $\vec{x}\,(u, v, 0)$. We will assume that the norm of the surface gradient of the initial eikonal on Γ is less than 1 at all points on Γ so that $\vec{p} = (u, v, 0)$ is a real vector over the entire surface Γ.

As we have seen in Chap. 2, the solution of the transport equation enables us to calculate the first term of the asymptotic expansion of the solution of the wave equation. This solution is proportional to the inverse of the square root of K, where K is given by

$$K = \frac{D(\vec{x}(t))}{D(\vec{x}(0))} = \frac{J(t)}{J(0)}, \tag{6.3}$$

and J is the Jacobian that relates the ray coordinates to the rectangular coordinates defined in Chap. 2. As a consequence, the solution derived from geometrical optics becomes infinite when J vanishes, that is, on the caustics. We recall that if Γ is a wavefront W_0, or in other words if the initial phase is constant, then the caustic is the surface of the centers of curvature of W_0.

In Chap. 3 we have provided the physical explanation of this infinite result and have shown how we can find a solution, which is valid close to the caustic, by using the boundary layer method. This method is simple as well as directly applicable at a regular point of the caustic C, that is, at a running point on one of the two sheets of C. Its use becomes somewhat cumbersome on the cusp lines of these sheets, and even more on the confluent points of two cusp lines and at the umbilical points of C. Maslov [Ma] has proposed a radically different approach to circumventing this problem, which enables us to treat the particular points of C in an easier manner than that provided by the boundary layer method. The basic idea is still to start with Eqs. (6.1a) and (6.1b), and not to work in the usual space (x_1, x_2, x_3) but in the phase space $P(x_1, x_2, x_3, p_1, p_2, p_3)$. Such a choice is suggested by Eq. (6.2), which tells us $\vec{p}$ that the pair of variables $(\vec{x}(t), \vec{p}(t))$ is simply related to the pair $(\vec{x}(0), \vec{p}(0))$ through a translation; consequently the Jacobian $\{D(\vec{x}(t), \vec{p}(t))\}/\{D(\vec{x}(0), \vec{p}(0))\}$ equals 1. Thus the volume element defined in the space P is conserved. Hence, the cancellations of J, and, consequently, the caustics, are generated by the projection of P on the usual space. More precisely, Maslov introduces a *Lagrangian manifold* (Sect. 6.1.1.2), and the caustics are obtained at the critical points of the projection of this manifold on the usual space (Sect. 6.1.1.3). This prompts us to define and study the geometry of this manifold in the next section.

6.1.1.2 Concept of Lagrangian manifold

In this section, we develop the concepts of the Lagrangian manifolds. To this end, we define on P the canonical differential form Ω, given by

$$\Omega = dp_i \times dx_i, \tag{6.4}$$

where the summation on i is carried out in accordance with Einstein's convention. At all points of the phase space P, this form is an alternate bilinear form, which can be represented by the matrix

$$\begin{bmatrix} 0 & I \\ -I & 0 \end{bmatrix},$$

where I is the three-dimensional identity matrix.

Let us define L as a simply connected, three-dimensional manifold [MS], which has the following parametric representation

$$M(\vec{x}, \vec{\nabla} S(\vec{x})).$$

(6.5)

On this manifold L the differential form Ω can be expressed as

$$\Omega (M) = d \left(\frac{\partial S}{\partial x_k} \right) \times dx_k.$$

(6.6)

By using the relationship

$$d \left(\frac{\partial S}{\partial x_k} \right) = \sum_j \frac{\partial^2 S}{\partial x_j \partial x_k} dx_j,$$

(6.7)

(6.6) can be re-written in the form

$$\Omega(M) = \frac{\partial^2 S}{\partial x_j \partial x_k} dx_j \times dx_k.$$

(6.8)

Next, by invoking the equality

$$\frac{\partial^2 S}{\partial x_j \partial x_k} = \frac{\partial^2 S}{\partial x_k \partial x_j},$$

(6.9)

and

$$dx_j \times dx_k = dx_k \times dx_j,$$

(6.10)

we can show that the terms $j \neq k$ cancel each other. Finally, the terms with $j = k$ vanish since $dx_k \times dx_k = 0$.

We have thus shown that Ω vanishes on the manifold L, which is called Lagrangian. In practice, this manifold is determined by two vectors representing the position and the impulse, that is, it admits the parametric representation

$$(\vec{x} (u, v, 0) + t \vec{p}(u, v, 0), \vec{p}(u, v, 0)).$$

(6.11)

We recall that (u, v) are the two parameters that define the position of a point on the surface Γ and t is the abscissa on a ray. L is then a two-parameter family (or congruence) of straight lines of the phase space. A particular straight line is obtained by fixing the values u and v and letting t run through R^+. The rays are the projections of these straight lines onto the configuration space.

6.1.1.3 Study of the Lagrangian manifold

The Lagrangian manifold L is three dimensional in nature. Let us now consider the projection Π of the phase space onto the configuration space R^3 restricted to L. It is given by

$$\Pi \begin{vmatrix} (\vec{x}, \vec{p}) \rightarrow \vec{x} \\ L \rightarrow R^3 \end{vmatrix}.$$

The projection Π, restricted to L, is a mapping of a three-dimensional manifold into another one of similar dimension. According to Sard's lemma, [GS] it is of rank-3 at a *generic* point. The points where the rank of Π is less than 3 constitute a submanifold of L called the singular submanifold VS of dimension 2.

The rank of Π at a point M is equal to the rank of its tangent mapping Π' at this point. According to (6.2), the tangent space to L at point M is spanned by the basis vectors $(\vec{e}_u, \vec{e}_v, \vec{e}_t)$

$$\vec{e}_u = \left(\left(\frac{\partial \vec{x}(0)}{\partial u} \right) + t \frac{\partial \vec{p}(0)}{\partial u}, \frac{\partial \vec{p}(0)}{\partial u} \right), \tag{6.12a}$$

$$\vec{e}_v = \left(\left(\frac{\partial \vec{x}(0)}{\partial v} \right) + t \frac{\partial \vec{p}(0)}{\partial v}, \frac{\partial \vec{p}(0)}{\partial v} \right), \tag{6.12b}$$

$$\vec{e}_t = (\vec{p}(0), \vec{0}). \tag{6.12c}$$

We can show that the vectors $(\vec{e}_u, \vec{e}_v, \vec{e}_t)$ form a basis set, except when the surface Γ is a set of rays. The projection Π will be of rank 3 if the vectors $(\Pi(\vec{e}_u), \Pi(\vec{e}_v), \Pi(\vec{e}_t))$ are linearly independent.

To simplify the calculations, let us consider the case in which Γ is a wavefront. Let (u, v) be the coordinate system of lines of curvature. For this case $(\partial \vec{x}(0)/\partial u) = \vec{t}_1$, $(\partial \vec{x}(0)/\partial v) = \vec{t}_2$

$$\vec{p}(0) = \vec{n}; \ \frac{\partial \vec{p}}{\partial u} = \frac{\vec{t}_1}{\rho_1}; \ \text{and} \ \frac{\partial \vec{p}}{\partial v} = \frac{\vec{t}_2}{\rho_2}, \tag{6.13}$$

where ρ_1 and ρ_2 are the principal radii of curvature, $\vec{i_1}$, $\vec{i_2}$ are the vectors tangent to the lines of curvature, and $\vec{n}$ is the vector normal to Γ at the point O from which the ray passing through $\Pi(M)$ emanates. The two last formulas in (6.13) are obtained from Rodrigue's formula, which provides the derivative of the normal vector with respect to the abscissa along a line of curvature. Using (6.12) and (6.13) we thus obtain

$$\Pi(\vec{e_u}) = \vec{i_1}\left(1+\frac{t}{\rho_1}\right),\tag{6.14a}$$

$$\Pi(\vec{e_v}) = \vec{i_2}\left(1+\frac{t}{\rho_2}\right),\tag{6.14b}$$

$$\Pi(\vec{e_t}) = \hat{n}.\tag{6.14c}$$

Consequently, Π is of rank 3, except if $t = -\rho_1$ or $t = -\rho_2$, that is, if $\Pi(M)$ is on the caustic. Thus, Π projects VS on the caustic.

At a point M of L whose projection on the caustic is where $\rho_1 \neq \rho_2$, in other words if M is not an umbilic, then Eq. (6.14) shows that the rank of Π equals 2.

Let us assume, for instance, that $t = -\rho_1$, or, in other words, that $\Pi(M)$ is on the first sheet of the caustic. In addition, let us assume that ρ_1 is finite. Let us choose $(\vec{i_1}, \vec{i_2}, \hat{n})$ as the coordinate axes in the phase space and consider the mixed projection Π_1, which projects the point $(x_1, x_2, x_3, p_1, p_2, p_3)$ onto the point $(0, x_2, x_3, p_1, 0, 0)$.

Accordingly, the projection of the basis vectors is given by $\Pi_1\ \vec{i_2}(\vec{e_u}) = (0, 0, 0, 1/\rho_1, 0, 0)$, $\Pi_1(\vec{e_v}) = (0, 1, 1+\frac{t}{\rho_2}, 0, 0, 0)$, $\Pi_1(\vec{e_t}) = (0, 0, 1, 0, 0, 0)$, so that the rank of Π_1 at point M equals 3. If M is an umbilic, we consider the mixed projection Π_{12} mapping the point $(x_1, x_2, x_3, p_1, p_2, p_3)$ to the point $(0, 0, x_3, p_1, p_2, 0)$. Accordingly,

$$\Pi_{12}(e_u) = (0, 0, 0, 1/\rho_1, 0, 0),$$

$$\Pi_{12}(e_v) = (0, 0, 0, 0, 1/\rho_2, 0),$$

$$\Pi_{12}(e_t) = (0, 0, 1, 0, 0, 0),$$

so that the rank of Π_{12} at point M is 3.

We now summarize the above results in the following:

(i) if $M \notin VS$, the projection of the Lagrangian manifold L on the configuration space is regular in a neighborhood of M

(ii) if $M \in VS$, and if its projection is a point where the two sheets of the caustic surface do not touch, there exists a mixed projection on a subspace of dimension-3 defined by one (impulse) and two spatial coordinates (x_1, x_2, x_3), which is regular in the neighborhood of M

(iii) if $M \in VS$, and if its projection is a point where the two sheets of the surface caustic are in contact, i.e., a point under an umbilic of a wavefront, there exists a mixed projection on a subspace of dimension-3 defined by two impulse coordinates and one position coordinate, regular in the neighborhood of M.

It is then possible to choose, at every point of L, a system of mixed coordinates regular on L. We note that this system of coordinates is not unique. For instance, it is possible to choose several systems of coordinates regular on L. We are now going to present a more rigorous demonstration of this important result [MS].

If $M \notin VS$, the coordinates (x_1, x_2, x_3) are regular on L, and the problem of finding a regular system of coordinates is solved.

Let us consider a point S of VS where the rank of Π equals 2. Thus Π projects the tangent space T to L on S in a plane Q. It is always possible (see Fig. 6.1) to choose two coordinates which are regular on this plane. Let us assume, for instance, that these coordinates are x_2 and x_3. It is logical to take p_1 as the impulse coordinate. We are going to show that the coordinate system (x_2, x_3, p_1) is regular in the neighborhood of S, which means that the projection Π_1 on the mixed subspace $M(0, x_2, x_3, p_1, 0, 0)$ is of rank-3 at S.

Let y be a vector of T belonging to the kernel of Π_1 and therefore having the form $y(x_1, 0, 0, 0, p_2, p_3)$. Since $y \in T$ it follows that its projection $\Pi(y) \in Q$. On the other hand, $\Pi(y) = (x_1, 0, 0)$ is then on the Ox_1 axis where O is the origin in the phase space Q. The intersection of Q and Ox_1 is the zero vector, and, hence, $x_1 = 0$.

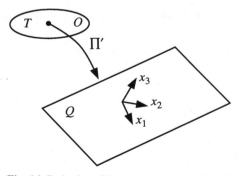

Fig. 6.1. Projection of the tangent space on plane Q

On the other hand, the second canonical form Ω vanishes on T, which means that if z is any vector of T, given by $z = (z_1, z_2, z_3, q_1, q_2, q_3)$, then

$$\Omega \,(y, z) = z_2 \, p_2 + z_3 \, p_3 = 0. \tag{6.15}$$

Equation (6.15) is valid for all (z_2, z_3); consequently, it implies that $p_2 = p_3 = 0$. Since it was demonstrated earlier that x_1 equals zero, it follows that $y(x_1, 0, 0, 0, p_2, p_3)$ must be zero. This leads us to conclude that the intersection of T and $Ker(\Pi_1)$, which is the null space of Π_1, is zero. The restriction of Π_1 to T is then an *isomorphism*, i.e., the coordinate system (x_2, x_3, p_1) on L is regular.

The case in which the rank of Π equals 1 is treated in the same manner. In this case, $\Pi(T)$ is a straight line D. We choose a coordinate axis which is not parallel to the straight line. Let us assume, for instance, that x_1 satisfies this condition. We can show that (x_1, p_2, p_3) represents a regular coordinate system on L at S. Note that the regular coordinate system is not unique. For instance, for the case in which the rank of Π, *viz.*, $rg(\Pi) = 1$, and if D is not parallel to any of the axes, it is possible to choose any of the coordinates x_1, x_2 or x_3.

We are now going to provide some examples which will help the reader to visualize the Lagrangian manifold.

Let us start with the simple case of one dimension. The general Lagrangian manifold is then a curve (a straight line if one considers the wave equation for an homogeneous medium), and VS is a set of points. The projection of this curve on x is regular, except at the collection of points that comprise VS, where the tangent to the curve becomes vertical. At these points, x is no longer a regular coordinate, although p is. Note that a point with abscissa x can correspond to several points of L. The manifold L, shown in Fig. 6.2, is the union of two curves $(x, p^+ (x))$ and $(x, p^- (x))$. We observe then that L possesses two branches and that they are attached to each other at the point S, which is the singular manifold in this particular case. It should be pointed out, however, that L has only one branch in the p-coordinate system.

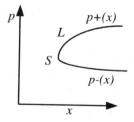

Fig. 6.2. Lagrangian manifold

In the n-dimensional configuration (or usual) space, the Lagrangian manifold L is a surface of dimension n inside a phase space of dimension $2n$. While it is impossible to represent the phase space completely once $n \geq 2$, the concepts presented earlier for the one-dimensional case continue to remain valid. There may exist, above the same point P of the configuration space, several points M of L. M is then a multivalued function of P, or, equivalently, L possesses several branches, which are connected on VS. In the configuration space, each point M corresponds to a ray passing through P. These rays coalesce on the projection of VS, i.e., on the caustic. It is possible to visualize L in two dimensions, by considering for instance the projection L' of L onto the space of dimension 3 described by (x_1, x_2, p_1). L' is the ruled surface defined as $(x_1(s) + tp_1(s), x_2(s) + tp_2(s), p_1(s))$ with $p_1^2(s) + p_2^2(s) = 1$. L' can display *folds* or *pleats*. Their projection on the configuration space will be, respectively, the caustic curve or the cusp points of this curve [ZD].

We now have the essential concepts necessary to understand the foundations of the Maslov method and we proceed next to the description of this method. We will begin with the case in which the rank of the projection-Π equals two. As we will see, it is then possible to obtain a representation of the solution in the form of a simple integral.

6.1.2 Representation by means of a simple integral

The Maslov method is based upon the Fourier transformation of spatial to spectral variables. In this section, we will treat the case in which the transformation is $x_1 \to p_1$. To simplify the formulas, Maslov redefines the Fourier transform as follows. Let the functions $u(x) = u(x_1, x_2, x_3)$ and $v(y) = v(p_1, x_2, x_3)$ be the Fourier transform pair. Then, we have

$$v(y) = \left(-\frac{ik}{2\pi}\right)^{1/2} \int u(x) \exp(-ikx_1p_1)\, dx_1, \tag{6.16a}$$

$$u(x) = \left(\frac{ik}{2\pi}\right)^{1/2} \int v(y) \exp(ikx_1p_1)\, dp_1. \tag{6.16b}$$

Equation (6.16b) provides an integral representation of u if v is known.

The Maslov method essentially entails the replacement of the function v in (6.16b) by an approximation obtained by invoking the stationary phase approximation to the integral in (6.16a), as shown below.

The geometrical optics approach, which is based on the Luneberg-Kline series and was discussed in Chap. 2, provides a solution of the diffraction problem in the form

$$u(x) = A(x) \exp (ikS(x)) + O(1/k). \tag{6.17}$$

Using (6.17) in (6.16a) we obtain

$$v(y) = \left(-\frac{ik}{2\pi}\right)^{1/2} \int A(x) \exp (ik(S(x) - x_1 p_1) \, dx_1, \tag{6.18}$$

where the terms neglected are $O(1/k)$. Next we compute (6.18) via the stationary phase integral method. The stationary phase points x_{1s} are given by

$$\frac{\partial S}{\partial x_{1s}} = p_1, \tag{6.19}$$

where x_{1s} is a function of $(p_1, x_2, x_3) = y$.

The second derivative of the phase, at the point where it is stationary, can be written as

$$k \frac{\partial^2 S}{\partial x_{1s}^2} = k \frac{\partial p_1}{\partial x_{1s}}. \tag{6.20}$$

By evaluating the integral in (6.18) using the stationary phase method, we obtain

$$v(y) \approx d\, A(x_s) \left|\frac{\partial x_{1s}}{\partial p_1}\right|^{1/2} \exp ik(S(x_s) - x_{1s} p_1) \tag{6.21}$$

where we have set $x_s = (x_{1s}, x_2, x_3)$ and $d = 1$ if $(\partial x_{1s}/\partial p_1)^{1/2} > 0$, and $d = -i$ if $(\partial x_{1s}/\partial p_1)^{1/2} < 0$.

This result has been obtained in an alternate way by Kravtsov [Kr], who, however, makes the distinction between the two cases associated with the sign of $(\partial x_1/\partial p_1)$. As a result, we must choose $\sqrt{-1} = -i$ if we want to recover the result given in (6.21) from Kravtsov's formula.

Equation (6.21) is exact to within the terms of $O(1/k)$. The second term in the right-hand side of (6.21) is then the first term v_0 of the asymptotic expansion of v in terms of the powers of $1/k$. Equation (6.21) thus provides v_0 as a function of u_0, and, hence, is designated the *asymptotic Fourier transform at zero order*. This transformation has been studied by Ziolkowski and Deschamps for zero and higher orders [ZD]. Thus, v_0 can be written in the form of $v_0(y) = d\, B(y)\, \exp(ikT(y))$. Additionally, we observe that:

(i) the phase $T(y)$ equals $S(x_s) - p_1\, x_{1s}$, which is the Legendre transform of the phase S of u_0 with $x_1 \to p_1$.

(ii) $B(y) = A(x_s)\, |(\partial x_{1s}/\partial p_1)|^{1/2}$. If A and B were two expressions of the same densities represented with the coordinates x and y, respectively, on the Lagrangian manifold, they would be related by $B(y) = A(x_s)\, (D(x_1, x_2, x_3)/D(p_1, x_2, x_3)) = A(x_s)\, |(\partial x_{1s}/\partial p_1)|$. Because of the square root appearing in the expression of B given above, we will refer to A and B as half-densities.

(iii) $A(x_s)$ is proportional to $J^{-1/2}$, where $J = (D(x_1, x_2, x_3)/D(\xi, \eta, t))$, and (ξ, η, t) are the ray coordinates (ξ, η characterize a ray, and t is the abscissa on this ray). Consequently, $B(x_s)$ is proportional to $K = (D(p_1, x_2, x_3)/D(\xi, \eta, t))^{-1/2}$, and it remains finite if (p_1, x_2, x_3) are regular coordinates on the Lagrangian manifold. We have seen above that it is always possible to choose such a coordinate system and then to obtain a bounded amplitude for v from (6.16).

In order to illustrate this aspect, we consider the coordinate system attached to the wavefront we have used in Sect. 6.1.1.3. In that system, the ray coordinates satisfy $\xi = s_1, \eta = s_2$, where s_1 and s_2 are the curvilinear abscissas on two curvature lines of the wavefront, chosen as the coordinate axes (see Fig. 6.3).

Let us assume that the two radii of curvature of the wavefront, viz., ρ_1 and ρ_2, are finite. Furthermore, let us choose the point 0 located on the wavefront as the origin of the coordinates such that the ray passing through the point M where one calculates the field emanates from 0. As we have seen in Sect. 6.1.1.3,

$$\frac{\partial \overrightarrow{OM}}{\partial s_1} = \hat{t}_1 + t\frac{\hat{t}_1}{\rho_1}, \quad \frac{\partial \overrightarrow{OM}}{\partial s_2} = \hat{t}_2 + t\frac{\hat{t}_2}{\rho_2}, \quad \frac{\partial \overrightarrow{OM}}{\partial t} = \hat{n},$$

so that

$$K = \frac{J(t)}{J(0)} = \left(1 + \frac{t}{\rho_1}\right)\left(1 + \frac{t}{\rho_2}\right).$$

From (6.19), and the expression of S in the vicinity of M, we obtain $\partial x_{1s}/\partial p_1 = \rho_1 + t$. The quantity B, which is proportional to $((1/K)\,(\partial x_{1s}/\partial p_1))^{1/2} = (\rho_1/1 + (t/\rho_2))^{1/2}$, then remains finite on the caustic $t = -\rho_1$.

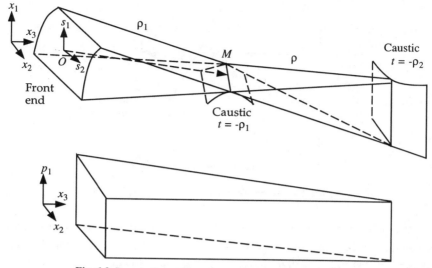

Fig. 6.3. Rays in the configuration space and in the mixed space

Figure 6.3 depicts the rays in the neighborhood of M, both in the configuration space and in the mixed space (p_1, x_2, x_3). We note that only the caustic at $x_3 = -\rho_2$ exists in this mixed space, while the caustic at $x_3 = -\rho_1$ is not present in this space, which explains why B remains bounded there.

Next, we observe that the axis Ox_1 is orthogonal to the caustic. Consequently, the practical rule is to choose as an axis Ox_1, the normal (or more generally a direction which is not tangent) to the caustic at the point in whose vicinity we wish to calculate the field. It is then possible, with the correct choice of the axis Ox_1, to obtain a finite value for v_0 given by (6.21). If we now insert (6.21) in (6.16b), we obtain the following integral representation of the solution, again to within $O(1/k)$

$$u(x) \approx \left(\frac{ik}{2\pi}\right)^{1/2} \int d \left|\frac{\partial x_{1s}}{\partial p_1}\right|^{1/2} A(x_s) \exp ik(S(x_s) - p_1 x_{1s} + p_1 x_1) \, dp_1. \qquad (6.22)$$

In this integral representation, x_{1s} and then x_s must be expressed as functions of p_1. In practice, according to (6.19) and for fixed value of x_2 and x_3, we determine the ray (or the rays) whose projection $\partial S/\partial x_1$ of the direction along the x_1 axis equals p_1. The ray passes through the point x_2, x_3, $x_{1s}(p_1)$, which determines x_{1s} as a function of p_1. Figure 6.4 illustrates the method in two dimensions.

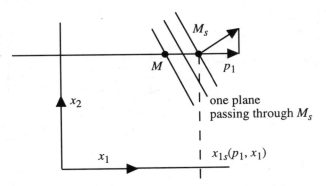

Fig. 6.4. Calculation of point M using the Maslov method

As pointed out by Arnold [Ar], it is possible to provide a very intuitive interpretation of the result in (6.22). Let us draw the straight line passing through the point $M(x_1, x_2, x_3)$ and parallel to the axis $0x_1$. The point $M_s(x_{1s}(p_1), x_2, x_3)$ is a point on this straight line such that the projection on $0x_1$ of the ray direction passing through M_s is p_1. The eikonal of the geometrical optics at the point M_s is $S(x_s)$, the gradient of the eikonal is $\vec{p}(p_1, p_2, p_3)$, and the amplitude is $A(x_s)$.

These three quantities define a plane wave passing through the point M_s. The field of this plane wave at point M is given by

$$u_p(x) = A(x_s) \exp(ik(S(x_s) + \vec{p} \cdot \overrightarrow{MM}_s)). \qquad (6.23)$$

But, since

$$\vec{p} \cdot \overrightarrow{MM}_s = p_1 x_1 - p_1 x_{1s}, \qquad (6.24)$$

we can write

$$u_p(x) = A(x_s) \exp ik(S(x_s) - p_1 x_{1s} + p_1 x_1), \qquad (6.25)$$

and one recognizes that the right-hand side of (6.25) appears under the integral sign in (6.22). In (6.22), the quantity $|(\partial x_{1s}/\partial p_1)|^{1/2}$ is a correction to the amplitude, while d is a phase correction which enables one to obtain the same phase for the ray whether or not it has crossed a caustic.

In physical terms, in the case of a one spectral parameter, the Maslov method consists of replacing by plane waves those rays that intersect a straight line passing through the point where the field is to be computed. Each point of the straight line gives rise to a plane wave; hence, we subsequently integrate along the straight line to obtain the total field. It is possible to show [Ar] that the application of the stationary phase method to the above integral recovers the geometrical optics field.

Before describing an application of (6.22) to the field computation near the caustic, we would like to make a few observations. First, the range of integration is never specified in the integrals appearing in (6.22). This problem is somewhat intricate as the range of p_1 is between -1 and 1, for real values of p_2 and p_3. The question we might ask is: Must we restrict the values of p_1 within the interval (-1 and 1), or extend the range of the integration on p_1 over the entire R by considering complex values of p_2 and p_3? This problem has not been completely resolved as yet. However, the value of (6.22) depends essentially on the amplitude and the phase of the integrand in a small neighborhood of the stationary phase points, defined by the directions of the rays of geometrical optics. In practice, we restrict the range of the integration in (6.22) to this neighborhood, for instance, by means of a regular function which has a *compact* support. Sometimes the expression *neutralizer* is used in the literature to indicate such a procedure.

Second, if (6.22) is limited to a bounded domain by means of a neutralizer, the integral exists as long as the integrand is bounded and, consequently, for all the points of the caustic surface, including the cusp edges and the intersections of the cusp edge (swallow tail). Thus, (6.22) provides a solution not only at the running point of the caustic, but also at all previous *singular* points. At the points where the two sheets of the caustic meet (i.e. the umbilics), it becomes necessary to employ an integral representation defined with two spectral parameters (see Sect. 6.1.3).

Third, if a number of rays of different origin pass through M, i.e., if L possesses several branches, then the result is a sum of several integrals of the same type as (6.22) in the coordinate system (p_1, x_2, x_3).

Let us now discuss a concrete example of the Maslov method. Using this method we are going to treat the case of the caustic in two dimensions, a problem that has already been addressed in Chap. 3 with the boundary layer method. As an illustration, let us assume that the caustic is circular and the amplitude is independent of the angle on the circle. We will first compute the various quantities appearing in (6.22).

We choose the coordinates as shown in Fig. 6.5 to obtain

$$p_1 = \sin \theta, \tag{6.26}$$

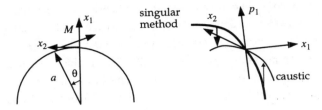

Fig. 6.5. Projection of the singular manifold on the caustic

$$x_{1s} = \frac{a}{\cos\theta} - a. \qquad (6.27)$$

Hence,

$$\frac{\partial x_{1s}}{\partial p_1} = a\frac{\sin\theta}{\cos^3\theta}, \qquad (6.28)$$

$$S(x_{1s}) = -a\theta + atg\,\theta \quad \text{and} \quad S-p_1 x_{1s} = a(\sin\theta - \theta), \qquad (6.29)$$

and $A(x_{1s}) = -(iC/\sqrt{tg\theta})$, if $\theta > 0$, since the ray has intersected the caustic, and $A(x_{1s})$ $= (C/\sqrt{|tg\theta|})$, if $\theta < 0$. The quantity C is a constant. However, d, which is a multiplicative factor of A in (6.22), equals 1 if $\theta > 0$ since $(\partial x_{1s}/\partial p_1) > 0$; and, it is equal to $-i$ if $\theta < 0$ since $(\partial x_{1s}/\partial p_1) < 0$. Hence, we can write

$$dA(x_{1s}) = -(iC/\sqrt{|tg\theta|}), \quad \text{for all } \theta. \qquad (6.30)$$

We see how d *compensates* for the phase shift at the caustic. Inserting (6.26)-(6.30) into (6.22), we obtain

$$u(M) \approx Ce^{-i\pi/4}\left(\frac{ka}{2\pi}\right)^{1/2} \int \cos^{-1}\theta \, \exp\, ik(a(\sin\theta - \theta)+ x_1 \sin\theta)\, d\theta. \qquad (6.31)$$

Next, we insert into (6.31) the small argument approximations of $\cos\theta$ and $\sin\theta$ to obtain

$$u(M) \approx Ce^{-i\pi/4}\left(\frac{ka}{2\pi}\right)^{1/2} \int \exp\, ik\left(-a\frac{\theta^3}{6} + x_1\theta\right) d\theta. \qquad (6.32)$$

After some simplification, the above reduces to

$$u(M) \approx 2\sqrt{\pi}\, e^{-i\pi/4}\left(\frac{ka}{2\pi}\right)^{1/6} Ai(-\nu)\, C, \qquad (6.33)$$

where $\nu = (kx_1/m)$ is the variable defined in Sect. 3.1.4 (refer to (3.8.a) of Sect. 3.1.4).

In (3.9), of Sect. 3.1.4, let us set $\rho = a$ and $B(s) = \sqrt{a}\ C$, implying that the caustic is circular and the amplitude is constant. If we insert the $A(s)$ obtained in (3.119) of Sect. 3.4, we obtain the expression in (6.33). Thus we observe that the Maslov method enables us to extract the result calculated with the boundary layer method in a direct manner.

The previous integral representation provides a solution which is valid everywhere on the surface of the caustic, including the cusp lines and their neighborhood. However, at points where the two sheets of the caustic surface intersect, the rank of Π, which projects L on the configuration space, is equal to 1, and we must resort to a representation of the solution in the form of a double integral with respect to two spectral variables. This is discussed further in the next section.

6.1.3 Representation by means of a double integral

To discuss the special case of contact points of the two sheets of a caustic surface, we will choose p_1 and p_2 as the spectral variables. We then follow an approach which is exactly the same as the one we have used in the previous section. We begin with the integral representation of the solution in the form

$$u(x) = \frac{ik}{2\pi} \int v(y) \exp(ik(x_1\, p_1 + x_2\, p_2))\, dp_1\, dp_2, \qquad (6.34)$$

where $y = (p_1, p_2, x_3)$ and v is the two-dimensional Fourier transform of u with $(x_1, x_2) \to (p_1, p_2)$. Next, we approximate v by replacing u with the first term of its asymptotic expansion in terms of the inverse powers of the wave number k to obtain

$$v(y) \approx -\frac{ik}{2\pi} \int A(x) \exp(ik(S(x) - p_1\, x_1 - p_2\, x_2))\, dx_1\, dx_2. \qquad (6.35)$$

The next step is to apply the method of stationary phase to (6.35). The stationary phase points x_{1s}, x_{2s} satisfy the conditions

$$\frac{\partial S}{\partial x_{is}} = p_i, \quad \text{where} \quad i = 1, 2, \qquad (6.36)$$

and x_{is} is a function of $(p_1, p_2, x_3) = y$.

At a stationary point, the matrix composed of the second derivative of the phase, called the Hessian, is given by

$$k \frac{\partial^2 S}{\partial x_{is} \partial x_{js}} = k \frac{\partial p_j}{\partial x_{is}}. \qquad (6.37)$$

Assuming that the stationary phase formula still applies, and by setting $x_s = (x_{1s}, x_{2s}, x_3)$, we obtain

$$v(y) \approx dA(x_s) \left| \frac{\partial x_{is}}{\partial p_j} \right|^{1/2} \exp ik(S(x_s) - x_{1s}p_1 - x_{2s}p_2), \qquad (6.38)$$

where $|(\partial x_{is}/\partial p_j)|$ is the Jacobian corresponding to the change of variables $x_i \rightarrow p_i$. Also, d equals 1, $-i$, or -1 depending upon whether the Hessian has two, one, or zero positive eigenvalues, respectively. As in the previous section, (6.38) leads to an integral representation

$$u(x) \approx \frac{ik}{2\pi} \int dA(x_s) \left| \frac{\partial x_{is}}{\partial p_j} \right|^{1/2} \exp ik(S(x_s) - x_{1s}p_1 - x_{2s}p_2 + x_{1s}p_1 + x_{2s}p_2) \, dp_1 \, dp_2.$$

$$(6.39)$$

The phase of the integrand is the Legendre transform of the phase given by geometrical optics, although with respect to two variables. In practice, for a fixed x_3, one determines the ray vector (or vectors) whose projection equals (p_1, p_2). It passes through the point $M_s(x_{1s}(p_1, p_2), x_{2s}(p_1, p_2), x_3)$ which determines x_{1s} and x_{2s} as functions of p_1, p_2. Figure 6.6 illustrates the method in three dimensions.

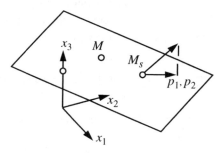

Fig. 6.6. Maslov method in three dimensions

The physical interpretation of the result is the same as given in the previous section. One replaces the ray passing through M_s by a plane wave propagating in the direction of the ray. As in the previous section, the application of the stationary phase in (6.39) recovers the geometrical optics result. These two remarks naturally suggest the following: Rather than working with a coordinate system passing through M, it must be possible to generate, directly from the ray result, an integral of the type in (6.39). This circumvents the need to introduce the rectangular coordinate system for deriving the field, as this system is not convenient in certain cases. This idea has been used by Arnold to prepare the so-called *spectral reconstruction* method.

Next, we describe the method of spectral reconstruction due to Arnold, which is described in the following section.

6.1.4 Method of spectral reconstruction

In practice, the application of the method of spectral reconstruction involves the steps outlined below. Consider the rays in the neighborhood of M where we wish to calculate the diffracted field. With each ray passing through N close to M, we associate a plane wave propagating in the direction $\hat{p}$ of the ray passing through N with the eikonal $S(\hat{p})$ of geometrical optics at point N, and an as yet undetermined amplitude $A(\hat{p})$ (see Fig. 6.7). We seek a solution at point M in the form

$$u(M) = \int_V A(\hat{p}) \exp(ik(S(\hat{p}) + \hat{p} \cdot \vec{NM})\, d\hat{p}. \qquad (6.40)$$

The integral is performed in the angular range in the neighborhood V of the direction of the ray passing through M.

plane wave passing through N

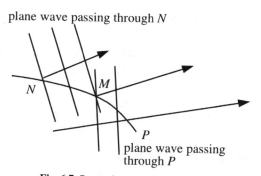

P
plane wave passing
through P

Fig. 6.7. Spectral reconstruction method

The points N can be located on a plane passing through M, and one reverts to the Maslov method, or more generally, on an arbitrary surface close to M which provides more flexibility to the method. $A(\hat{p})$ is then calculated by applying the method of stationary phase to (6.40) in a manner such that the solution becomes the same as that given by the geometrical optics. The justification of the method and some examples can be found in the publications by Arnold [Ar]. We now illustrate the method by means of an example for a point in the neighborhood of the cusp point R of the caustic line, in two dimensions.

Following Pearcey [Pe], we define the caustic C in the vicinity of R via the equation (with $x \to -x$ as compared to [Pe])

$$y = \left(\frac{8}{9\sigma}x^3\right)^{1/2} = 2\frac{\sqrt{2}}{3}\frac{x^{3/2}}{\sigma^{1/2}}. \tag{6.41}$$

The rays are tangent to C and the ray vector passing through $N(x, y(x))$ is $\hat{p}$

$$\hat{p}\left((1+2x/\sigma)^{-1/2}, \left(\frac{2x/\sigma}{1+2x/\sigma}\right)^{1/2}\right). \tag{6.42}$$

The eikonal at N equals s, where s is the curvilinear abscissa on C measured from R

$$s = \int_0^x dx\left(1+\frac{2x}{\sigma}\right)^{1/2}. \tag{6.43}$$

The eikonal $S(\hat{p})$ of the plane wave generated by the ray passing through N at the point $M(x_0, y_0)$ is given by

$$S(\hat{p}) = s + p_x(x - x_0) + p_y(y - y_0), \tag{6.44}$$

where (x, y, x_0, y_0) are small quantities of the same order. Next, we perform a Taylor expansion, retaining only the terms up to $O(x^2)$ to obtain

$$S = x_0 + \sqrt{\frac{2x}{\sigma}}\, y_0 - \frac{xx_0}{\sigma} + \frac{1}{6}\frac{x^2}{\sigma} + O(x^2). \tag{6.45}$$

The integral representation then takes the form

$$\exp(ik\, x_0) \int B(x) \exp ik\left(\sqrt{\frac{2x}{\sigma}}\, y_0 - \frac{xx_0}{\sigma} + \frac{1}{6}\frac{x^2}{\sigma}\right) dx. \tag{6.46}$$

We now introduce a change of the variables $t^4 = (k/6)\,(x^2/\sigma)$ to obtain the following expression for the phase φ in the integrand

$$\varphi = \left(\frac{24}{k\sigma}\right)^{1/4}(k\, y_0)\, t - \left(\frac{6}{k\sigma}\right)^{1/2}(k\, x_0)\, t^2 + t^4. \tag{6.47}$$

Setting

$$Y = \left(\frac{24}{k\sigma}\right)^{1/4} k\, y = \left(\frac{192\pi^3}{\sigma\lambda^3}\right)^{1/4} y, \tag{6.48}$$

$$X = -\left(\frac{6}{k\sigma}\right)^{1/2} k\, x = \left(\frac{12\pi}{\sigma\lambda}\right)^{1/2} x, \tag{6.49}$$

as in Pearcey [Pe], we obtain

$$\varphi = Yt + X\, t^2 + t^4, \tag{6.50}$$

in a deductive manner, whereas it was presented only heuristically in [Pe]. Depending upon the values of X and Y, there can be either one or three values of t that cancel $d\varphi/dt$, implying that there is either one or three stationary phase points in (6.46). In other words, the application of the method of the stationary phase to the integral representation in (6.46) will yield either one or three rays depending upon whether the point is outside or inside the beak. The difficulties in geometrical optics arise from its application in the neighborhood of the point R, where these three points coalesce at the point $t = 0$. The application of the boundary layer method to this problem would be rather involved, because of the coalescence of several layers (see Fig. 6.8).

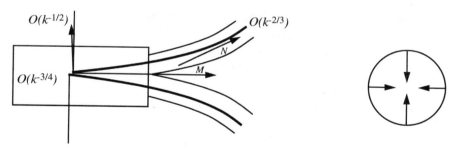

Fig. 6.8. Caustic beak   **Fig. 6.9.** Focus

Another simple example of the application is the case of the focus (see Fig. 6.9). The eikonal of the plane wave generated by the ray passing through N is given by

$$S = -r\, cos\, \theta, \tag{6.51}$$

which leads, if all the rays have the same amplitude, to an integral representation in the form

$$\int_0^\Pi exp\, (-ikr\, cos\, \theta)d\theta, \tag{6.52}$$

and we obtain a Bessel function $J_0(kr)$ for the characteristic function of the foci. If, on the other hand, only the rays emanating from an angular sector reach the focus, the field representation is an incomplete Bessel function.

6.1.5   An alternate approach to handling the caustic problem in the context of the Maslov method

Previously we have used the Fourier transform technique in conjunction with the method of stationary phase to handle the caustic problem. It is also possible to derive the same result by using a slightly different approach, which we will now discuss for the case where the spectral variable is p_1.

Let us Fourier transform the Helmholtz equation in the x_1 variable, with the transform variable p_1 as defined by (6.16.a), to obtain

$$(-k^2 p_1^2 + \nabla_1^2)\, v + k^2\, v = 0, \tag{6.53}$$

where ∇_1^2 denotes the *transverse* Laplacian involving only the variables x_2 and x_3.

Let us now attempt to solve (6.53) by expanding v in the form of a Lüneberg-Kline series

$$v = \exp\,(ik\,T(y))\!\left(v_0 + \frac{v_1}{ik} + \ldots\right). \tag{6.54}$$

Inserting (6.54) into (6.53), and ordering the result according to the powers of k, just as in Chap. 2, we are led to an *eikonal* equation

$$p_1^2 + (\nabla_1\,T)^2 = 1, \tag{6.55}$$

and a sequence of transport equations. The first one of these takes the form

$$2(\vec{\nabla}_1\,v_0) \cdot (\vec{\nabla}_1\,T) + v_0\,\nabla_1^2\,T = 0, \tag{6.56}$$

where $\vec{\nabla}_1$ denotes the transverse gradient.

The Legendre transform S_L of the eikonal S of geometrical optics with the transform pair, $x_1 \to p_1$, has the following partial derivatives

$$\frac{\partial S_L}{\partial p_1} = -x_1, \quad \frac{\partial S_L}{\partial x_2} = \frac{\partial S}{\partial x_2}, \quad \frac{\partial S_L}{\partial x_3} = \frac{\partial S}{\partial x_3}. \tag{6.57}$$

Hence, it satisfies the eikonal equation (6.55). The phase $S_L = S(x_1) - x_{1s} p_1$ of v_0 is given by (6.21), and we see by substituting S_L for T that, in view of (6.57), S_L satisfies the eikonal equation (6.55).

We now turn to the transport equation (6.56). Noting that p_1 is constant along a ray, and $\nabla_1 T = \nabla_1 S$, $\nabla^2_1 T = \nabla^2_1 S$, we can show that for fixed p_1, (6.56) then reduces to a transport equation in two dimensions expressed in terms of the variables x_2 and x_3. The solution of this equation is proportional to the $-1/2$ power of the Jacobian $(D(x_2, x_3)/D(\eta, t))$, where (p_1, η, t) is the ray coordinate system. But $p_1 = \xi$ is constant on a ray

$$(D(x_2, x_3)/D(\eta, t)) = (D(p_1, x_2, x_3)/D(p_1, \eta, t)).$$

We find by comparison that the Jacobian K is identical to that obtained in Sect. 6.1.2. If we use the coordinates (s_1, s_2, t) attached to a wavefront, we can show that (6.56) becomes

$$2\frac{dv_0}{dt} + \frac{v_0}{\rho_2 + t} = 0, \tag{6.58}$$

from which it follows that

$$v_0(t) = v_0(0)\sqrt{\frac{\rho_2}{\rho_2 + t}}, \tag{6.59}$$

where $\sqrt{\rho_2/(\rho_2 + t)}$ is the divergence in the mixed space (p_1, x_2, x_3) of the mixed space rays. The amplitude v_0 is then simply obtained by applying the laws of geometrical optics in mixed space. It remains to calculate the constant v_0 at $t = 0$, which can be achieved by calculating u with the method of the stationary phase and forcing the result to reproduce that given by the geometrical optics [Ar]. Indeed, we revert back to the same result as in Sect. 6.1.2 if we do this. Finally, the case of two spectral variables can also be handled in the same manner.

In conclusion, the Maslov method, based on the asymptotic Fourier transform discussed in Sects. 6.1.2 and 6.1.3, can also be interpreted as a generalization of the geometrical optics in a mixed space involving one or two spectral coordinates.

6.1.6 Extension to Maxwell's equations

In all of the discussion presented above, we have only been dealing with the wave equation although the method can be easily extended to the case of Maxwell's equations since each component of the $\vec{E}$ and $\vec{H}$ fields satisfies the wave equation. However, it is necessary to satisfy the divergence equations, viz., div $\vec{E}$ = div $\vec{H}$ = 0.

By transposing (6.22), to Maxwell's equations, we obtain for the case of a single spectral variable,

$$\vec{E}\ (x) \approx \left(\frac{ik}{2\pi}\right)^{1/2} \int d \left|\frac{\partial x_{1s}}{\partial p_1}\right| \vec{E}_0\ (x_s)\ \exp\ ik(S(x_s) - p_1 x_{1s} + p_1 x_1)\ dp_1. \qquad (6.60)$$

The term of highest order in k in div $\vec{E}$ is written as

$$div\,\vec{E} \approx \left(\frac{ik}{2\pi}\right)^{1/2} ik \int d \left|\frac{\partial x_{1s}}{\partial p_1}\right| \vec{E}_0\ (x_s)\ (\vec{\nabla}\ (S(x_s) - p_1 x_{1s} + p_1 x_1)$$

$$\exp\ ik(S(x_s) - p_1 x_{1s} + p_1 x_1)\ dp_1, \qquad (6.61)$$

but

$$\vec{\nabla}\ (S(x_s) - p_1\,x_{1s}) = \vec{\nabla}_1\,S(x_s), \qquad (6.62)$$

and

$$\vec{\nabla}\ (p_1 x_1) = p_1 \hat{e}_1 = \frac{\partial S}{\partial x_1}\,\hat{e}_1, \qquad (6.63)$$

so that

$$\vec{\nabla}\ (S(x_s) - p_1 x_{1s} + p_1 x_1) = \vec{\nabla}\,S(x_s). \qquad (6.64)$$

However, since

$$\vec{\nabla}\,S(x_s) \cdot \vec{E}_0(x) = 0,$$

we conclude that the highest-order term in k of div $\vec{E}$ vanishes.

Thus, (6.60) provides a solution that satisfies the divergence condition, viz., div $\vec{E} = 0$, with the same approximation as in geometrical optics. The demonstration that the representation in (6.39), which is in the form of a double integral, also satisfies the divergence condition follows along similar lines.

Thus, we have shown that the results established for the scalar wave equation extrapolate to the case of Maxwell's equations, and we need only use the vectorial solution of geometrical optics in the integral representations in (6.22) and (6.39).

We will now apply this result to the case of circular caustics, by assuming that the field $\vec{E}$ is in the plane of the caustic. The component of $\vec{E}$ along x_1 is still given by (6.32). However, a component of $\vec{E}$ now appears along x_2, and is proportional to

$$C e^{-i\pi/4} \left(\frac{ka}{2\pi} \right)^{1/2} \int \theta \, exp \, ik \left(-a \frac{\theta^3}{6} + x_1\theta \right) d\theta \,,$$

or

$$2\sqrt{\pi} \; e^{-i\pi/4} \left(\frac{ka}{2\pi} \right)^{-1/6} Ai(-v) \, C.$$

Therefore, a nonvanishing component of the electric field in the direction of the ray appears in the neighborhood of the caustic. The Maslov method provides a direct demonstration of this physical aspect of the field behavior in the neighborhood of the caustics.

6.1.7 Limitation of Maslov's method

As we have just seen, the Maslov method is very efficient for calculating the field in the neighborhood of the caustics, as long as they are located at finite distances. However, in the event that the caustics recede to infinity, i.e., if one of the radii of curvature, viz., either ρ_1 or ρ_2, of the wavefront becomes infinite, then the above method fails. As an example, let us consider the simple case of the reflection from a planar plate, located in the plane 0, x_1, x_2 (see Fig. 6.10).

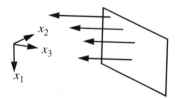

Fig. 6.10. Reflection from a plate

The space tangent to the Lagrangian manifold of the reflected field is generated by the three basic vectors $\hat{e}_1$, $\hat{e}_2$, $\hat{e}_3$. It is then of rank-3 and the method predicts that the result of geometrical optics is valid, which, however, is not true.

More generally, in the vicinity of a caustic, the choice of the coordinate p_1, normal to the caustic, leads to a factor $\sqrt{\rho_1}$ in the solution given in (6.22) in Sect. 6.1.2, where ρ_1 is the radius of curvature of the wavefront. Consequently, if the caustic is at infinity, i.e., if $\rho_1 = \infty$, the expression in (6.22) will yield an infinite result even at a finite distance.

6.2 Integration on a wavefront

As we saw in the previous section, Maslov's method provides an integral representation of the solution which is valid close to the caustics. In a homogeneous medium, a simpler method exists which has been available for some time and which enables one to construct the solution in the form of an integral. It is based on the concept that the geometrical optics provide the solution over the entire wavefront sufficiently far from the caustic and that the solution in the neighborhood of the caustic can be derived via an application of the Kirchoff formula to the wave equation on the wavefront. Thus, we obtain the solution in three dimensions, in the form of a double integral, which can be reduced to a simple integral in certain cases via an application of the stationary phase method. To illustrate this approach we will only consider the case of the scalar wave equations and merely mention that the case of Maxwell's equations can be treated in a similar manner and the results are qualitatively the same. We begin in Sect. 6.2.1 with an examination of the geometry of the caustic in a homogeneous medium, which is the surface of the centers of the wavefront. This will enable us to identify the particular points of the wavefront that correspond to the singular points of the caustic. Next, by asymptotically evaluating the integrals representing the field close to the caustic, we will show in Sect. 6.2.2 that these integrals can be expressed in terms of generic functions that appear in the *catastrophe theory*. For the sake of completeness we will provide, for the simplest cases, the arguments of these functions by means of the geometric parameters characterizing the wavefront.

6.2.1 Geometry of the surface of the centers of wavefront

To investigate the geometry of the wavefront center of curvature surface, we begin with the two-dimensional case for which a point N of the line of the centers LC is given by

$$\vec{ON} = \vec{OM} + R\vec{n} , \tag{6.65}$$

where R is the radius of curvature at point M of the wavefront W, and $\vec{n}$ is the normal to W at M. Let s be the curvilinear abscissa on W, then

$$\frac{d\vec{ON}}{ds} = \frac{dR}{ds}\vec{n} , \tag{6.66}$$

and the tangent to LC is $\vec{n}$. The velocity on LC is $(dR/ds) = R'$. Consequently, this velocity vanishes at the point where R reaches an extremum, which could either be its minimum or maximum, and which correspond to the cusp points of the caustic. In effect, at these points we have

$$\frac{d^2 \overrightarrow{ON}}{ds^2} = R'' \vec{n} \tag{6.67}$$

$$\frac{d^3 \overrightarrow{ON}}{ds^3} = R^{(3)} \vec{n} - 2R'' \frac{\vec{t}}{R}. \tag{6.68}$$

so that

$$\overrightarrow{N_0 N} = \left(\frac{s^2}{2} R'' + \frac{s^3}{6} R^{(3)} \right) \vec{n} - \frac{s^3}{3} \frac{R''}{R} \vec{t} + O(s^4). \tag{6.69}$$

Consequently, the caustic exhibits a *beak* which is described, if one chooses the coordinate x along $\vec{n}$ and the coordinate y along $\vec{t}$, with the parametric representation

$$y^2 = \frac{s^6}{9} \left(\frac{R''}{R} \right)^2, x^3 \approx \frac{s^6}{8} (R'')^3.$$

The equation describing LC in the neighborhood of the beak is then

$$y^2 \approx \frac{8}{9\sigma} x^3, \tag{6.70a}$$

where the notation of Pearcey and the definition,

$$\sigma = R'' R^2, \tag{6.70b}$$

have been used. Another way to obtain the previous result is to calculate the radius of curvature ρ of LC at the point N. The result is

$$\rho = R \frac{dR}{ds}. \tag{6.71}$$

The cusp points are the points at which ρ vanishes, that is, where $(dR/ds) = 0$. This condition defines some isolated points on the line of center of curvature LC.

For a generic wavefront the quantities, (d^2R/ds^2) and (dR/ds) do not vanish simultaneously; hence, $\sigma \neq 0$ and the singularities on the caustic are only of the beak (or cusp) type. However, for certain particular wavefronts, it is possible to have nongeneric singularities. When these quantities vanish simultaneously on the wavefront, we obtain, for instance, $y^2 = -(3/4R^{(3)})(x^4/R)$, instead of (6.70).

In three-dimension, the caustic surface C is the surface of the centers of curvature of the wavefront W. A point N on C is defined by

$$\vec{ON} = \vec{OM} + R\vec{n}, \tag{6.72}$$

where R is equal to one of the radii of curvature R or R', at point M on W, and where $\vec{n}$ is the normal to W at M. Each time a point N satisfies (6.72), we will say that it is below the point M. The surface is composed of two sheets, that are in contact, if M is an umbilic. In order to study the surface of the centers, it is convenient to relate W to a coordinate system of the lines of curvature. This system is regular outside the umbilics. Let us denote u (or v) as the curvilinear abscissa (or ordinates) on the line of curvature chosen as the axis of the abscissas (ordinates), and let $\vec{u} = (\partial \vec{OM}/\partial u)$, $\vec{v} = (\partial \vec{OM}/\partial v)$,

$$\frac{\partial \vec{ON}}{\partial u} = \frac{\partial R}{\partial u}\vec{n}, \tag{6.73}$$

so that $\vec{n}$ is tangent to C, which is then the envelope of the rays. The other tangent to C is given by

$$\frac{\partial \vec{ON}}{\partial v} = \left(1 - \frac{R}{R'}\right)\vec{v} + \frac{\partial R}{\partial v}\vec{n}. \tag{6.74}$$

Consequently, the plane tangent to C is generated by $\vec{v}$ and $\vec{n}$, and $\vec{u}$ is normal to C. If M runs along the u curvature line, N traverses a curve G of C whose tangent vector is $\vec{n}$. The curvilinear abscissa s on G satisfies,

$$ds = \frac{\partial R}{\partial u}du, \tag{6.75}$$

if we assume that $(\partial R/\partial u) > 0$. The derivative of the tangent vector $\vec{n}$ to G is given by

$$\frac{\partial \vec{n}}{\partial s} = -\vec{u}\left(R\frac{\partial R}{\partial u}\right)^{-1}. \tag{6.76}$$

It follows, therefore, that $(\partial \vec{n}/\partial s)$ is normal to C, and G is then a geodesic of C whose radius of curvature is $R\,(\partial R/\partial u)$. G is also an integral curve of the vector field defined by the directions of the rays tangent to C, i.e., a curve where the first approximation of the phase is iks. Thus, we recover the result established in Chap. 2 using an alternate approach, and discussed in Chap. 3 to investigate the solution in the vicinity of the point C. The result we refer to is that the integral curves of the rays are the geodesics of the caustic. The velocity on the geodesic G vanishes when $(\partial R/\partial u) = 0$. For this case, we have

$$\frac{\partial^2 \vec{ON}}{\partial u^2} = -\frac{1}{R}\frac{\partial^2 R}{\partial u^2}\vec{n} \tag{6.77}$$

Now, if $(\partial^2 R/\partial u^2) \neq 0$, $\vec{ON}\cdot\vec{n} = O(u^2)$ so that N is on the cusp line R of the caustic surface. The point N moves along R when the point M does the same along the line L on W defined by $(dR/du) = 0$. Let us now calculate the velocity of point N. The tangent to L is given by

$$d\left(\frac{\partial R}{\partial u}\right) = \frac{\partial^2 R}{\partial u \partial v}dv + \frac{\partial^2 R}{\partial u^2}du = 0. \tag{6.78}$$

When the point M travels along L, we have $(\partial \vec{ON}/\partial u) = 0$ from (6.73). Hence,

$$d\,\vec{ON} = \frac{\partial \vec{ON}}{\partial u}du + \frac{\partial \vec{ON}}{\partial v}dv = \frac{\partial \vec{ON}}{\partial v}dv, \tag{6.79}$$

becomes

$$d\,\vec{ON} = \left(\left(1-\frac{R}{R'}\right)\vec{v}+\frac{\partial R}{\partial v}\vec{n}\right)dv. \tag{6.80}$$

If $(\partial^2 R/\partial u^2) \neq 0$, (6.78) shows that $dv \neq 0$, and from (6.80) we have the result the velocity of N does not vanish on R; hence, we are at an ordinary point of R. If $(\partial^2 R/\partial u^2) = 0$, $dv = 0$, and L is tangent to the curvature line, $v = 0$. For this case, the velocity vanishes on R, which implies that R shows a cusp point and N is then located on a swallowtail (see Fig. 6.11).

The singularities of C are thus located below the special lines or special points of W. Specifically, the cusp lines R of the caustic C are below the lines defined by $(\partial R/\partial u) = 0$ on W, and the cusp points of these lines, which are the singularities of the swallowtail type, are underneath the points given by $(\partial R/\partial u) = 0$ and $(\partial^2 R/\partial u^2) = 0$.

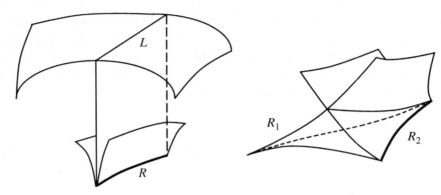

Fig. 6.11. Cusp line and swallowtail on the caustic

Evidently, these results are also valid on the second sheet of C provided that we replace R by R' and u by v. It is possible to extract other properties of C by studying the geometry of W, and we refer the interested reader to the works of Darboux [Da]. In particular, it is possible to express the product P of the radii of curvature of C as a function of the radii of curvature of W and of their derivatives as

$$P = - (R - R')^2 \, \frac{\partial R}{\partial u} \left(\frac{\partial R'}{\partial u} \right)^{-1} . \tag{6.81}$$

Equation (6.81) shows that $P = 0$ when $(\partial R/\partial u) = 0$, i.e., we are on the cusp line R. Thus, we find once again that N runs over R when $\partial R/\partial u = 0$.

So far we have considered the two sheets of the caustic C only separately. These two sheets are in contact with each other if the two radii of curvature R and R' are equal, i.e., under the umbilics of W. For the case of the umbilics, (6.81) simply yields $P = 0$, which means that the points of C below the umbilics of W are located on the cusp lines of C. This argument is not rigorous, however, since (6.81) was derived by using the coordinate system of the lines of curvature, which becomes singular close to the umbilics. A more rigorous demonstration of the fact that the points of C below the umbilics of W are located on the cusp lines has been given by Porteous [Po], and the geometry of the surface of the centers under the umbilics has been studied by Berry

[Be], who has shown that C is described locally by a surface of the elliptic or hyperbolic umbilic type (see Fig. 6.12).

We now recall the classification of the umbilics. The surface W in the neighborhood of the umbilic is described by

$$z = \frac{x^2 + y^2}{2R} + \frac{1}{6}(ax^3 + 3by^2 + cy^3).$$ (6.82)

The umbilic is classified as elliptic (or hyperbolic) depending upon the following inequalities

$$(b/a) + (c/2b)^2 < 0, \text{ for elliptic,}$$

and

$$(b/a) + (c/2b)^2 > 0, \text{ for hyperbolic.}$$

In the discussion presented above we have reviewed the nature and types of the singularities of C and we have associated them with the particular lines or points of C. Our next task will be to use the results derived above to calculate the field on C.

Fig. 6.12. Elliptic and hyperbolic umbilics

6.2.2 Field computation on the caustic C

The field $u(P)$ in the vicinity of center surface is written in the form of a radiation integral of the field on the wavefront W as follows

$$u(P) = \int_W \left(u(M) \frac{\partial G}{\partial n} - G \frac{\partial u(M)}{\partial n} \right) dS,$$ (6.83a)

where G is the Green function of the wave equation, and M is the running point of W. We are going to limit ourselves to the evaluation of the far field; consequently, we will replace the Green's functions appearing in the integral by their approximations for large arguments. Let us choose the origin at a point O of C and consider a point P close to O. The integrand of (6.83a) is an oscillating function of the phase φ

$$\varphi = \varphi(M, P) = k(\overrightarrow{PM}^2)^{1/2},\qquad (6.83b)$$

for which we will use a Taylor expansion. Since the point P is close to O, we truncate the above expansion by retaining only the first two terms as follows

$$\varphi \approx k\left(||\overrightarrow{OM}||-\overrightarrow{OP}\cdot\frac{\overrightarrow{OM}}{||\overrightarrow{OM}||}\right).\qquad (6.84)$$

In (6.84), $\overrightarrow{OP}\cdot\overrightarrow{OM}$ ($\overrightarrow{OM}\,||\overrightarrow{OM}||$) is a linear combination $\alpha x + \beta y + \gamma\,z$ of the coordinates (x, y, z) of the point P. We retain the leading term (beyond the constant) in the Taylor expansion of α, β, and γ in terms of the powers of u and v. In the same manner, we approximate $||\overrightarrow{OM}||$ by the leading terms of its Taylor expansion in terms of the powers of u and v. Our principal objective in this section is to recover the basic results for the field in the neighborhood of the caustics, especially the k dependence of the field, the special functions upon which it depends, and the arguments of these special functions.

6.2.2.1 Regular point of C, 2-D case (Fig. 6.13)

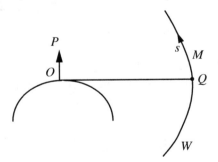

Fig. 6.13. Regular point of C

The third-order Taylor expansion of $\overrightarrow{OM}$ reads

$$\overrightarrow{OM} = -R\,\vec{n} + \left(s - \frac{s^3}{6R^2}\right)\vec{t} + \left(\frac{s^2}{2R} - \frac{s^3}{6R^2}\frac{dR}{ds}\right)\vec{n} + O(s^4),\qquad (6.85)$$

from which we can derive the result

$$\| \overrightarrow{OM} \| = R + \frac{s^3}{6R^2} \frac{dR}{ds} + O(s^4).$$

(6.86)

On the other hand, we can write

$$\overrightarrow{OP} \cdot \frac{\overrightarrow{OM}}{\| \overrightarrow{OM} \|} = x \frac{s}{R} + O(xs^3),$$

and use it to rewrite (6.84) as

$$\varphi = kR - kx \frac{s}{R} + k \frac{s^3}{6R^2} \frac{dR}{ds} + O(s^4, xs^3, x^2).$$

The two-dimensional Green's function is proportional to $k^{1/2}$, and the normal derivatives of G and $u(M)$ introduce a factor k; hence, the integral (I) is proportional to $k^{1/2}$. Accordingly, the field $u(P)$ can be written in the form

$$u(P) \approx k^{1/2} \int g(s) \exp ik \left(-x \frac{s}{R} + \frac{s^3}{6R^2} \frac{dR}{ds} + O(s^4, xs^3, x^2) \right) ds,$$

(6.87)

where g is a function of s and is proportional to the amplitude on the wavefront. Introducing the change of variable

$$t^3 = \frac{ks^3}{2R^2} \frac{dR}{ds},$$

(6.88)

and using (6.71), one is led to the expression

$$u(P) \approx k^{1/6} \int g(t) \exp \left(i \frac{t^3}{3} - it \, x \left(\frac{2k^2}{\rho} \right)^{1/3} + O(k^{-1/3}, x, kx^2) \right).$$

(6.89)

The first term of the asymptotic expansion of (6.89) is expressed by means of the Airy function $Ai \, (-x \, (2k^2/\rho)^{1/3}) = Ai(-\nu)$, where ν is the normal stretched variable introduced in Chap. 3. This result is valid if $x = O(k^{-2/3})$, i.e., in the boundary layer where we can neglect the term $O(k^{-1/3}, x, kx^2)$ of the phase which is $O(k^{-1/3})$. The above expression in terms of the Airy function of the normal stretched coordinate and the $k^{1/6}$ dependence is the same as that obtained in Chap. 3 by using the boundary layer method, although the integral approach provides these results in a more natural way. Let us apply the latter method to a case which is more difficult to treat with the boundary layer method, namely the cusp of the caustic.

6.2.2.2 Cusp of the caustic, 2-D case (Fig. 6.14)

We will follow the same method as in the previous subsection to determine the field at the cusp of the caustic. As a first step, we define the point P by the equation $\overrightarrow{OP} = x\, \vec{n} + y\, \vec{n}$. Next, by neglecting the terms s^5, we write the phase φ as

$$\varphi \approx k(R + x) + k\,\frac{ys}{R} - kx\,\frac{s^2}{2R^2} + k\,\frac{s^4}{24}\frac{R''}{R^2}. \qquad (6.90)$$

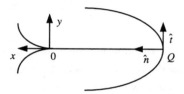

Fig. 6.14. Cusp of the caustic in the vicinity of a minimum of the radius of curvature R $(R'' > 0)$

The field $u(P)$ can then be written in the form

$$u(P) \approx k^{1/2} \int g(s)\, \exp ik\left(\frac{ys}{R} - \frac{xs^2}{2R^2} + \frac{s^4}{24}\frac{R''}{R^2}\right) ds. \qquad (6.91)$$

Next, we set $t^4 = k\,(s^4/24)\,(R''/R^2)$ in (6.84). The dominant contribution to the field is then expressed by means of the Pearcey function with the arguments

$$Y = -\left(\frac{24}{kR^2 R''}\right)^{1/4} ky = \left(\frac{24}{k\sigma}\right)^{1/4} ky\,, \qquad (6.92)$$

$$X = -\left(\frac{6}{kR^2 R''}\right)^{1/2} ky = -\left(\frac{6}{k\sigma}\right)^{1/2} kx\,, \qquad (6.93)$$

where we have used the notation $\sigma = R^2\, R''$ (formula (6.70b) Sect. 6.2.1). The sign of X does not modify the result since the Pearcey function is an even function of X. Note that we recover the $k^{1/4}$ dependence that characterizes the caustic beak.

Thus, by applying the integration technique on the wavefront, one can determine the form of the field in the vicinity of the caustic beak as derived in Sect. 6.1 with the Maslov method. Next, we will move onto the 3-D case and again discuss the problem of field computation at the caustics.

6.2.2.3 Regular point, 3-D case (Fig. 6.15)

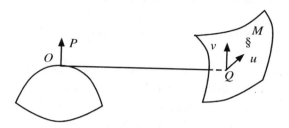

Fig. 6.15. Regular point, 3-D case

As in the 2-D case, we again use a Taylor expansion of the phase for the present 3-D problem. The point P is given by $\overrightarrow{OP} = x\,u$. We begin by calculating the Taylor expansion of $\overrightarrow{OM}^2$ in terms of the powers of u and v. The partial derivatives present in this expansion are obtained by using the formulas available from the theory of surfaces (see Appendix 2: Differential Geometry). The result is

$$\overrightarrow{OM}^2 \approx R^2 + v^2\left(1 - \frac{R}{R'}\right) + \frac{u^3}{3}\frac{1}{R}\frac{\partial R}{\partial u} + O(u^4, u^3v, uv^2), \qquad (6.94)$$

where the absence of the terms uv and u^2v is a consequence of the fact that the corresponding partial derivatives vanish.

For a sufficiently small x, and by employing (6.83) and (6.94), we obtain the following expression for the phase φ appearing in the integral defining the field

$$k^{-1}\,\varphi(u, v) = R - \frac{xu}{R} + \frac{v^2}{2R}\left(1 - \frac{R}{R'}\right) + \frac{u^3}{6}\frac{1}{R^2}\frac{\partial R}{\partial u} + O(u^4, u^3v, uv^2). \qquad (6.95)$$

As long as $R \ne R'$, i.e., if Q is not an umbilic, the phase is stationary at the point v defined by

$$2\frac{v}{R}\left(1 - \frac{R}{R'}\right) + a\,u^3 + b\,u^2\,v + \ldots = 0, \qquad (6.96)$$

and the solution is $v = O(u^3)$.

Let us apply the method of the stationary phase, with respect to the variable v, to the integral representation for the field $u(P)$, viz.,

$$u(P) \approx k \int g(u, v) \exp{(ik\,\varphi(u, v))}\,du\,dv. \tag{6.97}$$

In 3-D, the Green's function of the wave equation is $O(k^0)$ and the factor k results from the derivative with respect to the normal. The point at which the phase is stationary is $O(u^3)$; consequently, the terms containing v in the value of φ at this point will be $O(u^6)$ and, therefore, negligible. As a result, $u(P)$ is expressed as

$$u(P) \approx k^{1/2} \int h(u) \exp{\left(ik\left(R - x\frac{u}{R} + \frac{u^3}{3}\frac{1}{R^2}\frac{\partial R}{\partial u} + O(u^4) \right) \right)}\,du. \tag{6.98}$$

where $h(u) = g(u, 0)\,\exp(i\pi/4)(2\pi R)^{1/2}(1 - R/R')^{-1/2}$. The representation in (6.98) is similar to (6.87), which we derived in the two-dimensional case, and we can also perform the same change of variable as we did in the 2-D case. The dominant order in $u(P)$ can be written in terms of the Airy function with the argument $-(2k^2/\rho)^{1/3}x$, where $\rho = R\,(\partial R/\partial u)$ is the radius of curvature of the geodesic, i.e., is also the radius of curvature of the surface, in the direction of the ray passing through O. We recover, once again, the result obtained in Chap. 3 via the boundary-layer method.

6.2.2.4 Neighborhood of the cusp of a caustic, 3-D case

The procedure for computing the field in the neighborhood of a cusp of caustic in the 3-D case is similar to its 2-D counterpart. Again, after applying the stationary phase method with respect to the variable v, we arrive at the results that are similar to those found for the 2-D case. We can show that the result can be expressed in terms of the Pearcey function, with its arguments given by (6.92) and (6.93). The distance x is now measured along $\hat{u}$, and y along the vector orthogonal both to $\hat{u}$ and the tangent to the cusp line. We will not present the details of calculations leading to these results, but will only mention that they are expected and can be derived, from the physical point of view.

In the following section we will present some results about the swallowtail and the umbilics for which the following calculations become much more cumbersome. Since there is little practical interest for this case, we will restrict ourselves to only a brief discussion of this problem.

6.2.2.5 Neighborhood of a swallowtail

For computing the field in the neighborhood of the swallowtail, we still use a Taylor expansion (Appendix 2) to obtain

$$\vec{OM}^2 = R^2 + v^2\left(1 - \frac{R}{R'}\right) + \frac{u^5}{60}\frac{1}{R}\frac{\partial^3 R}{\partial u^3} + O(u^6, u^4 v, uv^2). \tag{6.99}$$

We will now find an estimate of the field at the point O. For sufficiently small u and v, the phase to be integrated is given by

$$\varphi(u, v) = \|\vec{OM}\| = R + \frac{v^2}{2R}\left(1 - \frac{R}{R'}\right) + \frac{u^5}{120R^2}\frac{\partial^3 R}{\partial u^3} + O(u^6, u^4 v, uv^2). \tag{6.100}$$

At the point O, the field is represented in terms of an integral of the type

$$u(O) \approx k \int g(u, v) \exp(ik\,\varphi(u, v))\, du\, dv. \tag{6.101}$$

We now evaluate this integral by using the method of stationary phase with respect to the variable v. The v coordinate of the stationary point is $O(u^4)$; hence, the terms in the phase φ that contain v are negligible at this point. Hence, the double integral (6.101) can be transformed into a simple one as follows

$$u(O) \approx k^{1/2} \int h(u) \exp\left(ik\left(R + \frac{u^5}{120R^2}\frac{\partial^3 R}{\partial u^3} + O(u^6, u^4 v, uv^2)\right)\right) du. \tag{6.102}$$

Introducing the variable

$$t^5 = \frac{u^5}{120R^2}\frac{\partial^3 R}{\partial u^3},$$

we obtain

$$u(O) \approx k^{3/10}(120R^2)^{1/5}\left(\frac{\partial^3 R}{\partial u^3}\right)^{-1/5}\exp(ikR)\,\Gamma(1/5)\frac{h(0)}{5}, \tag{6.103}$$

where

$$h(0) = g(0, 0)\exp(i\pi/4)(2\pi R)^{1/2}(1 - R/R')^{-1/2}. \tag{6.104}$$

From a physical point of view, (6.104) can be interpreted as being equivalent to a calculation of the field by applying geometrical optics, and taking into account only the confluence of rays with respect to v. Note that we recover the focusing factor of 3/10, which characterizes the field dependence in the vicinity of the swallowtail.

In all of the previous cases, we have transformed a double integral into a single one by using the method of stationary phase. Next we move on to the case of the umbilic, i.e., to a situation in which the two radii of curvature, viz., R and R', are equal.

6.2.2.6 Umbilic case

If (u, v, z) are now the usual Cartesian coordinates, then in the neighborhood of an umbilic, the wavefront can be described by the equation

$$z = -R + \frac{u^2 + v^2}{2} + \frac{1}{6}(au^3 + 3buv^2 + cv^3).$$ (6.105)

The phase function which enters into the definition of the integral representation of the field at point O is then given by

$$\varphi(u, v) = \|\overrightarrow{OM}\| = kR - \frac{k}{6}(au^3 + 3buv^2 + cv^3),$$ (6.106)

so that the field at the point O can be written as

$$u(O) \approx k \int g(u, v) \exp\left(i\,kR - \frac{k}{6}(au^3 + 3buv^2 + cv^3)\right) du\,dv.$$ (6.107)

Next we introduce the change of variable in (6.105) by letting $U = k^{1/3}u$, $V = k^{1/3}v$. We then obtain

$$u(O) \approx k^{1/3} \exp(i\,kR)\, g(0, 0) \int \exp\left(-\frac{1}{6}(aU^3 + 3bUV^2 + cV^3)\right) dU\,dV.$$ (6.108)

According to the type of umbilic, the integral in (6.107) can be expressed in terms of one of the two functions, namely the elliptic or hyperbolic umbilic function, occurring in the *catastrophy theory* and the field is found to be proportional to $k^{1/3}$. For more details on this subject the reader is referred to the two references [Be] and [GS].

6.2.3 Conclusions

We have derived an integral representation of the field in the neighborhood of the caustics by using the geometrical optics field on a wavefront located sufficiently far from the caustic. From this integral representation we can deduce, in a simple way, some of the results presented in the previous sections. Near the running point of the caustic surface, the field is expressed in terms of an Airy function and it is shown that it varies as $k^{1/6}$. In the neighborhood of the cusp lines of this surface, located under the lines of the wavefront where $(\partial R/\partial u)$ vanishes, the field is written in terms of the Pearcey function, and it varies as $k^{1/4}$. In the vicinity of the cusp points of these lines, i.e., the swallowtails of the caustic, located under the points on the wavefront where $(\partial^2 R/\partial u^2)$ vanishes, the field varies as $k^{3/10}$. The application of the stationary phase method enables us to replace the double integral in all of these cases by a single one which, in turn, is expressed by means of a standard function of the *catastrophe theory*. This is the case described in Sect. 6.1, where the Maslov method is employed to represent the field in terms of a single integral. In contrast, under the umbilics of the wavefront, the double integral cannot be transformed into a single integral, the field varies as $k^{1/3}$, and its expression uses umbilic functions of the *catastrophe theory*.

The method has enabled us to recover all the arguments of the Airy and Pearcey functions in the vicinity of the caustic. The calculation in the other cases is much more cumbersome and requires the knowledge of less well-known functions; hence, we will refrain from treating these cases in this work.

In conclusion, we have described two methods in this chapter which provide an integral representation of the field in the regions where the geometrical optics are not valid. The Maslov method, which is similar to the method of the Fourier integral operators, is mainly employed by the mathematicians and is also utilized for the calculation of seismic waves. In contrast, the second method, viz., the method of integration on a wavefront, is mainly employed in theoretical physics. Both methods give similar results and the choice between them is really a matter of convenience. Finally, it is also possible to devise other integral representations, which are all equivalent, to within $O(1/k)$, under the condition that one recovers the result of geometrical optics when applying the stationary phase method to them. It may be worthwhile to acknowledge at this point that despite the elegance of these approaches, they are not widely used in electromagnetic field computation.

In the next chapter we will present the physical theory of diffraction as devised by P. Ufimtsev, a technique that is indeed the most widely employed method for practical applications. In this method, the diffracted field is obtained from a radiation integral of the surface currents (or an approximation to these currents) on the object, an approach which is quite different from the ones we have discussed thus far.

References

[Ar] J. M. Arnold, "Spectral synthesis of uniform wave functions," *Wave Motion* **8**, 135-150 (1986).

[Be] M. Berry and C. Upstill, E. Wolf, Ed., *Progress in Optics* **18**, (1980).

[Da] G. Darboux, *Théorie Générale des Surfaces*, Chelsea, 1972.

[GS] V. Guillemin and S. Sternberg, *Geometric Asymptotics*, 1977.

[Kr] Y. Kravtsov, "Two new methods in the theory of inhomogeneous media," *Sov. Phys. Acoustics* **14(1)**, 1-17 (1968).

[KO] Y. Kravtsov and Y. Orlov, "Caustics, catastrophes and wave fields," *Sov. Phys. Usp.* **26**, 1039-1058 (Dec. 1983).

[Ma] V. Maslov, *Théorie des Perturbations et Méthodes Asymptotiques*, Dunod, 1972.

[MS] A. S. Mishchenko, V. E. Shalakov, and B. Y. Sternin, *Lagrangian Manifolds and the Maslov Operator*, Springer, 1990.

[Pe] T. Pearcey, *Phil. Mag.* **37**, 311-327 (1946).

[Po] Porteous, *J. Diff. Geometry*, 543 (1971).

[ZD] R. W. Ziolkowski and G. A. Deschamps, "Asymptotic evaluation of high frequency fields near a caustic, an introduction to Maslov's method," *Radio Science* **19(4)**, 1001-1025 (July-Aug. 1984).

7. Surface Field and Physical Theory of Diffraction

In Chap. 6, we presented several approaches to deriving integral representations of the diffracted field in the vicinity of the caustics, viz., Maslov's method in Sect. 6.1; method of integration over a wave front in Sect. 6.2; and, Molinet's method in Sect. 6.3.

Although these methods are relatively easy to apply to simple geometries, they can become quite involved for complex objects. As a consequence, these methods are seldom used for the calculation of the RCS of practical targets, e.g., aircrafts or missiles. An alternate approach to deriving an integral representation of the diffracted field is to determine, as a first step, the induced current on the diffracting object, and then compute the diffracted field by evaluating the radiation produced by these currents. It is well known that the induced currents on the object are simply related to the surface fields on the object; for instance, the induced electric current $\vec{J}$ is given by $\hat{n} \times \vec{H}$, where $\vec{H}$ is the surface magnetic field, and the surface magnetic current $\vec{M}$ is obtained from $-\hat{n} \times \vec{E}$ where $\vec{E}$ is the surface electric field, the latter being zero on a perfect conductor. The surface fields, in turn, can be calculated either by appealing to the Geometric Theory of Diffraction, discussed in Chap. 1, or by using the asymptotic expansion methods discussed in Chaps. 2 and 3. In fact, the computation of the surface fields can be regarded as a particular case of the more general problem of calculating the diffracted fields in the vicinity of an object, that was discussed in Chap. 3. In view of this, we will merely content ourselves in this chapter with presenting the results for the surface fields and refer the reader to earlier chapters for the details.

To compute the surface fields we follow Ufimtsev [U] and express the diffracted field as a sum of *uniform* (Sect. 7.1) and fringe fields (Sect. 7.2), the latter being associated with the discontinuities on the object. The Physical Theory of Diffraction (PTD), to be presented in Sect. 7.3, has approximate values for the uniform field and takes into account the fringe fields for sharp edges. As a consequence, although the PTD enables us to compute the fields reflected and diffracted by sharp edges, it does not include some of the other contributions to the diffracted fields that can, in fact, be significant at times. To correct this deficiency, various generalizations of the PTD have recently been proposed, and we will discuss them briefly in Sec. 7.4.

Let us begin by defining more precisely the essential notions of the uniform and fringe fields. Consider a convex scatterer whose surface is piecewise regular, i.e., it comprises portions of regular surface that are pieced together with linear or point discontinuities of the tangent, the curvature, or with discontinuities of higher order. We will now express the surface field on such an object as a sum of *uniform* and *fringe* fields in the context of Ufimtsev. The uniform field may be calculated on all of the regular portions of the surface by invoking the continuity expansion, and the fringe field may be derived from the discontinuities of the surface. In the next section, we will discuss the derivation of both the uniform and fringe fields.

7.1 Uniform field

When a plane wave with incident wave vector $\vec{k}$ illuminates an object Ω, we can distinguish points in the *lit zone* and the *shadow zone* by using the criteria $\hat{n} \cdot \vec{k} < 0$ and $\hat{n} \cdot \vec{k} > 0$, respectively, where $\hat{n}$ is normal to Ω. Thus, as a preamble to the computation of the surface fields, we divide up the surface into the lit and shadow regions with the *shadow boundary*, defined by $\hat{n} \; \vec{k} = 0$, separating the two regions. It should be realized, however, that rather than an abrupt change from one region to the other, there exists a *transition zone* in the vicinity of the separator, as discussed in Chap. 3. The approximate width of the transition zone is ρ/m, where ρ is the radius of curvature of the object, and m is the Fock parameter given by: $m = (k\rho/2)^{1/3}$ (Fig. 7.1).

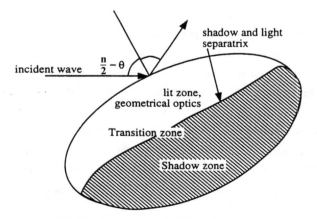

Fig. 7.1. Convex obstacle lit by a plane wave

7.1.1 Lit zone

As discussed in Chap. 2, in the lit zone the field is given by the Lüneberg-Kline series of which the first term is the geometrical optics field. The field on the surface can be written as

$$\vec{E} = \vec{E}^{\,i} + \vec{E}^{\,i}\,\underline{R},\tag{7.1}$$

where $\vec{E}^{\,i}$ is the incident field and $\underline{R}$ is the dyadic reflection coefficient defined in Sect. 1.3.4, and whose expression is given in (1.18) of Sect. 1.3.4. and also by (1.90) and (1.91) of Sect. 1.5.1. Let us introduce the vector $\hat{t}$ which is the intersection of the plane tangent to the surface of the object and the plane of incidence, and the vector $\hat{\alpha}$ such that $(\hat{t}, \hat{\alpha}, \hat{n})$ form an orthonormal, counter-clockwise triad of vectors. Then the tangential fields on the surface can be written as

$$\vec{E}_t = \vec{E}^{\,i} \cdot \hat{\alpha}\,\frac{2Z\sin\theta}{1+Z\sin\theta}\,\hat{\alpha} + (\vec{E}^{\,i} \cdot \hat{t})\,\frac{2Z}{Z+\sin\theta}\,\hat{t},\tag{7.2a}$$

$$\vec{H}_t = (\vec{H} \cdot \hat{\alpha})\,\frac{2\sin\theta}{Z+\sin\theta}\,\hat{\alpha} + (\vec{H}^{\,i} \cdot \hat{t})\,\frac{2}{1+Z\sin\theta}\,\hat{t}.\tag{7.2b}$$

Equation (7.2b) has been obtained from (7.1a) by replacing $\vec{E}^{\,i}$ with $\vec{H}^{\,i}$ and Z with $1/Z$. For the perfectly conducting case, we have the impedance $Z = 0$ and (7.1) and (7.2) simply become

$$\vec{E}_t = 0 \quad \text{and} \quad \vec{H}_t = 2\,\vec{H}_t^{\,i}.\tag{7.3}$$

This verifies that the magnetic field tangent to the surface is just twice the tangential incident field and the induced surface current, which is purely electric in nature, and is given by

$$\vec{J} = 2\,\hat{n} \times \vec{H}^{\,i}.\tag{7.4}$$

Equation (7.4) can be recognized as the well-known Physical Optics result.

7.1.2 Transition zone — lit region

The previous formulas are valid only in the lit region. The Lüneberg-Kline expansion is no longer applicable in the transition zone, and the appropriate formulas for this zone may be found in Sect. 3.8.2. Unfortunately, these formulas are not uniform, since they do not reduce to the results (7.2) and (7.3) of the previous section in the lit region.

Nonetheless, by interpreting the parameter σ in the Fock function [Fo] appearing in the expression for the field, it is possible to obtain a uniform representation of the field in a manner described by Pathak [Pa]. By neglecting the terms that are of order $O(k^{-1/3})$, we obtain the following expressions for the tangential magnetic and electric fields.

$$\vec{H}_t \approx (\vec{H}^{\,i} \cdot \hat{\alpha}) \, G_Z(-m \sin\theta) \, \hat{\alpha} + \frac{(\vec{H}^{\,i} \cdot \hat{n}')}{Z} G_{1/Z} (-m \sin\theta) \, \hat{t} \,, \qquad (7.5a)$$

$$\vec{E}_t \approx (\vec{E}^{\,i} \cdot \hat{\alpha}) \, G_{1/Z}(-m \sin\theta) \, \hat{\alpha} + Z \, (\vec{E}^{\,i} \cdot \hat{n}') \, G_Z (-m \sin\theta) \, \hat{t} \,. \qquad (7.5b)$$

In the above, $\hat{n}'$ is a vector such that ($\hat{k}$, $\hat{\alpha}$, $\hat{n}'$) form an orthonormal and counter-clockwise system, and G_Z is related to the Fock function F_Z by

$$G_Z(x) = \exp \, (ix^3/3) \, F_Z(x). \qquad (7.6)$$

By replacing F_Z and $F_{1/Z}$ with their asymptotic expansions for large negative values of the argument, it can be shown that (7.5) reduces to the geometrical optics formulas given in (7.2). On the other hand, when the incident field is a plane wave and θ is small, (7.5) reduces to the formulas (3.179) and (3.180) of Sect. 3.8.2.

For a perfectly conducting object $\vec{E}_t$ is zero and $\vec{H}_t$ is given by

$$\vec{H}_t \approx (\vec{H}^i \cdot \hat{\alpha}) \, G(-m \sin\theta) \, \hat{\alpha} + \frac{i}{m}(\vec{H}^i \cdot \hat{n}') \, F(-m \sin\theta) \, \hat{t} \,, \qquad (7.7)$$

(7.7) is only valid for the 2D case. For 3D case, we have not derived an equivalent formula. See however (7.13).

$$G(x) = \exp \, (ix^3/3) \, g(x) \quad \text{and} \quad F(x) = \exp \, (ix^3/3) \, f(x), \qquad (7.8)$$

where g and f are the magnetic and electric Fock functions, respectively. The expression for the field in (7.7) reduces to that in (7.3) when f and g are replaced by their asymptotic expansions, and to the formulas of Sect. 3.8.2 when θ is small.

Having presented the details of the field computation in the lit zone, we move on next to the discussion of shadow region fields in the following section.

7.1.3 Transition zone – Shadow region

The field computation in the shadow region is based upon the same principles as those presented in the previous section. A uniform result is again obtained from the non-uniform result of Sect. 3.8 by reinterpreting the parameter σ of the Fock function. The $\vec{H}$ field at point M is given, within $O(k^{-1/3})$, by

$$\vec{H} \approx \left\{ (\vec{H}^i \cdot \hat{\alpha}) \left(\frac{e_M(0)^{1/2}}{e_M(s)} \right) F_Z(\sigma_M) \, \hat{\alpha} + \frac{1}{Z} (\vec{H}^i \cdot \hat{n}_0) \left(\frac{e_E(0)}{e_E(s)} \right)^{1/2} F_{1/Z}(\sigma_E) \, \hat{s} \right\}$$

$$\exp{(iks)} \left(\frac{\rho(0)}{\rho(s)} \right)^{1/6} h^{-1/2}(s), \tag{7.9}$$

with

$$e_E(s) = \xi_E(s) + m^2 Z^{-2}(s), \tag{7.10}$$

$$e_M(s) = \xi_M(s) + m^2 Z^2(s), \tag{7.11}$$

$$\sigma_E = \frac{1}{\xi_E(0)} (k/2)^{1/3} \int_0^s \frac{\xi_E(s)}{\rho(s)^{2/3}} \, ds, \tag{7.12}$$

in (7.9), σ_M is obtained by replacing ξ_E by ξ_M in (7.12). The quantity s appearing in the above equations is the curvilinear abscissa measured from the separator along the creeping ray traversing through M, and $\hat{\alpha}_0$ and $\hat{n}_0$ are the vectors $\hat{\alpha}$ and $\hat{n}$ at the point with abscissa 0. The corresponding result for $\vec{E}$ is obtained by replacing $\vec{H}^i$ by $\vec{E}^i$; e_E by e_M; e_M by e_E; σ_E by σ_M; σ_M by σ_E; and, Z by $1/Z$ in (7.9).

In the transition zone, (7.9) reduces to the formula (3.170) of Sect. 3.8.2. On the other hand, for large values of σ, we can replace F_Z and $F_{1/Z}$ by their asymptotic expansions valid for large positive values of the argument. Upon doing this, we obtain the first creeping wave solution which is identical to that derived in Chap. 3. However, the higher-order creeping modes derived from the above procedure are not the same as those given in Chap. 3. Fortunately, in practice, the first creeping wave mode is quite dominant for a surface impedance Z whose real part is not too small compared to its imaginary part. Thus, despite the approximations, (7.9) often yields results in the shadow zone that are quite accurate. Nonetheless, the result in (7.9) is not very satisfactory from a theoretical point of view, and it can, at best, be regarded as only a temporary result.

For the case of a perfect conductor (7.9) becomes

$$\vec{H} \approx e^{iks} \left\{ (\vec{H}^i \cdot \hat{\alpha}_0)\, g(\sigma)\, \hat{\alpha} + \frac{i}{m} (\vec{H}^i \cdot \hat{n}_0)\, f(\sigma) \big((-\tau\rho)\hat{\alpha} + \hat{s} \big) \right\} \left(\frac{\rho(0)}{\rho(s)} \right)^{1/6} h^{-1/2}(s),$$

(7.13)

$$\sigma = (k/2)^{1/3} \int_0^s \frac{ds}{\rho(s)^{2/3}}.$$

(7.14)

We will show, in what follows, that (7.9) and (7.13) can be simplified in the deep shadow zone where the Fock functions can be replaced by their large argument approximations.

7.1.4 Deep shadow zone

The primary contribution in the deep shadow zone stems from the first creeping wave mode since the rest of the terms decay relatively quickly in this region. Thus, if we retain only this mode, (7.9) becomes

$$\vec{H} \approx \left\{ \frac{2i\sqrt{\pi}(\vec{H}^i \cdot \hat{\alpha}_0) \cdot \hat{\alpha}}{w_1(\xi_M(0))(e_M(0)e_M(s))^{1/2}} \, exp\left(i(k/2)^{1/3} \int_0^s \frac{\xi_M(s)}{\rho(s)^{2/3}} ds \right) \right.$$

$$\left. + \frac{2i\sqrt{\pi}(\vec{H}^i \cdot \hat{n}_0) \cdot \hat{n}}{Z w_1(\xi_E(0))(e_E(0)e_E(s))^{1/2}} \, exp\left(i(k/2)^{1/3} \int_0^s \frac{\xi_E(s)}{\rho(s)^{2/3}} ds \right) \right\}$$

$$exp\,(iks) \left(\frac{\rho(0)}{\rho(s)} \right)^{1/6} h^{-1/2}(s),$$

(7.15)

$\vec{E}$ is obtained from a similar formula by interchanging $\vec{H}^i$ with $\vec{E}$, ξ_E with ξ_M, e_E with e_M in (7.15). We note that when $Z \rightarrow 0$, $(e_E(0)\, e_E(s))^{1/2} \approx Z^{-2}$, so that the second term in (7.9) becomes negligible. Also, for the perfectly conducting case, the Fock function f becomes rapidly negligible in the deep shadow zone so that the first term in (7.13) provides the main contribution. Neglecting f and replacing g by its asymptotical expansion for large values of the argument, we obtain

$$\vec{H} \approx (\vec{H}^i \cdot \hat{\alpha}_0) \frac{\hat{\alpha}}{\beta A_i(-\beta)} \, exp\left(-\sqrt{\frac{3}{2}}\beta\sigma \right) exp\left(iks + i\frac{\beta}{2}\sigma \right) \left(\frac{\rho(0)}{\rho(s)} \right)^{1/6} h^{-1/2}(s),$$

(7.16)

where σ is given by (7.14), and $-\beta$ is the first zero of the derivative of the Airy's function ($\beta \approx 1,019$).

The previous formulas diverge if $h(s) = 0$, i.e., on the envelope of the geodesics followed by the creeping rays, i.e., at the caustics for the creeping rays. For a generic object, this caustic is a line with turning points. We can show [Bo1], by using the boundary layer method, that the caustic corrections valid for planar rays also apply to creeping rays up to the dominant order. Explicit formulas for the acoustic case, which can be directly generalized to the electromagnetic case have been given in [Bo2]. They involve Airy functions with the arguments of the type $(3/4(s^+ - s^-))^{2/3}$ for planar rays, where s^+ and s^- are the phases of the two creeping rays going through a point in the vicinity of the caustic.

7.2 Fringe field

The field scattered by an object is not described in its entirety by the uniform field alone, and it is necessary to augment it by using the so-called fringe field which we will discuss in this section. Let us consider the example of a plane TE wave with unit amplitude incident upon a perfectly conducting half-plane (see Fig 7.2).

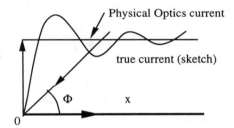

Fig. 7.2. Current on a half-plane illuminated by a plane wave

Fig. 7.3. Fringe current on a wedge with curved faces

On the lit side, the uniform field is given simply by (see (7.3) of Sect. 7.1.1)

$$\vec{H}^{u} = 2\vec{H}^{i} = 2\exp(-ikx\cos\varphi)\,\hat{z}. \tag{7.17}$$

For the half-plane, the exact solution can be expressed [BS] in terms of the Fresnel integral F as follows

$$\vec{H} = 2\,e^{-i\pi/4}\exp(-ikx\cos\theta)\,F(-\sqrt{2kx}\cos(\varphi/2))\,\hat{z}. \tag{7.18}$$

The fringe field H^f is defined as the difference between the actual field and the uniform field, as follows

$$\vec{H}^f = \vec{H} - \vec{H}^u. \tag{7.19}$$

Sufficiently far from the edge, and for large negative values of the argument, the function F can be replaced by the first terms of its asymptotic expansion. Introducing this approximation in (7.19), and using (7.18) we get

$$\vec{H}^f \approx -\frac{exp\ i(kx + \pi/4)}{\sqrt{2\pi kx}\ cos(\varphi/2)}. \tag{7.20}$$

The fringe field appears to be a wave traveling away from the edge, and it is $O(k^{-1/2})$ with respect to the uniform field at observation points sufficiently far from the edge.

Let us now consider the more general case of a wedge with curved, perfectly conducting faces illuminated by a plane wave (see Fig. 7.3).

We have seen in Chap. 3 that, to the dominant order in k, i.e., $O(1/k)$, the field in the vicinity of the edge is identical to that of a wedge with straight faces that are tangential to the curved faces at the tip of the wedge. In contrast to the half-plane case, there is no simple closed form expression similar to (7.18) for the case of the wedge. Borovikov [Bor] has shown by using the boundary layer method away from the neighborhood of the edge, that the fringe field can be expressed in terms of a Nicholson's function, and can be identified as a creeping wave at distances far enough from the edge, i.e., beyond the boundary layer described in Sect. 3.6. This is consistent with the launching of creeping rays by edges, discussed in Chap. 1.

Next, let us consider the case of oblique incidence for a perfectly conducting wedge for which the solution can be deduced from the corresponding result for normal incidence [BS]. In particular, we can again show that sufficiently far from the edge, the fringe field transforms into a creeping wave propagating along the direction of the creeping ray launched from the edge of the wedge, in accordance with the laws of GTD that were described in Chap. 1.

Let us now summarize the discussion given above. We have seen that at locations far from discontinuity the fringe field may be calculated by using the GTD, and that this field has the form of a creeping wave. Near the discontinuity, the fringe field can be derived from the solution of an associated canonical problem, e.g., a wedge with planar faces that are tangent to the curved surfaces of original wedge geometry near its tip.

In Sect. 7.1 we have seen how the uniform part of the surface field can be calculated, and in Sect. 7.2 the representation of the fringe field has been discussed. Therefore, in principle, it should be possible to compute the scattered field in the entire space by following the recipe we have provided above. However, this process can be very tedious because of the complex nature of the representations of the fringe field and of the uniform field in the shadow zone. The Physical Theory of Diffraction, presented in the next section, alleviates this problem by replacing the uniform currents with simpler, approximate currents, and by providing analytical expressions for the fields radiated by the fringe currents.

7.3 The Physical Theory of Diffraction

To help us explain the Physical Theory of Diffraction, we will introduce some useful preliminary notions of fringe waves in Sect. 7.3.1, of equivalent currents in Sect. 7.3.2, and of fringe equivalent currents in Sect. 7.3.3.

7.3.1 Fringe wave

The diffraction coefficient $\underline{D}$ of a discontinuity is the sum of the diffraction coefficient $\underline{D}^u$ associated with the uniform currents and $\underline{D}^f$ with the fringe currents. Therefore $\underline{D}^f$ can be derived simply by subtracting $\underline{D}^u$ from $\underline{D}$, which leads to

$$\underline{D}^f = \underline{D} - \underline{D}^u, \tag{7.21}$$

where all of these diffraction coefficients are dyads.

In the original version of the PTD proposed by Ufimtsev in the 1960s (see [U] for earlier references to Ufimtsev), the field radiated by the fringe currents, the so-called *fringe wave*, was calculated in the context of the GTD technique discussed in Chapter 1 by using the diffraction coefficient $\underline{D}^f$ given in (7.21). This technique allows us to suppress the divergence of the field computed by using the GTD at the incident and reflection shadow boundaries. However, in this method, the fringe wave is calculated as a ray field and, therefore, it tends to infinity in the vicinity of the caustics of the diffracted rays. An approach to circumventing this difficulty is to employ the equivalent current method introduced by Ryan and Peters [RP] and we will discuss this approach in the next section.

7.3.2 Equivalent current method

For two-dimensional problems, the method of equivalent currents was first introduced in the context of GTD by Ryan and Peters. They assumed that the field diffracted by a cylindrical discontinuity is associated with an imaginary line current located at the discontinuity. The complex amplitude of this line current is determined by equating the field diffracted by the discontinuity and that radiated by the equivalent line current. For instance, for an incident TM polarized wave with a unit amplitude, the equivalent current I is solved by using the equation

$$Z_0 I \sqrt{\frac{k}{8\pi r}} \, exp \, (ikr - i\pi/4) = -\frac{D}{\sqrt{r}} \, exp \, (ikr), \qquad (7.22)$$

where D is the diffraction coefficient of the discontinuity. This leads to the expression

$$I = -Z_0^{-1} \sqrt{8\pi /k} \, e^{i\pi/4} D. \qquad (7.23)$$

If the wave is TE-polarized, instead, the equivalent electric current is simply replaced by a magnetic current, but the rest of the procedure remains intact.

For a three-dimensional discontinuity, for instance the curved edge of a wedge, the discontinuity is locally replaced by its cylindrical or two-dimensional counterpart, e.g., the locally tangent wedge in our example. Ryan and Peters assumed that the equivalent currents on the edge can be approximated by the ones that would exist on the associated two-dimensional discontinuity. Thus depending on the polarization, we set up an array of electric or magnetic dipoles, with amplitudes $\vec{I}$ or $\vec{M}$, given by

$$\vec{I} = -Z_0^{-1} \sqrt{8\pi /k} \, e^{i\pi/4} \, D_s(\vec{E}^i \cdot \hat{t}) \, \hat{t} \,, \qquad (7.24a)$$

$$\vec{M} = -Z_0 \sqrt{8\pi /k} \, e^{i\pi/4} \, D_h(\vec{H}^i \cdot \hat{t}) \, \hat{t} \,, \qquad (7.24b)$$

where $Z_0 = 377 \, \Omega$ is the intrinsic impedance of free space, and $\hat{t}$ is the tangent vector to the edge. We should note at this point that although all of the discussion as well as the expressions, appearing above, are valid only for the case of normal incidence and observation, they can be easily generalized to the case of oblique incidence, with the direction of observation on the Keller cone (see Fig. 7.4).

Let us consider a line current whose distribution is given by $I \, exp \, (ikz \, \sin\beta)$, where z denotes the abscissa along the line. The field radiated by this line is still given by the left-hand side of (7.22), where r now denotes the distance from the observation point to a point on the line source that are connected by a ray subtending an angle β with respect to the line. We can interpret the radiation from the equivalent current source as though it radiates along a Keller cone whose aperture angle is determined by the phase variation along the line current source.

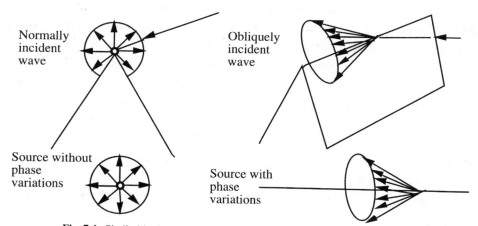

Fig. 7.4. Similarities between the source radiation and the diffraction by a wedge

We can notice the similarity (see Fig. 7.4) between the present problem to that of diffraction by a discontinuity. Hence, it is possible to regard, again for the oblique incidence case, the field diffracted by a discontinuity as though this field is radiated by a line current located at the discontinuity. The projection of $\vec{E}$ or $\vec{H}$ on the line of discontinuity has an amplitude given by $E^i\sin\beta$ or $H^i\sin\beta$ for the TM TE polarizations respectively. Hence, the equivalent currents can now be obtained from (7.24) upon dividing the previous result for the current by the factor $\sin\beta$, to get

$$I = - Z_0^{-1} \sqrt{8\pi/k}\; e^{j\pi/4}\; \frac{D_s}{\sin\beta}\; (\vec{E}^i \cdot \hat{t})\; \hat{t}. \qquad (7.25)$$

The field E^d diffracted by a line discontinuity L, is then calculated by computing the radiation for the above equivalent currents, located on L. When the point of observation d is located in the far field, $\vec{E}^d$ can be expressed by an integral on L as follows

$$\vec{E}^d \approx - \frac{ik}{4\pi d} \int_L (Z_0 I(\ell)\; \hat{d} \times (\hat{d} \times \hat{t}) + M(\ell)(\hat{d} \times \hat{t}))\; \exp(-ik\hat{d}\cdot\vec{OM})\; d\ell,$$
$$(7.26)$$

where M represents the source point with curvilinear abscissa ℓ that is located on the diffracting line L and $\hat{d}$ is the unit vector along the direction of observation. We can show by applying the stationary phase method to (7.26), that the diffracted field derived from the above integral is consistent with the GTD results presented in Chap. 1.

However, the advantage of (7.26) over the GTD formulas is that it provides finite results for the caustics as well, whereas the GTD displays a singular behavior there. The method of equivalent currents may thus be regarded as a simple means for obtaining an integral representation for the diffracted field that reduces to the familiar GTD field outside of the caustics region, but remains finite on the caustics.

At this point, it is worthwhile to draw the attention of the reader to one of the limitations of the equivalent current method we just presented above. Strictly speaking, the equivalence between the actual and equivalent current is limited to only along the direction of the Keller cone, because the derivation of the latter was based upon the equality of the radiated fields only on the surface of this cone. Of course, this does not prevent us from obtaining a valid integral representation of the diffracted field as long as (7.26) has a stationary phase point for the observation point situated precisely on the Keller cone. In fact, for a wave incident along $\hat{\imath}$ and for observation along the direction $\hat{d}$, the phase of the integrand in (7.26) is $k \, (\hat{\imath} - \hat{d}) \cdot \overrightarrow{OM}$ and the stationary phase points satisfy the equation

$$\frac{d}{ds}(\hat{\imath} - \hat{d}) \cdot \overrightarrow{OM} = (\hat{\imath} - \hat{d}) \cdot \hat{\imath} = 0, \tag{7.27}$$

the above implies that $\hat{d}$ and $\hat{\imath}$ have the same projection along $\hat{\imath}$; it follows, therefore, that $\hat{d}$ must be on the Keller cone.

The integral in (7.26) can be expressed in the form of a sum of terms evaluated at the stationary phase points, i.e., those points for which $\hat{\imath}$ and $\hat{d}$ are on the Keller cone and for which the Ryan and Peters' method is valid. However, when (7.26) has no stationary phase points, i.e., when there is no ray diffracted by the edge, we can no longer use (7.25), since it ceases to be valid on the Keller cone. In this case, to obtain the equivalent currents, it becomes necessary to calculate the field radiated by an infinitesimally small current strip on the diffracting object (see Fig. 7.5).

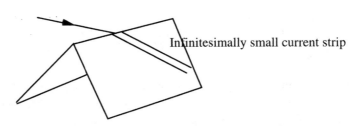

Infinitesimally small current strip

Fig. 7.5. Infinitesimally small fringe current strip

The equivalent current is then defined as the infinitesimally small current elements, located on D, that radiates the same field as the infinitely small strip. Michaeli [Mi] has applied this method to the case of a perfectly conducting wedge and Shore [SY] has generalized it to perfectly conducting discontinuities with planar sides, as for instance a narrow slot. The equivalent currents obtained in this manner depend upon the direction of incidence $\hat{i}$ as well as on the direction of observation $\hat{d}$. They reduce to the result derived previously when this direction of observation is located on the Keller cone.

Several observations can be made about these types of equivalent currents that are widely used in many practical applications. These are:

(i)   When a diffracted ray exists originating from D, the Ryan and Peters' equivalent currents provide a good approximation to the solution.

(ii) If D has extremities or edges (discontinuity of the tangent vector) the infinitely small current strip is not complete. Thus, it is incorrect to assume that this procedure provides the correct answer.

A rigorous proof asserting that Michaeli's approach yields a better result for practical problems, than does the simple method of Ryan and Peters, is not available. However, it is our experience that the use of Michaeli's method for case (ii), mentioned above, is typically more accurate. Explicit formulas for Michaeli's equivalent currents can be found in the original articles [Mi], and in the book by Knott et al. [KS] for the case of the conducting wedge. For the case of discontinuities on surfaces described by an impedance boundary condition, we are forced to return to Ryan and Peters' method, since the more general equivalent current approach of Yaghjian is restricted only to perfectly conducting structures. Finally, we note that the equivalent current method proposed by Ryan and Peters is extremely general and that it can be applied to any discontinuity problem of interest, so long as the diffraction coefficient of the discontinuity is known. It can, for instance, be employed using the diffraction coefficients in Chap. 1 but also using the uniform diffraction coefficients given in Chap. 5.

We have seen in the past that the equivalent currents enable us to calculate the fields on the caustics of the diffracted rays while the fringe wave approach solves the problem of divergence at the shadow boundaries. We now observe that the equivalent fringe currents combine the features of both of the above, and hence handle most of the singularity problems arising in GTD. Morover, one must use these fringe currents in PTD not to include twice the *po* diffracted field.

7.3.3 Equivalent fringe currents

The equivalent fringe currents are associated with the fringe diffraction coefficients D^f that are given by (7.21), in Ryan and Peters approach. They are derived by computing the radiation from the fringe currents in Michaeli's approach. Let us start with Ryan and Peters method, which is the only one that is available for the imperfectly conducting wedge. The fringe diffraction coefficient D^f is obtained by subtracting the total diffraction coefficient, given in Sect. 1.5, from the uniform diffraction coefficient D^u. For the case of a sharp edge, D^u is given, to the dominant order in k, by the diffraction coefficient D^{po}, which is obtained by calculating the radiation from the physical optic currents. For a wedge, D^{po} is the sum of the physical optics diffraction coefficient D^{po} for the two half-planes comprising the wedge. For backscattering calculations, D^{po} for the half-plane becomes

$$D^{po} = - \frac{e^{i\pi/4}}{2\sqrt{2\pi k}} R \, \mathrm{tg} \, \varphi, \tag{7.28}$$

where R is the reflection coefficient.

For the TM polarization, for which $\vec{E}$ is parallel to the edge, we have $R = R_s$. Likewise, for the TE case, R becomes replaceable by R_h. These reflection coefficients are given by

$$R_s = \frac{Z \sin \varphi - 1}{Z \sin \varphi + 1}, \qquad R_h = \frac{\sin \varphi - Z}{\sin \varphi + Z}, \tag{7.29}$$

where φ is the angle measured from the lit side.

Let us consider, for instance, a perfectly conducting half-plane. For the TM polarization, the diffraction coefficient for the half-plane, given in Sect. 1.5, reads

$$D_s = \frac{e^{i\pi/4}}{2\sqrt{2\pi k}} \frac{1 - \cos \varphi}{\cos \varphi}, \tag{7.30}$$

and

$$D_s^{po} = \frac{e^{i\pi/4}}{2\sqrt{2\pi k}} \, \mathrm{tg} \, \varphi, \tag{7.31}$$

since $R = -1$.

Both D_s and D^{po} diverge for the normal incidence case, i.e., for $\varphi = \pi/2$. By contrast, the coefficient D^f given by

$$D^f = \frac{e^{i\pi/4}}{2\sqrt{2\pi k}} \frac{1 - \cos\varphi - \sin\varphi}{\cos\varphi},\tag{7.32}$$

does not diverge when the incidence is normal, and tends instead to $-(e^{i\pi/4}/2\sqrt{2\pi k})$. The equivalent fringe current is an electric current in this case, and by using (7.25) and (7.32) , we find the following expression for this current

$$-\frac{i}{Z_0 k} \frac{1 - \cos\varphi - \sin\varphi}{\cos\varphi} (\vec{E}^i \cdot \hat{t})\,\hat{t}.\tag{7.33}$$

For the case of the imperfectly conducting wedge, the expressions are more complex than the one in (7.33). However, we can still show that the coefficient D^f, and hence the fringe current, does not diverge on the shadow boundaries.

Next, we discuss the case of discontinuity in the curvature. The uniform part of the surface field on the object is given by the Lüneberg-Kline series. The diffraction coefficient D^{LK}, associated with the second term of this series, is $O(k^{-3/2})$ as is the diffraction coefficient D^{po} due to the po currents. Therefore, this term must be taken into account when calculating the uniform diffraction coefficient. The fringe diffraction coefficient can be written as

$$D^u = D^{po} + D^{LK},\tag{7.34}$$

and

$$D^f = (D - (D^{po} + D^{LK})).\tag{7.35}$$

We can again show that D^f remains finite and even tends towards zero for the case of normal incidence [Bo3]. For discontinuities of order $n > 2$, that are perhaps not too important in practice since they only induce relatively small diffracted fields, D^u can be derived by taking into account the contribution of the first n terms of the Lüneberg-Kline series. In each case, the fringe current is obtained from D^f by using the formulas (7.24 a,b) for the TM and TE polarization, respectively.

The discussion above shows that the method of Ryan and Peters allows us to derive the fringe currents in a very systematic manner. It would be useful, at this point, to tabulate below the diffraction coefficients D^u or D^f for different objects. In the table below we present a summary of some results useful in practice.

Diffracting object	D^u or D^f
Half-plane with impedance surfaces, TE (the TM case is obtained by replacing Z by $1/Z$)	$D^u = -\dfrac{e^{i\pi/4}}{2\sqrt{2\pi k}} R_h \, \mathrm{tg}\, \varphi$
Wedge impedance	Sum of the D^u of the two half-planes comprising the wedge
Curvature discontinuity with TE impedance a_1: radius of curvature of the side that faces the incident wave a_2: radius of curvature of opposite side	$D^f = \dfrac{A(-sin^2\varphi + 2Z\,sin\,\varphi + Z^2)(sin\,\varphi - Z)\cos\varphi}{sin^2\varphi(sin\,\varphi + Z)^3}$ $A = -\dfrac{e^{i\pi/4}}{4k\sqrt{2\pi k}}\left(\dfrac{1}{a_1} - \dfrac{1}{a_2}\right)$

As we have seen in Sect. 7.3.2, the previous results are valid only on the Keller cone which becomes a plane for the backscattering case. However it is possible, using these results, to obtain an integral representation which, when evaluated using the stationary phase method, yields an expression for the fringe wave, and this expression remains finite on the caustics of the diffracted rays. For the case of perfectly conducting wedge, it is possible to obtain equivalent fringe diffraction coefficients by integrating the field radiated by the fringe currents. To this end we consider, in accordance with Michaeli [Mi] or Ufimtsev [U], an infinitesimally small current strip on a plane-faceted wedge in the direction $\hat{\alpha}$, subtending an angle $\hat{\alpha}$ with the edge of the wedge. The fringe current propagates along the intersection $\hat{c}$ of the Keller cone with the face of the wedge. If ℓ is the abscissa along the infinitely small strip, the phase of the integrated quantity will be $k\ell(\cos(\beta - \alpha) - \hat{c}\cdot\hat{\alpha})$. The resulting integral can be infinite if $\cos(\beta - \alpha) - \hat{c} \cdot \hat{\alpha} = 0$, that is to say if $\hat{c}$ is on the cone with an axis parallel to the strip and with half-angle $(\beta - \alpha)$. The most judicious choice is the one that avoids this singular case as much as possible, and this can be accomplished by choosing $\alpha = \beta$. Thus, the integration is carried out along a infinitely narrow strip which is the intersection of the Keller cone and the edge of the wedge. We will not reproduce the explicit formulas for the diffraction coefficients since they are rather involved, but the interested reader can locate them in Michaeli [Mi]. The equivalent currents obtained are singular only if the angle of incidence $\hat{\imath}$ is close to grazing and equal to the direction of observation $\hat{d}$, i.e., along the forward diffraction direction.

In particular, for the important practical case of backscattering, i.e., for $\hat{\imath} = -\hat{d}$, the equivalent fringe currents are never singular.

In this section, we have reviewed the notions of fringe wave equivalent currents and the equivalent fringe currents that are necessary for developing a good understanding of the PTD. We will now go on to discuss, in the next section, the evaluation of the diffracted field.

7.3.4 Calculation of the diffracted field by using the PTD

In this section we turn to the problem of computing the diffracted field in the context of the PTD. By examining the PTD procedure we find that in its conventional form it does not take into account the uniform surface currents defined in Sect. 7.1, but includes only the Physical Optic field, and moreover, it is restricted to the case of perfect conductors. Thus, on the lit surface we have (see (7.4) in Sect. 7.1)

$$\vec{J} = 2\hat{n} \times \vec{H}^{\,i}$$

Of course, this PO current should be augmented by the fringe current for the edges and, the diffracted field is the sum of $\vec{E}^{\,d}_{po}$, the diffracted far field by the Physical Optics currents, viz.,

$$\vec{E}^{\,d}_{po} = -\frac{ikZ_0}{4\pi d} \exp{(ikd)} \int_{s} (\hat{d} \times \hat{d} \times \vec{J}) \exp{(-ik\hat{d} \cdot \vec{OM})} \, ds, \qquad (7.36)$$

and the field radiated by the fringe currents given by (7.26). In practice, the usual version of the PTD takes into account only the equivalent currents on the sharp edges. We will now examine to what degree of accuracy these approximations obtain an accurate asymptotic representation of the field diffracted by an object.

7.3.5 PTD and GTD

In this section, we will show that by evaluating the integrals (7.26) and (7.36) via the stationary phase procedure, and adding the results of these evaluations, we obtain the PTD diffracted field, which is identical to that given by the Geometrical Theory of Diffraction (GTD) except for a supplemental term. This latter term stems from the non-physical jump of the Physical Optics field at the shadow boundaries.

To illustrate our point, we begin by considering the simple case of the diffraction of an incident TM wave by a circular cylinder (Fig. 7.6). Let the incident field be of unit amplitude given by $\vec{E}^{\,i} = \exp{(-ikx)}\hat{z}$.

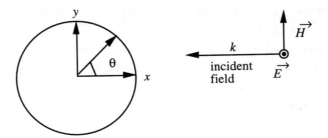

Fig. 7.6. Circular cylinder illuminated by a plane TM wave

We first obtain the surface field by using the formulas given in Sect. 7.1. The Physical Optics surface current is given by

$$\bar{J} = 2Y \cos \theta \exp (-ika \cos \theta) \, \hat{z},\tag{7.37}$$

which radiates the far field $\bar{E}^d = E^d \, \hat{z}$, where E^d is expressed as

$$E^d = \sqrt{\frac{k}{2\pi}} \frac{e^{-i\pi/4}}{\sqrt{r}} \int_{-\pi/2}^{\pi/2} \cos \theta \exp (-2ik \, a \cos \theta) \, a \, d\theta.\tag{7.38}$$

The integral in (7.38) has a stationary phase point at $\theta = 0$. Since the second derivative of the phase at this point is $2ka$, the contribution of this stationary phase point, say E^s, is given by

$$E^s \approx \frac{e^{i\pi/4}}{\sqrt{2ka}} \sqrt{2\pi} \sqrt{\frac{k}{2\pi}} \frac{e^{-i\pi/4}}{\sqrt{r}} \exp (-2ika) \, a,$$

which simplifies to

$$E^s \approx \sqrt{\frac{a}{2r}} \exp (-2ika).\tag{7.39}$$

We note that the above result is precisely that predicted by Geometrical Optics. The Equivalent Radar Length (ERL) of the specular point is given simply by

$$ERL = \lim_{r \to \infty} 2\pi r \, |E^s|^2 = \pi a.\tag{7.40}$$

We may also add the end point contribution E^f to the stationary phase contribution derived above (see the Appendix "Asymptotic Evaluation of Integrals"). The end point contribution is given by

$$E^f \approx \frac{k^{-3/2}}{\sqrt{2\pi r}} \frac{e^{-i\pi/4}}{4a^2}, \qquad (7.41)$$

and is found to be $O(k^{-3/2})$. Thus, it does not alter the validity of the PTD result, which, in this particular case, is a $O(k^{-1})$ approximation of the diffracted field.

Let us now discuss the more general case of a regular convex object, i.e., one without any sharp edges. The diffracted field calculated by the PTD, which reduces in this case to the Physical Optic field, is given by (7.36). For an incident plane wave propagating along $\hat{\imath}$, the phase of the integral in (7.36) is $k(\hat{\imath}-\hat{d}) \cdot \vec{OM}$, which is stationary if $k(\hat{\imath}-\hat{d})$ is parallel to the normal vector to the surface, i.e., if $\hat{\imath}$ and $\hat{d}$ are related by the laws of reflection. For the particular case of backscattering, $\hat{\imath} = -\hat{d}$, and we can find the specular point since the incidence is normal once again. The application of the stationary phase method to (7.36) then generates the following contribution to the diffracted field E^s (see Appendix "Asymptotic Evaluation of Integrals") when both the radii of curvature R_1 and R_2 at the specular point S are positive

$$\vec{E}^s \approx \frac{i \, 2\pi \sqrt{R_1 R_2}}{2k} \frac{-ikZ_0}{4\pi d} \hat{d} \times (\hat{d} \times \vec{J}) \exp(-2ik\hat{d} \cdot \vec{OS}),$$

or equivalently,

$$\vec{E}^s \approx - \frac{\sqrt{R_1 R_2}}{2d} \vec{E}^i(S) \exp(-2ik\hat{d} \cdot \vec{OS}). \qquad (7.42)$$

We can again verify that the above result is identical to that given by Geometrical Optics (see Chap. 1). In particular, the RCS contribution of these specular points is given by

$$RCS = \lim_{d \to \infty} 4\pi d^2 \frac{|\vec{E}^s|^2}{|\vec{E}^i|^2} = \pi R_1 R_2 \qquad (7.43)$$

For the case where R_1 is positive and R_2 is negative, (7.40) has to be multiplied by $e^{-i\pi/2}$. On the other hand, if R_1 and R_2 are both negative, i.e., the reflection is from a concave surface, (7.40) has to be multiplied by $e^{-i\pi} = -1$. In the first case, we only cross one caustic, and hence there is a $-\pi/2$ phase shift, while the second case has two caustic crossings, and, consequently, the phase shift is $-\pi$ (see Fig. 7.7).

a) Convex: no
caustic crossing

b) Convex-Concave:
crossing of one caustic

c) Concave:
crossing of two
caustics

Fig. 7.7. Reflection from a doubly-curved surface

Since the currents on the shadow boundary are truncated, the asymptotic expansion of (7.36) includes, in addition to these stationary points, the contributions from the points along the shadow boundary C, where the phase is stationary. At these points, if $\hat{\imath}$ is the vector tangent to C, then

$$(\hat{\imath} - \hat{d}) \cdot \hat{\imath} = 0, \qquad (7.44)$$

which implies that $\hat{\imath}$ and $\hat{d}$ are on a Keller cone whose axis is $\hat{\imath}$. Therefore, we find diffracted parasitical rays on the shadow boundary as shown in Fig. 7.8.

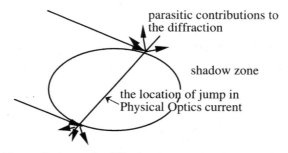

Fig. 7.8. Parasitic contributions to the diffraction due to the jump in the Physical Optics current

In the general case, the parasitical contribution of the shadow boundary can be of $O(k^{-1/2})$, which is similar to the contribution by other diffracted rays. However, it can be shown [GB] that this contribution remains of $O(k^{-3/2})$ for the backscattering case. Physically, this can be explained from the fact that the currents perpendicular to $\hat{s}$ are zero on the shadow boundary, and the field radiated by the currents parallel to $\hat{s}$ in the direction of $\hat{s}$ is zero. If, on the other hand, the diffracting object includes one (or several) linear edges, the integral (7.36) has critical points when the phase of the integrand stationary along the edge where (7.43) is satisfied, and these points become the origins of the diffracted rays. The asymptotical evaluation of these stationary phase contributions yields an expression which is similar to that of the GTD, but whose diffraction coefficient is that of the Physical Optics. In contrast, the asymptotic evaluation of the integral in (7.26), also by the stationary phase method, yields an edge-diffracted field with a fringe diffraction coefficient. By adding the above two contributions, we obtain a sum of the two diffraction coefficients viz., fringe and Physical Optics, and the result is identical to the GTD diffraction coefficients.

In summary, the PTD provides an integral representation of the diffracted field which, when evaluated asymptotically, yields the reflected as well as the GTD diffracted fields by sharp edges. Moreover, the PTD yields a finite result on all caustics as well as on shadow boundaries. However the PTD also has some drawbacks and it may be worthwhile, at this point, to list them. These are:

(i) The fictitious jump in the Physical Optics currents generates spurious bright points which can, except in the backscattering case, be $O(k^{-1/2})$, i.e., comparable to the contributions of sharp edges.

(ii) There are some contributions that are not included in PTD, as for instance, the creeping waves. Furthermore, for non-convex objects, the PTD does not incorporate the multiple reflection effects that are often significant.

(iii) The present version of the PTD only handles perfectly conducting objects.

However, it is possible to generalize the PTD to circumvent most of these drawbacks, and this is discussed in depth in the next section.

7.4 Generalizations of the PTD

7.4.1 Extension to objects described by an impedance condition

This extension to the case of objects whose surfaces are described by impedance type of boundary has been carried out by A. Pujols and S. Vermersch [PV]. The uniform currents used are the Physical Optics currents given by (7.2a) and (7.2b) in Sect. 7.1.1. The equivalent fringe currents for sharp edges, obtained by the method of Ryan and Peters, are given in Sect. 7.3.3.

7.4.2   Removal of the spurious contribution due to the fictitious jump discontinuity currents at the shadow boundary

The fictitious jump discontinuity in the PTD current at the shadow boundary can be removed by inserting an equivalent fictitious current located on the shadow boundary that radiates a field counteracting the one due to the jump. More specifically, one erects a locally tangent cylinder at every point on the shadow boundary of the diffracting object. Next, the radiation from the Physical Optics currents on this cylinder is asymptotically evaluated as in Sect. 7.3.4. The end point contribution is obtained for the TM polarization by using (7.41) of Sect. 7.3.4. A detailed exposition of this procedure, and useful formulas based upon the method, have been given by Gupta and Burnside [GB]. Yet another way to annihilate this contribution is to modify the currents in the shadow zone by using a neutralizer, which is a regular function that decreases from 1 in the lit zone to 0 in the shadow zone. This is a well-known technique, which is referred to as *tapering* in the antenna literature.

7.4.3 The use of more *realistic* uniform currents

A natural generalization of the PTD entails the replacement of the Physical Optics uniform current with the uniform current determined in Sect. 7.1. The above procedure has the following two advantages:

(i)   the current discontinuity at the shadow boundary is suppressed

(ii)   the integral representation of the diffracted field includes creeping rays.

At this point, we recall from Sect. 7.1.3 that, to a first approximation, the phase of the current in the shadow zone was ks, where s is the abscissa on a geodesic path traversed by a creeping ray. The phase of the field radiated by the currents in the vicinity of the geodesic in the direction of observation $\hat{d}$ is then

$$\varphi(M) = -k\,\hat{d}\,\cdot\,\vec{OM} + ks, \tag{7.45}$$

and φ is stationary when

$$\hat{d}\,\cdot\,\frac{\partial\vec{OM}}{\partial s} - 1 = \hat{d}\,\cdot\,\frac{\partial\vec{OM}}{\partial\alpha} = 0, \tag{7.46}$$

where s and α refer to the coordinates in a geodesic system whose origin is O (see Chap. 3). Equation (7.46) is only satisfied when $\hat{d}$ matches the vector $\hat{s}$ tangent to the geodesic, i.e., if the tangent to the creeping ray is aligned with the direction of observation. The procedure described above leads us, once again, to the GTD creeping rays. For a convex obstacle, Orlov [Oo] and Bouche [Bo1] have demonstrated, both for the perfectly conducting and the impedance surfaces, that an asymptotical evaluation of the radiation integral of the uniform currents leads to results that are the same as those obtained by applying the GTD.

The generalization of the PTD described in this section has been employed by Molinet [Mo] to predict the RCSs of two-dimensional objects. However, the procedure becomes rather involved, and quite tedious, for three-dimensional cases, since the derivation of the currents in the shadow zone not only requires the computation of the geodesics and the Fock functions, but also the implementation of the caustic corrections for the creeping rays. An alternative approach that has been used by Choi [CW] and Bouche et al. [BB] involves the use of equivalent currents, which are still derived by using the Ryan and Peters' technique, for calculating the creeping waves. As a next step, the creeping-wave contribution is expressed in the form of a line integral, where this line is usually the separator, although it is possible to employ alternate choices for the same. This area is still witnessing further development and we refer the reader to the recent publications [CW, BB] for further details.

To summarize this discussion, we point out that the use of the *real* uniform field rather than the physical optics currents in the context of the PTD has the advantage that it enables us to incorporate the contribution of the creeping waves. However, additional developments would be necessary before this approach could be applied to complex, three-dimensional objects.

7.4.4 Treatment of nonconvex objects

The physical optics approximation for the current distribution on an object does not include the contributions due to multiple reflections of the incident field on the object. Let us consider, for instance, the example of the right-angle corner reflector, shown in Fig. 7.9.

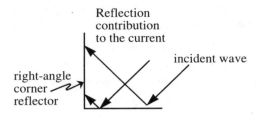

Fig. 7.9. Reflection contribution to the current distribution omitted by physical optics approximation for a right-angle corner reflector

For this geometry, the uniform current on each of the faces is derived by applying the method of images to the currents on an infinite wedge, and the reduced current derived in this manner is a sum of the physical optics currents due to the incident field on the face and the field reflected from the other face. For instance, for a TE-polarized incident wave of unit amplitude, the uniform current distribution on the horizontal side can be written

$$\bar{J} = (2\exp(-ikx\cos\theta) + 2\exp(ikx\cos\theta))\,\hat{x}. \qquad (7.47)$$

The first and second terms can be identified as the Physical Optics currents due to the incident and reflected fields, respectively. A natural generalization of the PTD would be to replace the Physical Optics current with the Geometrical Optics current that includes its multiple reflection term. In addition, it would also be necessary, in this case, to take into account the fringe currents due to multiple reflections of the diffracted fields. Of course, it should be realized that there may be some difficulties in this procedure if the caustics or the shadow boundaries of these fields were to cross the diffracting object. In this event, it would become necessary to resort to the uniform formulas of Chap. 5 to calculate the multiply-reflected field on the object. Yet another approach would be to calculate the field radiated by the Physical Optics currents, regarding this field as though it were an incident field for the second reflection, and using it to derive the currents due to the double reflections. Subsequently, by repeating the process just outlined, one can also compute the currents due to multiple reflections, as shown, for instance, in [CB]. We should note, however, that while the generalization of the PTD to non-convex objects is theoretically possible, the implementation of the steps just outlined may be very tedious for many practical situations.

We conclude this section by pointing out there are several possible ways available to us for improving the PTD approach presented above. First, we can extend to objects described by an impedance condition (Sect. 7.4.1). Second, we can enhance the accuracy of the PTD by replacing the Physical Optics currents by a more realistic one (Sects. 7.4.2-7.4.4). These generalized versions of the PTD yield integral representations of the field which, when asymptotically evaluated, not only yield results that are consistent with the GTD, but also remain finite in the problem regions where the GTD fails. We also hasten to point out that the discussion of the topic of PTD is by no means closed, and investigations into the practical and efficient computation of the *exact* uniform current (Sect. 7.4.3) and the generalization to non convex objects still remain active areas of investigation. Incidentally, the generalizations described above are only representative examples, and other possibilities for improving the PTD still exist. For instance, we can obtain an integral representation of the solution by calculating the fields on any surface enclosing the object rather than computing them on the object itself. Finally, it is useful to note that in computing the integral representation of the diffracted field, a portion of it can be calculated via the GTD or the equivalent current method, while the balance may be derived by using the PTD.

Before closing this chapter, it would be useful to provide some examples illustrating the applications of the PTD and we will do this in the following section.

7.5 Applications of PTD: Illustrative examples

7.5.1 The strip

The simplest example which illustrates the advantages of the PTD over the GTD is the perfectly conducting strip illuminated by a TM plane wave (see Fig. 7.10) given by

$$\vec{E}^i = \exp\left(-ik(x\cos\theta + y\sin\theta)\right)\hat{z}. \tag{7.48}$$

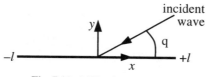

Fig. 7.10. Diffraction by a strip

We first obtain the *uniform* surface current on the strip by using the Physical Optics. It takes the form

$$\vec{J} = 2\hat{n} \times \vec{H}^i = 2\frac{\sin\theta}{Z_0}\exp\left(-ikx\cos\theta\right)\hat{z}. \tag{7.49}$$

Next we augment this current with the fringe currents, launched by the edges of the strip. These fringe currents are the same as those on a half-plane (see Sect. 7.2). In this case, the PTD replaces these currents by the equivalent fringe currents, or simply by the fringe waves with amplitudes given by (7.32) of Sect. 7.3.3. Hence, the back-scattered field $\vec{E}^d = E^d \, \hat{z}$ is the sum of the field E_{po}^d radiated by the Physical Optic currents, viz.,

$$E_{po}^d = - \sqrt{\frac{k}{2\pi r}} \, e^{-i\pi/4} \int_{-\ell}^{\ell} \sin\theta \, \exp(-2ikx\cos\theta) \, dx, \qquad (7.50)$$

and E_f, the fringe waves due to the edges, given by

$$E_f(\pm \ell) = \mp \frac{e^{-i\pi/4}}{2\sqrt{2\pi kr}} \frac{1 \pm \cos\theta - \sin\theta}{\cos\theta}. \qquad (7.51)$$

The expression in (7.50) can be written in a closed-form as follows

$$E_{po}^d = \frac{e^{i\pi/4}}{2\sqrt{2\pi k}} \, \mathrm{tg}\,\theta \, (\exp(2ik\ell\cos\theta) - \exp(-2ik\ell\cos\theta)). \qquad (7.52)$$

We note that (7.52) resembles the GTD result provided we recognize that the diffraction coefficients have been calculated via the Physical Optics, and are given by (7.31) in Sect. 7.3.3. The first and second terms in (7.52) may be identified as the fields diffracted by the edges that are singular $x = \ell$ and $x = -\ell$, respectively. We might recall that one of the drawbacks of the GTD is that when $\theta = \pi/2$, it yields diffracted fields that are singular. However, the advantage of the PTD construction is that these infinite fields exactly annihilate each other by virtue of the subtraction in (7.52). In fact, (7.52) can also be written as

$$E_{po}^d = \frac{e^{-i\pi/4}}{2\sqrt{2\pi kr}} \sin\theta \, \frac{\sin(2k\ell\cos\theta)}{2k\ell\cos\theta} \, 2k\ell, \qquad (7.53)$$

and it is evident that (7.53) is totally free of any singularities when $\theta = \pi/2$.

By adding the fringe waves, given in (7.51), to the result in (7.52), we find the GTD results with the *correct* diffraction coefficients which do not exhibit any fictitious singularities when added to the expression in (7.52). We observe, therefore, that the PTD provides a bounded representation of the field, regardless of the incidence angle and that this representation reduces to the GTD results away from the shadow boundaries when the diffraction is simple, i.e., not multiple in nature.

For the backscattering case, it is useful to note that the multiple diffractions can be calculated by using the GTD, since they do not exhibit any singularities, except for the case of grazing incidence.

In the next section, we consider a somewhat more complex example of the application of the generalized PTD, viz., the diffraction by a cone with an impedance surface.

7.5.2 Diffraction by a sharp-tipped cone with an impedance surface

A three-dimensional high frequency code called the 3DHF code, has been generated recently by Vermersch and Pujols [PV]. It uses the PTD in its generalized form, applies to objects whose surfaces are described, in general, by impedance conditions, and handles double-diffraction effects.

The results generated by this code will now be compared with those computed by using the integral equation code SHF 89 devised by P. Bonnemason and B. Stupfel [BSt] in order to establish the accuracy of the 3DHF code.

Figures 7.10 and 7.11 compare the backscattering in RCS computed by using the above two codes and plot them as functions of the incident angle. We note that the agreement between the two results is quite good except for the following situations:

(i)   in the vicinity of the axis where there exists a caustic of the doubly-diffracted rays, and

(ii) for incident angles between 120° and 150° where the 3DHF code does not include the creeping rays. Although, for the impedance value chosen in the example, the intensities of the creeping waves are relatively low, they do become noticeable for angles for which RCS is weak.

7.6 Conclusions

The PTD is a powerful method for calculating the diffracted field which, in contrast to the GTD, yields bounded results for all angles of observation. It is useful for practical application and is widely used for industrial applications. However, it also has some disadvantages as compared to the GTD, viz., (i) it requires the evaluation of an integral; and, (ii) it omits a portion of the field contributions, e.g., those due to the creeping rays or the multiple reflections. However, it is possible to overcome these limitations of the PTD by deriving generalized variants of the method and then using it to calculate the RCS of complex objects with a good accuracy.

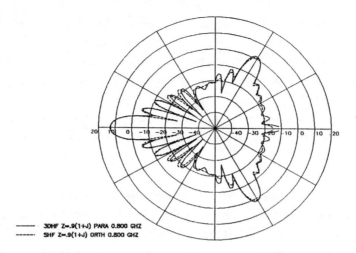

Fig. 7.11. Diffraction by an impedance cone; electric field in the plane of incidence

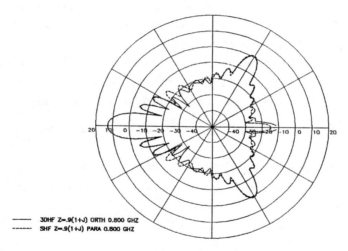

Fig. 7.12. Diffraction by an impedance cone; electric field orthogonal to the plane of incidence

References

[BB]   D. Bouche, J. J. Bouquet, M. Pierronne, and R. Mittra, "Diffraction by low observable axisymmetric objects at high frequency," *IEEE AP*, **40**(10), 1165-1174, 1992.

[Bo1] D. Bouche, "La méthode des courants asymptotiques," Ph.D Dissertation, University of Bordeaux, 1992.

[Bo2] D. Bouche, "Calcul du champ à la surface d'un obstacle convexe vérifiant une condition d'impédance par une méthode de développement asymptotique," *Journal d'Acoustique*, **5**, 507-530, 1992.

[Bo3] D. Bouche, "Courant sur un obstacle cylindrique parfaitement conducteur présentant une discontinuité de courbure," *Annales des Télécomm*, **47**, 391-399, 1992.

[Bor] V. A. Borovikov, "Diffraction by a wedge with curved faces," *Sov. Phys. Acoust.* **25**(6), Nov.-Dec. 1979.

[BS]   J. J. Bowman, T. B. A. Senior, and P. L. Uslenghi, "Acoustic and electromagnetic scattering by simple shapes," *Hemisphere*, 1987.

[BSt] P. Bonnemason and B. Stupfel, "Modeling high-frequency scattering by axisymmetric perfectly or imperfectly conducting scatterers," *Electromagnetics,* **13**, 111-129, 1993.

[CB]   Cavelier and Balabanne, Exposé aux journées Maxwell, Bordeaux, May 1993.

[CW]   Choi, N. Wang, and L. Peters, "Near-axial backscattering from a cone-sphere," *Radio Science,* **25**, 427-434, July-August 1990.

[Fo]   V. Fock, *Electromagnetic Diffraction and Propagation Problems*, Pergamon Press, 1965.

[GB]   I. J. Gupta and W. D. Burnside, "A PO correction for backscattering from curved surface," *IEEE Trans. Ant. Prop.*, **AP-35**, 553-561, May 1987.

[KS]   E. F. Knott , E. Shaeffer, and M. Tuley, *Radar Cross Section*, Artech House, 1992.

[Mi]   A. Michaeli, "Elimination of infinities in equivalent edge currents," *IEEE Trans. Ant. Prop.*, **AP-34**, 912-918, July 1986 and 1034-1037, Aug. 1986.

[Mit] K. M. Mitzner, "Incremental length diffraction coefficients," report AFAL-TR 73-26, Northrop Corporation, Aircraft Division, April 1974.

[Mo]   Molinet, "Contribution au benchmark," *JINA,* Nov. 1992.

[Oo]   Orlova and Orlov, "Scattering of waves by smooth convex bodies of large electrical dimensions," *Radiotechnika et Electronika*, 31-40, 1975.

[Pa]   P. H. Pathak, "An asymptotic analysis of the scattering of plane waves by a smooth convex cylinder," *Radio Sci.*, **14**(3), 419-435, 1979.

[PV]   A. Pujols and S. Vermersch, "La condition d'impédance dans le code 3DHF," rapport interne CEA/CESTA.

[RP]   C. E. Ryan, Jr., and L. Peters, Jr., "Evaluation of edge diffracted field including equivalent currents for the caustic region," *IEEE Trans. Antennas Propagat.*, **AP-17**, 292–299, May 1969.

[SY]   R. A. Shore and A. D. Yaghjian, "Incremental diffraction coefficients for planar surfaces," *IEEE Trans. Ant. Prop.*, **AP-34,** 55-70, Jan. 1988.

[Uf]   P. Ya Ufimtsev, "Elementary edge waves and the Physical Theory of Diffraction," *Electromagnetics*, **11**, 125-160, 1991.

8. Calculation of the Surface Impedance, Generalization of the Notion of Surface Impedance

In all of the previous chapters, we have dealt with objects whose surfaces are described by an impedance Z, which is defined via the relationship

$$\vec{E} - \hat{n}\ (\hat{n} \cdot \vec{E}) = Z\ \hat{n} \times \vec{H}\ .$$

The impedance condition enables us to describe, under certain conditions, either a conducting object coated with a dielectric material or an imperfectly-conducting, penetrable object. In Sect. 8.1, we will present the mathematical foundations of the impedance condition along with the formulas for computing this impedance.

Under certain circumstances the impedance condition is not an accurate description of the relationship between the electric and magnetic fields on the surfaces of the object. As we will see in Sect. 8.2, it is possible to either treat the problem directly without resorting to the impedance boundary condition or to generalize it by means of various techniques as explained in Sect. 8.3. However, since these latter methods are employed much less frequently than the impedance condition, we will restrict ourselves to presenting only a brief review of these approaches.

As we have done in the past (with the exception of Sect. 4.4.1), we will assume in this chapter that the materials constituting the object are lossy such that the surface waves can be neglected.

8.1 Mathematical foundations and determination of the surface impedance

The concept of surface impedance was introduced by Rytov [R] and Léontovitch [L] to treat the problem of diffraction by lossy dielectric objects with high refractive index. Their analysis was based upon an asymptotic expansion where the small parameter is the inverse of the index $n = \sqrt{|\varepsilon\mu|}$. This approach is global and requires no assumptions with regard to the nature of the field which can be either Geometrical Optical or diffracted field. The Rytov-Léontovitch boundary condition has been improved and a more rigorous foundation for this condition has been established by Artola, Cessenat and Cluchat [AC, C], and we will review their work in Sect. 8.1.1. Yet another approach to be discussed in Sect. 8.1.2, entails the use of high frequency approximations, with the small parameter being the inverse of the wave number k, and the impedance then depends upon the physical nature of the field.

8.1.1 Surface impedance for lossy materials with high index

For the case of lossy materials with a high index, the field cannot penetrate the material deeper than one skin depth.

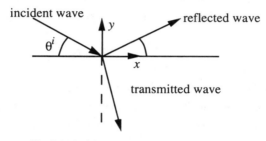

Fig. 8.1. Incident wave on a material half-space

Let us consider a plane wave incident at an angle θ^i upon a half-space of material ε, μ. The wave vector k_t of the transmitted wave is given by the expression (see Fig. 8.1)

$$\vec{k}_t = k \cos \theta^i \, \hat{x} - k \sqrt{\varepsilon\mu - \cos^2 \theta^i} \, \hat{y} \qquad (8.1)$$

where the square root is chosen in such a way that the wave attenuates in the negative y-direction. From (8.1) we see that if $n >> 1$, then $k_t \approx k \sqrt{\varepsilon\mu} \, \hat{y}$. The transmitted wave depends upon the stretched variable (or *fast* variable) $Y = n\,y$, which can also be written as $Y = y/\eta$ after setting $\eta = 1/n$.

Artola, Cessenat and Cluchat [AC] have shown that this behavior continues to be valid for a general object, regardless of whether the object is a lossy dielectric body or a conductor coated with dielectric layers. The reader is referred to [C] for the demonstration of this result. For a general object, the variable y is then the distance to the surface of the object.

We now seek a solution in the interior of the object in the form of an asymptotic expansion in terms of the integer powers of η. In practice, the asymptotic series is typically truncated after three terms and the Ansatz for the $\vec{E}$ and $\vec{H}$ fields then reads

$$\vec{E} = \vec{E}_0(\vec{r}, Y) + \eta\, \vec{E}_1(\vec{r}, Y) + \eta^2\, \vec{E}_2(\vec{r}, Y) + O(\eta^2), \qquad (8.2a)$$

$$\vec{H} = \vec{H}_0(\vec{r}, Y) + \eta\, \vec{H}_1(\vec{r}, Y) + \eta^2\, \vec{H}_2(\vec{r}, Y) + O(\eta^2), \qquad (8.2b)$$

where $\vec{r}$ is the position inside the object. It is convenient to employ a curvilinear coordinate system adapted to the problem by choosing the axes of the system as follows. We choose, for instance, a grid of orthogonal curves on the surface S of the object for the axes 1 and 2, and the curves initially normal to the surface for axis 3 such that the coordinate system is orthogonal. For the case where the coordinate curves on S are the lines of curvature, the normal curves are just the straight lines normal to S as shown in Fig. 8.2. This coordinate system is regular in a neighborhood $O(min(R_1, R_2))$ of S, even for large Y.

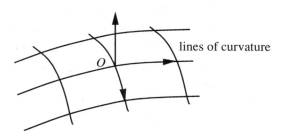

Fig. 8.2. Coordinate system for the lines of curvature

Next, we insert Eqs. (8.2a) and (8.2b) into the Maxwell's equations expressed in the curvilinear coordinate system. Then, using the same procedure as in Chap. 3, we order the terms in accordance to the powers of η, which, in turn, enables us to extract the equations satisfied by $\vec{E}_i$ and $\vec{H}_i$. These equations are then supplemented with various conditions depending upon the nature of the object. Thus, for a lossy dielectric we add the condition that the field decreases in the interior of the body; for a perfect conductor, we impose the criterion that the tangential component of the field vanishes on the surface; and, finally, for the coated case, we use the transmission conditions at the interfaces. The equations and the boundary conditions provide the relationships between the fields $\vec{E}$ and $\vec{H}$. By restricting ourselves to the first non-vanishing term in the representation of $\vec{E}$ and $\vec{H}$, we can show that we recover the Léontovitch condition [L], with an impedance given by

$$Z = \sqrt{\frac{\mu}{\varepsilon}} \quad \text{for the lossy dielectric,} \tag{8.3}$$

and

$$Z = -i \sqrt{\frac{\mu}{\varepsilon}} \; tg \, (k \, \sqrt{\varepsilon\mu} \; d), \tag{8.4}$$

for a conductor coated with a dielectric layer of thickness d. We see that the impedances can be calculated for a normally-incident plane wave on a half-space with material parameters (ε, μ), as in (8.3), and, for a half-plane coated with a layer of material (ε, μ) with thickness d. This result also still holds true for the case of a conductor coated with several layers of dielectrics.

The above discussion leads us to the following important result: As a first approximation, the surface impedance of a dielectric object with a large refractive index, or a conductor coated with dielectric layers, is the same as that of the equivalent multi-layered dielectric at normal incidence. Although this result has long been employed in practice, it has never been demonstrated rigorously. Cluchat [C] has calculated the O(1) correction to the above impedance, and has found that the higher-order impedance is anisotropic, which is a diagonal matrix when expressed in the coordinate system of the lines of curvature. Its expression takes the form

$$Z = \begin{pmatrix} Z_1 & 0 \\ 0 & Z_2 \end{pmatrix}, \tag{8.5}$$

with

$$Z_{1,2} = \sqrt{\frac{\mu}{\varepsilon}} \left(1 \pm \frac{h_1 - h_2}{2ik\sqrt{\varepsilon\mu}} \right), \tag{8.6}$$

for the homogeneous case. This result has also been given in [L], except for a misprint of a factor of 1/2.

For the case of one layer, the impedances are given by

$$Z_{1,2} = -i \sqrt{\frac{\mu}{\varepsilon}} \; tg(k\sqrt{\varepsilon\mu} \; d) \left(1 \mp tg(k\sqrt{\varepsilon\mu} \; d) \frac{(h_1 - h_2)}{2k\sqrt{\varepsilon\mu}} \right). \tag{8.7}$$

In both of the expressions given in (8.6) and (8.7), h_1 and h_2 are the curvatures along the directions 1 and 2.

The corrections thus obtained are small, especially at high frequencies, where k is large; hence, the correction terms can usually be neglected.

8.1.2 Surface impedance at high frequencies

At high frequencies, and for indices that are not necessarily large, it is still possible to define a surface impedance condition of the dielectrics as lossy and the object as smooth. We distinguish between the lit and shadow zones, where the reflection phenomenon and the creeping rays dominate, respectively. The procedure employed in the literature begins with the solution of the canonical problem of the cylinder, coated with one or several dielectric layers and the solution is expressed in an integral form. For a cylinder illuminated with a plane wave TE (Fig. 8.3), the total field at point M (ρ, θ) takes the form

$$H(M) = \sum_{m=-\infty}^{\infty} \int_{-\infty}^{+\infty} \left(J_\nu(k\rho) - \frac{J_\nu'(kb) + iZ_\nu J_\nu(kb)}{H_\nu'^1(kb) + iZ_\nu H_\nu^1(kb)} H_\nu^1(k\rho) \right) e^{i\nu\psi_m} \, d\nu , \quad (8.8)$$

where $\psi_m = |\theta + 2m\pi| - \pi/2$ and Z_ν is the impedance relative to the vacuum of the mode associated with $\exp(i\nu\theta)$ and b is the external radius of the coated cylinder.

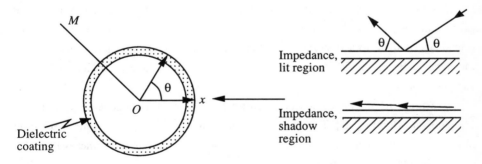

Fig. 8.3. Calculation of the surface impedance of a coated cylinder illuminated by a plane wave

Next, Z_ν is derived by requiring that the integrand satisfy the transmission conditions at the interfaces and the normal derivative vanish on the surface of the conductor.

For the case of a layer with constitutive parameters (ε_1, μ_1) and with thickness $(b - a)$ coating a conducting cylinder of radius a, we obtain, for instance

$$Z_v = i \sqrt{\frac{\mu_1}{\varepsilon_1}} \frac{J'_v(k_1 a) N'_v(k_1 b) - N'_v(k_1 a) J'_v(k_1 b)}{J'_v(k_1 a) N_v(k_1 b) - N'_v(k_1 a) J_v(k_1 b)}. \tag{8.9}$$

Kim and Wang [KW] have calculated the asymptotic expansion of the integral (8.8) for the case of a single layer. In the lit zone, we obtain the Geometrical Optics result, with a Fresnel reflection coefficient of a plane surface coated with a layer whose thickness e $= b - a$ is the same as that coating the cylinder. In the shadow zone, the field is expressed in the form of a series of creeping modes. When the layer is thin and lossy, these creeping modes are characterized by propagation constants v close to kb. If k_1 is sufficiently different from k, the Bessel functions in (8.9) can be replaced by their Debye asymptotic expansion. The impedance Z_v then takes the form

$$Z_v \approx -i \sqrt{\frac{\mu_1}{\varepsilon_1}} \sqrt{1 - \frac{v^2}{k_1^2 b^2}} \, tg\left(k_1 e \sqrt{1 - \frac{v^2}{k_1^2 b^2}}\right), \tag{8.10}$$

since $v \approx kb$, we have $\sqrt{1 - (v^2 / k_1^2 b^2)} \approx \sqrt{1 - (1/n^2)} = \cos\theta_1$, where θ_1 is the sin of the angle inside the material layer at grazing incidence (see Fig. 8.3).

Thus, we recognize in (8.10) the impedance of a plane conductor coated with a layer of the same material and thickness as that for the cylinder and illuminated at grazing incidence. For the case of a lossy layer whose index is not too close to 1, the impedance is then, to a first approximation, the same as that of a plate illuminated at the local angle of incidence for the lit region, and at the grazing incidence for the shadow zone.

Similar results have been obtained for the case of a homogeneous cylinder by Langlois and Boivin [LB]. To a first approximation, it is logical to assume that for an object coated by a lossy multilayer dielectric, the impedance is the same as that of a conducting plane covered with layers with the same thicknesses as those on the original coatings on the object. The impedance is calculated by illuminating the coated plane with the local angle of incidence for the lit zone and at grazing incidence for the shadow zone. In passing we note that, when the indices of the layers are large, then the wave propagates almost along the normal direction, and we can then use, with good approximation, the impedance derived from the normal incidence case over the entire object. We can then retrieve the result obtained in the previous section by means of an expansion in terms of the inverse powers of the index.

8.1.3 Treatment of the diffraction by edges and discontinuities

Up to this stage, we have only considered the case of smooth objects. For the case of a general object, we will then be able to treat, using the impedance condition, the contributions arising from the reflection and the creeping rays. The case of the edges, and more generally of the discontinuities of the coating, is much more involved and cannot be analyzed rigorously. In practice, we can still introduce an impedance type of description for the faces of the edge, or for each side of the discontinuity. This impedance is calculated as though the surface and the coating were continuous, i.e. the perturbation due to the discontinuity is ignored. This approach is probably valid for large indices and for lossy materials.

8.1.4 Summary

In summary, the impedance condition is applicable to smooth objects coated with lossy and/or large index materials regardless of the frequency, or to objects coated with lossy materials at high frequencies. In most practical cases, this enables us to calculate the contributions arising from the reflections and diffractions. However, we must resort to accurate solutions for low loss coatings, for those with only a moderately large index, or for the case of discontinuities for which the analyses presented previously no longer apply. Alternatively, we can use a direct procedure detailed in Sect. 8.2 below, to treat the material coating case. It relies upon the canonical problems of the coated plane and the cylinder.

8.2   Direct treatment of the material

The impedance condition gives, in most interesting practical cases, a satisfactory answer to the problem of calculating the RCS. It is also possible to take into account the material coating on an object in a more direct manner.

8.2.1 Reflected rays

As we have seen in Sect. 8.1., in analyzing the behavior of reflected rays from a coated object, one employs the reflection coefficient of an equivalent, planar multilayer sheet.

8.2.2 Transition zone and shadow zone of a smooth obstacle

It is possible to generalize the Fock functions defined for a surface impedance (see Chap. 3 and Appendix 5), to the coating case. Let us consider a circular cylinder coated with a multilayer dielectric for the case of the TE incidence. The total field for this case is given by (8.8) of Sect. 8.1.2, and we make the assumption, which is valid for lossy materials, that the principal contribution to the integral in (8.8) comes from the region $v = kb + m\tau$, where $m = (kb/2)^{1/3}$, and where τ is of O(1). The Bessel functions with argument kb can then be approximated by their Watson expansion in terms of the Fock-Airy functions, and by following the same procedure as that employed in Chap. 5 we obtain uniform asymptotic expansions of the solution. These expansions are similar to those we derived earlier by using the impedance condition except that the Fock functions are now replaced by the generalized ones, and that the impedance Z is no longer a constant but is replaced by the impedance Z_v of the mode associated with $\exp(iv\theta)$. For example, for the case of a layered material, Z_v is calculated by means of Eq. (8.9) of Sect. 8.1.2, with $v = kb + m\tau$, and, consequently, the impedance Z_v depends upon τ.

In order to provide a little more detail of the procedure we consider the example of the Pekeris function, which is given by the integral

$$P(\xi) = \frac{e^{-i\pi/4}}{\sqrt{\pi}} \int_{-\infty}^{+\infty} \frac{v'(\tau) - imZ(kb + m\tau)\, v(\tau)}{w'_1(\tau) - imZ(kb + m\tau)\, w_1(\tau)}\, d\tau, \tag{8.11}$$

with the constant impedance of the Pekeris function replaced by an impedance which is a function of the integration variable τ. The same is also true for other Fock functions occurring in the expressions for either the surface field or the field inside the boundary layer.

Finally, in the shadow zone, these Fock functions are expressed in terms of series of residues which can be interpreted as creeping rays. In particular, these series of residues enable us to calculate not only the detachment coefficients D but the attenuation constants α of the creeping waves as well.

To summarize, the canonical problem of the coated cylinder can be interpreted within the framework of the Geometric Theory of Diffraction, in exactly the same way as in the case of the canonical problem of the cylinder with an impedance type of surfaces. Thus, the results for a smooth obstacle can be derived from the formulas given in Chap. 1 (standard GTD) by replacing the detachment coefficients and attenuation constants calculated for a surface impedance with those derived for a coating, and also by replacing the conventional Fock function formulas of Chap. 5 (uniform version) with the generalized Fock functions. However, we note that this approach is heuristic, since it is based on the extrapolation of canonical solutions in the case of general obstacles. The direct treatment of the material, enables one to refine the evaluation of the contributions arising from the reflections and creeping rays. Having discussed the problem of scattering by a smooth object, we now proceed to discuss the case of diffraction by an edge.

8.2.3 Diffraction by a wedge coated with material coating

The problem of the diffraction by a wedge, defined $-\Phi < \theta < \Phi$ (see Fig. 8.4), with impedances Z_+ and Z_- on the faces $\theta = \Phi$ and $-\Phi$, respectively has been solved by Maliuzhinets in 1958 [Ma]. He began with a representation of the solution in the form of a plane wave spectrum along the Sommerfeld-Maliuzhinets contour D (see Chap. 4)

$$u(r,\ \theta) = \frac{1}{2\pi i}\int_D s(\alpha + \theta)\ exp(-ikr\ cos\ \alpha)\ d\alpha. \tag{8.12}$$

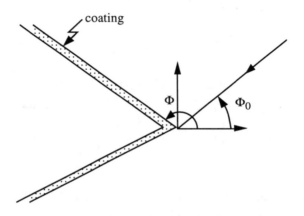

Fig. 8.4. Diffraction by a coated wedge

By enforcing the impedance condition on u given by (8.12) on each face of the wedge, Maliuzhinets derived two equations determining the function s, viz.,

$$s(\Phi + \alpha) = -R_+\ s\ (\Phi - \alpha), \tag{8.13a}$$

$$s(-(\Phi + \alpha)) = -R_-\ s(-\Phi + \alpha), \tag{8.13b}$$

where R_+ and R_- are the reflection coefficients of the faces $\theta = \varphi$ and $-\varphi$ respectively, that are described by the surface impedance Z_+ and Z_-.

After defining R_+ as $(\sin \alpha - Z_+/\sin \alpha + Z_+)$ and R_- as $(\sin \alpha - Z_-/\sin \alpha - Z_-)$ for the TE case, Michaeli [Mi] proposes to take the reflection coefficients for the coating on the wedge instead of those for a constant impedance. He succeeds, then, to calculate the function s(α) following the same procedure, as described by Maliuzhinets. For a plane wave, which is incident from the direction $\theta = \Phi_0$, Michaeli finds

$$s(\alpha) = \sigma(\alpha)\ \psi(\alpha)/\psi(\Phi_0) \tag{8.14}$$

where the function $\sigma(\alpha)$, which does not depend upon the coating, is given by

$$\sigma(\alpha) = \frac{1}{n}\cos\frac{\Phi_0}{n}\Big/\left(\sin\frac{\alpha}{n} - \sin\frac{\Phi_0}{n}\right), \tag{8.15}$$

$$n = 2\Phi/\pi. \tag{8.16}$$

On the other hand, $\psi(\alpha)$, which does take into account the properties of the coating, is given by

$$\psi(\alpha) = \psi_+(\alpha + \Phi)\ \psi_-(\alpha + \Phi). \tag{8.17}$$

In the above formula the of functions ψ_+ and ψ_- are defined by

$$\psi_\pm(\alpha) = \exp\left[-\frac{1}{2\pi}\int_0^\infty \ell n\left(\frac{cos\left(\frac{y}{n}\right) + ch\left(\frac{y}{n}\right)}{ch\left(\frac{y}{n}\right) + 1}\right)\frac{R'_\pm(iy)}{R_\pm(iy)}\,dy\right]. \tag{8.18}$$

The asymptotic evaluation of (8.18) with the stationary phase method enables one to extract the following diffraction coefficient of the coated wedge

$$D(\theta, \Phi_0) = \frac{\psi(\theta - \pi)\ \sigma(\theta - \pi) - \psi(\theta + \pi)\ \sigma(\theta + \pi)}{\psi(\Phi_0)}\ \frac{e^{i\pi/4}}{\sqrt{kr}}. \tag{8.19}$$

The solution of Michaeli is found to be more accurate than that obtained by using a constant impedance. However, we note that it relies upon the assumption that the impedance of the coated infinite plane continues to apply all the way up to the tip of the wedge, which is not completely correct. Indeed, near the tip of the wedge, the impedance is perturbed by the corner.

8.2.4 Summary of the procedure for the direct treatment of material

To summarize the discussion given above, it is possible to treat, directly, the case of a material-coated object by using:

(i)   the reflection coefficient of the equivalent plane multilayer for the reflected rays,

(ii) the generalized Fock functions, as deduced from the canonical problem of a cylinder coated with the same layered medium as that coating the object, for the creeping rays,

(iii) Michaeli's solution for the coated wedge to treat the diffracted rays by edges.

We should point out that although the procedure described above is very promising, it has not yet been employed extensively in the literature. In addition, we note that the procedure relies upon an extrapolation of the solution of an associated canonical problem to the case of a general obstacle, and this is precisely the initial approach of the GTD as presented in Chap. 1. Consequently this treatment still appears as based on heuristic argument.

Furthermore, the case of the edges is solved only approximately since one assumes that the impedance of the coating on an infinite plane applies up to the tip of the edge which means that the effect of the tip on the impedance is neglected. An other approach to treat a coated object, which follows a strategy halfway between the impedance boundary condition and the direct treatment of materials, consists in using a more general surface condition which is known as the generalized impedance condition. We are now going to review this method in the next section.

8.3 Generalized surface impedance

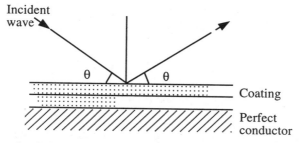

Fig. 8.5. Geometry for deriving the generalized impedance boundary condition with a plane wave incident upon a coated surface at an angle θ

The notion of generalized surface impedance has been introduced by Karp and Karal [KK], and it has been developed, in particular, by Senior and Volakis [SV] and Rojas [Ro1]. These authors start in general from a coating on a conducting plane. For the sake of simplification let us restrict ourselves to the bidimensional case. We consider a plane wave at incidence characterized by the angle θ. The surface impedance, defined as the ratio Etg/Htg calculated at the surface of the coating, depends on θ, i.e., $Z = Z(\theta)$. Using a symmetry argument we conclude that $Z(\pi - \theta) = Z(\theta)$, or $Z = Z(\sin \theta)$. But, for $\theta \in [0, \pi]$, $\sin \theta = \sqrt{1 - \cos^2 \theta}$, and it appears that Z is a function of $\cos^2 \theta$. Let us assume that this function is analytical in such a way that it can be expanded as an entire series

$$Z = a_0 + a_1 \cos^2 \theta \ldots + a_n \cos^{2n} \theta + \ldots . \qquad (8.20)$$

It then becomes possible to extract this surface impedance dependent on θ thanks to a local boundary condition. In fact, let us consider the following boundary condition for $y = 0$

$$\frac{\partial u}{\partial Y} = b_0 u + b_1 \frac{\partial^2 u}{\partial X^2} + \ldots b_n \frac{\partial^n u}{\partial X^n}, \qquad (8.21)$$

where $X = kx$, $Y = ky$.

The incident and reflected waves are given by $u^i = \exp(-ik(x \cos \theta + y \sin \theta))$ and $u^r = R \exp(-ik(x \cos \theta - y \sin \theta))$ where R is the reflection coefficient. Since the total field satisfies (8.21), we are led to the equation

$$- i \sin \theta + R\, i \sin \theta = (b_0 - b_1 \cos^2 \theta + b_2 \cos^4 \theta \ldots)\,(1 + R), \qquad (8.22)$$

which yields

$$R = \frac{\sin\theta - i(b_0 - b_1 \cos^2 \theta + b_2 \cos^4 \theta + \ldots)}{\sin\theta + i(b_0 - b_1 \cos^2 \theta + b_2 \cos^4 \theta + \ldots)}. \qquad (8.23)$$

Thus the impedance can be written as

$$Z = i\,(b_0 - b_1 \cos^2 \theta + b_2 \cos^4 \theta + \ldots), \qquad (8.24)$$

and the case of plane coating with impedance given by (8.20) can be handled by setting

$$b_0 = - ia_0,\ b_1 = ia_1,\ \ldots b_n = (-1)^{n+1}\, ia_n. \qquad (8.25)$$

The conventional impedance condition consists in keeping only the first term in the series (8.24), whereas the generalized impedance condition is derived by retaining a finite number of terms in the series. Alternatively, one can express Z as a rational fraction, as opposed to a series, which lead other forms of generalized impedance boundary condition. For instance, the boundary condition

$$\frac{\partial}{\partial Y}\left(c_0 u + c_1 \frac{\partial^2 u}{\partial X^2} + c_2 \frac{\partial^2 u}{\partial X^4} + \dots\right) = b_0 u + b_1 \frac{\partial^2 u}{\partial X^2} + b_2 \frac{\partial^2 u}{\partial X^4} + \dots, \quad (8.26)$$

is equivalent to the surface impedance

$$Z = i\, \frac{b_0 - b_1 \cos^2\theta + b_2 \cos^4\theta + \dots}{c_0 - c_1 \cos^2\theta + c_2 \cos^4\theta + \dots}. \quad (8.27)$$

As far as the computation of the reflected rays and the creeping rays via the use of an asymptotic method are concerned, the generalized impedance condition does not improve the solution over that obtained by a direct treatment of the material, either in terms of simplicity or accuracy. The situation is different for the case of the discontinuities of a material. Consider, for example, the diffraction by a plane satisfying the following generalized impedance condition (Fig. 8.6)

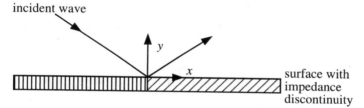

Fig. 8.6. Diffraction by a planar discontinuity in generalized impedance

$$\frac{\partial u}{\partial y} = a_- u + b_- \frac{\partial^2 u}{\partial x^2} \quad \text{for } x < 0, \quad (8.28a)$$

$$\frac{\partial u}{\partial y} = a_+ u + b_+ \frac{\partial^2 u}{\partial x^2} \quad \text{for } x > 0. \quad (8.28b)$$

For $y > 0$, the total field satisfies the wave equations

$$(\nabla^2 + k^2)\, u = 0. \quad (8.29)$$

Multiplying (8.29) by a test function υ and integrating by parts, one obtains [Ly] (the function v was assumed to be continuous)

$$- \int_{y>0} (\vec{\nabla} u \cdot \vec{\nabla} \upsilon) \, dx \, dy + k^2 \int_{y>0} u\upsilon \, dx \, dy - \int_{-\infty}^{+\infty} a_{\pm} u\upsilon \, dx +$$

$$+ \int_{-\infty}^{+\infty} b_{\pm} \frac{\partial u}{\partial x} \frac{\partial v}{\partial x} \, dx - \left(b_+ \frac{\partial u}{\partial x}(0_+) - b_- \frac{\partial u}{\partial x}(0_-) \right) \upsilon(0) = 0. \tag{8.30}$$

We now take $\bar{u}$, the conjugate of u, as the test function, υ. We also assume that $\mathrm{Im}\ k^2 > 0$, which is the case if the half-space $y > 0$ is filled with a lossy medium. The first term of (8.29) has a vanishing imaginary part; the second has a positive imaginary part; the third has a positive imaginary part if $\mathrm{Im}\ a_{\pm} < 0$, which corresponds to a passive impedance; and the fourth has a positive imaginary part if $\mathrm{Im}\ b_{\pm} > 0$. We assume that the above conditions are satisfied. If the imaginary part of the fifth term is also positive, the solution of the homogeneous equation (8.29), satisfying the homogeneous boundary conditions given by (8.28), is unique. To this end, one define a linear relation, called the contact condition

$$b_+ \frac{\partial u}{\partial x}(0_+) - b_- \frac{\partial u}{\partial x}(0_-) = c\, u(0), \tag{8.31}$$

with

$$\mathrm{Im}\ c > 0. \tag{8.32}$$

The choice of c is arbitrary if (8.32) is satisfied, which provides us with another degree of freedom with which to take into account the effect of the discontinuity. The generalized impedance method has been applied in particular to the case of the coated wedge by Bernard [Be] and to a number of practical geometries by Senior, Volakis and their collaborators (for example [SV]) and also by Rojas [Ro2]. The problem of deriving the quantity c is still being investigated, and it remains a research topic.

8.4 Conclusions

The surface impedance approach permits us to treat, either at high frequencies or for large indices, the case of lossy coatings on smooth objects. The calculation of the impedance relies, in particular in the second case, upon a rigorous mathematical demonstration. It consists in calculating, as a first step, the impedance at the surface of a plane multilayer geometry illuminated by a plane wave. The impedance representation is a reliable approximation and is convenient to use in most practical cases.

However, to obtain a better accuracy, it is possible, to derive the contributions of the reflection and creeping rays, by treating the material problem directly, and this is done by extrapolating the results of canonical problems. The diffraction by discontinuities coated with material media still continues to be a research topic. A number of approximations are available for handling the coated body case and are based upon simple or generalized impedance condition or the calculation of the exact reflection coefficient of the coating. However, the perturbation due to the discontinuity is not completely accounted for in the above approaches.

References

[AC] M. Artola and M. Cessenat, "Diffraction d'une onde électromagnétique par un obstacle à perméabilité élevée," CRAS, t 314, série 1, 349-354, 1992.

[C] T. Cluchat, "Etude de validité de la condition d'impédance," Ph.D Dissertation, University of Bordeaux, 1992.

[Be] J. M. Bernard, "Diffraction by a metallic wedge covered with a dielectric material," *J. Wave Motion*, 9, 543-561, 1987.

[KK] S. Karp and F. Karal, "Generalized impedance boundary conditions with applications to surface wave structures," in E. Brown (Ed.) *Electromagnetic Wave Theory*, Pergamon Press, 1965.

[KW] H. T. Kim and N. Wang, "UTD solution for electromagnetic scattering by a circular cylinder with a thin coating," *IEEE Trans. Ant. Prop.*, **AP-37**, 1463-1472, Nov. 1989.

[L] M. A. Léontovich, "Investigation of radio wave propagation," Partie II, *Izd. AN SSSR.*, Moscow, 1948.

[LB] P. Langlois and A. Boivin, *Can Journal of Physics*, **61**, 332, 1983.

[Ly] M. A. Lyalinov, "Electromagnetic scattering by coated surfaces," Journées de diffraction, St. Pétersbourg, 1993.

[Mi] A. Michaeli, "Extension of Malhiuzinets solution to arbitrary enabletivity and permeability of coatings on a perfect conducting wedge," *Electron. Lett.*, **24**, 1291-1294, Erratum, p. 1521, 198.

[Ma] Maliuzhinets, *Sov. Phys. Dokl.*, **3**, 752, 1958.

[Ro1] R. Rojas, "Generalized impedance boundary conditions for electromagnetic scattering problems," *Electron. Lett.*, **24**(17), 1093-1094, 1987.

[Ro2] R. Rojas and L. Chou, "Diffraction by a partially coated PEC half-plane," *Radio Sci.*, **25**(2), 175-188, 1990.

[R]   S. M. Rytov, *Zh. Eksp.Teor. Fiz.*, **10**, 180, 1940.

[SV]   T. Senior and S. Volakis, "Derivation and application of a class of GIBC," *IEEE AP*, **37**, 1566-1572, Dec. 1989.

Appendix 1. Canonical Problems

In this appendix we consider some canonical problems, whose solutions may be regarded as the cornerstones of the GTD. We will restrict ourselves to three representative canonical problems, viz., (i) plane; (ii) cylinder; and (iii) wedge, that are useful, respectively, for studying reflection, creeping wave and diffraction phenomena. Much of the literature on the subject of scattering from canonical geometries is devoted to perfectly conducting cases. However, we will extend our discussion to the more general case of scatterer surfaces described by an impedance boundary condition for the cylinder and the plane. For the case of the wedge we will begin, for the sake of simplicity, with perfectly conducting scatterers and extend it later to treat the surface impedance case.

For a TE (TM) incident field, the problem can be reduced to a scalar problem for a wave function $u = H$ or E, satisfying the following boundary condition on the scatterer

$$\frac{\partial u}{\partial n} + i \, k \, Z \, u = 0,$$

when u represents the magnetic field, and

$$\frac{\partial u}{\partial n} + i \, \frac{k}{Z} \, u = 0,$$

when $u = E$.

A1.1 Reflection of a plane wave by a plane

Let us consider a plane wave whose polarization is TE and incident angle is θ. Let it impinge on a plane satisfying an impedance condition, namely

$$\frac{\partial u}{\partial n} + i \, k \, Z \, u = 0. \tag{A1.1}$$

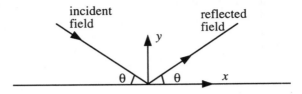

Fig. A1.1. Reflection of a plane wave by a plane

We seek the solution in the form of a sum of the incident wave

$$u^i = exp \, (ik(x \, cos \, \theta - y \, sin \, \theta)),\qquad (A1.2)$$

and a reflected wave

$$u^r = R \, exp \, (ik(x \, cos \, \theta + y \, sin \, \theta)),\qquad (A1.3)$$

where R is unknown.

In order to determine R, we insert (A1.2) and (A1.3) into (A1.1), which leads to

$$R = \frac{sin \, \theta - Z}{sin \, \theta + Z},\qquad (A1.4)$$

which is precisely the reflection coefficient R_{TE} (see Eq. (1.90) of Sect. 1.5.1).

For an incident wave with a TM polarization, the impedance condition reads $\partial u / \partial n + i \, k/Z \, u = 0$, and the reflection coefficient takes the form

$$R_{TM} = \frac{Z \, sin \, \theta - 1}{Z \, sin \, \theta + 1}.$$

We see, then, that the canonical problem of the plane enables us to calculate the TE and TM reflection coefficients. Next, we consider the canonical problem of the cylinder whose solution will yield the propagation constants of the creeping rays and their detachment coefficients.

A1.2 Diffraction by a circular cylinder whose surface impedance is constant

A1.2.1 General solution of the cylinder problem

Consider an electric or magnetic type of line source current whose intensity is unity and whose direction is parallel to the generatrix of an infinite cylinder with radius a

and a constant surface impedance Z (Fig. A1.3). The problem we wish to solve is the calculation of the total field in the space exterior to the cylinder.

Let Oxyz be a triad with the Oz axis as the axis of revolution of the cylinder. Let us denote ρ and θ as the polar coordinates of the intersection point of the linear source and the plane xy. As mentioned earlier, the problem can be reduced to the solution of the following scalar problem:

$$(\nabla^2 + k^2)\, u(r,\, \theta) = -\frac{1}{r}\, \delta(r-\rho)\, \delta(\theta), \quad r > a, \tag{A1.3}$$

$$\frac{\partial u}{\partial r} + i\, k\, \zeta u = 0, \quad r = a, \tag{A1.4}$$

where $\zeta = Z$, $u = H_z$ for a wave TM and $\zeta = 1/Z$, $u = E_z$ for a wave TE.

The Eqs. (A1.3) and (A1.4) must be supplemented by the Sommerfeld radiation condition at infinity, which is stated as

$$\lim_{r\to\infty} r^{1/2}\left(\frac{\partial u}{\partial r} - iku\right) = 0. \tag{A1.5}$$

The solution to the scattering problem, as stated above, is straightforward and leads to the following expression for the expression for u in the range r > a

$$u = \frac{i}{4} \sum_{m=-\infty}^{+\infty} e^{im\theta}\left[J_m(kr_<) - \frac{\Omega J_m(ka)}{\Omega H_m^{(1)}(ka)} H_m^{(1)}(kr_<) \right] H_m^{(1)}(kr_>), \tag{A1.6}$$

where we have set

$$\Omega = \frac{\partial}{\partial r} + ik\zeta,$$

$$r_< = Inf(r,\rho),\ r_> = sup(r,\rho).$$

The series in (A1.6) converges very slowly when the radius of the cylinder is large, i.e., when ka >> 1, and also for points far away from the cylinder. However, as we will now show, we can apply the Watson transform to the series and convert it into one which is rapidly convergent, at least in the region of space that corresponds to the deep shadow zone.

Let S denote the following series

$$S = \sum_{n=1}^{\infty} f(n).$$

Each term of this series can be viewed as the residue of an integral along a close contour in the complex plane. If we denote $g(v)$ as an analytical function such that $g(n) = 0$ for n = 1, 2, 3, ..., then using the residue theorem we can write

$$S = -\frac{1}{2\pi i} \int_C f(v) \frac{\frac{dg(v)}{dv}}{g(v)} dv,$$

where C is a contour in the complex v-plane, which surrounds the Re v axis as shown in Fig. A1.2 below.

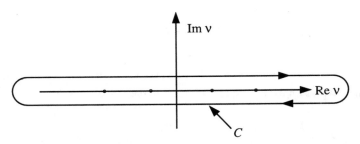

Fig. A1.2. Contour C for the Watson transformation

If we choose the function $g(v) = \sin v\pi$, the right-hand side of (A1.6) can be written in the form

$$U = -\frac{1}{8} \int_C \frac{e^{iv(\theta-\pi)}}{\sin v\pi} \left[J_v(kr_<) - \frac{\Omega J_v(ka)}{\Omega H_v^{(1)}(ka)} H_v^{(1)}(kr_<) \right] H_v^{(1)}(kr_>) \, dv. \quad (A1.7)$$

Next, replacing v by $-v$ on the part of the contour C for which $Imv < 0$, and taking into account the fact that the integrand of (A1.7) is odd except for the factor $\exp[iv(\theta - \pi)]$, the integral in (A1.7) can be rewritten as

$$U = -\frac{1}{4} \int_D \frac{\cos v(\theta-\pi)}{\sin v\pi} \left[J_v(kr_<) - \frac{\Omega J_v(ka)}{\Omega H_v^{(1)}(ka)} H_v^{(1)}(kr_<) \right] H_v^{(1)}(kr_>) \, dv, \quad (A1.8)$$

where D is a contour in the upper half-plane of v, parallel to the real v axis. When the observation point is located in the geometric shadow of the source, the contour D can be closed by the circle at infinity in the upper half plane and the integral can be evaluated by using the calculus of residues, after noting that the poles v_n of the integrand are the zeros of $\Omega H_v^{(1)}$.

Next, using the relationship

$$J_v(x) = \frac{1}{2}\left[H_v^{(1)}(x) + H_v^{(2)}(x)\right],$$

we can obtain the following representation for U from (A1.8)

$$U = \frac{\pi i}{4} \sum_{n=1}^{\infty} \frac{\cos v_n(\theta - \pi)}{\sin v_n \pi} \frac{\Omega H_{v_n}^{(2)}(ka)}{\left[\frac{\partial}{\partial v}\Omega H_v^{(1)}(ka)\right]_{v=v_n}} H_{v_n}^{(1)}(k\rho)\, H_{v_n}^{(1)}(kr). \quad \text{(A1.9)}$$

We now proceed to derive an asymptotic representation of U from the expression in (A1.9). To this end, we first note that

$$\frac{\cos v_n(\theta - \pi)}{\sin v_n \pi} = -\frac{e^{iv_n\theta} + e^{iv_n(2\pi-\theta)}}{1 - e^{2\pi i v_n}}\, i\,,$$

and that for large values of $k\rho$ and kr, the Hankel functions $H_{v_n}^{(1)}(k\rho)$ and $H_{v_n}^{(1)}(kr)$ can be replaced by their Debye asymptotic expansions

$$H_v^{(1)}(x) = \frac{\sqrt{2}}{\sqrt{\pi x \sin\gamma}} e^{ix(\sin\gamma - \gamma\cos\gamma)}\, e^{-i\,\pi/4},\ \cos\gamma = \frac{v}{x},|v - x| > O(|v|^{1/3})\,.$$

Inserting these expressions into (A1.9) we obtain

$$U = -i\,\frac{e^{ik\left[(r^2-a^2)^{1/2} + (\rho^2-a^2)^{1/4}\right]}}{2k(r^2 - a^2)^{1/4}(\rho^2 - a^2)^{1/4}}$$

$$\sum_{n=1}^{\infty} \frac{e^{iv_n\theta} + e^{iv_n(2\pi-\theta)}}{1 - e^{2\pi i v_n}} e^{-iv_n[\text{Arc}\cos(v_n/k\rho) + \text{Arc}\cos(v_n/kr)]} \times \frac{\Omega H_{v_n}^{(2)}(ka)}{\left[\frac{\partial}{\partial v}\Omega H_v^{(1)}(ka)\right]_{v=v_n}}.$$

$$\text{(A1.10)}$$

The reason why the expression in (A1.10) is valid only if the observation point is located in the geometric shadow of the source is as follows. The terms

$$e^{i\nu_n[\theta - Arc\,cos(\nu_n/k\rho) - Arc\,cos(\nu_n/kr)]}, \quad e^{i\nu_n[2\pi - \theta - Arc\,cos(\nu_n/k\rho) - Arc\,cos(\nu_n/kr)]},$$

are evanescent only if θ falls in the range

$$Arc\,cos\left(\frac{\nu_n}{k\rho}\right) + Arc\,cos\left(\frac{\nu_n}{kr}\right) < \theta < 2\pi - Arc\,cos\left(\frac{\nu_n}{k\rho}\right) - Arc\,cos\left(\frac{\nu_n}{kr}\right).$$

We will see later that, $\nu_n \approx ka \gg 1$. Consequently, the condition on θ given above reads

$$\Phi_1 + \Phi_2 < \theta < 2\pi - (\Phi_1 + \Phi_2),$$

with

$$\Phi_1 = Arc\,cos\left(\frac{a}{\rho}\right), \quad \Phi_2 = Arc\,cos\left(\frac{a}{r}\right).$$

Using the above notations, the field in the shadow region, given by (A1.10), can be written as a sum of two terms, viz.,

$$e^{i\nu_n[\theta - \Phi_1 - \Phi_2]}, \quad e^{i\nu_n[2\pi - \theta - \Phi_1 - \Phi_2]},$$

which can be interpreted as the creeping rays. More precisely, (A1.10) takes the form

$$U = -i\frac{exp\,ik(\rho^2 - a^2)^{1/2}}{(2k)^{1/2}(\rho^2 - a^2)^{1/4}}\frac{exp\,ik(r^2 - a^2)^{1/2}}{(2k)^{1/2}(r^2 - a^2)^{1/4}}$$

$$\sum_{n=1}^{\infty}\frac{1}{1 - e^{2\pi i\nu_n}}\left(e^{i\nu_n[\theta - \Phi_1 - \Phi_2]} + e^{i\nu_n[2\pi - \theta - \Phi_1 - \Phi_2]}\right)\frac{\Omega H_{\nu_n}^{(2)}(ka)}{\left[\frac{\partial}{\partial\nu}\Omega H_\nu^{(1)}(ka)\right]_{\nu=\nu_n}}.$$

$$\tag{A1.11}$$

We will now present a physical interpretation of the above representation for U. Referring to Fig. A1.3, we can identify the following three stages through which the ray undergoes: (i) attachment; (ii) propagation on the surface of the cylinder as a creeping ray; and, finally, (iii) detachment of the ray. The first term in (A1.11) corresponds to the propagation in free space of the incident field produced by the source S. Let us denote P_1 as a point on the cylinder such that SP_1 is tangent to its surface and, hence, the distance SP_1 equals $\sqrt{\rho^2 - a^2}$. Next, let us assume that the point is located at a distance sufficiently large from the surface S such that it is possible to replace the Hankel function, describing the incident field U^i by the first term of its asymptotic expansion. Then the incident field reads

$$U^i(P_1) = \frac{e^{i\pi/4}}{\sqrt{8\pi k}} \frac{exp\, i(\rho^2 - a^2)^{1/2}}{(\rho^2 - a^2)^{1/4}}.$$

The propagation of the detachment point P_2 of the creeping ray at the observation point yields a phase term $exp\, ik(r^2 - a^2)^{1/2}$ and an amplitude term whose geometric divergence is $(r^2 - a^2)^{1/4}$. The angle $\theta_1 = [\theta - \Phi_1 - \Phi_2]$ corresponds to the distance P_1P_2 traversed by the creeping ray on the upper surface of the circular cylinder, while the angle $\theta_2 = [2\pi - \theta - \Phi_1 - \Phi_2]$ is associated with the distance $P'_1 P'_2$ traveled by the creeping ray underneath the cylinder. Thus, the creeping rays propagate on a cylinder with a propagation constant ν_n/a, where ν_n is one the zeros of $\Omega H_\nu^{(1)}(ka)$, which will be determined in Sect. 2. Finally, the term $(1 - e^{2\pi i\nu_n})^{-1}$ can be expressed in the form of the series $(1 + e^{2\pi i\nu} + e^{4\pi i\nu} + ...)$, and the terms of this series can be associated with the creeping rays circumnavigating the cylinder one or more times. We note, however, that the term $(1 - e^{2\pi i\nu_n})^{-1}$ is always very close to unity.

The GTD interpretation of the field at the point P finally leads to the following expression for U

$$U = U^i(P_1) \sum_{n=1}^{\infty} \sum_{m=1}^{\infty} D_n^2 (exp(i\, m\nu_n\theta_1) + exp(i\, m\nu_n\theta_2)) \frac{(exp\, ik(r^2 - a^2)^{1/2}}{(r^2 - a^2)^{1/4}}.$$

$$(A1.12)$$

The double summation is carried out over the index n of the creeping ray mode and m, the number of trips the ray takes around the cylinder. A comparison of (A1.12) with (A1.11) enables us to determine the square of the detachment coefficient, and it is given by

$$D^2 = e^{-3i\pi/4}(2\pi/k)^{1/2}\Omega H_{\nu_n}^{(2)}(ka) \Big/ \Big[\frac{\partial}{\partial\nu}\Omega H_\nu^{(1)}(ka)\Big]_{\nu=\nu_n}.$$

In summary, the solution to the canonical problem of the circular cylinder illuminated by a point source can be interpreted in terms of creeping rays and the fact that this solution expressed in an explicit form helps us extract the propagation constants and the detachment coefficients of these rays. We will present some results for the propagation constants and the detachment coefficients in the next section and in Sect. 3, respectively.

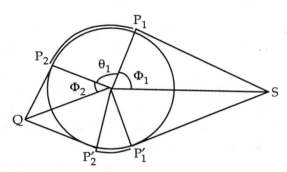

Fig. A1.3. Creeping rays on a circular cylinder

A1.2.2 Expressions for the zeros of $\Omega H_\nu^{(1)}(ka)$ and the propagation constants

The zeros of $H_\nu^{(1)}(ka)$ and $H_\nu^{\prime(1)}(ka)$ are located in the first and fourth quadrant of the complex plane of ν. If ka is real, which is always the case when the cylinder is in free space, the zeros of $H_\nu^{(1)}(ka)$ are located on the curves as shown in Fig. A1.4 below.

These curves, which are symmetrical with respect to the origin, are also called *Stokes lines* of the function $H_\nu(ka)$. The contour D, introduced previously, is located in the upper half-plane, and passes below the first zero of $H_\nu^{(1)}(ka)$ in the first quadrant. Since the contour is closed at infinity with a circle located in the upper half-plane, only the zeros with a positive imaginary part contribute to the series in (A1.9) and (A1.10). Consequently, we only show the zeros in the first quadrant in the figure below.

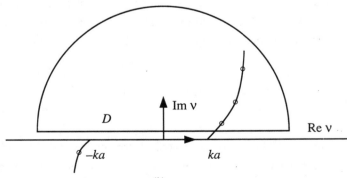

Fig. A1.4. Location of the zeros of $H_\nu^{(1)}(ka)\,\nu_n$ and the contour D in the complex plane

The values of ν_n appearing in the series in (A1.9) and (A1.10) are the zeros of $\Omega H_\nu^{(1)}(ka)$. Thus they satisfy the equation

$$\left[\frac{\partial H_{\nu_n}^{(1)}(k\rho)}{\delta\rho}\right]_{\rho=a} + ik\zeta H_{\nu_n}^{(1)}(ka) = 0 ,$$

which, by setting $ka = x$, can be rewritten as

$$\frac{\partial H_{\nu_n}^{(1)}(x)}{\partial x} + i\zeta H_{\nu_n}^{(1)}(x) = 0 . \tag{A1.13}$$

The asymptotic expansions (of Watson) of Hankel functions and of their derivatives with respect to the argument and the index are valid in the vicinity of the Stokes line. These expansions are given by

$$H_{\nu_n}^{(1)}(ka) = 2^{4/3}(\nu_n)^{-1/3} e^{-i\pi/3} Ai(y)$$

$$\left[\frac{\partial}{\partial x} H_{\nu_n}^{(1)}(x)\right]_{x=ka} = 2^{4/3}(\nu_n)^{-1/3} e^{-i\pi/3} Ai'(y)\left(\frac{\partial y}{\partial x}\right)_{x=ka} ,$$

$$\left[\frac{\partial}{\partial \nu} H_\nu^{(1)}(ka)\right]_{\nu=\nu_n} = 2^{4/3}(\nu_n)^{-1/3} e^{-i\pi/3} Ai'(y)\left(\frac{\partial y}{\partial \nu}\right)_{\nu=\nu_n}$$

with

$$y = 2^{1/3} e^{-i\pi/3} x^{-1/3} (x - \nu),$$

$$\frac{\partial y}{\partial x} = -2^{1/3} e^{-i\pi/3} 1/3 \, x^{-4/3}(x - \nu) + 2^{1/3} e^{-i\pi/3} x^{-1/3} \text{ Å } 2^{1/3} e^{-i\pi/3} x^{-1/3},$$

$$\frac{\partial y}{\partial \nu} = -2^{1/3} e^{-i\pi/3} x^{-1/3},$$

where $Ai(y)$ is the Airy function defined by (see Abramowitz and Stegun [AS] p. 447)

$$Ai(y) = \frac{1}{\pi} \int_0^\infty \cos\left(\frac{1}{3}\tau^3 + y\tau\right) d\tau ,$$

which satisfies the equation

$$Ai''(y) = yAi(y).$$

Inserting these expansions into (A1.13), we can derive the following equation for y

$$Ai'(y) - m \, \zeta \, e^{-i\pi/6} \, Ai(y) = 0. \tag{A1.14}$$

We can identify (A1.14) as (1.117) of Sect. 1.5.4, which yields the attenuation constants of the creeping ray. In the above expression, $y = -q_h^n(Z)$, and $y = -q_s^n(Z)$, for the TE and TM polarizations, respectively.

For $\zeta = 0$, the problem becomes a Neuman type. All of the zeros of the derivative of the Airy function are solutions of (A1.12). The first zero is approximately equal to -1.019. In contrast, it is a Dirichlet problem for $\zeta = \infty$, and all of the zeros of the Airy function are solutions of (A1.14).

There exists an infinite number of zeros of the Airy function and of its derivative, and, consequently, the number of solutions is infinite as well. Each solution corresponds to a mode with an index n. This is also true for the general case of an arbitrary ζ.

Once $y(\zeta)$ has been determined, the propagation constant can be found from $\nu/a = k - e^{i\pi/3} ym/a$, where m = $(ka/2)^{1/3}$ is the Fock parameter. It is the sum of the wave number k in vacuum and of the attenuation constant $a_h^n(Z)$ introduced in Chap. 1. The attenuation constant is given by $a_h^n(Z) = -e^{i\pi/3} ym/a = -e^{i\pi/3} q_h^n(Z)$ m/a for the TE polarization; thus, we recover the result in (1.115), given in Sect. 1.5.4. We note, therefore, that the canonical problem of the circular cylinder with a surface impedance condition provides the attenuation constants of the creeping rays. Before concluding this section, we will briefly present a method for solving (A1.12) and discuss the effect of the impedance on the propagation constant.

To determine $y(\zeta)$ as a function of the parameter ζ, we have to solve (A1.14) in the complex plane [GB]. This problem can be reduced to the numerical solution of a differential equation. The initial solution, defined at $\zeta = 0$, is a zero of the derivative of the Airy function. It is also possible to solve (A1.13) directly for ν_n, by using the Newton method. Figure A1.5 shows the evolution of the $\nu_n(\zeta)$ in the complex plane as ζ traverses the complex plane with an angle θ with respect to the origin. For surface impedances that are negative and purely imaginary, the attenuation is lower than that of the perfectly conducting case, and these surface waves can play an important role in the diffraction of fields from smooth surfaces with an impedance type of boundary condition. For surface impedances whose real part is sufficiently large, the attenuation is more important than for a perfect conductor. We thus observe that the surface impedance provides a way to modulate the attenuation of the creeping rays.

A1.2.3 Diffraction coefficient Đ(n, ζ, ka)

The diffraction coefficient D(n, ζ, ka) is derived from an asymptotic expansion of $D^2 = e^{-3i\pi/4}(2\pi/k)^{1/2}R_n$, where

$$R_n = \frac{\Omega H_{\nu_n}^{(2)}(ka)}{\left[\frac{\partial}{\partial\nu}\Omega H_\nu^{(1)}(ka)\right]_{\nu=\nu_n}} = \frac{H_{\nu_n}'^{(2)}(ka) + i\zeta H_{\nu_n}^{(2)}(ka)}{\frac{\partial}{\partial\nu}\left[H_\nu'^{(1)}(ka) + i\zeta H_\nu^{(1)}(ka)\right]_{\nu=\nu_n}}. \qquad (A1.15)$$

The Wronskian of $H_{\nu_n}^{(1)}$ and $H_{\nu_n}^{(2)}$ takes the form

$$-H_{\nu_n}^{(1)}(ka)H_{\nu_n}'^{(2)}(ka) + H_{\nu_n}'^{(1)}(ka)H_{\nu_n}^{(2)}(ka) = \frac{4i}{\pi ka}.$$

Using $H_{\nu_n}'^{(1)}(ka) + i\zeta H_{\nu_n}^{(1)}(ka) = 0$, we obtain

$$H_{\nu_n}'^{(2)}(ka) + i\zeta H_{\nu_n}^{(2)}(ka) = \frac{-4i}{\pi ka H_{\nu_n}^{(1)}(ka)}.$$

Hence (A1.15) can be simplified as

$$R_n = \frac{-4i}{\pi ka H_{\nu_n}^{(1)}\left[H_{\nu_n}'^{(1)}(ka) + i\zeta H_{\nu_n}^{(1)}(ka)\right]}, \qquad (A1.16)$$

where

$$H_{\nu_n}'^{(1)}(ka) = \left[\frac{\partial}{\partial\nu}H_\nu'^{(1)}(ka)\right]_{\nu=\nu_n}.$$

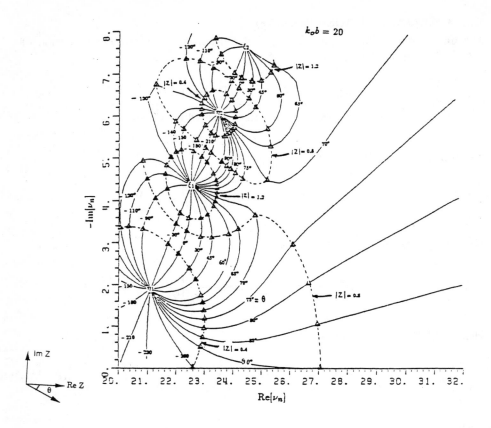

Fig. A1.5. Propagation constants of the creeping rays on a cylinder as a function of the impedance (extracted from H. T. Kim, Ph.D. Thesis, Ohio State University, 1986)

If we replace ζ in (A1.16) by its expression given in (A1.13), we get

$$\zeta = \frac{iH_{\nu_n}^{\prime(1)}(ka)}{H_{\nu_n}^{(1)}(ka)} \, .$$

This, in turn, leads to the following expression

$$R_n = \frac{-4i}{Q_n}, \tag{A1.17}$$

where

$$Q_n = \pi \, ka\Big[H_{\nu_n}^{(1)}(ka) H_{\nu_n}^{\prime(1)}(ka) - H_{\nu_n}^{\prime(1)}(ka) H_{\nu_n}^{(1)}(ka) \Big] . \tag{A1.18}$$

Inserting the asymptotic expansions of different Hankel functions in the vicinity of the Stokes lines in (A1.18), we get

$$Q_n = \pi k a \left(2^{10/3} v_n^{-2/3} e^{2\pi i/3} (ka)^{-2/3} \left\{ [Ai'(y_n)]^2 - y_n [Ai(y_n)]^2 \right\} \right). \quad (A1.19)$$

Combining all of these, we obtain the final expression for R_n

$$R_n = -\frac{2i}{\pi 2^{7/3}} (ka)^{1/3} e^{-2\pi i/3} \frac{1}{[Ai'(y_n)]^2 - y_n [Ai(y_n)]^2}. \quad (A1.20)$$

Using the notation introduced by Levy and Keller, which we adopt in this book, the diffraction coefficient D_n can be written as

$$D_n^2 = \frac{e^{i\pi/12}}{2^{5/6} \pi^{1/2} (ka)^{1/6}} \frac{a^{1/2}}{[Ai'(y_n)]^2 - y_n [Ai(y_n)]^2}, \quad (A1.21)$$

with

$$y_n = - 2^{1/3} e^{-i\pi/3} (ka)^{-1/3} (n_n - ka). \quad (A1.22)$$

We observe that (A1.21) is identical to the formula for the diffraction coefficient given in Sect. 1.5.

Interpreting the solution to the problem of a point source radiating in the presence of a circular cylinder enables us, once again, to find the propagation constants and the detachment coefficients of the creeping rays. To a first order, the quantities thus obtained are the same regardless of the geometry of the canonical problem considered, whether they be circular or elliptic cylinders, or a sphere. Thus we regard the discussion of the canonical problem of the circular cylinder as being complete and go on to discuss the canonical problem of diffraction by a perfectly conducting wedge whose solution will yield the diffraction coefficient on an edge.

A1.3 Diffraction by a wedge

Consider a wedge with planar faces which make the angles $\pm \Phi$ with the x axis. Let the wedge be illuminated by an incident plane wave.

$$u^{inc} = exp\,(-ik\rho cos(\varphi - \Phi_0)) = exp\,(-ikx cos\,\Phi_0 - iky sin\,\Phi_0), \quad (A1.23)$$

where u^{inc} denotes the electric field for the TM polarization and the magnetic field when the polarization is TE. We seek a solution to the diffraction problem

$$(\nabla^2 + k^2)u = 0, \ for -\Phi < \varphi < \Phi,$$

$$u = 0 \ for \ \varphi = \ \pm\Phi \ (TM), \ \frac{\partial u}{\partial n} = 0 \ for \ \varphi = \ \pm\Phi \ (TE),$$

such that energy associated with the field solution is bounded even though the field itself or its derivative can be singular at the tip of the wedge. We begin by representing the solution in the form of a plane wave spectrum (see Chap. 4).

$$u(\rho,\varphi) = \frac{1}{2\pi i} \int_D exp(-ik\rho\cos\alpha) \ p(\alpha+\varphi)d\alpha , \tag{A1.24}$$

where p is an unknown weight function and D is the Sommerfeld or Malhiuzinets contour shown in Fig. A1.6.

The weight function $p(\alpha)$ is such that $p(\alpha) - (\alpha - \varphi_0)$ is regular in the range $|Re \ \alpha| < \Phi + \varepsilon$. We will see later how the presence of a pole enables us to extract the incident field from the solution and thereby find the desired representation for the diffracted field.

The application of the boundary conditions on the faces $\varphi = \pm\Phi$ leads to the equations

$$\int_D exp(-ik\rho\cos\alpha) \ p(\alpha \pm \Phi)d\alpha = 0, \ for \ TM , \tag{A1.25}$$

$$\int_D exp(-ik\rho\cos\alpha) \ (sin\alpha) \ p(\alpha \pm \Phi)d\alpha = 0, \ for \ TE . \tag{A1.26}$$

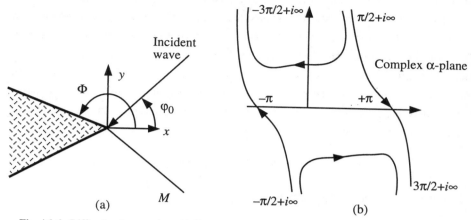

Fig. A1.6. Diffraction by a wedge: (a) Geometry of problem; (b) Contours in the complex α-plane

In view of the symmetry of the contour D with respect to the origin, it can be shown [Ma] that (A1.25) and (A1.26) are satisfied if and only if

$$p(\alpha \pm \Phi) = p(-\alpha \pm \Phi), \text{ for TM,} \qquad (A1.27)$$

and

$$p(\alpha \pm \Phi) = -p(-\alpha \pm \Phi), \text{ for TE.} \qquad (A1.28)$$

It is possible to provide a simple interpretation of the last two equations. Each spectral component of the wave is reflected by the appropriate face of the edge, with the reflection coefficient given by –1 and +1 for TM and TE polarization, respectively. Hence this approach is referred to in the literature [Va] as the *method of generalized reflection*.

The solutions to Eqs. (A1.27) and (A1.28) may be written as

$$p(\alpha) = \frac{1}{2n}(cot(\alpha - \varphi_0)/2n + tg(\alpha + \varphi_0)/2n) = \frac{cos(\varphi_0/n)}{n(sin(\alpha/n) - sin(\varphi_0/n))} \text{ for TM,}$$

$$(A1.29)$$

$$p(\alpha) = \frac{1}{2n}(cot(\alpha - \varphi_0)/2n - tg(\alpha + \varphi_0)/2n) = \frac{cos(\alpha/n)}{n(sin(\alpha/n) - sin(\varphi_0/n))} \text{ for TE,}$$

$$(A1.30)$$

where $n = 2\Phi/\pi$, according to the notations chosen in Sect. 1.5 and we note that these solutions are particular solutions of (A1.27) and (A1.28). We proceed next to evaluate (A1.24) asymptotically at a point $M(\rho,\varphi)$ for a large $k\rho$. To this end we deform the contour D into the steepest descent paths passing through $\pm \pi$. The asymptotic expansion of (A1.24) consists of two parts, the first of which is a set of contributions of the poles of the integrand located between the two contours, and the second arising from the contributions of the two steepest descent paths, computed by using the Laplace method (see the method of steepest descent in the appendix "Asymptotic expansion of integrals"). The poles of the weight functions are real. Only the poles located between the two contours, i.e., those between $-\pi$ and π are to be considered. They satisfy the condition

$$sin(\alpha + \varphi/n) = sin(\varphi_0/n),$$

which leads to the following alternatives

$$\alpha + \varphi = \varphi_0,$$
$$\alpha + \varphi = n\pi - \varphi_0 = 2\Phi - \varphi_0,$$
$$\alpha + \varphi = -n\pi - \varphi_0 = -2\Phi - \pi_0.$$

In fact, the other poles are never between the two contours. The conditions for the poles to be situated between $-\pi$ and π are as follows. p For the first pole, $\varphi > \varphi_0 - \pi$, and the point M is in the lit zone; for the second pole, $\varphi > 2\Phi - \varphi_0 - \pi$, and M is in the zone of the field reflected by the upper face of the wedge; finally, for the last pole, $\varphi < \pi - 2\Phi - \varphi_0$, M is in the zone of the field reflected by the lower face of the wedge (see Fig. A1.7)

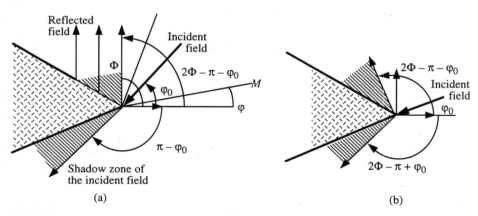

Fig. A1.7. Interpretation of the contributions of the poles — incident and reflected fields: (a) single reflected field; (b) two reflected fields

The evaluation of the contribution u_p of the various poles with the residue method finally leads to the following result (H denotes the Heaviside function)

$$u_p = H(\varphi - (\varphi_0 - \pi))u^i + H(\varphi - (2\Phi - \varphi_0 - \pi))\,u^r_{sup} + H(-(2\Phi + \varphi_0 - \pi + \varphi))\,u^r_{inf}\,,$$

(A1.31)

where u^i is the incident plane wave, and u^r_{sup}, u^r_{inf} represent the plane waves reflected by the upper and lower faces of the wedge, respectively.

Thus, u_p is the geometrical optics contribution of the total field as illustrated in Fig. A1.7. Let us turn now to the contributions u_s of the saddle points at $\pm\pi$. By applying the method of steepest descent (Appendix 4), we obtain

$$u_s = -e^{i\pi/4}\big(p(\pi + \varphi) - p(-\pi + \varphi)\big)\frac{e^{ikr}}{\sqrt{2\pi kr}}\,.$$

(A1.32)

After some manipulations, this formula can be reduced to the form

$$u_s = \frac{D}{\sqrt{r}}\,,$$

(A1.33)

where D is the diffraction coefficient of the wedge given in Sect. 1.5, for an incident angle $(\Phi - \varphi_0)$ and an observation angle $(\Phi - \varphi)$, these two angles being measured with respect to the upper face of the wedge. The total field in presence of the wedge is then the sum of the geometrical optics field and the field diffracted by the wedge, this separation being valid only if the poles are not too close to the saddle points. For the situation when this condition is not met, one has to resort to a uniform asymptotic estimation as explained in Chap. 5. Thus, when suitably interpreted, the canonical problem of the wedge enables us to calculate the diffraction coefficient of an edge.

We next turn our attention to the more complex case of a wedge described with a surface impedance condition $\partial u/\partial n + ikZ_\pm u = 0$. Here, Z_+ and Z_- denote the surface impedances on the upper and lower faces of the wedge, respectively. We still seek a solution in the form of a wave plane spectrum of the type (A1.24), and by using the impedance condition in place of (A1.26), we now get

$$\int_D exp(-ik\rho cos\alpha)\,(sin\alpha \pm Z_\pm)\,p(\alpha \pm \Phi)d\alpha = 0\,.$$

(A1.34)

This equation is satisfied if the weight function p satisfies

$$p(\alpha \pm \Phi) = \frac{-\sin\alpha \pm Z_\pm}{\sin\alpha \pm Z_\pm} p(-\alpha \pm \Phi).$$ (A1.35)

We recognize the appearance of the reflection coefficient of the plane with a surface impedance. The functional Eq. (A1.35) has been solved by Maliuzhinets [Ma] and the solution, which shows a pole at φ_0, is given by

$$p(\alpha) = p_0(\alpha)\frac{\Psi(\alpha)}{\Psi(\varphi_0)},$$ (A1.36)

where $p_0(\alpha) = cos(\varphi_0/n) / n(sin(\alpha/n) - sin(\varphi_0/n))$ is the weight function of the TM case given in (A1.29), and Ψ is the Maliuzhinets function. The asymptotic expansion of the integral, that leads to the desired result, is obtained in the same manner as in the past. The field is the sum of the contribution of the saddle point, which gives the diffracted field with the diffraction coefficient of Sect. 1.5, and those of the poles of the integrand. For the impedance case, not only are there contributions of the poles of $p_0(\alpha)$ that contribute to the GO field as they do in the conducting case, but there may be additional contributions from the poles of Ψ, which are associated with the surface waves propagating on the faces of the wedge. Let us recall that in this book we have treated the case where the ReZ is not negligible compared to ImZ, implying that the waves are rapidly attenuated.

Thus, we observe that the method of the canonical problems not only enables us to interpret the diffraction phenomenon but to also calculate the diffraction coefficients that are indispensable for the application of GTD. We should mention, however, that solving the canonical problem is the difficult step in GTD and, although a whole host of canonical problems have been solved in the past, it is not easy to enlarge the list with new additions. Finally, we add the remark that although we have restricted ourselves to only one representative problem for each of the diffraction phenomena, the interested reader can find more details on the canonical problems for the conducting case in the very complete work of Bowman, Senior, and Uslenghi [BS].

References

[AS] Abramowitz and Stegun, "Handbook of Mathematical Functions," Dover Publications.

[BS] J. J. Bowman, T. B. A. Senior, and P. L. Uslenghi, "Acoustic and electromagnetic scattering by simple shapes," *Hemisphere*, 1987.

[GB] J. Gay and D. Bouche, Internal Report (not available).

[Ma] G. D. Maliuzhinets, "Excitation, reflection and emission of surface waves from a wedge with given face impedances," *Sov. Phys. Dokl.* **3**, 752, 1958.

[Va] V. G. Vaccaro, "The generalized reflection method in electromagnetism," AEU Band 34, Heft 12, pp. 493-500, 1980.

Appendix 2. Differential Geometry

Differential geometrical concepts were employed mainly in Chaps. 3 and 6 in connection with a number of applications. In this appendix we are going to provide some examples based on differential geometry that illustrate the calculation of the length of the rays which was discussed in Chap. 3. In addition, we present an expression for the phase of an incident plane wave expressed in the (s, n) coordinate system. Finally, we present some information on the geodesics, the geodesic coordinate systems, the (s, α, n) coordinate system, and the coordinate system of the lines of curvature that are discussed in Chaps. 6 and 8.

A2.1 Calculation of the ray lengths

We choose the ray coordinates on a cylinder in order to facilitate the demonstration of the formulas in (3.35) in Sect. 3.1.8. We might approach this task in two possible ways:

(i) start with an arbitrary cylinder and perform a Taylor expansion on the formula
(ii) replace the arbitrary cylinder by a circular cylinder and use trigonometric formulas to derive the ray lengths.

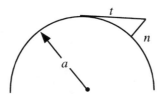

Fig. A2.1. Ray (t) and normal (n) coordinates for a circular cylinder

Although both of these procedures lead to the same result, and the second procedure is actually less rigorous, we have chosen to use it because it is much less time-consuming to apply. Let us consider a circle with radius a (see Fig. A2.1) and a point located at the distance n from this circle, and let us draw the tangent at this point.

The length t of this tangent is also the length of the diffracted ray given below.

$$t^2 = (a + n)^2 - a^2 = 2an + n^2 \approx 2an. \qquad (A2.1)$$

Hence, we have the result

$$Y = \left(m \frac{t}{a} \right)^2 \approx 2 \frac{m^2}{a} n = \frac{2^{1/3} k^{2/3}}{a^{1/3}} n = v, \qquad (A2.2)$$

which is the first formula (3.35) of Sect. 3.1.8. For the case of a circle, the first formula in (3.35) is exact. For the general case, we can express t as a series of $\sqrt{n}$. Then we can write

$$t^2 = 2\rho n + O(n^{3/2}). \qquad (A2.3)$$

Using $n^{3/2} = O(k^{-1})$ in (A2.2), we can obtain the following expression for Y

$$Y = v + O(m^2 k^{-1}) = v + O(k^{-1/3}). \qquad (A2.4)$$

It is also possible to find an explicit form of the second term, i.e., $O(k^{-1/3})$ term in (A2.4) and to show that it is proportional to the derivative of the radius of curvature.

We will return to (3.35) in Sect. 3.1.8 and present a proof of the second equation in (3.35). We begin by noting that the $\ell - s$ is the difference between the cord and the circular arc. It has the form

$$\ell - s = a \, (\mathrm{tg} \, \theta - \theta) = a \frac{\theta^3}{3} + O(a \, \theta^5). \qquad (A2.5)$$

Using $\theta = \dfrac{t}{a} + (t^3)$, we get

$$\ell - s = \frac{t^3}{3a^2} = \frac{2}{3k} Y^{3/2}. \qquad (A2.6)$$

The terms neglected are of $O(t^4)$ or $O(k^{-4/3})$ for the case of the circle.

The other formulas in (3.35) can be derived by noting that s and s^r differ only by a quantity $k^{-1/3}$. The phase difference between two rays passing through a point near the caustic may be calculated in the same manner. For the sake of simplicity, let us consider a circular caustic with radius a, as shown in Fig. A2.2.

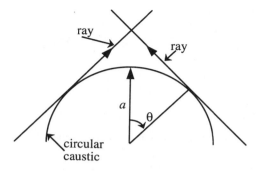

Fig. A2.2. Circular caustic

The phase difference δ between the ray receding from the caustic and the ray approaching it, is twice the difference between the lengths of the cord and the arc, as calculated above. Thus, we can write

$$\delta = 2ka\,(\text{tg}\,\theta - \theta) \approx 2k\,\frac{t^3}{3a^2} \approx \frac{2^{5/2}}{3}\,k\,n^{3/2}\,a^{-1/2}. \tag{A2.7}$$

The phase difference is $O(k\,n^{3/2})$ and is of $O(1)$ in a neighborhood of $O(k^{-2/3})$ of the caustic.

Let us now consider the case of the grazing ray which is discussed in Sect. 3.7, and calculate the phase difference between the direct (incident) ray passing through the cylinder and the creeping ray passing through the point whose coordinates are (s, n) (see Fig. A2.3). Let us assume that the phase reference or the origin is the contact point of the ray and the cylinder.

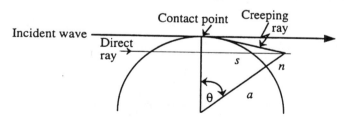

Fig. A2.3. Ray at grazing incidence on a circular cylinder

The phases of the direct and creeping ray are given, respectively, by

$$k(n + a)\sin\frac{s}{a} \approx k\left(s - \frac{s^3}{6a^2} + \frac{ns}{a}\right), \tag{A2.8}$$

and

$$k(s - a\,\theta + a\,\text{tg}\,\theta) \approx k\left(s + a\frac{\theta^3}{3}\right), \qquad (A2.9)$$

with

$$\theta \approx \sqrt{\frac{2n}{a}}. \qquad (A2.10)$$

The phase difference δ between the two rays then becomes

$$\delta \approx k\left(-\frac{s^3}{6a^2} + \frac{ns}{a} - \frac{2}{3}n\sqrt{\frac{2n}{a}}\right), \qquad (A2.11)$$

which is (3.143) of Sect. 3.7.

The path difference between the rays or, equivalently, their phase difference, can be calculated by using simple calculus. They determine (see Chap. 3) the stretching of coordinates that are useful for deriving uniform solutions.

We will now provide the expression for the phase of the incident field using the (s, n) coordinate system.

A2.2 Phase of the incident field expressed in terms of the coordinates (s, n) or (s, α, n)

A2.2.1 Two-dimensional case

Let the incident field be a plane wave given by exp(ikx). The problem at hand is to express the scalar product $x = \hat{x} \cdot \vec{OM}$ in the vicinity of O using the coordinates (s, n). We choose the coordinate system as shown in Fig. A2.4 and define M_0 as the projection on the cylinder of the point M.

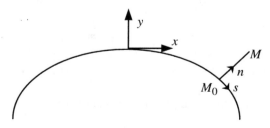

Fig. A2.4. (s, n) coordinate system for an arbitrary cylinder

The successive derivatives of $\overrightarrow{OM}_0$ with respect to s at point O, can be expressed as

$$\frac{d\overrightarrow{OM}_0}{ds} = \hat{x}, \quad \frac{d^2\overrightarrow{OM}_0}{ds^2} = -\frac{\hat{y}}{\rho}, \quad \frac{d^3\overrightarrow{OM}_0}{ds^3} = -\frac{\rho'}{\rho}\hat{y} - \frac{\hat{x}}{\rho^2}, \qquad (A2.12)$$

which leads to the following Taylor expansion of $\hat{x} \cdot \overrightarrow{OM}_0$.

$$\hat{x} \cdot \overrightarrow{OM}_0 = s - \frac{s^3}{6\rho^2} + O(s^4), \qquad (A2.13)$$

$$\overrightarrow{M_0 M} = n\,\hat{n} \qquad (A2.14)$$

Similarly, the Taylor expansion of $\hat{n}$ is

$$\hat{n} = \hat{y} + s\,\frac{\hat{x}}{\rho} + O(s^2). \qquad (A2.15)$$

Using the above two expansions we derive the desired result

$$\hat{x} \cdot \overrightarrow{OM} = s - \frac{s^3}{6\rho^2} + \frac{ns}{\rho} + O(s^4, s^2 n). \qquad (A2.16)$$

We note that we recover the expression (A2.8) given in 2.1, which was derived for the circular cylinder geometry. The terms neglected in the above are of $O(s^4, s^2 n)$, i.e., $O(k^{-4/3})$.

A2.2.2 General surface in $\mathbf{R}^3$

For the general surface, the approach followed for the derivation of an expression for $\hat{x} \cdot \overrightarrow{OM}$ is identical to that employed in the two-dimensional case. We begin by choosing the origin at a point O on the separatrix and carry out the expansion in the vicinity of this point, recognizing that this time there are two coordinates, viz., s and α on the surface (see Fig. A2.5).

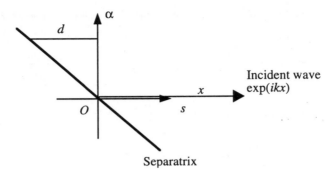

Fig. A2.5. (s, α) coordinate system in the vicinity of a point O on the separatrix

We now express the Taylor series expansion of $\hat{x} \cdot \overrightarrow{OM}$ using the coordinates (s, α, n). The wave vector $\hat{x}$ of the incident wave at the point O is directed along the s axis, the vector $\hat{y}$ along the α axis, and $\hat{z}$ along the normal to the surface. The coordinates s and α are assumed to be of $O(k^{-1/3})$ and the coordinate n of $O(k^{-2/3})$.

The equations of Gauss-Weingarten, which take a simple form when expressed in the geodesic coordinates (see Sect. A2.3 of this appendix), enable us to express the successive derivatives of $\overrightarrow{OM}$ with respect to s and α. We then carry out a Taylor expansion, similar to the one for the cylinder, up to O(3) with respect to s and α for OM_0, and up to O(1) with respect to s and α for $\hat{n}$. The calculations are straightforward though somewhat cumbersome, and we only report the final result below, viz.,

$$\hat{x} \cdot \overrightarrow{OM} \approx s - \frac{b_{ss}^2}{6}\left(s + \frac{b_{s\alpha}}{b_{ss}}\alpha\right)^3 + b_{ss}\left(s + \frac{b_{s\alpha}}{b_{ss}}\alpha\right)n .$$

$$(A2.17)$$

In common with the case of the cylinder, the terms we have neglected are $O(k^{-4/3})$. Also, we observe that the pertinent variable is not s, but

$$s + \frac{b_{s\alpha}}{b_{ss}}\alpha .$$

A simple physical interpretation of this variable is as follows. Let us express the equation of the light-shadow separatrix $\hat{n} \cdot \hat{x} = 0$, in the vicinity of O, by using a Taylor expansion. We then obtain

$$b_{ss}\, s + b_{s\alpha}\, \alpha \approx 0.$$

Next, we consider the distance $d(\alpha)$ between the separatrix and a point on the α-axis (see Fig. A2.5), and express it as

$$d(\alpha) \approx \frac{b_{s\alpha}}{b_{ss}} \alpha, \tag{A2.18}$$

where $d > 0$ for $\alpha > 0$, and $d < 0$ for $\alpha < 0$, as can be seen from the figure. We can then interpret the quantity $b_{s\alpha}/b_{ss} \, \alpha$ in the vicinity of O, as the distance traveled along the geodesic by the creeping or the direct ray (according to the sign of h) between the separatrix and the α-axis.

Thus, in view of (A2.17), and the fact that $b_{ss} = 1/\rho$, where ρ is the curvature radius of the geodesic, we obtain, after setting

$$s' = s + d(\alpha), \tag{A2.19}$$

the result

$$\hat{x} \cdot \overrightarrow{OM} \approx s'^3 - \frac{s'}{6\rho^2} + \frac{s'n}{\rho}. \tag{A2.20}$$

The error is $O(s'^{\,4}, s'^{\,2}n)$, i.e., $O(k^{-4/3})$. Thus we find the same result for an arbitrary surface as we did for the cylinder, using the expression of s' as provided by (A2.19).

The physical interpretation of this result is straightforward. The pertinent quantity that governs the attenuation of a creeping wave is the distance it travels from the separatrix along the geodesic. The quantity $b_{s\alpha}/b_{ss}$ may also be interpreted as follows by expressing $b_{s\alpha}/b_{ss} = -\tau\rho$, where τ is the torsion and ρ the curvature radius of the geodesic tangent to $\hat{x}$ at point O. The function $d(\alpha)$ then takes the form $-(\tau\rho)\alpha$.

A2.3 Geodesic coordinate system and applications

A2.3.1 The geodesic coordinates

We will now provide some information on the geodesic coordinates that were introduced in Chap. 3. Let us recall that the geodesics are defined as the lines which minimize the distance between two points on the surface and they satisfy a system of second order differential equations. There exists one and only one geodesic passing through a point with a given tangent at that point. A system of geodesic coordinates can be defined as long as we have a one-parameter family of geodesics on a surface. In our case this family is the family of creeping rays initiated at the shadow boundary. We define the trajectories orthogonal to this family of geodesics. From these trajectories we choose one curve as the α-axis. The coordinate α is the curvilinear abscissa along this curve. The length of the segment cut by two orthogonal trajectories on the geodesics is constant (see Fig. 1.14 of Chap. 1).

The arc length of the curve (ds, dα) on the surface will be ($ds^2 + h^2\,d\alpha^2$), where h measures the broadening (if $h > 1$) of the pencil of geodesics between the α-axis and the running point on the surface.

A2.3.2 The surface in geodesic coordinates

The first quadratic form, written in the geodesic coordinates, takes the following form.

$$ds^2 + g_{\alpha\alpha}\,d\alpha^2 = d^2s + h^2\,d\alpha^2.$$

The quantity h, which is a function of s and α, completely defines the metric on the surface. For the second quadratic form we use the following definition (the sign is opposite to that used by Struik)

$$b_{ss} = \frac{\partial \vec{OM}}{\partial s}\cdot\frac{\partial \hat{n}}{\partial s},\; b_{\alpha\alpha} = \frac{\partial \vec{OM}}{\partial \alpha}\cdot\frac{\partial \hat{n}}{\partial \alpha},\; b_{s\alpha} = \frac{\partial \vec{OM}}{\partial \alpha}\cdot\frac{\partial \hat{n}}{\partial s},$$

where $\hat{n}$ is the unit normal vector pointing outward from the surface. This choice leads to positive values of b_{ss} and $b_{\alpha\alpha}$ for a convex object. The equations providing the derivative of $\hat{n}$ (the so-called Weingarten's equations) are written as

$$\frac{\partial \hat{n}}{\partial s} = b_{ss}\frac{\partial \vec{OM}}{\partial s} + \frac{b_{s\alpha}}{h^2}\frac{\partial \vec{OM}}{\partial \alpha},$$

$$\frac{\partial \hat{n}}{\partial \alpha} = b_{s\alpha}\frac{\partial \vec{OM}}{\partial s} + \frac{b_{\alpha\alpha}}{h^2}\frac{\partial \vec{OM}}{\partial \alpha}.$$

The vector $\hat{n}$, normal to the surface, is opposite the normal vector to the geodesic, $\partial \vec{OM}/\partial s$ is the tangent vector, and $1/h\,\partial \vec{OM}/\partial s$ is the binormal vector to the geodesic. Thus, we also have

$$\frac{\partial \hat{n}}{\partial s} = \frac{1}{\rho}\frac{\partial \vec{OM}}{\partial s} - \frac{\tau}{h}\frac{\partial \vec{OM}}{\partial \alpha},$$

where ρ is the radius of curvature and τ the torsion of the geodesic at the point M. Thus, we can write

$$b_{ss} = \frac{1}{\rho} \ , \ b_{s\alpha} = -h\tau,$$

where $b_{\alpha\alpha}/h^2$ is the inverse of the normal radius of curvature of the curve orthogonal to the geodesic at the point M, which is denoted here as ρ_t.

The Christoffel symbols of the surface vanish except for

$$\Gamma^\alpha_{s\alpha} = \frac{1}{h}\frac{\partial h}{\partial s}, \ \Gamma^s_{\alpha\alpha} = -h\frac{\partial h}{\partial s}, \ \Gamma^\alpha_{\alpha\alpha} = \frac{1}{h}\frac{\partial h}{\partial \alpha}.$$

These symbols help us in expressing the second derivative with the help of the Gauss equations which can be written in the form

$$\frac{\partial^2 \overrightarrow{OM}}{\partial s^2} = -\frac{\hat{n}}{\rho},$$

$$\frac{\partial^2 \overrightarrow{OM}}{\partial s\partial\alpha} = \frac{1}{h}\frac{\partial h}{\partial s}\frac{\partial \overrightarrow{OM}}{\partial s} + h\tau\hat{n},$$

$$\frac{\partial^2 \overrightarrow{OM}}{\partial\alpha^2} = -h\frac{\partial h}{\partial s}\frac{\partial \overrightarrow{OM}}{\partial s} + \frac{1}{h}\frac{\partial h}{\partial s}\frac{\partial \overrightarrow{OM}}{\partial \alpha} - \frac{h^2}{\rho_t}\hat{n}.$$

The above equations, combined with those of Weingarten, provide the successive derivatives of $\overrightarrow{OM}$ and consequently, the Taylor expansion of $\overrightarrow{OM}$. It enables us to establish Eq. (A2.20) of A2.2 of this appendix, which provides the Taylor expansion of the phase of the incident wave.

A point in space is located by its coordinates (s, α) of its projection on the surface, and its distance n to the surface. This curvilinear coordinate system (s, α, n) is not orthogonal. Thus we would need to know the metric matrix of this system, which we will discuss below.

A2.3.3 Calculation of the metric matrix of the coordinate system (s, α, n)

To compute the metric matrix of the (s, α, n) coordinate system, we set $\overrightarrow{OM} = \overrightarrow{OP} + n$ $\hat{n}$, where P is on the surface. The vector $\overrightarrow{OP}$ is a function of both of the coordinates (s, α) of the semi-geodesic coordinate system.

We are going to calculate the metric coefficients of the coordinate system (s, α, n), that are scalar products of the partial derivatives of $\vec{OM}$. We have $g_{xy} = \partial \vec{OM}/\partial x \cdot \partial \vec{OM}/\partial y$, where x (or y) is equal to s, α or n, $\partial \vec{OM}/\partial n = \hat{n}$, so that $g_{sn} = g_{\alpha n} = 0$ and $g_{nn} = 1$. The terms that differ from unity are g_{ss}, $g_{\alpha\alpha}$ and $g_{s\alpha}$.

We will now calculate g_{ss}. To this end, we write

$$\frac{\partial \vec{OM}}{\partial s} = \frac{\partial \vec{OP}}{\partial s} + n\frac{\partial \hat{n}}{\partial s} = (1 + b_{ss}n)\frac{\partial \vec{OP}}{\partial s} + \frac{b_{s\alpha}}{h^2}n\frac{\partial \vec{OP}}{\partial \alpha}, \qquad (A2.21)$$

where we have used the notation of Chap. 3. Then, we have

$$g_{ss} = \left(\frac{\partial \vec{OM}}{\partial s}\right)^2 = (1 + b_{ss}\,n)^2 + \frac{b_{s\alpha}^2}{h^2}n^2. \qquad (A2.22)$$

Next, we turn to $g_{\alpha\alpha}$ and write

$$\frac{\partial \vec{OM}}{\partial \alpha} = \frac{\partial \vec{OP}}{\partial \alpha} + n\frac{\partial \hat{n}}{\partial \alpha} = \left(1 + \frac{b_{\alpha\alpha}}{h^2}n\right)\frac{\partial \vec{OP}}{\partial \alpha} + b_{s\alpha}n\frac{\partial \vec{OP}}{\partial s}, \qquad (A2.23)$$

and we get

$$g_{\alpha\alpha} = \left(\frac{\partial \vec{OM}}{\partial \alpha}\right)^2 = h^2\left(1 + \frac{b_{\alpha\alpha}}{h^2}n\right)^2 + b_{s\alpha}^2\,n^2. \qquad (A2.24)$$

Finally, we calculate $g_{s\alpha}$ as follows

$$g_{s\alpha} = \frac{\partial \vec{OM}}{\partial s} \cdot \frac{\partial \vec{OM}}{\partial \alpha} = b_{s\alpha}n\,(1 + b_{ss}n) + b_{s\alpha}n\left(1 + \frac{b_{\alpha\alpha}}{h^2}n\right), \qquad (A2.25)$$

$$g_{s\alpha} = 2b_{s\alpha}n + b_{s\alpha}n^2\left(b_{ss} + \frac{b_{\alpha\alpha}}{h^2}\right). \qquad (A2.26)$$

The metric matrix of the coordinate system (s, α, n) is then given by

$$g_{ij} = \begin{pmatrix} (1+b_{ss}n)^2 + \dfrac{b_{s\alpha}^2}{h^2}n^2 & 2b_{s\alpha}n + b_{s\alpha}n^2\left(b_{ss} + \dfrac{b_{\alpha\alpha}}{h^2}\right) & 0 \\[3mm] 2b_{s\alpha}n + b_{s\alpha}n^2\left(b_{ss} + \dfrac{b_{\alpha\alpha}}{h^2}\right) & h^2\left(1 + \dfrac{b_{\alpha\alpha}}{h^2}n\right)^2 + b_{s\alpha}^2 n^2 & 0 \\[3mm] 0 & 0 & 1 \end{pmatrix}.$$

However, we have seen in the previous section that $b_{ss} = 1/\rho$, $b_{s\alpha} = -h\tau$ and $b_{\alpha\alpha}/h^2 = 1/\rho_t$. Using these we can rewrite the metric matrix as

$$g_{ij} = \begin{pmatrix} \left(1 + \dfrac{n}{\rho}\right)^2 + \tau^2 n^2 & -h\tau\left(2n + n^2\left(\dfrac{1}{\rho} + \dfrac{1}{\rho_t}\right)\right) & 0 \\[3mm] -h\tau\left(2n + n^2\left(\dfrac{1}{\rho} + \dfrac{1}{\rho_t}\right)\right) & h^2\left(\left(1 + \dfrac{n}{\rho_t}\right)^2 + \tau^2 n^2\right) & 0 \\[3mm] 0 & 0 & 1 \end{pmatrix},$$

which is the result we were seeking.

A2.4 Coordinate system of the lines of curvature

The lines of curvature represent a network of orthogonal curves on the surface. The coordinate system of the curvature lines is defined and regular, except at the umbilics, the points where the two radii of curvature are equal. The coordinate axes are the two lines of curvature passing through the origin. The coordinates are the curvilinear abscissas denoted on the two axes by u and v. This system can be completed by adding the normal distance as the third coordinate, and the coordinate system so defined is then orthogonal, in contrast to the geodesic coordinate system. In fact, if $\hat{n}$ is the inner normal vector to the surface, we obtain for a convex surface (Rodrigues's Theorem)

$$\frac{d\hat{n}}{du} = -\frac{1}{R}\frac{\partial \overrightarrow{OM}}{\partial u}. \tag{A2.27}$$

Hence

$$g_{uv} = \frac{d}{du}(\overrightarrow{OM} + n\,\hat{n}) \cdot \frac{d}{dv}(\overrightarrow{OM} + n\,\hat{n}) = 0. \tag{A2.28}$$

Let a surface be described in terms of the coordinate system of the lines of curvature and let the point Q on the surface be the origin of this coordinate system. We choose the origin O of the space coordinate as the center of curvature, associated with the line of curvature defining the abscissa axis. Then O is one of the two points of the caustic surface below the point Q, and

$$\vec{OQ} = -R\,\hat{n}\,. \qquad\qquad\qquad\qquad (A2.29)$$

It is possible to calculate the partial derivatives of $\vec{OM}\,^2$, used in Sect. 6.2, to calculate the phase of the integral of the field on a wavefront.

Next, let us calculate the first few partial derivatives of $\vec{OM}$. Note that the derivatives $\partial\vec{OM}/\partial u = \hat{u}$ and $\partial\vec{OM}/\partial u = \hat{v}$ at point Q define the unit vectors. The first quadratic form at point Q then reads $Edu^2 + Gdv^2 = du^2 + dv^2$. Furthermore, since u is the curvilinear abscissa on the u-axis, this property still holds true if $v = 0$. We then deduce that $E = 1$ if $v = 0$, and, consequently, all the derivatives of E with respect to u vanish. We can similarly show that the partial derivatives of G with respect to v also vanish. These observations make it easier for us to calculate the partial derivative of $\vec{OM}\,^2$ and the results are expressed as

$$\frac{\partial\vec{OM}\,^2}{\partial u} = 2\,\vec{OM}\cdot\hat{u} = -2R\,\hat{n}\cdot\hat{u} = 0,$$

$$\frac{1}{2}\frac{\partial^2\vec{OM}\,^2}{\partial u^2} = \hat{u}\cdot\hat{u} + \vec{OM}\,\frac{\partial\hat{u}}{\partial u},$$

$$= E - R\,\hat{n}\,\frac{\partial\hat{u}}{\partial u},$$

$$= 1 + R\,\frac{\partial\hat{n}}{\partial u}\cdot\hat{u},$$

$$= 1 + R\left(-\frac{1}{R}\right),$$

$$= 0.$$

For higher-order derivatives we find

$$\frac{1}{2}\frac{\partial^3\vec{OM}\,^2}{\partial u^3} = -R\,\frac{\partial(1/R)}{\partial u},$$

$$\frac{1}{2}\frac{\partial^4 \overrightarrow{OM}^2}{\partial u^4} = -R\,\frac{\partial^2(1/R)}{\partial u^2}\,,$$

$$\frac{1}{2}\frac{\partial^5 \overrightarrow{OM}^2}{\partial u^5} = -R\,\frac{\partial^3(1/R)}{\partial u^3}\,.$$

For the partial derivatives with respect to v the result is

$$\frac{1}{2}\frac{\partial^2 \overrightarrow{OM}^2}{\partial v} = \overrightarrow{OM}\cdot\hat{v} = 0,$$

$$\frac{1}{2}\frac{\partial^2 \overrightarrow{OM}^2}{\partial v^2} = \hat{v}\cdot\hat{v} + \overrightarrow{OM}\,\frac{\partial\hat{v}}{\partial v}\,,$$

$$= 1 - \frac{R}{R'}\,.$$

Finally for the mixed derivatives, we can obtain

$$\frac{1}{2}\frac{\partial^2 \overrightarrow{OM}^2}{\partial u\partial v} = \hat{v}\cdot\hat{u} + \overrightarrow{OM}\cdot\frac{\partial \overrightarrow{OM}}{\partial u\partial v}\,,$$

$$= 0.$$

Similarly

$$\frac{\partial^2 \overrightarrow{OM}}{\partial u^2\partial v} = 0.$$

The only non-vanishing mixed derivative is given by

$$\frac{\partial^3}{\partial u\partial v^2}\overrightarrow{OM}^2 = R\,\frac{\partial}{\partial u}\left(\frac{1}{R'}\right) \neq 0.$$

These partial derivatives enable us to calculate $\overrightarrow{OM}^2$, which is needed in Sect. 6.2.

Reference

Struik, D. J., *Lectures on Classical Differential Geometry,* Dover, 1988.

Appendix 3. Complex Rays

As we have seen in Chap. 1, the most common applications of the ray theories are based on the theory of real rays. However, in some cases it is useful to introduce the concept of complex rays, and, as an example, we refer the reader to Sect. 1.6 where we have used this notion to derive the solution in the shadow region of a caustic. The theory of complex rays continues to be an active area of research and it is difficult to be very comprehensive in discussing this topic. Thus we will restrict ourselves to the presentation of some of the main concepts underlying this theory, but will provide a bibliography on this subject to assist the reader desiring additional information. Finally, we will only consider the two-dimensional case.

A3.1 Complex solutions of the eikonal equation, complex rays

We recall that the eikonal equation in a homogeneous medium reads

$$(\vec{\nabla} S)^2 = 1. \tag{A3.1}$$

In Chap. 2 we have studied the real solutions of this equation; however, it is also possible to search for complex rays as solutions of (A3.1). We do this by inserting a solution of the form $S = (R + iI)$ into (A3.1), which leads to the following coupled set of equations

$$(\vec{\nabla} R)^2 - (\vec{\nabla} I)^2 = 1, \tag{A3.2a}$$

$$\vec{\nabla} R \cdot \vec{\nabla} I = 0. \tag{A3.2b}$$

A geometrical optics type of solution of the wave equations is written in the form

$$A \exp(ikS) = A \exp(ikR) \exp(-kI).$$

The two functions R and I are related to the phase and the attenuation of the ray, respectively. From (A3.2a) it follows that $\vec{\nabla} R^2 > 1$, and Eq. (A3.2b) reveals that, in two dimensions, the equi-phase contours are $R =$ constant, while the equi-amplitude contours are $I =$ constant; together, they constitute a network of orthogonal curves. If we seek the solutions to (A3.1) of the type $\vec{k} \cdot \vec{r} / k$, we obtain real plane waves for real $\vec{k}$, while these waves are the complex waves, as in Chap. 4, if $\vec{k}$ is complex.

For the case of a real eikonal, we have seen in Chap. 2 that Eq. (A3.1), with initial conditions on a line, is solvable by means of the method of characteristics. For the complex case, let us assume that we know the two functions, R and I, on a line L. The following two methods have been proposed for solving (A3.2). The first of these is to seek a solution in the real space. This is the Evanescent Wave Tracking (EWT) approach of Felsen and his collaborators [F], which is not altogether robust for the general case. The second approach is to solve the problem by using the method of complex rays [WD]. At each point M_0 of L, the phase S, and consequently the projection $\hat{t} \cdot \vec{\nabla} S$ of its gradient onto L, are known. Let us set $\hat{t} . \vec{\nabla} S = \cos \theta$. Then the norm of $\vec{\nabla} S$ equals 1; hence, $\vec{\nabla} S = \cos \theta \, \hat{t} + \sin \theta \, \hat{n}$. If $\cos \theta$ is real and less than 1, then $\vec{\nabla} S$ is real; otherwise, it is complex. We define the complex ray R originating from M_0 as the following set of points M of $\mathbb{C} \times \mathbb{C}$, where $\mathbb{C}$ is the complex space.

$$R = \{M \mid M = M_0 + \ell \, \vec{\nabla} S \, (M_0), \, \ell \in \mathbb{C}\}.$$

We note that the two complex rays emanate from M_0 since $\sin \theta$ is, in fact, defined except for its sign.

For the real rays, it is easy to distinguish between the incoming and the outgoing solutions; however, this is no longer true for complex rays. To our knowledge, no unambiguous and rigorous method has been proposed to date that helps discriminate between the two solutions. Additionally, we have to consider not only the points M_0 of L, but the points M_0 of the complex extension complex $\tilde{L}$ of L as well. Furthermore, the analytical continuation of L is not stable, and other problems also appear in defining $\tilde{L}$ correctly when L is a line with two end-points [WD].

Let us assume that these problems have been solved and that L has been parameterized by its curvilinear abscissa s. A complex ray emanates from each point of $\tilde{L}$, and its direction vector is given by

$$\hat{d} = \cos \theta \, \hat{t} + \sin \theta \, \hat{n}. \tag{A3.3}$$

As a real ray, the eikonal at point M with complex abscissa ℓ on the ray is then

$$S(M) = S(M_0) + \ell\,(\cos\theta\;\hat{\imath} + \sin\theta\;\hat{n}),\tag{A3.4}$$

which is the result we were seeking.

A3.2 Solution of the transport equations

For complex rays, the transport equations are solved in the same manner as they are done for real rays. Let R be the radius of curvature of $\tilde{L}$ at the point M_0. Then we can write

$$A(M) = A(M_0)\left(\frac{-\cos\theta}{-\cos\theta + \ell\left(\frac{d\theta}{ds}+\frac{1}{R}\right)}\right)^{1/2}.\tag{A3.5}$$

We can thus derive the solution of the eikonal and transport equations with the method of complex rays for given initial conditions on L. These solutions pertain to the complex space $\mathbb{C}\times\mathbb{C}$ and the task remains to extract the field in the real space $\mathbb{R}\times\mathbb{R}$.

A3.3 Field calculation in real space

Consider a point M in real space. The complex ray which passes through this point satisfies

$$M_0(s) + \ell\,\hat{d}\,(s) = M,$$

which yields two equations for the two complex unknowns. Thus, one or more complex rays may pass through a point M. Each complex ray produces a field

$$A(M)\exp\,(ikS(M)),$$

at the point M, where S is defined by (A3.4) and the amplitude of A by (A3.5). The computation of the field via of the method of complex rays entails the following steps.

(i) Choosing the line L on which the field is given and calculation of the field on this line. For a diffraction problem, L is simply the diffracting object. The field on L is calculated via Geometrical Optics (see Chap. 2).

(ii) Extending L to $\tilde{L}$.

(iii) Determining the complex rays emanating from $\tilde{L}$.

(iv) Calculating the eikonal along a ray using (A3.4).

(v) Calculating the amplitude along a ray using (A3.5).

(vi) Determining the ray passing through the point where the field has been calculated.

In spite of the fact that the above method appears to be quite direct, it may be difficult to implement in practice, especially the step involving the analytical continuation. In many applications, it is possible to circumvent this difficulty by introducing a class of objects that are defined by using a single parameter. The diffracting object corresponds to a value of the parameter for which the rays are complex, although they are real for other values. The procedure consists of finding an explicit expression for the real rays and performing an analytical continuation of this expression to derive the result for the complex ones. We will provide an example of this procedure in the next section. An alternate procedure, which we believe to be more reliable than the analytic continuation procedure, is to employ an integral representation for the rays. The rays are real when the stationary phase point of this representation is real and, otherwise, they are complex. In the latter case, the ray field can be evaluated with the method of steepest descent.

A3.4   Calculation of the reflected field with the method of complex rays

Let $\exp(-iky)$ $\hat{z}$ represent the field incident on a perfectly conducting scatterer, where y is given by the expression $y = (x^3/3a) - bx$. When $b > 0$, this scatterer has two specular reflection points located at $x = \mp\sqrt{ab}$, $y = \pm 4/3b \sqrt{ab}$ (see Fig. A3.1). The radius of curvature of the scatterer at these points is $R_c = 1/2 \sqrt{ab}$. The far field u $\exp(iky)/\sqrt{y}$ reflected by the scatterer at the point $(0,y)$ in the $\hat{y}$ direction is due to two reflected rays. This field is calculated by using the geometrical optics and it is given by the following expression

$$u = -\frac{1}{2}\left(\frac{a}{b}\right)^{1/4}\left(\exp\left(-i\frac{4}{3}k\,b\,\sqrt{ab}\right) - i\,\exp\left(i\frac{4}{3}k\,b\,\sqrt{ab}\right)\right). \tag{A3.6}$$

The first and second terms correspond to the points S_1 and S_2 shown in Fig. A3.1. The second term is multiplied by $-i$ because of the phase-shift suffered by the ray as it passes through the caustic formed by the reflected rays.

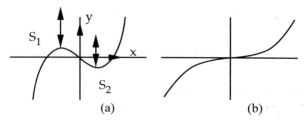

Fig. A3.1. Reflection from a cubic profile: (a) Two reflection points; (b) No reflection point

When $b < 0$, the scatterer no longer has a specular point of reflection. However, (A3.6) can be analytically continued for negative values of b, although an ambiguity arises as to the choice of the signs of $\sqrt{b}$ and $b^{1/4}$, the choices being $\sqrt{b} = \pm i\sqrt{-b}$; and, $b^{1/4} = \pm (-b)^{1/4}$ or $\pm i(-b)^{1/4}$. With all of these possible choices, the continuation of (A3.6) for the negative values of b leads to a term of the type $\exp(-4/3\, kb\sqrt{-ab})$, which we must reject because it is exponentially increasing and, consequently, does not satisfy the radiation condition. However, we retain the following term which decreases exponentially

$$u = \frac{e^{-i\pi/4}}{2}\left(\frac{a}{-b}\right)^{1/4} \exp(4/3\, kb\sqrt{-ab}),\qquad (A3.7)$$

and which is defined within a factor ± 1 or $\pm i$. The exact determination of u via the method of complex rays requires us to follow, carefully, the rays in the complex space and to determine their contact with the complex caustics – and this is not always easy. An alternate way to derive the reflected field entails the use of the surface field to calculate the radiated field, which leads to the result

$$u = - e^{-i\pi/4}\sqrt{\frac{k}{2\pi}}\int \exp\left(2ik\left(\frac{x^3}{3a} - bx\right)\right) \sin\theta(x)\, dx,\qquad (A3.8)$$

where $\theta(x)$ is the angle between the tangent to the obstacle and the incident ray. For a sufficiently small b, we can use the approximation $\sin\theta(x) = 1$, and recognize that we can rewrite (A3.8) as

$$u \approx - e^{-i\pi/4}\,\pi^{1/2}\,2^{1/6}\,k^{1/6}\,Ai(-(2k)^{2/3}\,a^{1/3}\,b),\qquad (A3.9)$$

where Ai is the Airy function. When $b > 0$, the asymptotic expansion for large negative values of the argument of Ai enables us to recover the expression given in (A3.6). For large positive values of the argument, which correspond to $b < 0$, we obtain

$$u \approx - \frac{e^{-i\pi/4}}{2} \left(\frac{a}{-b} \right)^{1/4} \exp(4/3 \, kb \sqrt{-ab}). \tag{A3.10}$$

This is the expression in Eq. (A3.7) which we have now derived in a systematic manner.

This example illustrates the most important features of the method of complex rays, viz., that the complex rays are associated with complex stationary points of the integral representation of the field radiated by surface fields. Their contributions can be obtained through a direct analytical continuation of the geometrical optics result, either by using the complex version of geometrical optics or an asymptotic evaluation of the radiation integral, the latter being the most reliable procedure. In addition, it also has the advantage that it provides a uniform result, while the method of complex rays leads to infinities on the caustic. The contribution of the complex rays decreases exponentially as a function of the frequency, and, as a consequence, only the rays that are almost real contribute significantly at high frequencies.

A3.5 Other applications of complex rays

The complex rays can also be used to represent, in a simple manner, the fields that are difficult to describe using the real rays. In particular, the radiated field by a source located in the complex space is represented by a Gaussian beam. Thus, if we know the solution of the diffraction problem of a point source by a scatterer, it is possible to calculate the diffraction of a Gaussian beam by the scatterer using analytical continuation [F].

Finally, we mention that it is also possible to treat the diffraction by cavities using complex rays. A number of other applications have been reported by Ruan [R].

References

[EF]   P. D. Einziger and L. B. Felsen, "Evanescent waves and complex rays," *IEEE. Trans. Ant. Prop.,* **AP-30**(4), 594-605, 1982.

[F]   L. B. Felsen, "Geometrical theory of diffraction: Evanescent waves, complex rays and Gaussian beams," *Geophys. J. R. Astr. Soc.,* **79**, 77-88, 1984.

[R]   Y. Ruan, *Application of complex ray theory in E.M. scattering and RCS analysis*, Rapport FTD-ID(RS)T-0535-90, 1990.

[WD]   W. D. Wang and G. A. Deschamps, "Application of complex ray tracing to scattering problems," *Proc. IEEE,* **62**(11), 1541-1551, Nov. 1974.

Appendix 4. Asymptotic Expansion of Integrals

In Chaps. 6 and 7, we have shown how we can derive an integral representation of the field which is valid on the caustics. In Chap. 4, we have discussed how to represent the field in the form of a plane wave spectrum. In all of these cases, the field is described in the form of the following simple or double integral

$$I = \int A(x) \exp\left(ikS(x)\right) dx, \tag{A4.1}$$

where the domain of integration may be either infinite or finite.

The function $A(x)$ is the complex amplitude and $S(x)$ is the phase which is generally real. Of course, it is always possible to compute, numerically, the integrals of the form (A4.1). However, this can be costly because the large parameter k present in the phase induces rapid oscillations in the integrand. Moreover, the numerical integration does not provide much insight into the physical aspects of the problem. In the majority of cases, it is possible to express (A4.1) in the form of an asymptotic expansion when k is large. The theory of asymptotic expansions of oscillating integrals has been the subject of many mathematical investigations and it has been demonstrated that this theory can be developed in a totally rigorous manner. In this appendix, we will restrict ourselves to the most useful formulas pertaining to asymptotic expansion, and pay particular attention to the physical interpretation of the results.

The asymptotic expansion of (A4.1) can be written as a sum of contributions associated with the critical points. These are: the points where $\vec{\nabla} S = 0$; a singularity (pole or branching point); and, a discontinuity of S or A. We will begin by discussing the methods for calculating the contributions associated with isolated critical points which are not too close. Following this, we will briefly present the case where the critical points coalesce, i.e., come close enough to merge.

A4.1 Evaluation of the contributions of isolated critical points

A4.1.1 The method of the stationary phase

Let us first treat the case of a simple integral. We consider the integrand of (A4.1) at a point where $S'(x) \neq 0$. In the vicinity of this point the phase varies as k $S'(x)$, i.e., very rapidly since k is large, while the amplitude varies relatively slowly as $A'(x)$. Accordingly, the integrand behaves as an oscillating function whose modulus is constant, and, hence, its integral essentially vanishes. This result can be justified mathematically in the following manner. Let us assume that A is a regular function defined in a compact domain and that $S(x)$ is a regular monotonous function whose derivative S' does not vanish. Let us integrate (A4.1) by parts to get

$$I = -\frac{1}{ik} \int \left(\frac{A}{S'}\right)' \exp\,(ik\,S(x))\,dx, \tag{A4.2}$$

where the factor $1/k$ is multiplied in front of the integral. Next, by repeating this procedure n times we obtain the expression

$$I = \left(-\frac{1}{ik}\right)^n \int B(x) \exp\,(ik\,S(x))\,dx, \tag{A4.3}$$

hence

$$|I| < C_n\,k^{-n}. \tag{A4.4}$$

We conclude that I decreases faster than any power of k.

If the derivative of S vanishes, then the essential contributions to the integral I come from the neighborhood of the points where $S'(x) = 0$. This condition defines the so-called stationary phase points. Let us assume that the stationary phase point is not degenerate, which implies that $S''(x) \neq 0$. In the neighborhood of the stationary phase point we can write

$$S(x) - S(x_s) \approx \frac{1}{2}S''(x_s)\,(x - x_s)^2, \tag{A4.5}$$

$$A(x) \approx A(x_s). \tag{A4.6}$$

Accordingly, the integral becomes

$$I \approx A(x_s) \exp{(ik\, S(x_s))} \int_{-\infty}^{+\infty} exp\frac{ik}{2}\, S''(x_s)\, (x - x_s)^2,$$

$$I \approx \sqrt{\frac{2\pi}{k|S''(x_s)|}}\, A(x_s) \exp\left(ik\, S(x_s) + i\frac{\pi}{4}\, sgn\, S''(x_s)\right). \tag{A4.7}$$

The argument presented above is not rigorous, although (A4.7) can be proven if A and S satisfy a number of conditions that we will define precisely at the end of this section. The integral I is of $O(k^{-1/2})$, the relative error when using (A4.7) is of $O(k^{-1})$.

It is also possible to provide a complete asymptotic expansion of (A4.1) in terms of the powers of $1/k$. We refer the reader to the references for additional information on this topic.

Let us now study the case of a double integral having a stationary point x_s, such that $\vec{\nabla} S(x_s) = 0$. Again, we perform an expansion of the phase around the point x_s to get

$$S(x) \approx {}^t(x - x_s)\, H(x_s)\, (x - x_s), \tag{A4.8}$$

where H denotes the matrix of the second derivatives of S (or Hessian). After inserting this expansion into the integral we obtain

$$I \approx A(x_s) \exp{(ik\, S(x_s))} \int_{-\infty}^{+\infty} exp\left(\frac{ik}{2}\, {}^t(x - x_s)\, H(x_s)\, (x - x_s)\right) dx, \tag{A4.9}$$

$$I \approx \frac{2\pi}{k}\, |\det H(x_s)|^{-1/2}\, A(x_s) \exp\left(ik\, S(x_s) + i\frac{\pi}{4}\, sgn\, H(x_s)\right), \tag{A4.10}$$

where sgn H represents the difference between the number of positive and negative eigenvalues of H. sgn H can take the values of 2, 0, or –2. Equations (A4.7) and (A4.10) are used frequently, and, in particular, they enable us to find the phase-shift at the caustics. In two dimensions, we can compute the field at a point M at distance r from the wavefront as an integral on the convex wavefront. In front of the caustic $r < R$, where R is the curvature radius of the wavefront taken at the projection of M on this front, so that $S''(x_s) > 0$. On the other hand, behind the caustic, $r > R$ and $S''(x_s) < 0$. According to (A4.7), the jump of the phase is $-\pi/2$ in the direction of the ray.

In Chap. (7) we can find an application of (A4.10), viz., that the asymptotic evaluation of the field radiated by the physical optics currents flowing on a doubly-curved surface recovers the geometrical optics result with the correct phase-shift at the caustics. In general, the application of the stationary phase to the calculation of the integrals occurring in diffraction problems provides the same result as that obtained with the ray method. In fact, the ray method can be viewed as an application of the stationary phase to an integral representation of the solution, and the limits of validity of the stationary phase method are the same as those of the ray method. We will now review these limits in the following.

(i) S is replaced by the approximation (A4.5) and the limit of integration is extended to the interval $-\infty$ to $+\infty$. This assumes that (A4.5) is valid for $k(x - x_s)^2 \gg 1$, and in particular, that S has no other stationary points. It also assumes that there are no singularities in the neighborhood $O(k^{-1/2})$ of x_s, and that the domain of integration contains a neighborhood $O(k^{-1/2})$ of x_s.

(ii) A is replaced by a constant. This approximation leads to errors if A has a singularity or a discontinuity in a neighborhood $O(k^{-1/2})$ of the stationary point.

Globally, the method of the stationary phase yields good results only if the limiting process described in (A4.5) and (A4.6) are well-justified in a neighborhood $O(k^{-1/2})$ of x_s, i.e., if the critical (stationary) points, the singularities, and the discontinuities of $A(x)$ and $S(x)$ are sufficiently far apart so that they can be treated separately. If this is not the case, then we must resort to formulas that are uniform and we will discuss this further in Sect. A4.2 below. Before we do this, however, we will present the techniques for calculating the contributions arising from the isolated critical points.

A4.1.2 The method of steepest descent

In certain situations, it is necessary to evaluate, asymptotically, simple integrals of the type (A4.1) which possess no stationary points. Let us assume that A and S are analytical functions and I can be considered as an integral in the complex plane by choosing a particular contour of integration along the real axis. According to the Liouville theorem, $S'(z)$ has at least one zero, unless it is a constant.

Let z_s be such a point and let us assume that $S''(z_s) \neq 0$, which implies that, $u = Re(S - S(z_s))$ and $v = - Im(S - S(z_s))$ have simple saddle points at z_s. Then, the steepest descent method consists in deforming the real axis into a contour of integration D, so-called steepest descent path, passing through z_s, on which u is constant, and $v < v\ (z_s)$.

The vector $\vec{\nabla} u$ is normal to D and $\vec{\nabla} v$ is tangent to it so that D is the contour where v decreases the fastest on each side of z_s. Along D, the quantity exp (ikS) = exp(iku) exp (kv) will then have a constant phase and will decrease rapidly on both sides of z_s. The behavior of the integral will then be determined essentially by the neighborhood of z_s and it is possible to determine D in the vicinity of z_s. In fact, we can write

$$S(z) - S(z_s) \approx \frac{1}{2} S''(z_s) \, (z - z_s)^2, \tag{A4.11}$$

$$u(z) \approx \frac{1}{2} |S''(z_s)| \, |z - z_s|^2 \cos{(Arg \, S''(z_s) + 2 \, Arg \, (z - z_s))}, \tag{A4.12}$$

and, consequently, u is constant near the point z_s, for

$$Arg \, (z - z_s) = \frac{\pi}{4} - \frac{1}{2} Arg \, S''(z_s) \mod \left(\frac{\pi}{2} \right). \tag{A4.13}$$

Equation (A4.13) defines two orthogonal straight lines, which make the angle $\pm \, \pi/4 - 1/2 \, Arg \, S''(z_s)$, with the x-axis.

On the straight line, which is at an angle $- \pi/4 - 1/2 \, Arg \, S''(z_s)$ with the x-axis, we have

$$v(z) \approx \frac{1}{2} |S''(z_s)| \, |z - z_s|^2, \tag{A4.14}$$

and, hence, $v \geq v \, (z_s)$. Thus the above line is not tangent to D.

On the straight line T, which forms the angle $- \pi/4 - 1/2 \, Arg \, S''(z_s)$, we have

$$v \, (z) \approx -\frac{1}{2} |S''(z_s)| \, |z - z_s|^2. \tag{A4.15}$$

We then have u = constant and $v \leq v \, (z_s)$. Thus, T is the tangent to D at z_s. Along D, the integral I_D has the form

$$I_D = \exp{(ikS(z_s))} \int_D A(z) \exp{(-kv(z))} \, dz, \tag{A4.16}$$

which we can rewrite as follows after denoting t as the curvilinear abscissa D

$$I_D = \exp{(ikS(z_s))} \int_D A(z) \, \frac{dz}{dt} \exp{(-kv(t))} \, dt. \tag{A4.17}$$

Equation (A4.17) is no longer an oscillating integral. Hence, it is possible to evaluate it numerically without difficulty and, alternatively, the Laplace method can be employed to obtain an explicit estimation of the same. In view of the rapid decay of the exponential in the vicinity of z_s, we can make the following approximations

$$v(t) \approx \frac{1}{2} |S''(z_s)| \, |t - t_s|^2,$$
(A4.18)

$$A(t) \frac{dz}{dt} \approx A(z_s) \exp\left(i \left(\frac{\pi}{4} - \frac{1}{2} Arg \, S''(z_s) \right) \right),$$
(A4.19)

and extend the domain of integration in (A4.17) to $-\infty$ to $+\infty$. We then obtain the result

$$I_D \approx A(z_s) \exp\left(ikS(z_s) \right) \sqrt{\frac{2\pi}{kS''(z_s)}} \exp i\frac{\pi}{4}.$$
(A4.20)

We still need to specify the choice of $1/2 \, Arg \, (S''(z_s))$, that is to say the choice of the square root of $S''(z_s)$, or, equivalently the choice of the direction of integration on D (Fig. A4.1). The choice of the direction of integration on D is imposed by the sense of integration on the real axis, which says that the tangent to D at point z_s must point in the direction of $x > 0$. Thus, $1/2 \, Arg \, (S''(z_s)) = Arg \, \sqrt{S''(z_s)}$ must be between $-\pi/4$ and $3\pi/4$.

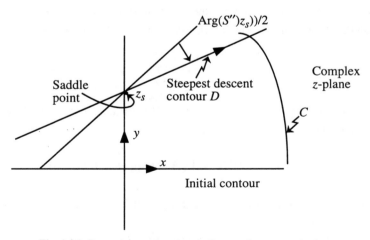

Fig. A4.1. Steepest descent contour in the complex $z = x + iy$ plane

If there are no singularities between the real axis and D, and if one can neglect the integrals on the semi-circular contour at infinity in Fig. A4.1, then $I = I_D$, and (A4.20) provides an approximation of I. If singularities exist between the two contours, then we must account for them by including supplementary contributions.

We note that for z and $S(z)$ real, (A4.20) recovers the expression in (A4.7), which is derived via the stationary phase method. In effect, if $S''(z) > 0$, $1/2\,Arg\,(S''(z)) = 0$, and if $S''(z) < 0$, $1/2\,Arg\,(S''(z)) = \pi/2$, and we obtain (A4.7) upon inserting these results into (A4.20). However, we should point out that the spirits of these two methods are different from each other. The method of stationary phase uses a contour on which the imaginary part of the phase is constant and the real part oscillates rapidly, while the contour employed by the steepest descent method is one where the real part of the phase is constant and the imaginary part decreases rapidly on each side of the stationary phase point. In fact, these two contours intersect each other at an angle of 45°. Also, the steepest descent method is somewhat more general than the method of the stationary phase, and the application of the former to an integral of type-1 enables us to not only retrieve the real rays, as is possible by using the stationary phase method, but the complex rays as well (see Appendix 3 "Complex Rays"). On the other hand, the steepest descent method forces us to assume that the functions A and S are analytical, which is not required for the application of the stationary phase method.

The steepest descent method can be generalized to apply to the case of a multiple integral [F]. For the case of a double integral we can obtain the result

$$I \approx A(z_s) \exp\,(ikS(z_s)) \frac{2\pi}{k} (det\ S''(Z_S))^{-1/2} \exp\left(i\frac{\pi}{4}\right). \qquad (A4.21)$$

However, the choice of the square root in (A4.21) is somewhat intricate and is not too obvious in Fedoriuk [F].

So far we have discussed the integrals over the entire $\mathbb{R}$-space ($\mathbb{R}$ = real line) and we are now going to treat the contribution of the boundaries of the integration for integrals on bounded intervals.

A4.1.3 Integration by parts

The method of integration by parts enables us to obtain the contribution arising from the boundaries of integration, or more generally from the discontinuities of A. Consider the integral

$$I = \int_a^b A(x)\, exp(ikS(x))\, dx,\qquad\qquad (A4.22)$$

where $S'(x) \neq 0$ on [a, b]. Let us assume that A and S are defined on $\mathbb{R}$.

$$I = \left(\int_a^{+\infty} + \int_{-\infty}^b - \int_{-\infty}^{+\infty} \right) exp\,(ikS(x))\, dx. \qquad (A4.23)$$

The interval of integration in the last integral is the entire space $\mathbb{R}$ and, consequently, it can be evaluated with the techniques presented above. Let us perform the first integral by assuming that $S'(x) \neq 0$ for $x \geq a$. This yields the result

$$I_a = \int_a^{+\infty} A(x)\, exp(ikS(x))\, dx,$$

$$I_a = \int_a^{+\infty} \left(\frac{A(x)}{ikS'(x)} \right) (exp(ikS(x)))'\, dx,$$

$$I_a = -\frac{A(a)}{ikS'(a)} exp\,(ikS(a)) - \frac{1}{ik} \int_a^{+\infty} \left(\frac{A(x)}{S'(x)} \right)' exp(ikS(x))\, dx.$$

$$(A4.24)$$

Equation (A4.24) is purely formal and the conditions of its validity are given in [B]. The second term in (A4.24) is an integral of the same type as (A4.22) multiplied by $(ik)^{-1}$, and, consequently, it is of lesser order in k. Thus, we can write

$$I_a \approx -\frac{A(a)}{ikS'(a)} exp\,(ikS(a)), \qquad\qquad (A4.25)$$

by neglecting the term which is of $O(1/k^2)$.

If either A or S has a discontinuity, the technique applies to each part of I, and the contribution is of $O(1/k)$. If, on the other hand, A is continuous up to the order $(n-2)$, and exhibits a jump in its derivative of order $(n-1)$, we can repeat the procedure outlined above n times, and the contribution resulting from this discontinuity is of $O(1/k^n)$.

Finally the method described above can be generalized to double integrals. If the integral is to be performed on a bounded domain, or has a line of discontinuity, an integration by parts will give rise to a contribution in the form of an integral on the edge of the domain or on the line of discontinuity. This integral can, in turn, be evaluated asymptotically to obtain the desired result.

If a corner is present in the domain of integration, the integral takes the form

$$I = \int_a^{+\infty} \int_b^{-\infty} A(x, y) \, exp \, ik(S(x, y)) \, dx \, dy \,, \tag{A4.26}$$

and we can obtain an estimation of the contribution of the corner by carrying out two integrations by parts to get

$$I_c \approx \frac{1}{k^2} A(a, b) \left(\frac{\partial S}{\partial x}(a, b) \frac{\partial S}{\partial y}(a, b) \right)^{-1} exp \, ikS(a, b). \tag{A4.27}$$

These results enable us to retrieve, in a simple manner, the k-dependence of the reflected and diffracted fields by an object derived from the radiation integral of the currents on the object. Since, in $\mathbb{R}^3$, the Green's function is proportional to k, we can show from (A4.10) that the contribution from a specular point of reflection behaves as k^0. Likewise, from (A4.25) we can arrive at the result that the diffraction by a discontinuity of order n behaves as $k^{-n-1/2}$. Finally, according to (A4.27), the contribution from a tip behaves as k^{-2}.

We now proceed below to provide the domains of validity of the various asymptotic evaluation methods that we have just discussed above.

A4.1.4 Limitations of the previous methods

The methods discussed above provide an asymptotic expansion of an integral of type (A4.1), expressed as a sum of contributions associated with the critical points. They assume that these critical points are sufficiently isolated, and the methods fail when they are too close to each other. For instance, if a stationary phase point coincides with an extremity, then the expression in (A4.25) leads to an infinite result. More generally, if the separation distance between two critical points is too small, (less than $k^{-\alpha}$, where $\alpha > 0$), the formulas given above do not apply. This can be rather inconvenient, especially when the integrals to be calculated depend upon a parameter, as for instance the distance between two critical points, which can take all real positive values. To handle this case it becomes necessary to resort to the methods of uniform estimation which we are going to present in the next section.

A4.2 Coalescence of critical points, uniform expansions points

The integral

$$J = \int_{a}^{+\infty} \exp{(ikx^2/2)} \, dx \,, \tag{A4.28}$$

is a simple example illustrating the problem associated with the coalescence of critical points. In accordance with the prescriptions given in Sects. A4.1.1 and A4.1.3 of this appendix, we obtain

(i)   a contribution J_s from the stationary phase point at $x = 0$, if $a < 0$. As $S(0) = 0$, $S''(0) = 1$, $A(0) = 1$, this contribution, according to (A4.7), is given by

$$J_s \approx \sqrt{\frac{2\pi}{k}} \, e^{i\pi/4}. \tag{A4.29}$$

(ii)   a contribution from the extremity, $S(a) = a^2/2$, $S'(a) = a$, $A(a) = 1$. Following (A4.25) we can write this contribution, viz., J_a, as

$$J_a \approx \frac{i}{ka} e^{ia^2/2}. \tag{A4.30}$$

We note, however, that J_a becomes infinite when $a = 0$, while it is obvious that J remains finite. The problem stems from the fact that $S'(a) = 0$ and we have derived the contributions from the two critical points as though they were independent, and yet they coalesce for $a = 0$.

As a general observation, the uniform expansions of the integrals are always derived by following the same approach. First, we seek a regular transformation, viz., $x \to s$, of the neighborhood of the stationary points onto a neighborhood of the origin in a way such that S is transformed into a polynomial $P(s)$ with the same configuration of stationary points as S. More precisely, S is transformed into one of the elementary polynomials of the theory of catastrophes and I is written as

$$I = \int B(s) \exp{(ikP(s))} \, ds. \tag{A4.31}$$

Since the transformation is regular, $B(s) = A(x) \, dx/ds$ exhibits the same singularities as those of $A(x)$. However, only a neighborhood of the origin contributes to I. B will therefore be replaced by a simple function showing the same singularities. I will be resolved into a *canonical* integral.

The concepts of asymptotic expansion techniques are relatively simple, but a complete derivation of these expansions can be a formidable task. Here we will be content with providing, in the table below, the associated canonical integral for only a simple integral.

Coalescence type	Associated canonical integral
2 stationary points	Airy function
3 stationary points	Pearcey function
4 stationary points	Swallowtail function
1 stationary point and one pole	Fresnel function
1 stationary point and 1 extremity point	Fresnel function
2 stationary points and 1 extremity point	Incomplete Airy

We recognize that the functions useful for expressing the fields are: Airy and Pearcey on the caustics; Fresnel on the shadow boundaries; and, incomplete Airy on the truncated caustics. As we have seen in Chaps. 5 and 6, these zones correspond, in effect, to the coalescence of critical points.

We will conclude this brief review of the methods of asymptotic expansion for integrals with these observations. We refer the interested reader to the literature on this subject and especially to the references below for more complete and rigorous results.

References

[BH]   Bleistein and Handelsman, *Asymptotic expansions of integrals*, Dover, 1986.

[F]   Fedoriuk, *Asymptotic Methods in Analysis*, Springer, 1990, (Encyclopedia of Mathematical Science).

[FM]   Felsen and Marcuvitz, Chap. 4 in *Radiation and Scattering of Waves*, Prentice Hall, 1973.

Appendix 5. Fock Functions

A5.1 Utilization of the Fock functions

We have seen in Chaps. 3 and 5, that the Fock functions are used to express the field radiated by a source in the presence of a smooth surface. More precisely, they describe the transition between the illuminated and shadow regions, these regions being associated with large negative and positive arguments of these functions, respectively. A very comprehensive study of the Fock functions has been carried out by Logan, and some of the results of his investigation have been reported in [Lo]. However, it is somewhat difficult to follow the presentation in the above work because the results are not always demonstrated explicitly. The principal objective of this appendix is to extract the most useful formulas from the above-mentioned reference.

There are three types of Fock functions, and they are listed below.

(i)   Nicholson function, or the coupling function   (N)
(ii) Fock function, or the radiation function   (F)
(iii) Pekeris function, or the reflection coefficient function   (P)

The location of the source and observation points determine the nature of the function to employ, as indicated in the table below.

Table A5.1

Source Point	Observation Point	Function
on the surface	on the surface	Nicholson
on the surface	at infinity	Fock
at infinity	on the surface	Fock
at infinity	at infinity	Pekeris

If one of the points is located at a finite distance, then it becomes necessary to resort to other functions that are more complicated, such as the Fock V function with two arguments (Eq. (3.160) of Sect. 3.7), whose properties can be found in [Lo]. We will define the Fock functions for large positive and negative arguments in Sects. 4 and 5.

A5.2 Definition of the Fock functions

The Fock functions are defined for $q = \mathrm{im}\, Z = \text{constant}$, in [Lo], from which we have extracted the following formulas.

$$N(x) = \frac{exp\left(-i\frac{\pi}{4}\right)}{2}\sqrt{\frac{x}{\pi}}\int_{-\infty}^{\infty} exp(ixt)\frac{w_1(t)}{w_1'(t) - qw_1(t)}\,dt\,, \qquad (A5.1)$$

$$F(x) = \frac{1}{\sqrt{\pi}}\int_{-\infty}^{\infty} exp(ixt)\frac{1}{w_1'(t) - qw_1(t)}\,dt\,, \qquad (A5.2)$$

$$P(x) = \frac{exp\left(i\frac{\pi}{4}\right)}{\sqrt{\pi}}\int_{-\infty}^{\infty} exp(ixt)\frac{v'(t) - qv(t)}{w_1'(t) - qw_1(t)}\,dt\,, \qquad (A5.3)$$

where v and w_1 are the Airy functions in accordance with Fock's notation. For $q = \infty$, which corresponds to the case of a perfect conducting cylinder and electric polarization, the functions N and F associated with the surface electric field vanish, and we define two new functions, viz., the so-called electric or soft functions. The first one of these is denoted U, and is given by

$$U(x) = \frac{x^{3/2}e^{-3i\pi/4}}{\sqrt{\pi}}\int_{-\infty}^{+\infty} exp(ixt)\frac{w_1'(t)}{w_1(t)}\,dt\,.$$

The second function, denoted as f (see [Lo]), has the expression

$$f(x) = \frac{1}{\sqrt{\pi}}\int_{-\infty}^{+\infty} exp(ixt)\frac{1}{w_1(t)}\,dt\,.$$

All of these expressions contain the Airy functions whose asymptotic behaviors will be discussed in the following section.

A5.3 Asymptotic behaviors of Airy functions

Except for a constant factor, Fock functions are the Fourier transforms of ratios of combination of Airy functions and their derivatives. This is not entirely surprising since we construct them by performing a Fourier transformation (see Sect. 3.7). In order to derive the asymptotic expansions of the Fock functions we deform the original path of integration, which is along the real axis, into the complex plane.

The asymptotic behavior of the Airy functions v and w_1 is given in the figure below. It determines the regions in the complex plane where it is permissible to deform the contour. The shaded areas are the regions where the amplitude of the function decreases and the bold straight lines indicate the directions of the most rapid growth.

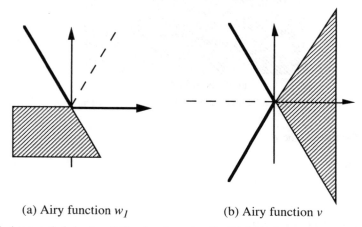

(a) Airy function w_1 (b) Airy function v

Fig. A5.1. Asymptotic behavior of Airy functions. Amplitude of this function decreases in the shaded region. Directions of most rapid growth denoted by bold lines.

Now that we have discussed the asymptotic behaviors of the Airy functions, we will go on to evaluate the integrals (A5.1) to (A5.3) that define the Fock functions for large positive values of their argument, i.e., in the shadow zone.

A5.4 Behavior of the Fock functions for large positive x

a. Nicholson function

The integrand of the Nicholson function behaves as $|t|^{-1/2} e^{ixt}$, for large $|t|$, which means that, if x is positive, the integral converges for Im $t > 0$. Then it is possible to close the contour in the upper half-plane and subsequently to calculate the integral via the residue calculus technique.

Using the relations

$$w_1'(\xi_p + h) - q w_1(\xi_p + h) = h(\xi_p w_1(\xi_p) - q w_1'(\xi_p)) - O(h^2),$$

and

$$w_1'(\xi_p) = q w_1(\xi_p).$$

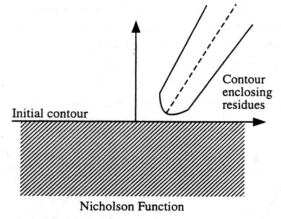

Nicholson Function

Fig. A5.2. Evaluation of the Nicholson function

We obtain for the residue R

$$R = \frac{exp\, i\xi_p x}{\xi_p - q^2}.$$

So that, $N(x)$ is given by (see [Lo])

$$N(x) = exp\left(i\frac{\pi}{4}\right)\sqrt{\pi x}\, \sum_{p=0}^{\infty} \frac{exp\, i\xi_p x}{\xi_p - q^2}, \text{ where } w_1'(\xi_p) = q w_1(\xi_p).$$

The function $U(x)$ is calculated in a similar manner and its expression is given in Table A5.2.

b. Fock function

The integrand of the Fock function F behaves as $e^{i\xi t}/w_1'$ (t) for t large. It is therefore possible to deform the contour of integration in the unshaded sector, i.e., in a region where this quantity decreases as $|t| \to +\infty$, especially in the upper half plane. The integral can thus be expressed as a sum of residues. The same is also true for the function f and the relevant formulas appear in Table A5.2.

c. Pekeris function

The integrand of the Pekeris function behaves as $exp(ixt)\, v'(t)/w_1'$ (t) when t is large. For $|t| \to +\infty$, the sector in which the integrand decreases is the same as it is for the Fock function

Table A5.2

Function	$x \approx 0$ for N and U $x \ll 0$ for F and P	$x \gg 0$
$N(x)$	≈ 1 (for $Z = 0$)	$\sqrt{\pi x}\, exp\left(i\dfrac{\pi}{4}\right)\dfrac{exp\, i\xi x}{\xi + m^2 Z^2}$
U	≈ 1	$2x\sqrt{\pi x}\, exp\left(i\dfrac{\pi}{4}\right) exp\, i\xi_E x$
F	$\dfrac{2x}{x - mZ}\, e^{-i\frac{x^3}{3}}$	$2i\sqrt{\pi}\, \dfrac{exp\, i\xi x}{(\xi + m^2 Z^2)w_1(\xi)}$
f	$2ix\, e^{-i\frac{x^3}{3}}$	$2i\sqrt{\pi}\, \dfrac{exp\, i\xi_E x}{w_1'(\xi)}$
P	$\dfrac{x + 2mZ}{x - 2mZ}\, \dfrac{\sqrt{-x}}{2}\, e^{-i\frac{x^3}{3}}$	$2e^{3i\pi/4}\sqrt{\pi}\, \dfrac{exp\, i\xi x}{(\xi + m^2 Z^2)w_1^2(\xi)}$

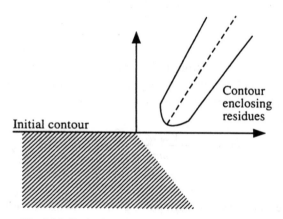

Initial contour

Contour
enclosing
residues

Fig. A5.3. Evaluation of the Fock and Pekeris functions

(unshaded sector of Fig. A5.3). The integral is calculated by using the residue calculus method in the same manner as it is done for the Fock function and the result is given in Table A5.2.

d. Conclusions

For $x > 0$, i.e., in the shadow zone, the Fock functions can be expressed as a series of residues, of which each is associated with a creeping ray. The convergence improves as

x gets larger. In practice, once x is on the order of 2, a good approximation of the function can be obtained with just the first term, which corresponds to the first creeping ray.

Next, we will study the behavior of the Fock functions in the lit zone, i.e., for large negative arguments.

A5.5 Behavior of the Fock functions for large $x < 0$

a. Nicholson function

The Nicholson function is only defined for $x > 0$.

b. Fock function

To obtain the Fock function, the integral in (A5.2) is calculated via the stationary phase method, by replacing the Airy functions in the integrand by their respective asymptotic expansions. The stationary phase point t (< 0) exists only when $x < 0$, and it is given by the expression $(-t)^{1/2} + x = 0$. Then, the stationary phase evaluation leads to the approximation

$$F(x) \approx \frac{2x}{x - mZ} e^{-ix^3/3}$$

c. Pekeris function

The Pekeris function is derived from the integral in (A5.3) which is evaluated in a manner similar to the Fock function, although the stationary point $t'(< 0)$ is now given by $2(-t')^2 + x = 0$. The stationary phase evaluation leads to the result (see [Lo])

$$P(x) \approx \frac{-x + 2mZ}{-x - 2mZ} \frac{\sqrt{-x}}{2} \exp{-i \frac{x^3}{12}}. \qquad (A5.4)$$

d. Conclusions

For large $x < 0$, we can extract the asymptotic approximations of the Fock and Pekeris functions via the stationary phase method and, as we have seen in Sect. 5.5, these expressions enable us to retrieve the Geometrical Optics results in the lit zone.

A5.6 Behavior of the Nicholson functions for $x \approx 0$

To investigate the behavior of the Nicholson function for $x \approx 0$, we perform a change of variable, viz., $xt = u$, in the integral (A5.1). The Airy functions have arguments u/x, that are large for $u \neq 0$, and, consequently, we can use their asymptotic expansions.

Here, we restrict ourselves to the perfectly conducting case for which a considerable amount of simplification results. The results are

$$N(x) \approx \frac{e^{-i\pi/4}}{2} \sqrt{\frac{x}{\pi}} \int_{-\infty}^{+\infty} x^{-1/2} u^{-1/2} \, exp(iu) \, du \approx 1,$$

and

$$U(x) \approx \frac{e^{-3i\pi/4} x^{3/2}}{\sqrt{\pi}} \int_{-\infty}^{+\infty} x^{-3/2} u^{1/2} \, exp(iu) \, du \approx 1.$$

A5.7 Conclusions

Table A5.2 appearing below, delineates the various approximations of the Fock functions discussed in this appendix. We have used the notations ξ for the first zero of $w_1'(\xi) - imZw_1(\xi)$ and ξ_E for the first zero of $w_1(\xi)$. Also, we have only retained the first term of the series for $x \gg 0$. In fact, in the deep shadow zone, retaining only a single creeping ray is found to be an accurate representation of the Fock functions. Similarly, in the lit zone where $x \ll 0$, the first term of the stationary phase integral gives a fair approximation of these functions. In the transition zone, the previous formulas are no longer valid and one has to resort either to a numerical calculation of the integrals defining the Fock functions, or to tabulated values. In Logan's reference [Lo], one will find tables of values and more accurate asymptotic expansions. The results for perfect conductors are available in other references.

References

[Lo]   N. Logan and K. Yee, *Electromagnetic waves*, R. E. Langer, Ed., 1962.
[BSU]   J. J. Bowman, T. B. A. Senior, and P. L. Uslenghi, *Acoustic and electro-magnetic scattering by simple shapes*, Hemisphere, 1987.

Appendix 6. Reciprocity Principle

The general form of the reciprocity principle reads

$$\int_V \vec{J}_1 \cdot \vec{E}_2 - \vec{M}_1 \cdot \vec{H}_2 \, dv = \int_V \vec{J}_1 \cdot \vec{E}_1 - \vec{M}_2 \cdot \vec{H}_2 \, dv, \qquad \text{(A6.1)}$$

where $\vec{J}_1$, $\vec{M}_1$ and $\vec{J}_2$, $\vec{M}_2$ are the electric and magnetic currents, that radiate the fields $(\vec{E}_1, \vec{H}_1)$ and $(\vec{E}_2, \vec{H}_2)$, respectively, and V is a volume that contains the sources.

Equation (A6.1) is obtained from the identity [FM]

$$\nabla \cdot (\vec{E}_1 \times \vec{H}_2 - \vec{E}_2 \times \vec{H}_1) = (\vec{J}_1 \cdot \vec{E}_2 - \vec{M}_1 \cdot \vec{H}_2) - (\vec{J}_2 \cdot \vec{E}_1 - \vec{M}_2 \cdot \vec{H}_1)$$

$$+ \vec{D}_1 \cdot \vec{E}_2 - \vec{D}_2 \cdot \vec{E}_1 + \vec{B}_1 \cdot \vec{H}_2 - \vec{B}_2 \cdot \vec{H}_1,$$

which leads to the result

$$\nabla \cdot (\vec{E}_1 \times H_2 - \vec{E}_2 \times \vec{H}_1) = (\vec{J}_2 - \vec{E}_2 - \vec{M}_1 \cdot \vec{H}_2) - (\vec{J}_2 \cdot \vec{E}_1 - \vec{M}_2 \cdot \vec{H}_1),$$
$$\text{(A6.2)}$$

provided that the permittivity and permeability tensors are symmetrical.

We now integrate (A6.2) on a sphere with radius R. We apply the divergence theorem to the left-hand side, let R tend to infinity, and then employ the radiation condition. We find that the left-hand side vanishes and yields (A6.1). We note that the integration volume containing the source can be arbitrary since the integrand vanishes outside the sources.

The reciprocity principle is valid for conducting objects that are coated with dielectric or isotropic magnetic materials, and, consequently, for objects described by a surface impedance type of boundary conditions, of the type we have discussed in this book. The reciprocity principle is not satisfied, for instance, in media that have non-symmetric permeability tensors, e.g., magnetized ferrites. A comprehensive discussion of the validity of this principle has been provided by Felsen and Marcuvitz [FM]. This principle is also called the Rumsey reaction principle. The result (A6.1) is then interpreted as the equality of the reactions between the fields with subscript 2 and the sources of these fields with subscript 1, or vice versa.

A very useful form of the reciprocity principle is obtained by considering the particular case of point dipole sources, e.g., electric ones. For this case we obtain the following equality from (A6.1)

$$\vec{J}_1 \cdot \vec{E}_2 = \vec{J}_2 \cdot \vec{E}_1. \tag{A6.3}$$

Using (A6.3), we can calculate the electric far-fields radiated by a source on a surface. For this, it is sufficient to know the electric field induced on the surface by a source located far from the surface (see Chap. 1). In particular, we see that on a perfectly conducting surface, where the tangent electric field vanishes, the field radiated by a source tangent to the surface is zero.

Reference

[FM]   L. B. Felsen and N. Marcuvitz, *Radiation and Scattering of Waves*, Prentice Hall, 1973.

Subject Index

Printing: Mercedesdruck, Berlin
Binding: Buchbinderei Lüderitz & Bauer, Berlin